SPACE KINEMATICS AND LIE GROUPS

SPACE KINEMATICS AND LIE GROUPS

ADOLF KARGER
Charles University, Prague, Czechoslovakia
JOSEF NOVÁK
Technical University, Prague, Czechoslovakia

Translated by Michal Basch

GORDON AND BREACH SCIENCE PUBLISHERS
New York London Paris Montreux Tokyo

P.O. Box 161, 1820 Montreux 2, Switzerland

Gordon and Breach Publishers

P.O. Box 786
Cooper Station
New York, NY 10276
United States of America

P.O. Box 197
London WC2E 9PX
England

58, rue Lhomond
75005 Paris
France

14-9 Okubo 3-chome
Shinjuku-ku
Tokyo 160
Japan

Gordon and Breach Science Publishers S.A., Montreux
in co-edition with
SNTL - Publishers of Technical Literature, Prague

Originally published in Czech as *Prostorová kinematika a Lieovy grupy,* by SNTL Prague, Czechoslovakia. © 1978 Karger, Novák

LIBRARY OF CONGRESS CATALOGING IN PUBLICATION DATA

KARGER, ADOLF 1929 –
Space Kinematics and Lie Groups,

Translation of: Prostorová kinematika a Lieovy grupy.
1. Kinematics. 2. Lie Groups. 1. Novák, Josef.
1905 –. 11. Title
QA841.K3713 1985 531′.112 84-29064
ISBN 2-88124-023-2

ISBN: 2-88124-023-2. Printed in Czechoslovakia.

Contents

Preface

Kinematic geometry and the kinematics of space motion form a traditional mathematical discipline. Its origins go back to the last century. Although the theory of continuous groups emerged at the beginning of the present century we, can first trace its influence on space kinematics in the works of W. Blaschke and H. R. Müller. This influence manifests itself mainly in the application of the so-called moving frame method of E. Cartan to the investigation of motions, especially multiparametric motions. The moving frame method is easily applicable to kinematic geometry. We shall not discuss this method because it is theoretically so difficult. Its advantages show themselves mainly when investigating multiparametric motions, which are not studied in this book.

In the course of development of space kinematics, several formal tools were introduced which can be used for the description of motion, e.g., quaternions, dual quaternions, matrices, dual vectors, helix calculus, etc. The aim of this book is to show that all these formal approaches arise quite naturally within the scope of the theory of the Lie group of congruences of the space or the sphere, and that to every representation or realization of this group we obtain the corresponding formal apparatus. For instance, the cross product of dual vectors is nothing else than the bracket operation in the Lie algebra of the group of congruences. The application of dual numbers also follows naturally from the fact that the extension of the group SO(3) to dual numbers yields the group of congruences of the space. Even such "classical" concepts as pseudospherical space and Klein's quadric are determined as sets in the Lie algebra of the group of congruences of the space, on which the invariant forms vanish. The adjoint representation then induces on these sets the same as obtained when applying dual quaternions. It is possible to find more such examples.

The book is intended not only for readers oriented to theoretical work but also for those who deal with concrete problems in the theory of mechanisms and gearings or in the construction of cutting tools. In view of the two-fold orientation of the book it is organized into relatively independent chapters.

The first chapter is of a rather auxiliary nature. It is largely meant for the reader who wishes to gain deeper insight into the foundations of the theory presented in the second and third chapters. For the reader who is interested in only the applications of the theory, a detailed study of the first chapter is not necessary. If the reader's interest is oriented only to the applications of space motion he need not read the second chapter devoted to spherical motion.

In any case, we suggest that the reader choose some concrete example of spherical or space motion (e.g. an analogy to II, 2.7 or III, 4.7). Also, that he carry out all the considerations of the second and third chapters in parallel on his chosen concrete case of motion. The authors' opinion is that the computation of a single concrete case gives the reader deeper insight into the problems than detailed study of the theory. This is why the reader will also find in this book a considerable number of examples and exercises.

Examples of medium difficulty were deliberately selected, even though the resulting expressions are sometimes of considerable complexity. In fact, we fear that the reader cannot gain much insight from examples that are "nice" to solve. Using the examples, we have tried to show that even relatively simple motions lead to rather complicated expressions, be they represented with the aid of whatever formalism. Consequently, matrix calculus is used, as a rule, since it is our belief that its foundations are familiar generally. Nevertheless, some conversion relations are given for the sake of completeness. Naturally, the reader who is not interested in these problems need not pay attention to them.

In view of the complexity of the expressions obtained, e.g., when computing motions determined by rather simple mechanisms (e.g., Bennet's mechanism), or when computing the characteristics of envelope surfaces, it is obvious that concrete results can be obtained only with the aid of computers. We take account of this fact throughout the book and adapt all the formulae into forms applicable for computer processing. The advantages of computer solution are shown for some examples for which the results were obtained as graphic representations of the computer output with the aid of an automatic graph plotter. This also indicates the possibilities of computer graphics in the field.

The examples take up a considerable part of the contents of the book. Obviously, this has a necessarily negative influence on the clarity of the exposition. Therefore, we give a brief outline of the contents of the book and add some comments on its overall arrangement in what follows.

The first and second sections of Chapter I comprise a brief survey of the concepts and relations used in the sequel.

The third and fifth sections of Chapter I are devoted to the discussion of the concepts of differentiable manifold and Lie group. The exposition goes into considerable details, as these concepts are not used currently in kinematic literature. However, detailed study is recommended only to the reader who wishes to pursue deeper theoretical investigation of kinematics. If the reader is content with the discovery that space motion is given by a matrix whose elements are differentiable functions of one variable (i.e., functions with a sufficient number of derivatives), he need not read these sections.

The fourth section of Chapter I is devoted to the differential geometry of curves and ruled surfaces. The knowledge of the differential geometry of curves is necessary for reading Chapter II, while the knowledge of the entire section is necessary for reading Chapter III. Special and detailed attention should be paid to this section. We recommend that the reader works out a number of examples (particularly the Frenet formulae for the helix, for the helical ruled surface and the hyperboloid of revolution, the parametric representation of the helical ruled surface and the hyperboloid of revolution in Plücker coordinates).

Chapter II is devoted to "classical" spherical kinematic geometry and kinematics. It was included, in particular, since spherical motion is a special case of space motion. This enabled us to present the theory of Chapter III as applied to a simpler case first. Also, this case is well known and described in detail in the literature.

In the first section of Chapter II one finds the definition of spherical motion as a curve on the group of orthogonal transformations. Also, the object of investigation of kinematics and kinematic geometry is defined. The considerations of this section are of an entirely general nature and the fact that spherical motion is discussed is not essential. Therefore, the reader will encounter the same considerations, in abbreviated form, repeated for space motion in Chapter III.

The second section describes the relations between different representations of spherical motion. The connection between spherical motion and quaternions has been known for a long time. The connection between the groups $O(3)$ and $SU(2)$ is also familiar. However, it has not been applied to the investigation of spherical motion before.

In the third section, directing cones of motion are introduced. The concept of the directing cone of motion differs from the concept of the directing cone of a ruled surface from Chapter I. The name was chosen because there actually are some similarities. The concept of the directing cones of motion is obtained by the formal generalization of the concept of the spherical image of a space curve. The spherical image of a space curve is obtained by parallel translation of all its tangent unit vectors to a fixed chosen point. Here we proceed similarly. Motion is a curve on a group. At all its points tangent vectors are constructed (unit in a certain sense) and "parallel" translated (i.e. by left or right translation on a group) to a fixed point – the unit of the group. Further, it is shown that directing cones are of decisive significance for the investigation of motion, i.e. they contain all the information necessary for this investigation – e.g. the directing cone determines the instantaneous centres of revolution. Finally, the invariants of the motion (curvature) are found. They are functions which uniquely determine the motion, independently of the randomly chosen initial position of the moving and fixed space (e.g. the Euler angles also determine a motion, but the same motion can be given different functions).

The fourth section discusses the trajectory of a point. Here the exposition is traditional, being merely a transcription of a certain part of plane kinematic geometry, as the reader familiar with plane kinematic geometry will certainly note (however, familiarity is not necessary to understand the text). The section also includes the theory of double-rotation motions which occur most frequently in practical problems. All the formulae given here are obtained as special cases of formulae for double-helical motions in Chapter III. The derivation of the formulae for double-rotation motions was included here especially to facilitate comprehension of the considerations concerning double-helical motions in Chapter III. In Examples 4.15 and

4.16 the reader will find formulae for the approximate computation of the trajectory of a point. The formula in 4.15 is suitable for practical computations; the formula in 4.16 is suitable for the investigation of the properties of the trajectory which are of higher order (hyperosculation, poles of higher orders etc.). We wish to draw attention to the fact that the same formulae can also be derived for general space motion, only in the case of 4.16 it is necessary to apply the respective Frenet formulae.

The fifth section is devoted to the traditional exposition of the kinematics of spherical motion. It includes the necessary theoretical base for this investigation of "relative" motions by applying the methods discussed in this book. The corresponding considerations for space motion are exactly the same. Therefore, we did not repeat this topic in Chapter III.

The considerations of Chapter III are similar to those of the previous chapter, but they are a bit more complicated. In the first section the formal tools necessary for the further exposition are introduced. The second section elucidates the relations between the straight line in Euclidean space and all helical motions which preserve it, or – which is the same – between a vector from the Lie algebra of the group of congruences and the Plücker coordinates of a straight line.

The third section is devoted to the representation of concrete space motion. The Bennet mechanism is used as an example to show how it is necessary to proceed in the case of motion determined by a multiple-member mechanism, originating as the composition of several simple motions.

In the fourth section the reader will find the theory of space motion. The considerations are similar to those given in Chapter II.

In the fifth section the Frenet formulae and the formulae for the computation of invariants are derived. The reader should pay attention to Example 5.11, where the formulae are directly in the form suitable for computer solution. These are obtained by suitable substitutions.

In the sixth section space motion is characterized as the identification of the Frenet frames of two ruled surfaces – axoids.

The seventh section characterizes the trajectory of a point for general space motion. Here, the situation is much more complicated than in the case of spherical or plane motion. The space curve of inflection points is described (several results from the theory of quadrics were applied here; it is not necessary, however, to bother the reader with the details). The dependence of the first and second curvatures of a trajectory on the position of an investigated point is found. However, it is too complicated for us to present its simple description. The remainder of the section is devoted to the problem of the so-called synthesis of motion, namely to the determination of all space motions which have a straight-line trajectory. A system of algebraic equations is obtained, the solution of which has to satisfy a certain system of differential equations. The solution is easy to perform as far as the computations are concerned if the invariants of the motion are constant, i.e. in the case of helical motion. The method of the solution of the given problem is applicable also to the solution of other similar problems.

In the eighth section double-helical motions are discussed, i.e. motions composed of two helical motions about fixed axes. Here, it was assumed that the displacement can be an arbitrary function (not necessarily linear) of the angle of rotation. The ratio of the angular velocities need not be constant; thus, more general motions are obtained than the helical motions usually considered (axoids need not be ruled helical surfaces). The corresponding theory for helical motions is then obtained easily by mere specialization.

The character of the ninth section is theoretical. The so-called dual numbers which are used for the description of space motion are introduced there. Further, the relation between the formulae for space motion and the "dualized" formulae for spherical motion is pointed out. This relation enables the derivation of the formulae for space motion by substitution of the dual numbers into the corresponding formulae for spherical motion.

The kinematics of space motion is discussed in the tenth section, i.e. the analysis of the vector fields of velocity and of acceleration. The analysis of the velocity vector field is traditional – the null complex of all the normals of the trajectories and the quadratic complex of the tangents of the trajectories are defined, their equations are found, and those properties necessary to the applications are described. Further, it is shown that the vector field of acceleration is of a different character from that of the vector field of velocity; its analysis requires the solution of third-order equations. Consequently, it is preferable to decompose acceleration into two components – the radial component corresponding to rotation about the instanteneous axis, and the tangential component. Here, the tangential components constitute the field of the velocities of helical motion for a suitable axis and parameter.

In the eleventh section kinematically generated envelope surfaces are studied. Various possibilities of the construction of the characteristic are shown. In this connection the conjugate polars of a linear complex are investigated in more detail.

The twelfth section is devoted to applications of ruled surfaces to the construction of cutting tools and gearings. The application of computers, including computer graphics, to the solutions of concrete problems of cutting tool construction is pointed out.

The interdependencies of the individual sections are outlined in the following scheme.

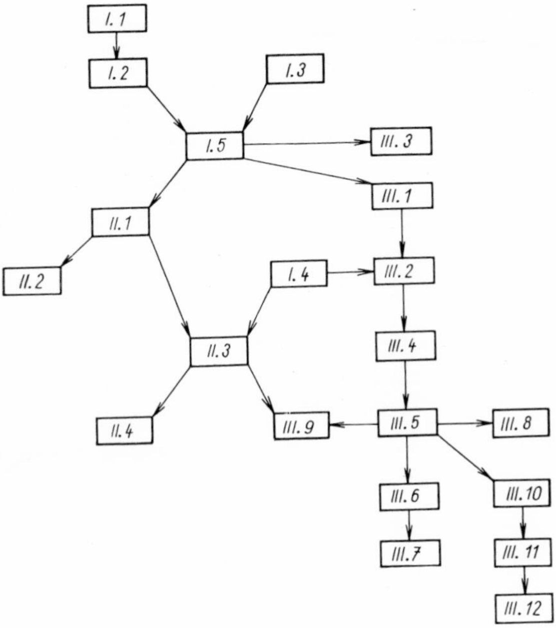

List of symbols and notations

Symbol	Application	Meaning
( )	(a^i_j)	matrix a
	$(\boldsymbol{u}, \boldsymbol{v})$	inner product of the vectors $\boldsymbol{u}$ and $\boldsymbol{v}$
	$(\boldsymbol{p}; \bar{\boldsymbol{p}})$	Plücker coordinates of a straight line
[ ]	$[x^1, x^2]$	coordinates of point X
	$[\boldsymbol{X}, \boldsymbol{Y}]$	Lie bracket of vectors $\boldsymbol{X}$ and $\boldsymbol{Y}$ in $\mathfrak{E}$
| |	$\lvert\lambda\rvert$	absolute value of real number λ
	$\begin{vmatrix} a^1_1, & a^1_2 \\ a^2_1, & a^2_2 \end{vmatrix}$	determinant of matrix a
	$\lvert\boldsymbol{u}, \boldsymbol{v}, \boldsymbol{w}\rvert$	mixed product of vectors $\boldsymbol{u}$, $\boldsymbol{v}$, $\boldsymbol{w}$
‖ ‖	$\lVert\boldsymbol{u}\rVert$	length of vector $\boldsymbol{u}$
{ }	$\{\boldsymbol{f}_1, \boldsymbol{f}_2, \boldsymbol{f}_3\}$	base (orthonormal)
	$\{A, \boldsymbol{f}_1, \boldsymbol{f}_2, \boldsymbol{f}_3\}$	frame
×	$\boldsymbol{u} \times \boldsymbol{v}$	cross product of vectors $\boldsymbol{u}$ and $\boldsymbol{v}$
	$M \times N$	Cartesian product of sets M and N
→	$X \to Y$	mapping of set X into set Y
∘	$a \circ b$	binary operation on set M; $a, b \in M$

Notation	Meaning
$\boldsymbol{\alpha}$	quaternion, dual number
δ_{ij}, δ^i_j	Kronecker delta
ad$g$(t)	adjoint motion to given motion g(t)
$\boldsymbol{u}_i$	vectors of orthonormal base $\boldsymbol{u}_1$, $\boldsymbol{u}_2$, $\boldsymbol{u}_3$
$g(t)$	matrix determining motion
$\boldsymbol{r}$	fixed directing conical surface
$\bar{\boldsymbol{r}}$	moving directing conical surface
$\boldsymbol{u}$	vector
$\boldsymbol{u}_{AB}$	representative of vector $\boldsymbol{u}$
$\boldsymbol{v}_0$	reduced height of thread or parameter of helical motion
A	point or position vector of point
A	dual vector
E_3	three-dimensional Euclidean space
$\mathscr{E}$	Lie group of all congruences in E_3
$\mathfrak{E}$	Lie algebra of group $\mathscr{E}$
K($\boldsymbol{X}$, $\boldsymbol{Y}$)	Killing bilinear form in $\mathfrak{E}$
$\mathscr{K}$($\boldsymbol{X}$, $\boldsymbol{Y}$)	Klein bilinear form in $\mathfrak{E}$
$\mathscr{N}$	zero complex
O(3)	orthogonal group of dimension 3
R	set of all real numbers
$\mathscr{R}$	frame
$\boldsymbol{R}$	fixed directing cone
$\bar{\boldsymbol{R}}$	moving directing cone
$\mathscr{S}$	congruence
Tr a	trace of matrix a
$\boldsymbol{X}$	element of Lie algebra $\mathfrak{E}$
$\mathfrak{A}$	fixed axoid
$\bar{\mathfrak{A}}$	moving axoid

Chapter I

Preliminaries

1. SELECTION OF BASIC CONCEPTS FROM ALGEBRA

In this introductory section we review the material assumed to be known in the remainder of the book. The moment the reader realizes that he is not renewing previous knowledge he should turn to the recommended literature [I],[X], [XI] for more detailed information.

A. *Matrices and Determinants*

1.1 Definition. An array consisting of mn real numbers a^i_j $(i = 1, 2, \ldots, m, j = 1, 2, \ldots, n)$, arranged in m rows and n columns

$$\begin{pmatrix} a^1_1, a^1_2, \ldots, a^1_n \\ a^2_1, a^2_2, \ldots, a^2_n \\ \ldots\ldots\ldots\ldots \\ a^m_1, a^m_2, \ldots, a^m_n \end{pmatrix},$$

is called an $m \times n$ *matrix* or *matrix of type* $m \times n$. The number a^i_j is the *element* of the matrix in its i-th row and j-th column. The brief notation for a matrix is $a = (a^i_j)$.

1.2 Definition. An $m \times n$ matrix is called a

row matrix	if $m = 1$ while $n > 1$,	
column matrix	if $m > 1$ while $n = 1$,	
square matrix of order n	if $m = n$.	

A square matrix is called

symmetric if $a^i_j = a^j_i$,
skew-symmetric if $a^i_j = -a^j_i$,

diagonal	if $a^i_j = 0$ for $i \neq j$,	
unit e	if $a^i_j = \delta^i_j$,	
zero 0	if $a^i_j = 0$,	

for all $i, j = 1, \ldots, n$ where $\delta^i_j = \delta_{ij} = \begin{cases} 1 & \text{for } i = j \\ 0 & \text{for } i \neq j \end{cases}$

is the so-called *Kronecker delta function.*

Interchanging the rows and columns (without changing their order) of an $m \times n$ matrix $a = (a^i_j)$ we obtain the *transpose* a^T of matrix a. The matrix $a^T = (\bar{a}^i_j)$ is an $n \times m$ matrix, and $\bar{a}^i_j = a^j_i$ holds for its elements.

The elements $a^1_1, a^2_2, \ldots$ constitute the *main diagonal* of a matrix. The sum of the elements of the main diagonal of a square matrix a is called the trace of the matrix. We denote it by $\operatorname{Tr} a$. Thus, we have $\operatorname{Tr} a = a^1_1 + a^2_2 + \ldots + a^n_n$ if a is a matrix of order n.

1.3 Definition. Matrices a and b are *equal* if and only if they are of the same type and if

$$a^i_j = b^i_j \quad \text{is true for all } i \text{ and } j\,.$$

The *sum* of two matrices a, b of the same type is the matrix

$$c = a + b \quad \text{for which} \quad c^i_j = a^i_j + b^i_j \quad \text{holds for all } i \text{ and } j\,.$$

The *product* of a real number λ and a matrix a is the matrix

$$b = \lambda a \quad \text{for which} \quad b^i_j = \lambda a^i_j \quad \text{holds for all } i \text{ and } j\,.$$

For $\lambda = -1$ we put $-1a = -a$.

The product of an $m \times n$ matrix (a) with an $n \times p$ matrix (b) is the $m \times p$ matrix

$$c = ab\,.$$

For the elements of this matrix we have

$$c^i_j = a^i_1 b^1_j + a^i_2 b^2_j + \ldots + a^i_n b^n_j\,.$$

From the given formula we see that the element c^i_j is the sum of the products of the respective elements of the i-th row of matrix a and the j-th row of matrix b.

1.4 Remark. In the sequel, summation will be used to simplify the notation. Unless otherwise stated, the following convention will be kept: If a product has two identical indices it is understood to be the symbol for summation with respect to this index. For $k = 1, 2, 3$ we have, for instance,

$$a^i_k b^k_j = a^i_1 b^1_j + a^i_2 b^2_j + a^i_3 b^3_j ,$$
$$a^k_k = a^1_1 + a^2_2 + a^3_3 .$$

The commutative law is not valid for matrix multiplication, in general. Therefore, it is necessary to pay careful attention to the order of the matrix factors in a product. For a product ab we say that matrix b is *left-multiplied* by matrix a, or that matrix a is *right-multiplied* by matrix b. Matrices for which $ab = ba$ hold are called *commutative* matrices.

1.5 Remark. Wherever the types of matrices involved in matrix operations will not be specified in what follows, it will be assumed that they satisfy Definition 1.3.

1.6 Theorem. *For the multiplication of matrices a, b, c, the associative and distributive laws hold*:

$$(ab)\,c = a(bc) ,$$
$$(a + b)\,c = ac + bc .$$

1.7 Theorem. *For matrices a, b and their respective transposes a^{T}, b^{T} we have*

$$(a + b)^{\mathrm{T}} = a^{\mathrm{T}} + b^{\mathrm{T}} , \quad (ab)^{\mathrm{T}} = b^{\mathrm{T}} a^{\mathrm{T}} , \quad (a^{\mathrm{T}})^{\mathrm{T}} = a .$$

Let us pay attention to the matrix product

$$\begin{pmatrix} e, & 0 \\ f, & 0 \end{pmatrix} \begin{pmatrix} 0, & 0 \\ g, & h \end{pmatrix} = \begin{pmatrix} 0, & 0 \\ 0, & 0 \end{pmatrix} = 0 .$$

We see that the product of two nonzero matrices can be a zero matrix. Consequently, the equalities

$$ab = 0 , \quad \text{and} \quad ab = ac$$

do not imply, in general,

$$a = 0 \quad \text{or} \quad b = 0 , \quad \text{and} \quad a = 0 \quad \text{or} \quad b = c ,$$

respectively.

1.8 Definition. The *derivative* of a matrix $a(\mathrm{t})$, whose elements are differentiable functions of the variable t on an interval **I**, is the matrix

$$a'(\mathrm{t}) = \frac{\mathrm{d}a(\mathrm{t})}{\mathrm{dt}} = \left(\frac{\mathrm{d}a_j^i(\mathrm{t})}{\mathrm{dt}}\right).$$

For the derivative of a matrix product we have

$$\frac{\mathrm{d}(ab)}{\mathrm{dt}} = a\frac{\mathrm{d}b}{\mathrm{dt}} + b\frac{\mathrm{d}a}{\mathrm{dt}}.$$

Before introducing the concept of determinant, let us recall the definition of permutation and inversion. By a *permutation* of n different elements we understand any ordered n-tuple of those elements. The number of all possible permutations is $P_n = 1.\ 2.\ 3 \ldots (n-1)\, n = n!$.

Consider a permutation $(m_1, m_2, \ldots, m_n)$ of positive integers $1, 2, \ldots$ $\ldots, n$. The numbers m_i and m_j, for which $i < j$ and $m_i > m_j$ hold, constitute an *inversion* in this permutation. A permutation with an even, or odd, number of inversions is called *even*, or *odd*, respectively. For instance, the permutation (5, 3, 1, 2, 4) is even since it contains 6 inversions. If p is the number of inversions in permutation $\pi = (m_1, m_2, \ldots, m_n)$, then the number $(-1)^p$ is called the *sign* of this permutation. It is denoted by π.

1.9 Definition. The *determinant of order n* of a square matrix

$$a = \begin{pmatrix} a_1^1, a_2^1, \ldots, a_n^1 \\ \cdots\cdots\cdots\cdots \\ a_1^n, a_2^n, \ldots, a_n^n \end{pmatrix}$$

is the number

$$\sum \operatorname{sgn} \pi a_{m_1}^1 a_{m_2}^2 \ldots a_{m_n}^n,$$

where addition is performed over all permutations $\pi = (m_1, m_2, \ldots, m_n)$ of numbers $1, 2, \ldots, n$. We denote the determinant by

$$\det a = \begin{vmatrix} a_1^1, a_2^1, \ldots, a_n^1 \\ a_1^2, a_2^2, \ldots, a_n^2 \\ \cdots\cdots\cdots\cdots \\ a_1^n, a_2^n, \ldots, a_n^n \end{vmatrix}.$$

1.10 Definition. The determinant obtained from the determinant det a by the omission of its i-th row and j-th column is called the *minor* of det a corresponding to the element a^i_j. This subdeterminant multiplied by the number $(-1)^{i+j}$ is called the *cofactor* of the element a^i_j in the determinant det a. We denote it $\det^i_j a$.

Before presenting several fundamental theorems on determinants, we recall that the properties of determinants related to their rows are valid for their columns as well, and vice versa.

1.11 Theorem. *A determinant does not change if to one of its rows we add a linear combination of its other rows.*

A determinant changes its sign if two of its rows are interchanged.

A determinant is equal to zero if any of its rows is a linear combination of the remaining rows. In particular this is true if a determinant has a zero row, or two identical rows, or two proportional rows.

1.12 Theorem. *A determinant is multiplied by a number λ if all the elements of one of its rows are multiplied by this number.*

1.13 Theorem. *The determinant of a product of matrices of order n is equal to the product of the determinants of the individual factors:*

$$\det ab = \det a \,.\, \det b \,.$$

The determinants of a matrix and of its respective transpose a^{T} are equal:

$$\det a = \det a^{\mathrm{T}} \,.$$

1.14 Theorem. *For a determinant of order n we have*

$$\det a = a^i_1 \det {}^i_1 a + a^i_2 \det {}^i_2 a + \ldots + a^i_n \det {}^i_n a$$

(the summation convention is not applied here).

This theorem on the so-called *expansion of a determinant with respect to the elements of its i-th row* enables the computation of a determinant by expanding this determinant, of order n, into a sum of determinants, of order $n - 1$.

1.15 Remark. The *derivative of a determinant* det $a(\mathrm{t})$ *of order* n, whose elements are differentiable functions of the variable t on an interval **I**, is equal to the sum of n determinants obtained from the given determinant by the successive substitution of all elements of the individual rows by their derivatives:

$$\frac{\mathrm{d}(\det a(\mathrm{t}))}{\mathrm{dt}} = \begin{vmatrix} \frac{\mathrm{d}a_1^1}{\mathrm{dt}}, & \frac{\mathrm{d}a_2^1}{\mathrm{dt}}, & \dots, & \frac{\mathrm{d}a_n^1}{\mathrm{dt}} \\ a_1^2, & a_2^2, & \dots, & a_n^2 \\ \dots & \dots & \dots & \dots \\ a_1^n, & a_2^n, & \dots, & a_n^n \end{vmatrix} + \begin{vmatrix} a_1^1, & a_2^1, & \dots, & a_n^1 \\ \frac{\mathrm{d}a_1^2}{\mathrm{dt}}, & \frac{\mathrm{d}a_2^2}{\mathrm{dt}}, & \dots, & \frac{\mathrm{d}a_n^2}{\mathrm{dt}} \\ \dots & \dots & \dots & \dots \\ a_1^n, & a_2^n, & \dots, & a_n^n \end{vmatrix} + \dots.$$

Let us return to matrices again:

1.16 Definition. A square matrix a is called *regular* or *singular* if $\det a \neq 0$ or $\det a = 0$, respectively.

1.17 Definition. The *inverse* matrix of a square matrix a of order n is the matrix a^{-1} of the same order for which

$$a^{-1}a = aa^{-1} = e\,,$$

Here e is the unit matrix of order n.

1.18 Theorem. *The inverse matrix a^{-1} of a matrix a exists if and only if the matrix a is regular. We have*

$$a^{-1} = (1/\det a)\begin{pmatrix} \det {}_1^1a, & \det {}_1^2a, & \dots, & \det {}_1^na \\ \det {}_2^1a, & \det {}_2^2a, & \dots, & \det {}_2^na \\ \dots & \dots & \dots & \dots \\ \det {}_n^1a, & \det {}_n^2a, & \dots, & \det {}_n^na \end{pmatrix}.$$

For the determinant of the inverse matrix a^{-1}, we have

$$\det a^{-1} = 1/\det a\,.$$

1.19 Theorem. *If a, b are regular matrices of the same order, we have*

$$(ab)^{-1} = b^{-1}a^{-1}\,, \quad (a^{-1})^{-1} = a\,, \quad (a^{-1})^{\mathrm{T}} = (a^{\mathrm{T}})^{-1}\,.$$

1.20 Definition. An *orthogonal* matrix is a matrix a such that

$$a^{\mathrm{T}} = a^{-1}\,.$$

1.21 Theorem. *A square matrix a is orthogonal if for all its elements*

$$a_i^j a_k^j = \delta_{ik} \quad \text{or} \quad a_j^i a_j^k = \delta_{ik}\,.$$

If a matrix is orthogonal, then all its elements satisfy both conditions given above.

Consequently, in an orthogonal matrix the sum of the products of all elements of its arbitrary row (column) with the corresponding elements of a different row (column) is equal to zero, while the sum of the squares of all the elements of an arbitrary row (column) is equal to one.

1.22 Theorem. *The determinant of an orthogonal matrix is equal to* 1 *or* -1. *The product of orthogonal matrices is an orthogonal matrix. The inverse matrix to an orthogonal matrix is an orthogonal matrix.*

1.23 Theorem. *The matrices* a, a^{T} *which satisfy the equations*

$$aa^{\mathrm{T}} = a^{\mathrm{T}}a = e$$

are orthogonal matrices.

1.24 Definition. We say that an $m \times n$ matrix has *rank* h if there exists at least one of its minors of order h (i.e. a determinant of order h consisting of those elements at which arbitrary h rows and h columns of matrix a intersect) which differs from zero while all its minors of higher order are equal to zero.

B. *Vectors in Three-dimensional Euclidean Space*

First, we deal with vectors in the three-dimensional Euclidean space E_3 from the purely geometrical point of view.

The segment AB, whose one endpoint, e.g. A, is designated as the *initial* point while the other endpoint B as the *terminal* point, is called an *oriented segment* AB. By the *length* of an oriented segment we understand the distance of its endpoints A, B. The zero segment for which $A \equiv B$ is also considered to be an oriented segment.

Let AB and CD be parallel oriented segments, let AD' be the segment obtained by parallel translation of CD. Then AB and CD have *the same* (*opposite*) orientation if D' lies (does not lie) on the half line AB, respectively.

1.25 Definition. The set of all oriented segments which have the same length and orientation is called a *vector*. Each element of this set is called a *representative of the vector*. By the *length of a vector* we understand the length of its arbitrary representative. If this length is equal to 1 or to 0 we speak of a *unit vector* or *zero vector*, respectively.

Vectors are denoted by lower case semibold type letters, e.g. by $\boldsymbol{u}$ – the only exception being the zero vector $\boldsymbol{0}$. Their representatives, e.g. AB, are denoted by $\boldsymbol{u}_{AB}$. The lengths of a vector and of its representative are denoted by $\|\boldsymbol{u}\|$ and $\|\boldsymbol{u}_{AB}\|$, respectively. Sometimes, if there is no danger of misunderstandings, we denote by the same letter the representative of a vector and the vector itself.

It is obvious that at an arbitrary point $A \in E_3$ every vector has a single representative with initial point A. This representative is called the *bound vector* at the point A.

1.26 Definition. The vectors $\boldsymbol{u}$ and $\boldsymbol{v}$ are *equal*, $\boldsymbol{u} = \boldsymbol{v}$, if we have $B \equiv C$ for their representatives $\boldsymbol{u}_{AB}$ and $\boldsymbol{u}_{AC}$.

The *sum* of two vectors $\boldsymbol{u}$ and $\boldsymbol{v}$ is the vector $\boldsymbol{w} = \boldsymbol{u} + \boldsymbol{v}$ whose representative is $\boldsymbol{w}_{AC}$, where $\boldsymbol{u}_{AB}$ and $\boldsymbol{v}_{BC}$ are representatives of $\boldsymbol{u}$ and $\boldsymbol{v}$, respectively.

The *product* of a real number λ and a vector $\boldsymbol{u}$ is the vector $\boldsymbol{v} = \lambda\boldsymbol{u}$ whose length is $\|\boldsymbol{v}\| = |\lambda| \cdot \|\boldsymbol{u}\|$. Here the representatives $\boldsymbol{u}_{AB}$ and $\boldsymbol{v}_{AC}$ have the same (opposite) orientation for $\lambda > 0$ (for $\lambda < 0$). We say that the vectors $\boldsymbol{u}$ and $\boldsymbol{v}$ are *colinear*. If $\lambda = -1$, the vector $\boldsymbol{v} = -\boldsymbol{u}$ is called the *inverse* vector to vector $\boldsymbol{u}$; the representatives of these vectors are $\boldsymbol{u}_{AB}$, $\boldsymbol{v}_{BA}$.

The *difference of vectors* $\boldsymbol{u}$ and $\boldsymbol{v}$ (denoted by $\boldsymbol{u} - \boldsymbol{v}$) is the vector $\boldsymbol{w} = \boldsymbol{u} + (-\boldsymbol{v})$.

1.27 Theorem. *For vector operations we have*

$$\begin{aligned}
&\alpha)\ \boldsymbol{u} + \boldsymbol{v} = \boldsymbol{v} + \boldsymbol{u},\ \boldsymbol{u} + (\boldsymbol{v} + \boldsymbol{w}) = (\boldsymbol{u} + \boldsymbol{v}) + \boldsymbol{w}\,,\\
&\beta)\ \boldsymbol{u} + \boldsymbol{0} = \boldsymbol{u},\ \boldsymbol{u} + (-\boldsymbol{u}) = \boldsymbol{0},\ 1 \,.\, \boldsymbol{u} = \boldsymbol{u}\,,\\
&\gamma)\ \lambda(\mu\boldsymbol{u}) = (\lambda\mu)\,\boldsymbol{u},\ (\lambda + \mu)\,\boldsymbol{u} = \lambda\boldsymbol{u} + \mu\boldsymbol{u},\ \lambda(\boldsymbol{u} + \boldsymbol{v}) = \lambda\boldsymbol{u} + \lambda\boldsymbol{v}\,,
\end{aligned}$$

where λ, μ are arbitrary real numbers.

1.28 Definition. The *angle* between two noncolinear (two colinear) nonzero vectors is the acute (zero eventually straight) angle between their representatives with a common initial point.

1.29 Definition. The *inner* product of nonzero vectors $\boldsymbol{u}$ and $\boldsymbol{v}$ is the real number

$$(\boldsymbol{u}, \boldsymbol{v}) = \|\boldsymbol{u}\| \, \|\boldsymbol{v}\| \cos \varphi \, .$$

Here φ $(0 \leqq \varphi \leqq \pi)$ is the angle between the two vectors. If one of the vectors is a zero vector, then their inner product is $(\boldsymbol{u}, \boldsymbol{v}) = 0$.

1.30 Theorem. *For the inner product of vectors we have*

$$(\boldsymbol{u}, \boldsymbol{v}) = (\boldsymbol{v}, \boldsymbol{u}) \, , \quad (\boldsymbol{u} + \boldsymbol{v}, \boldsymbol{w}) = (\boldsymbol{u}, \boldsymbol{w}) + (\boldsymbol{v}, \boldsymbol{w}) \, ,$$
$$(\lambda \boldsymbol{u}, \boldsymbol{v}) = (\boldsymbol{u}, \lambda \boldsymbol{v}) = \lambda(\boldsymbol{u}, \boldsymbol{v}) \, ,$$

where λ is an arbitrary real number.

In case of the equality of vectors $\boldsymbol{u} = \boldsymbol{v}$ Definition 1.29 implies:

1.31 Theorem. *The length of the vector $\boldsymbol{u}$ is*

$$\|\boldsymbol{u}\| = [(\boldsymbol{u}, \boldsymbol{u})]^{1/2} \, .$$

1.32 Theorem. *Nonzero vectors $\boldsymbol{u}$ and $\boldsymbol{v}$ are perpendicular if and only if* $(\boldsymbol{u}, \boldsymbol{v}) = 0$.

Let us construct, from m vectors $\boldsymbol{u}_1, \boldsymbol{u}_2, \ldots, \boldsymbol{u}_m$ and m real numbers $\lambda_1, \lambda_2, \ldots, \lambda_m$, the vector

$$\boldsymbol{u} = \lambda_1 \boldsymbol{u}_1 + \lambda_2 \boldsymbol{u}_2 + \ldots + \lambda_m \boldsymbol{u}_m \, .$$

This vector is called a *linear combination* of vectors $\boldsymbol{u}_i$. The numbers λ_i are called the *coefficients of* this *linear combination.*

1.33 Definition. The vectors $\boldsymbol{u}_1, \boldsymbol{u}_2, \ldots, \boldsymbol{u}_m$ are *linearly dependent* if they satisfy the equation

$$\lambda_1 \boldsymbol{u}_1 + \lambda_2 \boldsymbol{u}_2 + \ldots + \lambda_m \boldsymbol{u}_m = \boldsymbol{0}$$

under the assumption that at least one of the coefficients λ_i is different from zero. In the opposite case the vectors are called *linearly independent.*

On the linear dependence of vectors in E_3 the following theorem holds:

1.34 Theorem. *Two or three vectors are linearly dependent if and only if they are colinear or coplanar, respectively.* (Coplanar means that their representatives with common initial point lie in one plane.) *Four and more vectors are always linearly dependent.*

1.35 **Definition.** An ordered triad of linearly independent vectors $\boldsymbol{f}_1, \boldsymbol{f}_2, \boldsymbol{f}_3$ is called a *base* and it is denoted $\{\boldsymbol{f}_1, \boldsymbol{f}_2, \boldsymbol{f}_3\}$. An ordered triad of bound vectors at a point A which represent the base vectors $\boldsymbol{f}_1, \boldsymbol{f}_2, \boldsymbol{f}_3$ is called a *frame* and it is denoted $\mathscr{R} = \{A, \boldsymbol{f}_1, \boldsymbol{f}_2, \boldsymbol{f}_3\}$. The point A is the *origin* of the frame. Frame vectors will be denoted $\boldsymbol{f}_{A1}, \boldsymbol{f}_{A2}, \boldsymbol{f}_{A3}$, or sometimes $\boldsymbol{f}_1, \boldsymbol{f}_2, \boldsymbol{f}_3$ only. If the base vectors are mutually perpendicular, or unit and mutually perpendicular, we say that the base and its respective frame are *orthogonal*, or *orthonormal*, respectively.

Before introducing the concept of the orientation of a base or of a frame a convention is made concerning the orientation of the space E_3 by an ordered pair of noncolinear bound vectors at a common point. Let $\boldsymbol{u}_{AB}, \boldsymbol{v}_{AC}$ be such a pair of vectors forming an angle φ $(0 < \varphi < \pi)$. The plane determined by these vectors is the boundary plane of two halfspaces. The halfspace from which the rotation of the halfline AB into the halfline AC about A through the angle φ appears to be positive, i.e. counterclockwise oriented, will be positively oriented. The other halfspace is negatively oriented.

1.36 **Definition.** The frame $\{A, \boldsymbol{f}_1, \boldsymbol{f}_2, \boldsymbol{f}_3\}$ is *right-handed*, or *left-handed*, respectively, if the bound vector $\boldsymbol{f}_{A3}$ lies in the positive, or negative,

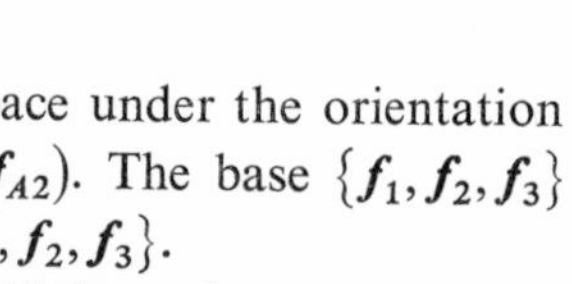

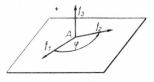

Fig. 1.

halfspace under the orientation of the space by the pair of bound vectors $(\boldsymbol{f}_{A1}, \boldsymbol{f}_{A2})$. The base $\{\boldsymbol{f}_1, \boldsymbol{f}_2, \boldsymbol{f}_3\}$ is oriented in conformity with the frame $\{A, \boldsymbol{f}_1, \boldsymbol{f}_2, \boldsymbol{f}_3\}$.

Unless otherwise stated we will always consider a right-handed frame (see Fig. 1). Note that a right-handed, or left-handed, frame $\{A, \boldsymbol{f}_1, \boldsymbol{f}_2, \boldsymbol{f}_3\}$ can be illustrated quite well by representing the fixed vectors $\boldsymbol{f}_{A1}, \boldsymbol{f}_{A2}, \boldsymbol{f}_{A3}$

by the thumb, forefinger, and middlefinger of the right hand, or left hand, respectively.

1.37 Definition. The *cross product* of noncolinear vectors $\boldsymbol{u}$ and $\boldsymbol{v}$, denoted by $\boldsymbol{u} \times \boldsymbol{v}$, is the vector $\boldsymbol{w}$ with the following properties:

1. $\boldsymbol{w}$ is perpendicular to the vectors $\boldsymbol{u}$ and $\boldsymbol{v}$,
2. $\boldsymbol{w}$ constitutes a right-handed base $\{\boldsymbol{u}, \boldsymbol{v}, \boldsymbol{w}\}$ together with the vectors $\boldsymbol{u}$ and $\boldsymbol{v}$,
3. the length of $\boldsymbol{w}$ is $\|\boldsymbol{w}\| = \|\boldsymbol{u}\| \, \|\boldsymbol{v}\| \sin \varphi$, where φ is the angle between the vectors $\boldsymbol{u}$ and $\boldsymbol{v}$.

The cross product of two colinear vectors is the zero vector.

1.38 Theorem. *For the cross product we have*

$$\boldsymbol{u} \times \boldsymbol{v} = -\boldsymbol{v} \times \boldsymbol{u}\,, \quad \boldsymbol{u} \times (\boldsymbol{v} + \boldsymbol{w}) = \boldsymbol{u} \times \boldsymbol{v} + \boldsymbol{u} \times \boldsymbol{w}\,,$$
$$\boldsymbol{u} \times (\boldsymbol{v} \times \boldsymbol{w}) = (\boldsymbol{u}, \boldsymbol{w})\, \boldsymbol{v} - (\boldsymbol{u}, \boldsymbol{v})\, \boldsymbol{w}\,,$$
$$\lambda \boldsymbol{u} \times \boldsymbol{v} = \lambda(\boldsymbol{u} \times \boldsymbol{v})\,, \quad \textit{where } \lambda \textit{ is an arbitrary real number}\,,$$
$$(\boldsymbol{u} \times \boldsymbol{v}, \boldsymbol{w} \times \boldsymbol{t}) = (\boldsymbol{u}, \boldsymbol{w})(\boldsymbol{v}, \boldsymbol{t}) - (\boldsymbol{u}, \boldsymbol{t})(\boldsymbol{v}, \boldsymbol{w})\,,$$
$$\|\boldsymbol{u} \times \boldsymbol{v}\| = [(\boldsymbol{u}, \boldsymbol{u})(\boldsymbol{v}, \boldsymbol{v}) - (\boldsymbol{u}, \boldsymbol{v})^2]^{1/2}\,.$$

1.39 Theorem. *For the vectors of an orthonormal base* $\boldsymbol{u}_1, \boldsymbol{u}_2, \boldsymbol{u}_3$ *we have*

$$\boldsymbol{u}_1 \times \boldsymbol{u}_2 = \boldsymbol{u}_3\,, \quad \boldsymbol{u}_2 \times \boldsymbol{u}_3 = \boldsymbol{u}_1\,, \quad \boldsymbol{u}_3 \times \boldsymbol{u}_1 = \boldsymbol{u}_2\,,$$
$$\boldsymbol{u}_1 \times \boldsymbol{u}_1 = \boldsymbol{0}\,, \quad \boldsymbol{u}_2 \times \boldsymbol{u}_2 = \boldsymbol{0}\,, \quad \boldsymbol{u}_3 \times \boldsymbol{u}_3 = \boldsymbol{0}\,.$$

1.40 Definition. The *mixed product* of an ordered triad of vectors $\boldsymbol{u}, \boldsymbol{v}, \boldsymbol{w}$, denoted by $|\boldsymbol{u}, \boldsymbol{v}, \boldsymbol{w}|$, is the real number

$$|\boldsymbol{u}, \boldsymbol{v}, \boldsymbol{w}| = (\boldsymbol{u}, \boldsymbol{v} \times \boldsymbol{w})\,.$$

1.41 Theorem. *Three vectors are coplanar if and only if their mixed product is equal to zero.*

1.42 Theorem. *For the mixed product we have*

$$|\boldsymbol{u}, \boldsymbol{v}, \boldsymbol{w}| = |\boldsymbol{v}, \boldsymbol{w}, \boldsymbol{u}| = |\boldsymbol{w}, \boldsymbol{u}, \boldsymbol{v}|\,,$$
$$|\boldsymbol{u}, \boldsymbol{v}, \boldsymbol{w}| = -|\boldsymbol{v}, \boldsymbol{u}, \boldsymbol{w}| = -|\boldsymbol{u}, \boldsymbol{w}, \boldsymbol{v}| = -|\boldsymbol{w}, \boldsymbol{v}, \boldsymbol{u}|\,,$$
$$|\lambda \boldsymbol{u}, \boldsymbol{v}, \boldsymbol{w}| = \lambda |\boldsymbol{u}, \boldsymbol{v}, \boldsymbol{w}|\,, \quad \textit{where } \lambda \textit{ is an arbitrary real number}\,.$$

Now we express vectors by algebraic objects and describe the basic vector operations in this mode of expression:

1.43 Theorem. *Let* $\{f_1, f_2, f_3\}$ *be a base in* E_3. *Then there is a one to one correspondence between vectors* $\boldsymbol{u} \in E_3$ *and ordered triads* u^1, u^2, u^3 *of real numbers such that*

$$\boldsymbol{u} = u^1 \boldsymbol{f}_1 + u^2 \boldsymbol{f}_2 + u^3 \boldsymbol{f}_3 . \tag{1.1}$$

1.44 Definition. The coefficients of the linear combination (1.1) are called *coordinates* of vector $\boldsymbol{u}$ while the vectors $u^1\boldsymbol{f}_1$, $u^2\boldsymbol{f}_2$, $u^3\boldsymbol{f}_3$ are *components* of vector $\boldsymbol{u}$, in the base $\{\boldsymbol{f}_1, \boldsymbol{f}_2, \boldsymbol{f}_3\}$, or in the frame $\{A, \boldsymbol{f}_1, \boldsymbol{f}_2, \boldsymbol{f}_3\}$.

Further we consider the coordinates of the vector $\boldsymbol{u}$ in the orthonormal frame $\mathcal{R} = \{A, \boldsymbol{u}_1, \boldsymbol{u}_2, \boldsymbol{u}_3\}$, and use the notation $\boldsymbol{u} = [u^1, u^2, u^3]$ for the coordinate form of the vector $\boldsymbol{u}$. To the frame $\mathcal{R} = \{A, \boldsymbol{u}_1, \boldsymbol{u}_2, \boldsymbol{u}_3\}$ we assign a Cartesian coordinate system (A, x^1, x^2, x^3) whose origin is at the point A while the unit point on the x^i axis is the terminal point of the vector u_i of the frame $\mathcal{R}$ $(i = 1, 2, 3)$.

1.45 Theorem. *The coordinates of vector* $\boldsymbol{u}$ *in an orthonormal frame satisfy the equations*

$$u^i = c^i - b^i , \quad i = 1, 2, 3 ,$$

where c^i *and* b^i *are the Cartesian coordinates of the terminal and initial points of an arbitrary representative* $\boldsymbol{u}_{BC}$ *of the vector* $\boldsymbol{u}$.

If the representative $\boldsymbol{u}_{AU}$ of the vector $\boldsymbol{u}$ is bound at the origin of the frame it is called the *position vector* or *radius vector* of its terminal point U. As follows from Theorem 1.45 the coordinates of the vector $\boldsymbol{u}$ are identical with the Cartesian coordinates of the point U.

Relations between the coordinates of equal vectors, of the sum of two vectors and of the product of a vector and a real number follow from equation (1.1).

1.46 Theorem. *For vectors and their coordinates we have*

$$\boldsymbol{u} = \boldsymbol{v} \Leftrightarrow u^i = v^i ,$$

$$\boldsymbol{u} + \boldsymbol{v} = \boldsymbol{w} \Leftrightarrow u^i + v^i = w^i ,$$

$$\lambda \boldsymbol{u} = \boldsymbol{v} \Leftrightarrow \lambda u^i = v^i , \quad \textit{where } \lambda \textit{ is an arbitrary real number} ,$$

$$i = 1, 2, 3 .$$

1.47 Theorem. *In an orthonormal base the inner product of vectors* $\boldsymbol{u}, \boldsymbol{v}$

$$(\boldsymbol{u}, \boldsymbol{v}) = u^1 v^1 + u^2 v^2 + u^3 v^3$$

while the length of vector $\boldsymbol{u}$ *is*

$$\|\boldsymbol{u}\| = [(u^1)^2 + (u^2)^2 + (u^3)^2]^{1/2} .$$

The angle φ *between non-zero vectors* $\boldsymbol{u}, \boldsymbol{v}$ *satisfies the equation*

$$\cos \varphi = \frac{(\boldsymbol{u}, \boldsymbol{v})}{\|\boldsymbol{u}\| \, \|\boldsymbol{v}\|} .$$

Further, equation (1.1) and Theorems 1.38 and 1.39 imply:

1.48 Theorem. *In the orthonormal base* $\boldsymbol{u}_1, \boldsymbol{u}_2, \boldsymbol{u}_3$ *the cross pro*[c]
vectors $\boldsymbol{u}, \boldsymbol{v}$ *is*

$$\boldsymbol{w} = \boldsymbol{u} \times \boldsymbol{v} =$$
$$= (u^2 v^3 - u^3 v^2)\, \boldsymbol{u}_1 + (u^3 v^1 - u^1 v^3)\, \boldsymbol{u}_2 + (u^1 v^2 - u^2 v^1)\, \boldsymbol{u}_3$$

Coordinates of the cross product can be expressed in brief a

$$w^i = \varepsilon_{ijk} u^j v^k , \quad \text{where} \quad i, j, k = 1, 2, 3$$

and

$$\varepsilon_{ijk} = \begin{cases} 1 \\ -1 \\ 0 \end{cases} \text{for} \begin{cases} \text{an even permutation of the elements 1} \\ \text{an odd permutation of the elements 1} \\ \text{at least two equal indices.} \end{cases}$$

For the mixed product of vectors $\boldsymbol{u}, \boldsymbol{v}, \boldsymbol{w}$ we have, by De
and Theorem 1.48,

$$|\boldsymbol{u}, \boldsymbol{v}, \boldsymbol{w}| = \varepsilon_{ijk} u^i v^j w^k$$

which yields further, by Definition 1.9 of the determinant,
theorem:

1.49 Theorem. *In an orthonormal base the mixed pro*
$\boldsymbol{u}, \boldsymbol{v}, \boldsymbol{w}$ *is*

$$|\boldsymbol{u}, \boldsymbol{v}, \boldsymbol{w}| = \begin{vmatrix} u^1, & u^2, & u^3 \\ v^1, & v^2, & v^3 \\ w^1, & w^2, & w^3 \end{vmatrix} .$$

1.50 Definition. A rule which assigns to every real number t from an interval **I** exactly one vector $\boldsymbol{v}$ is called a *vector function of one variable.* It is denoted by $\boldsymbol{v} = \boldsymbol{v}(t)$, or in coordinate form by

$$\boldsymbol{v} = [v^1(t), v^2(t), v^3(t)],$$

where $v^i(t)$, $i = 1, 2, 3$, are functions of one real variable defined on the common interval **I**.

1.51 Definition. If the functions $v^i(t)$ have finite derivatives at the point $t = t_0$,

$$\dot{v}^i(t_0) = \frac{dv^i(t_0)}{dt}, \quad i = 1, 2, 3,$$

then by *derivative of the vector function* $\boldsymbol{v}(t)$ at the point $t = t_0$ we understand the vector

$$\boldsymbol{v}^{\cdot}(t_0) = [\dot{v}(t_0), \dot{v}^2(t_0), \dot{v}^3(t_0)].$$

If a vector function $\boldsymbol{v}(t)$ has a finite derivative at every point $t \in \mathbf{I}$, then the vector function

$$\boldsymbol{v}^{\cdot}(t) = [\dot{v}^1(t), \dot{v}^2(t), \dot{v}^3(t)]$$

is called the derivative of the vector function $\boldsymbol{v}(t)$ on the interval **I**.

1.52 Theorem. *For derivatives of the vector functions* $\boldsymbol{u}(t)$, $\boldsymbol{v}(t)$ *on the interval* **I** *we have*

$$(\boldsymbol{u} \pm \boldsymbol{v})^{\cdot} = \boldsymbol{u}^{\cdot} \pm \boldsymbol{v}^{\cdot},$$

$$(\boldsymbol{u}, \boldsymbol{v})^{\cdot} = (\boldsymbol{u}^{\cdot}, \boldsymbol{v}) + (\boldsymbol{u}, \boldsymbol{v}^{\cdot}),$$

$$(\boldsymbol{u} \times \boldsymbol{v})^{\cdot} = \boldsymbol{u}^{\cdot} \times \boldsymbol{v} + \boldsymbol{u} \times \boldsymbol{v}^{\cdot}.$$

A vector function $\boldsymbol{v}(t)$ is sometimes briefly called a *vector* $\boldsymbol{v}(t)$. Similarly we define the vector function of two variables

$$\boldsymbol{v}(t, w) = [v^1(t, w), v^2(t, w), v^3(t, w)],$$

where $v^i(t, w)$, $i = 1, 2, 3$, are functions of two real variables defined on a common domain. Under the assumption of the differentiability of the

functions $v^i(t, w)$, $i = 1, 2, 3$, we then define the partial derivatives

$$\frac{\partial v}{\partial t} = \left[\frac{\partial v^1}{\partial t}, \frac{\partial v^2}{\partial t}, \frac{\partial v^3}{\partial t}\right], \quad \frac{\partial v}{\partial w} = \left[\frac{\partial v^1}{\partial w}, \frac{\partial v^2}{\partial w}, \frac{\delta v^3}{\partial w}\right].$$

C. *Vector Spaces*

In Definition 1.26 the operations of the sum of two vectors and of the product of a vector and a real number were introduced. Their outcomes are vectors again. The properties of these operations were given in Theorem 1.27. However, one frequently encounters sets of mathematical objects for which the two operations are also defined, satisfying the properties mentioned above. The reader can verify this easily by considering the following examples, in which operations upon elements of sets are the usual ones. Consider the sets: (a) of all real numbers (this set will be denoted by **R** in the sequel), (b) of all polynomials with real coefficients, (c) of all matrices of a given type, (d) of all triads of real numbers which are the coordinates of vectors in a given base.

The quoted examples show the heterogeneity of the investigated objects and indicate the desirability of generalization. Therefore we will not discuss concrete objects and the form of the operations applied to them, but we shall present a system of axioms which the operations with given objects should satisfy. From these axioms additional properties of the objects will then be derived. The advantage of this axiomatic approach lies in the fact that the consequences following from the axioms can be elaborated once and for all and then applied to all sets of elements which satisfy the considered system of axioms.

1.53 Definition. A set V of elements $\boldsymbol{u}, \boldsymbol{v}, \ldots$ is called a *vector space over the set of all real numbers* **R** if:

1) the *sum* of two elements is defined in V,
2) the *product* of a real number and an element from V is defined in V,
3) an element from V, denoted by $\boldsymbol{0}$, exists such that the relations (α), (β) and (γ) of Theorem 1.27 are satisfied for arbitrary $\lambda, \mu \in \mathbf{R}$ and for arbitrary elements $\boldsymbol{u}, \boldsymbol{v}, \boldsymbol{w}$ from V, $-\boldsymbol{u} = (-1)\,\boldsymbol{u}$.

The elements of V will be called *vectors* while the real numbers will be called *scalars*.

Note that the concept of vector is used here in a broader sense than in Part B.

(α) and (β) of Theorem 1.27 imply that in a vector space there exists only one zero element and just one inverse element to a given element.

1.54 Exercise. Decide whether

a) all unit vectors in E_3
b) all polynomials of degree at most n
c) all polynomials of degree n

constitute a vector space.

Using the operations introduced above it is possible to generate a linear combination of the vectors $\boldsymbol{u}_i$

$$\boldsymbol{u} = \lambda_1 \boldsymbol{u}_1 + \lambda_2 \boldsymbol{u}_2 + \ldots + \lambda_m \boldsymbol{u}_m \quad (\lambda_i \in \mathbf{R};\ i = 1, \ldots, m)$$

and then define linear dependence and independence of vectors.

1.55 Definition. Vectors $\boldsymbol{u}_1, \boldsymbol{u}_2, \ldots, \boldsymbol{u}_m$ are *linearly dependent* if they satisfy the equation

$$\lambda_1 \boldsymbol{u}_1 + \lambda_2 \boldsymbol{u}_2 + \ldots + \lambda_m \boldsymbol{u}_m = 0$$

under the assumption that at least one of the coefficients $\lambda_i \in \mathbf{R}$ differs from zero. In the opposite case the vectors are *linearly independent*.

1.56 Definition. A vector space V is called *n-dimensional*, and is denoted V_n, if the following axioms are satisfied:

C1. There exist n linearly independent vectors in V.
C2. Every set of $n + 1$ vectors of the space V is linearly dependent.

The positive integer n, called the *dimension of the vector space* V_n, determines thus the maximal number of linearly independent vectors from the space V_n. If such a number exists, we speak of a *finite dimensional vector space*.

1.57 Exercise. Decide whether all matrices u of type $(n + 1) \times 1$, n positive integer, for which $u_1^1 = 0$ and the operations of addition and multiplication by a scalar are defined in Definition 1.3, constitute a vector space. In the affirmative case, determine its dimension.

1.58 Definition. An ordered n-tuple of linearly independent vectors $f_1, f_2, \ldots, f_n$ of an n-dimensional vector space is called its *base* and we denote it by

$$\{f_1, f_2, \ldots, f_n\}.$$

1.59 Theorem. *Every vector $u \in V_n$ can be expressed uniquely as a linear combination of the vectors of the base $\{f_1, f_2, \ldots, f_n\}$ of the space V_n*

$$u = u^1 f_1 + u^2 f_2 + \ldots + u^n f_n.$$

The numbers $u^1, u^2, \ldots, u^n$ are called *coordinates* of the vector u in the base $f_1, f_2, \ldots, f_n$, the vectors $u^1 f_1, u^2 f_2, \ldots, u^n f_n$ are components of the vector u in the base $f_1, f_2, \ldots, f_n$.

1.60 Theorem. *Let f_i, $i = 1, 2, \ldots, n$, be linearly independent vectors of the vector space V, and let every vector from V be their linear combination. Then the vectors f_i constitute a base of the vector space V whose dimension is n.*

1.61 Example. Let us show that all ordered n-tuples of real numbers constitute an n-dimensional vector space. Define the operations of addition of two n-tuples $(x^1, \ldots, x^n)$ and $(y^1, \ldots, y^n)$ and of multiplication of an n-tuple $(x^1, \ldots, x^n)$ by a real number λ by

$$(x^1, x^2, \ldots, x^n) + (y^1, y^2, \ldots, y^n) = (x^1 + y^1, x^2 + y^2, \ldots, x^n + y^n),$$
$$\lambda(x^1, x^2, \ldots, x^n) = (\lambda x^1, \lambda x^2, \ldots, \lambda x^n).$$

It is obvious that all the axioms 1.27 are satisfied, i.e. all ordered n-tuples of real numbers constitute a vector space. Choose n vectors of this space

$$(1, 0, \ldots, 0), (0, 1, \ldots, 0), \ldots, (0, 0, \ldots, 1)$$

and form their linear combinations ($\lambda_i \in \mathbf{R}$):

$$\lambda_1(1, 0, \ldots, 0) + \lambda_2(0, 1, \ldots, 0) + \ldots + \lambda_n(0, 0, \ldots, 1)$$
$$= (\lambda_1, \lambda_2, \ldots, \lambda_n).$$

The vector $(\lambda_1, \lambda_2, \ldots, \lambda_n)$ is a zero vector if and only if all λ_i, $i = 1, \ldots, n$, are equal to zero. This implies that the chosen vectors are linearly independent.

Conversely, if $(\lambda_1, \ldots, \lambda_n)$ is an arbitrary n-tuple of real numbers it can be written as a linear combination of the chosen vectors. Consequently, the chosen vectors constitute a base and the dimension is n.

1.62 Example. Let us consider the set of all polynomials in one variable $\mathbf{x}$ with real coefficients whose degree is at most n. According to Exercise 1.54 these polynomials constitute a vector space V. Every vector of this space is a linear combination of $n + 1$ vectors $\mathbf{x}^n, \mathbf{x}^{n-1}, \ldots, \mathbf{x}^0$ which are obviously linearly independent. According to Theorem 1.60 the vectors $\mathbf{x}^n, \mathbf{x}^{n-1}, \ldots$ $\ldots, \mathbf{x}^0$ thus constitute a base of the vector space V of all the considered polynomials, and this space is $(n + 1)$-dimensional.

Note that the operations of vector addition and of multiplication of a vector by a scalar lead to the same operations, applied to the coordinates of these vectors, as defined in Example 1.61 for ordered n-tuples of real numbers. Thus, the n-dimensional vector space is, in a certain sense, constructed in the same way as the space of all ordered n-tuples of real numbers. Consequently, the construction of all other n-dimensional vector spaces is essentially the same. Such spaces are called *isomorphic*. This idea is expressed precisely in the following definition:

1.63 Definition. The vector spaces V and V' are called *isomorphic* if there exists a one-to-one correspondence between their elements

$$u \leftrightarrow u', \quad v \leftrightarrow v', \quad u, v \in V, \quad u', v' \in V',$$

such that

$$u + v \leftrightarrow u' + v',$$

$$\lambda u \leftrightarrow \lambda u', \quad \lambda \in \mathbf{R}.$$

1.64 Theorem. *Two vector spaces are isomorphic if and only if they have the same dimension.*

In concluding we discuss the concepts of linear and bilinear form on a vector space V.

1.65 Definition. A *linear form* $\mathbf{g}$ on a vector space V is a rule which assigns to every vector $\boldsymbol{u} \in V$ exactly one real number $\mathbf{g}(\boldsymbol{u})$ such that

$$\mathbf{g}(\boldsymbol{u} + \boldsymbol{v}) = \mathbf{g}(\boldsymbol{u}) + \mathbf{g}(\boldsymbol{v}),$$
$$\mathbf{g}(\lambda \boldsymbol{u}) = \lambda\, \mathbf{g}(\boldsymbol{u})$$

where $\boldsymbol{u}, \boldsymbol{v} \in V$ and $\lambda \in \mathbf{R}$.

1.66 Exercise. Ascertain that the mixed product $|\boldsymbol{u}, \boldsymbol{v}, \boldsymbol{w}|$ of vectors $\boldsymbol{u}, \boldsymbol{v}, \boldsymbol{w} \in E_3$, where the vectors $\boldsymbol{v}, \boldsymbol{w}$ are constant while $\boldsymbol{u}$ is a variable vector, is a linear form.

1.67 Exercise. Show that the function Tr a defined in 1.2 is a linear form on the vector space of all square matrices of order n.

1.68 Definition. A *bilinear form* $\mathbf{g}(\boldsymbol{u}, \boldsymbol{v})$ on a vector space V is a rule which assigns to every pair $\boldsymbol{u}, \boldsymbol{v} \in V$ a real number $\mathbf{g}(\boldsymbol{u}, \boldsymbol{v})$ in such a way that $\mathbf{g}(\boldsymbol{u}_0, \boldsymbol{v})$ $(\mathbf{g}(\boldsymbol{u}, \boldsymbol{v}_0))$ is a linear form on V for fixed $\boldsymbol{u}_0 \in V$ $(\boldsymbol{v}_0 \in V)$, respectively.

1.69 Theorem. *$\mathbf{g}(\boldsymbol{u}, \boldsymbol{v})$ is a bilinear form if and only if the equations*

$$\mathbf{g}(\boldsymbol{u} + \boldsymbol{v}, \boldsymbol{w}) = \mathbf{g}(\boldsymbol{u}, \boldsymbol{w}) + \mathbf{g}(\boldsymbol{v}, \boldsymbol{w}),$$
$$\mathbf{g}(\boldsymbol{u}, \boldsymbol{v} + \boldsymbol{w}) = \mathbf{g}(\boldsymbol{u}, \boldsymbol{v}) + \mathbf{g}(\boldsymbol{u}, \boldsymbol{w}),$$
$$\mathbf{g}(\lambda \boldsymbol{u}, \boldsymbol{v}) = \lambda\, \mathbf{g}(\boldsymbol{u}, \boldsymbol{v}),$$
$$\mathbf{g}(\boldsymbol{u}, \lambda \boldsymbol{v}) = \lambda\, \mathbf{g}(\boldsymbol{u}, \boldsymbol{v})$$

are true for vectors $\boldsymbol{u}, \boldsymbol{v}, \boldsymbol{w} \in V$ and an arbitrary real number λ.

1.70 Example. Let $\mathbf{g}(\boldsymbol{u}), \mathbf{h}(\boldsymbol{v})$ be linear forms. Let us prove that $\mathbf{f}(\boldsymbol{u}, \boldsymbol{v}) = \mathbf{g}(\boldsymbol{u})\, \mathbf{h}(\boldsymbol{v})$ is a bilinear form: we have

$$\mathbf{f}(\boldsymbol{u} + \boldsymbol{w}, \boldsymbol{v}) = \mathbf{g}(\boldsymbol{u} + \boldsymbol{w})\, \mathbf{h}(\boldsymbol{v})$$
$$= [\mathbf{g}(\boldsymbol{u}) + \mathbf{g}(\boldsymbol{w})]\, \mathbf{h}(\boldsymbol{v}) = \mathbf{f}(\boldsymbol{u}, \boldsymbol{v}) + \mathbf{f}(\boldsymbol{w}, \boldsymbol{v})$$

and further

$$\mathbf{f}(\lambda \boldsymbol{u}, \boldsymbol{v}) = \mathbf{g}(\lambda \boldsymbol{u})\, \mathbf{h}(\boldsymbol{v}) = \lambda\, \mathbf{g}(\boldsymbol{u})\, \mathbf{h}(\boldsymbol{v}) = \lambda\, \mathbf{f}(\boldsymbol{u}, \boldsymbol{v}).$$

Similarly we prove the validity of the conditions from Theorem 1.69 for the second argument, which accomplishes the proof.

If u^i and v^i are coordinates of vectors $\boldsymbol{u}$ and $\boldsymbol{v}$ in the base $\{\boldsymbol{f}_1, \boldsymbol{f}_2, \ldots, \boldsymbol{f}_n\}$ of the vector space V, then

$$g(\boldsymbol{u}, \boldsymbol{v}) = g(u^i \boldsymbol{f}_i, v^j \boldsymbol{f}_j), \quad i, j = 1, 2, \ldots, n .$$

The properties of the bilinear form imply

$$g(\boldsymbol{u}, \boldsymbol{v}) = u^i v^j g(\boldsymbol{f}_i, \boldsymbol{f}_j) = u^i v^j \lambda_{ij}, \quad i, j = 1, 2, \ldots, n, \quad \text{where}$$
$$\lambda_{ij} = g(\boldsymbol{f}_i, \boldsymbol{f}_j) .$$

A square matrix of n-th order formed by the coefficients (l^i_j), where $l^i_j = \lambda_{ij}$, is called the *matrix of the bilinear form* $g(\boldsymbol{u}, \boldsymbol{v})$ *in the base* $\{\boldsymbol{f}_1, \ldots, \boldsymbol{f}_n\}$. A bilinear form is called *symmetric*, or *skew-symmetric*, if

$$g(\boldsymbol{u}, \boldsymbol{v}) = g(\boldsymbol{v}, \boldsymbol{u}), \quad \text{or} \quad g(\boldsymbol{u}, \boldsymbol{v}) = -g(\boldsymbol{v}, \boldsymbol{u}),$$

respectively, for all vectors $\boldsymbol{u}, \boldsymbol{v} \in V$. For the coefficients of a symmetric, or skew-symmetric, bilinear form we have

$$\lambda_{ij} = \lambda_{ji}, \quad \text{or} \quad \lambda_{ij} = -\lambda_{ji}, \quad i, j = 1, \ldots, n,$$

respectively.

1.71 Example. The inner product $(\boldsymbol{w}, \boldsymbol{v})$ is a symmetric bilinear form on the vector space of all vectors in E_3.

1.72 Definition. A vector space V is called an *algebra* if a further operation is given on V which assigns, to every pair of vectors $\boldsymbol{u}, \boldsymbol{v}$ from V, a vector $\boldsymbol{u} * \boldsymbol{v}$ from V such that

$$(\boldsymbol{u} + \boldsymbol{v}) * \boldsymbol{w} = \boldsymbol{u} * \boldsymbol{w} + \boldsymbol{v} * \boldsymbol{w}, \quad \boldsymbol{u} * (\boldsymbol{v} + \boldsymbol{w}) = \boldsymbol{u} * \boldsymbol{v} + \boldsymbol{u} * \boldsymbol{w},$$
$$(\lambda \boldsymbol{v}) * \boldsymbol{w} = \boldsymbol{v} * (\lambda \boldsymbol{w}) = \lambda(\boldsymbol{v} * \boldsymbol{w})$$

for arbitrary $\boldsymbol{u}, \boldsymbol{v}, \boldsymbol{w}$ from V and λ from $\mathbf{R}$.

An algebra V is called *associative* if

$$\boldsymbol{u} * (\boldsymbol{v} * \boldsymbol{w}) = (\boldsymbol{u} * \boldsymbol{v}) * \boldsymbol{w},$$

and *commutative* if $\boldsymbol{u} * \boldsymbol{v} = \boldsymbol{v} * \boldsymbol{u}$, for all $\boldsymbol{u}, \boldsymbol{v}, \boldsymbol{w}$ from V.

Let V_n be an algebra of dimension n, let $\{\boldsymbol{f}_1, \ldots, \boldsymbol{f}_n\}$ be an arbitrary base in V_n. Then the products $\boldsymbol{f}_i * \boldsymbol{f}_j$ determine multiplication in V_n uniquely.

In fact, if $\boldsymbol{u} = x^i \boldsymbol{f}_i$, $\boldsymbol{v} = y^j \boldsymbol{f}_j$ are two arbitrary vectors from V_n, then $\boldsymbol{u} * \boldsymbol{v} = x^i y^j \boldsymbol{f}_i * \boldsymbol{f}_j$.

For the operation from Definition 1.72 we often use a different symbol than the asterisk, as will be seen below.

1.73 Example. Square matrices of order n constitute an associative algebra if we define $a * b = a \,.\, b$ for matrices a, b. This algebra is not commutative.

1.74 Example. The vector space of all vectors in E_3 constitutes an algebra when multiplication is given by $\boldsymbol{u} * \boldsymbol{v} = \boldsymbol{u} \times \boldsymbol{v}$. This algebra is not associative.

1.75 Example. The *algebra of quaternions* Q is a vector space of dimension 4 with base $\{1, \mathbf{i}, \mathbf{j}, \mathbf{k}\}$ where multiplication is determined by the formulae

$$\mathbf{ij} = -\mathbf{ji} = \mathbf{k}\,, \quad \mathbf{jk} = -\mathbf{kj} = \mathbf{i}\,, \quad \mathbf{ki} = -\mathbf{ik} = \mathbf{j}\,,$$
$$\mathbf{i}^2 = \mathbf{j}^2 = \mathbf{k}^2 = -1\,.$$

1 possesses the properties of a unit element, i.e. $1 \,.\, \boldsymbol{\alpha} = \boldsymbol{\alpha}$ for all $\boldsymbol{\alpha}$ from Q. With multiplication defined in this manner quaternions constitute an associative algebra which is not commutative. By the *norm of a quaternion* $\boldsymbol{\alpha} = a_0 + a_1\mathbf{i} + a_2\mathbf{j} + a_3\mathbf{k}$, $a_i \in \mathbf{R}$, $i = 0, \ldots, 3$, we understand the number

$$|\boldsymbol{\alpha}| = (a_0^2 + a_1^2 + a_2^2 + a_3^2)^{1/2}\,.$$

A quaternion is called a *unit quaternion* if its norm is equal to one, or *pure imaginary* if $a_0 = 0$.

D. *Mappings of Sets and Vector Spaces*

In this section the simplest properties of mappings will be derived and congruences and linear mappings will be discussed.

1.76 Definition. Let us have two sets X and Y. The rule φ which assigns to every element $x \in X$ precisely one element $y \in Y$ is called a *mapping of the set X into the set Y* and it is denoted by $\varphi: X \to Y$ or $\varphi: X \to Y: x \to y$. The element $y = \varphi(x)$ is called the *image* of x under φ.

A mapping $\varphi: X \to Y$ where to every $y \in Y$ there exists at least one element $x \in X$ such that $y = \varphi(x)$ is called a *mapping of the set X on to the set Y.*

A mapping $\varphi: X \to Y$ is called *one-to-one* if we have $\varphi(x_1) \neq \varphi(x_2)$ for every two distinct elements $x_1, x_2 \in X$.

A mapping of a set X into the set X where every element is identical with its image is called the *identity mapping*, and it is denoted by id_X: $X \to X$.

If φ is a one-to-one mapping of a set X onto a set Y, then there exists exactly one mapping $\varphi': Y \to X$ such that $x = \varphi'(y)$ if $\varphi(x) = y$. The mapping φ' is called the *inverse* mapping of the mapping φ, and it is denoted by φ^{-1}.

1.77 Example. Let $\mathbf{R}$ be the set of all real numbers. Then the mapping $\varphi: \mathbf{R} \to \mathbf{R}: x \to x^2$ is a mapping of the set $\mathbf{R}$ into the set $\mathbf{R}$. The mapping $\lambda: \mathbf{R} \to \mathbf{R}^+: x \to x^2$, where $\mathbf{R}^+ \equiv \langle 0, \infty)$, is a mapping of the set $\mathbf{R}$ on to $\mathbf{R}^+$ since $\varphi(x) \geqq 0$. However, this mapping is not one-to-one. Finally the mapping $\tau: \mathbf{R} \to \mathbf{R}: x \to 2x$ is a one-to-one mapping of the set $\mathbf{R}$ on to $\mathbf{R}$ and its inverse mapping is $\tau^{-1}: \mathbf{R} \to \mathbf{R}: x \to \frac{1}{2}x$.

1.78 Definition. A one-to-one mapping φ of the set X onto X is called a *transformation of the set X*. If X is a finite set, i.e. a set with a finite number of elements, the mapping φ is called a *permutation* of the set X.

Now, we turn our attention to the successive application of several mappings.

1.79 Definition. If two mappings $\psi: X \to Y$ and $\varphi: Y \to Z$ are given, then by their *product* or *composition* denoted by $\varphi \circ \psi$ we understand the mapping

$$\varphi \circ \psi: X \to Z$$

defined by the relation

$$(\varphi \circ \psi)(x) = \varphi(\psi(x)).$$

This definition implies that the product of two mappings $\varphi \circ \psi$ is obtained by the successive application of the two mappings, first ψ and then φ. We call this successive application a *superposition*.

1.80 Theorem. *Let mappings* $\kappa: X \to Y$, $\psi: Y \to Z$ *and* $\varphi: Z \to T$ *be given. Then we have*

$$(\varphi \circ \psi) \circ \kappa = \varphi \circ (\psi \circ \kappa): X \to T.$$

1.81 Theorem. *Let us have a mapping* $\varphi: X \to Y$; *then*

$$\varphi \circ \mathrm{id}_X = \mathrm{id}_Y \circ \varphi = \varphi.$$

Further we consider mappings of points of the three-dimensional Euclidean space. Until now this space was viewed as the space of our conception and its points as basic geometrical figures. Introducing the Cartesian coordinate system it becomes possible to assign, in one-to-one correspondence, to each point an algebraic object, namely the ordered triad of its Cartesian coordinates.

1.82 Definition. The *Euclidean space* E_3 is the set of all ordered triads of real numbers $[z^1, z^2, z^3]$ to whose any two elements $X = [x^1, x^2, x^3]$, $Y = [y^1, y^2, y^3]$ the real number

$$\varrho(X, Y) = [(x^1 - y^1)^2 + (x^2 - y^2)^2 + (x^3 - y^3)^2]^{1/2}$$

is assigned. Its elements are called *points from* E_3 while the number $\varrho(X, Y)$ is the *distance* of the points X, Y.

Note that the distance of two points from Definition 1.82 yields the customary distance of two points with Cartesian coordinates $[x^1, x^2, x^3]$ and $[y^1, y^2, y^3]$.

1.83 Definition. Let M and N be two non-empty sets of points from E_3. A one-to-one mapping $\mathscr{S}$ of the set M on to the set N which preserves distances

$$\varrho(X, Y) = \varrho(\mathscr{S}(X), \mathscr{S}(Y))$$

for every two points $X, Y \in M$ is called a *congruence* of the set M on to the set N. If $M = N$, we speak of the *congruence of the set M*.

It is obvious that any congruence of the Euclidean space E_3 maps a bound vector into a bound vector, with the length of vectors as well as the magnitude of angles between pairs of corresponding vectors preserved since $\|\boldsymbol{u}_{AB}\| = \varrho(A, B)$ and $(\boldsymbol{u}, \boldsymbol{v}) = \frac{1}{2}[\|\boldsymbol{u} + \boldsymbol{v}\|^2 - \|\boldsymbol{u}\|^2 - \|\boldsymbol{v}\|^2]$. An orthonormal frame $\mathscr{R} = \{A, \boldsymbol{u}_1, \boldsymbol{u}_2, \boldsymbol{u}_3\}$ is then mapped into an orthonormal frame $\mathscr{R}' = \{A', \boldsymbol{u}'_1, \boldsymbol{u}'_2, \boldsymbol{u}'_3\}$ again. By Theorem 1.43 the vectors of this frame satisfy the equation

$$\boldsymbol{u}'_i = a_i^m \boldsymbol{u}_m , \tag{1.2}$$

where the indices take the values 1, 2, 3. If the frames are orthonormal, i.e. if

$$(\boldsymbol{u}_m, \boldsymbol{u}_k) = \delta_{mk} , \quad (\boldsymbol{u}'_i, \boldsymbol{u}'_j) = \delta_{ij} , \tag{1.3}$$

then the coordinates a_i^m of the vectors $\boldsymbol{u}'_i$ in the frame $\mathscr{R}$ satisfy the conditions which follow from relations (1.2) and (1.3):

$$\delta_{ij} = a_i^m a_j^k (\boldsymbol{u}_m, \boldsymbol{u}_k) = a_i^m a_j^k \delta_{mk} = a_i^m a_j^m . \tag{1.4}$$

The reader will easily prove that these conditions are also sufficient for the vectors $\boldsymbol{u}'_i$ from equation (1.2) to constitute an orthonormal frame $\mathscr{R}'$.

If we form a matrix a of third order so that the elements of its i-th column are the coordinates of the vector $\boldsymbol{u}'_i$ in the frame $\mathscr{R}$, then relation (1.4) and Theorem 1.21 imply that the matrix a is orthogonal. According to Theorem 1.47 its elements satisfy

$$a_i^m = \cos \varphi_i^m ,$$

where φ_i^m is the angle between the vectors $\boldsymbol{u}'_i$ and $\boldsymbol{u}_m$.

In what follows we assume both frames $\mathscr{R}$ and $\mathscr{R}'$ to be right-handed, i.e.

$$\boldsymbol{u}_1 \times \boldsymbol{u}_2 = \boldsymbol{u}_3 , \quad \boldsymbol{u}'_1 \times \boldsymbol{u}'_2 = \boldsymbol{u}'_3 .$$

Hence, upon substitution from relation (1.2), it follows that

$$a_1^m a_2^k (\boldsymbol{u}_m \times \boldsymbol{u}_k) = a_3^p \boldsymbol{u}_p .$$

Finally, comparing the coefficients of both linear combinations of vectors $\boldsymbol{u}_i$ we obtain

$$a_3^i = \det {}_3^i a .$$

By Theorem 1.14 (for the expansion of the determinant with respect to the elements of its third column) we thus have

$$\det a = a_3^i a_3^i = 1 .$$

Consequently, the above condition is necessary for the frame $\mathscr{R}'$, the vectors $\boldsymbol{u}'_i$ of which are expressed by equation (1.2), to be right-handed. As an exercise prove that this condition is sufficient as well. Then the following theorem holds:

1.84 Theorem. *Let* $\{\boldsymbol{u}_1, \boldsymbol{u}_2, \boldsymbol{u}_3\}$ *be a right-handed orthonormal base. The vectors*

$$\boldsymbol{u}'_i = a_i^m \boldsymbol{u}_m$$

constitute a right-handed orthonormal base $\{\boldsymbol{u}'_1, \boldsymbol{u}'_2, \boldsymbol{u}'_3\}$ *if and only if the matrix* $a = (a^i_j)$ *is orthogonal and if* $\det a = 1$.

Thus, the frame $\mathscr{R}' = \{A', \boldsymbol{u}'_1, \boldsymbol{u}'_2, \boldsymbol{u}'_3\}$ is determined by the orthogonal matrix a with $\det a = 1$ and by the coordinates of its origin $A' = [b^1, b^2, b^3]$ in the frame $\mathscr{R}$, i.e. in the coordinate system corresponding to the frame $\mathscr{R}$.

1.85 Remark. A congruence is called direct, or indirect, if the frames $\mathscr{R}$ and $\mathscr{R}'$ have the same, or opposite, orientation (i.e. if $\det a = 1$, or $\det a = -1$), respectively. In what follows only direct congruences will be used.

Consider points from E_3 as terminal points of position vectors in the frame $\mathscr{R}$, or $\mathscr{R}'$. The coordinates of a point and of its corresponding position vector are then equal. Let the congruence $\mathscr{S}$ of the Euclidean space E_3 carry a point X into a point X', where X and X' have coordinates $[x^1, x^2, x^3]$ and $['x^1, 'x^2, 'x^3]$ in the frames $\mathscr{R}$ and $\mathscr{R}'$, respectively. Let us investigate these coordinates in more detail. Since the angles between the position vector $\boldsymbol{x}_{AX}$ of the point X and the vectors $\boldsymbol{u}_i$ of the frame $\mathscr{R}$ must be equal to the angles between the position vector $'\boldsymbol{x}_{A'X'}$ of the point X' and the vectors $\boldsymbol{u}'_i$ of the frame $\mathscr{R}'$ and since we have, further,

$$\|\boldsymbol{u}_i\| = \|\boldsymbol{u}'_i\|, \quad \|\boldsymbol{x}_{A'X'}\| = \|'\boldsymbol{x}_{A'X'}\|,$$

it is true that

$$(\boldsymbol{u}_i, \boldsymbol{x}) = (\boldsymbol{u}'_i, '\boldsymbol{x}).$$

This implies

$$x^i = 'x^i,$$

so that the coordinates of the point X' in the frame $\mathcal{R}'$ are equal to the coordinates of the point X in the frame $\mathcal{R}$. Now, let us determine the relations between the coordinates of the points $X = [x^1, x^2, x^3]$ and $X' = [x'^1, x'^2, x'^3]$ in a fixed chosen frame $\mathcal{R}$. For the position vector $\boldsymbol{x}'_{AX'}$ of the point X' we have

$$\boldsymbol{x}'_{AX'} = \boldsymbol{b}_{AA'} + {}'\boldsymbol{x}_{A'X'} . \tag{1.5}$$

The vector $'\boldsymbol{x}$ has coordinates x^j in the frame $\mathcal{R}'$ so that $'\boldsymbol{x} = x^j\boldsymbol{u}'_j = x^j a^i_j \boldsymbol{u}_i$. This implies that $a^i_j x_j$ are coordinates of the vector $'\boldsymbol{x}$ in the frame $\mathcal{R}$.

From equation (1.5) we obtain the relation between the coordinates x^i of the point X (u^i of the vector $\boldsymbol{u}$) and the coordinates x'^i of its image X' (u'^i of its image $\boldsymbol{u}'$):

$$x'^i = a^i_j x^j + b^i , \quad u'^i = a^i_j u^j . \tag{1.6}$$

Here b^i are the coordinates of the vector $\boldsymbol{b}_{AA'}$ in the frame $\mathcal{R}$.

Conversely we easily verify that the mapping given by (1.6), where $a = (a^i_j)$ is an orthogonal matrix, is a congruence in E_3. Equation (1.6) can be written in matrix form as

$$\begin{pmatrix}1\\ x'^1\\ x'^2\\ x'^3\end{pmatrix} = \begin{pmatrix}1, & 0, & 0, & 0\\ b^1, & a^1_1, & a^1_2, & a^1_3\\ b^2, & a^2_1, & a^2_2, & a^2_3\\ b^3, & a^3_1, & a^3_2, & a^3_3\end{pmatrix}\begin{pmatrix}1\\ x^1\\ x^2\\ x^3\end{pmatrix}, \quad \begin{pmatrix}0\\ u'^1\\ u'^2\\ u'^3\end{pmatrix} = \begin{pmatrix}1, & 0, & 0, & 0\\ b_1, & a^1_1, & a^1_2, & a^1_3\\ b_2, & a^2_1, & a^2_2, & a^2_3\\ b_3, & a^3_1, & a^3_2, & a^3_3\end{pmatrix}\begin{pmatrix}0\\ u^1\\ u^2\\ u^3\end{pmatrix}, \tag{1.7}$$

where we have assigned to the point X and to the vector $\boldsymbol{u}$, with coordinates x^1, x^2, x^3 and u^1, u^2, u^3, the column matrices

$$\begin{pmatrix}1\\ x^1\\ x^2\\ x^3\end{pmatrix} \text{ and } \begin{pmatrix}0\\ u^1\\ u^2\\ u^3\end{pmatrix}, \tag{1.8}$$

respectively.

1.86 Definition. Let V_n and V_m be vector spaces of dimensions n and m, respectively. The mapping $\boldsymbol{A}: V_n \to V_m$ is called *linear* if

$$\boldsymbol{A}(\lambda \boldsymbol{u} + \mu \boldsymbol{v}) = \lambda\, \boldsymbol{A}(\boldsymbol{u}) + \mu\, \boldsymbol{A}(\boldsymbol{v})$$

for all $\boldsymbol{u}$ and $\boldsymbol{v}$ from V_n and $\lambda, \mu \in \mathbf{R}$.

1.87 Definition. Let A and B be linear mappings from V_n into V_m, $\lambda \in \mathbf{R}$. We define:

$$(A + B)(u) = A(u) + B(u), \quad A(\lambda u) = \lambda A(u).$$

1.88 Exercise. Show that the set of all linear mappings from V_n into V_m constitutes a vector space under the operations from Definition 1.87.

1.89 Definition. Let V_n and V_m be vector spaces of dimensions n and m, respectively, with bases $\mathscr{B} = \{f_1, \ldots, f_n\}$ and $\mathscr{C} = \{g_1, \ldots, g_m\}$. If $A: V_n \to$ $\to V_m$ is a linear mapping, then the $n \times m$ matrix a, where

$$\{A(f_1), \ldots, A(f_n)\} = \{g_1, \ldots, g_m\}\, a \tag{1.9}$$

is called the *matrix of the linear mapping A in the bases $\mathscr{B}$ and $\mathscr{C}$*. The product on the right side of (1.9) means the formal product of the row $\{g_1, \ldots, g_n\}$ with the matrix a.

1.90 Exercise. Show that the matrix of a linear mapping determines this mapping uniquely.

1.91 Exercise. Show that the vector space of all linear mappings from V_n into V_m and the vector space of all $n \times m$ matrices are isomorphic.

1.92 Exercise. Investigate how does the matrix of a linear mapping change with a change of the bases $\mathscr{B}$ and $\mathscr{C}$.

1.93 Exercise. Show that congruence in E_3 determines a linear mapping of the vector space of all vectors from E_3. Further show that the matrix of this linear mapping is expressed, in an orthonormal base, by an orthogonal matrix.

1.94 Exercise. If V is an algebra with the operation $*$, show that the mapping $L_u: V \to V$, determined by the rule $L_u(v) = u * v$ for every $v \in V$, is linear (u is a fixed chosen vector from V).

1.95 Example. Find the matrix a of the mapping L_{u_1} in the algebra V_3 of vectors from E_3 with the operation of cross product, in the orthonormal base $\mathscr{B} = \{u_1, u_2, u_3\}$. By definition we have

$$L_{u_1}(u_1) = 0, \quad L_{u_1}(u_2) = u_1 \times u_2 = u_3, \quad L_{u_1}(u_3) = u_1 \times u_3 = -u_2.$$

Thus, 1.9 yields

$$\{0, \boldsymbol{u}_3, -\boldsymbol{u}_2\} = \{\boldsymbol{u}_1, \boldsymbol{u}_2, \boldsymbol{u}_3\} \cdot a\,, \quad \text{i.e.} \quad a = \begin{pmatrix} 0, & 0, & 0 \\ 0, & 0, & -1 \\ 0, & 1, & 0 \end{pmatrix}.$$

2. BASIC CONCEPTS FROM GROUP THEORY

On some non-empty set M let an operation be defined which assigns to every ordered pair of elements $a, b \in M$ an element $c \in M$ uniquely. Such an operation is called a *binary operation* on the set M and we denote it by a circle

$$a \circ b = c\,.$$

For instance, addition and multiplication are binary operations on the set of integers; however, this is not true for division. A different example of a binary operation is the addition of matrices on the set of all matrices of the same type.

If a binary operation on a set satisfies certain conditions, we arrive at the important concept of group (see also [I]):

2.1 Definition. A non-empty set G on which a binary operation is defined is called a *group* if the following axioms are satisfied:

G1. For any three elements $a, b, c \in G$ we have

$$(a \circ b) \circ c = a \circ (b \circ c)\,.$$

G2. In the set G a so-called *unit element* e exists such that for every element $a \in G$ we have

$$a \circ e = e \circ a = a\,.$$

G3. To every element $a \in G$ a so-called *inverse element* $a^{-1} \in G$ exists for which we have

$$a \circ a^{-1} = a^{-1} \circ a = e\,.$$

Several important consequences follow from the presented axioms. In every group G there exists precisely one unit element, and every element of G has precisely one inverse element. Further, for all $a, b, c \in G$ we have

$$a \circ b = a \circ c \Rightarrow b = c\,,$$

$$b \circ a = c \circ a \Rightarrow b = c\,,$$

$$(a^{-1})^{-1} = a\,, \quad e^{-1} = e\,, \quad (a \circ b)^{-1} = b^{-1} \circ a^{-1}\,.$$

Finally, the equations

$$a \circ x = b\,, \quad y \circ a = b$$

have, for all $a, b \in G$, a single solution in the group G,

$$x = a^{-1} \circ b\,, \quad y = b \circ a^{-1}\,.$$

The binary operation on a group, the so-called *group operation*, is usually written either as a product ab or as a sum $a + b$. According to the choice of notation we then speak either of a *multiplicative* group or of an *additive* group; simultaneously we change the symbolism. In a multiplicative group we replace the unit element by the symbol 1 and call it the unit. In an additive group we replace the unit element by the symbol 0 and call it zero. The inverse element to an element b is denoted by $-b$. Note that multiplication as a group operation of a multiplicative group does not necessarily mean the arithmetic product but a binary operation on a group.

2.2 Example. We shall prove that the set M of all positive rational numbers constitutes a group whose group operation is the arithmetic product.

We have to prove that: 1. multiplication of numbers from M is a binary operation on M, 2. all group axioms are satisfied. The first assertion is obvious since the product of positive rational numbers is again a positive rational number. Moreover the group axioms are also satisfied since multiplication of numbers from M is associative, the number 1 is the unit, and the inverse element to the number $a \in M$ is the positive rational number $1/a$.

2.3 Exercise. Decide whether the set of all real numbers constitutes

a) an additive group,

b) a multiplicative group,

if the group operations are the arithmetic (a) sum, (b) product, respectively.

2.4 Definition. A group is called *commutative* or an *Abelian* group if its every two elements a, b satisfy the axiom

G4. $a \circ b = b \circ a$

2.5 Exercise. Prove that the set of all regular square matrices of order n constitutes a noncommutative group whose group operation is matrix multiplication.

2.6 Exercise. Verify whether the numbers $1, -1$ constitute a commutative group under the group operation of multiplication of numbers.

2.7 Example. Every vector space is a commutative group whose group operation is vector addition. This follows from the fact that the axioms of vector addition in 1.27 are simultaneously the group axioms G1 to G4.

The example which follows was chosen to show, together with the previous examples and exercises, how heterogeneous can the elements of different groups be.

2.8 Example. Consider the set M of all rotations of an equilateral triangle about its space axis (i.e. a line perpendicular to the plane of the triangle which passes through its center) such that the vertices of the rotated triangle coincide with the vertices of its original position. When speaking of such an operation we say that it maps a triangle into itself. For instance, after rotation of the triangle through the oriented angle $2\pi/3$ the new position of the vertices $1', 2', 3'$ coincides with the original position of the vertices 2, 3, 1.

Precisely the same result is obviously obtained also for the rotation $u \in U = \{2\pi/3 \pm 2\pi k;\ k = 0, 1, \ldots\}$. Thus, the element u is the representative of the set U which moves the triangle vertices so that to a vertex of the original position a vertex of the new position is assigned:

$$\begin{Bmatrix} 1 \to 3' \\ 2 \to 1' \\ 3 \to 2' \end{Bmatrix}.$$

The primes to the numbers of the second column as well as the arrowheads will be omitted in the exposition which follows.

Similarly we obtain the sets

$$V = \{4\pi/3 \pm 2\pi k;\ k = 0, 1, \ldots\},$$
$$W = \{\pm 2\pi k;\ k = 0, 1, \ldots\},$$

whose elements carry the vertices of the triangle as follows

$$v = \begin{pmatrix} 1, & 2 \\ 2, & 3 \\ 3, & 1 \end{pmatrix},$$
$$w = \begin{pmatrix} 1, & 1 \\ 2, & 2 \\ 3, & 3 \end{pmatrix}.$$

Then the set M is the union of the sets U, V, W. Each of the three mutual positions of the triangle before rotation and after it can be characterized by the representatives u, v, w of the sets U, V, W.

On the set M it is possible to define the binary operation

$$x \circ y = t$$

where $t \in M$ stands for rotation obtained as the outcome of the superposition of rotations x and y. Recall that rotation y has to be performed first, then x. The condition of a binary operation is satisfied since, e.g.,

$$u \circ v = \begin{Bmatrix} 1, & 3 \\ 2, & 1 \\ 3, & 2 \end{Bmatrix} \circ \begin{Bmatrix} 1, & 2 \\ 2, & 3 \\ 3, & 1 \end{Bmatrix} = \begin{Bmatrix} 1, & 1 \\ 2, & 2 \\ 3, & 3 \end{Bmatrix} = w.$$

Similarly we find out that

$$v \circ u = w, \quad u \circ u = v, \quad v \circ v = u, \quad w \circ w = w,$$
$$w \circ u = u \circ w = u, \quad w \circ v = v \circ w = w.$$

We can prove easily that superposition on M is an assotiative binary operation

$$x \circ (y \circ t) = (x \circ y) \circ t.$$

Recalling the definition of a group the outlines of our target become gradually clear from the obtained results: to show that the set M of considered rotations of a triangle constitutes a group. If we choose superposition on M to be the group operation it suffices to show that the group axioms are satisfied. These axioms are satisfied since superposition on M is associative, the unit element is $w = e$, and the inverse elements are $u^{-1} = v$, $v^{-1} = u$, $w^{-1} = w$.

Thus the set M is a group.

2.9 Exercise. Let a rhombus be given. Reflections in its axes, reflection in its centre and identity carry the rhombus into itself. Prove that the set of considered operations constitutes a group with superposition as the group operation.

In a multiplicative group G (with the unit element 1) it is of advantage to introduce *powers* of the element $g \in G$:

$$g^0 = 1\,, \quad g^m = gg^{m-1}\,, \quad g^{-m} = (g^m)^{-1}\,, \quad m \text{ is a positive integer}\,.$$

For these powers we have

$$g^p g^q = g^{p+q}\,, \quad (g^p)^q = g^{pq}\,, \quad p, q \text{ integers}\,.$$

2.10 Example. The set of all roots of the equation $x^4 - 1 = 0$, i.e. $\{1, -1, \mathbf{i}, -\mathbf{i}\}$, upon which complex number multiplication is defined as the group operation constitutes a commutative group. This group consists of the powers of the number $\mathbf{i}$.

2.11 Example. Let us show that the set of all orthogonal matrices of order n constitutes a group whose group operation is the product of matrices. First of all, Theorem 1.22 implies that the product of orthogonal matrices is a binary operation on the investigated set. The fulfilment of all three group axioms follows from Theorems 1.6 and 1.22 and from the fact that the unit matrix is an orthogonal matrix.

2.12 Definition. The group of all orthogonal matrices of order n is called the *orthogonal group* of order n and denoted by $O(n)$.

2.13 Definition. *Left* and *right translation on the group* G determined by an element $a \in G$ are the mappings

$$L_a\colon G \to G\colon x \to ax\,, \quad \text{and} \quad R_a\colon G \to G\colon x \to xa\,,$$

respectively, for all $x \in G$.

2.14 Exercise. Construct left translation L_i on the group of Example 2.10.

2.15 Definition. A subset H of elements of a group G is called a *subgroup* of group G if it constitutes a group which has common group operation with G.

It can be easily shown that the subset H of a group G is a subgroup if and only if gh and h^{-1} belong to H for all g and h from H.

2.16 Exercise. Show that the set of all regular matrices of order n constitutes a group under the operation of matrix multiplication.

2.17 Definition. The group from Exercise 2.16 is called the *general linear group of order n* and it is denoted by $G\,L(n)$.

2.18 Exercise. Show that the subset of all matrices from $G\,L(n)$ which have the positive determinant is a subgroup of the group $G\,L(n)$. Show also that $O(n)$ is a subgroup of $G\,L(n)$.

Every group is its own subgroup and it further contains the subgroup which consists of the unit element only. These subgroups are called *trivial* subgroups, the other subgroups are called *proper subgroups*.

2.19 Exercise. Show that all non-zero quaternions constitute a multiplicative group under the group operation of quaternion multiplication. Further, show that the set of all unit quaternions as well as the set $H = \{1, -1, \mathbf{i}, \mathbf{j}, \mathbf{k}\}$ constitute its subgroup.

2.20 Exercise. Show that all matrices of the form $g = \begin{pmatrix} a, & 0 \\ b, & c \end{pmatrix}$, where $a \in G\,L(n)$, $c \in G\,L(m)$, b is of type $m \times n$, 0 is an $n \times m$ zero matrix, constitute a subgroup of the group $G\,L(n + m)$.

Note that operations with matrices whose elements are matrices again are defined in the same way as is the case for ordinary matrices in 1.3. Naturally, we should not forget that each operation must make sense.

2.21 Exercise. Prove that the intersection of two subgroups of a group G is a subgroup of G again, and that every subgroup of a commutative group is commutative.

2.22 Definition. A one-to-one mapping φ of a group G on to a group G' is called *isomorphism* of groups G and G' if

$$\varphi(gh) = \varphi(g) \circ \varphi(h) \quad \text{for all} \quad g, h \in G .$$

The groups G and G' are then called *isomorphic*.

It is easily proved that an isomorphism carries the unit element into the unit element and the inverse element into the inverse element.

2.23 Example. Consider the group $G = \{e, u, v\}$ of Example 2.8 and the group $G^1 = \{1, s, \bar{s}\}$, where $s = \frac{1}{2}(-1, + \mathbf{i}\sqrt{3})$, $\bar{s} = \frac{1}{2}(-1 - \mathbf{i}\sqrt{3})$, with the group operation of complex number multiplication. If we put $\varphi(e) = 1$, $\varphi(u) = s$, $\varphi(v) = \bar{s}$, we find out easily that φ is an isomorphism of G on G'. For instance, we have $s^2 = \bar{s}$, $u^2 = v$, so that $\varphi(u^2) = \varphi(v) = \bar{s}$, $\varphi(u)$. $\varphi(u) = s \,.\, s = \bar{s}$. Thus, $\varphi(u^2) = \varphi(u) \,.\, \varphi(u)$. A similar result holds for the other elements.

2.24 Example. We will show that a group can be isomorphic with its proper subgroup. Let G be the additive group of all integers and let H be the set of all even numbers, thus a subgroup of the group G. We prove that the mapping $\varphi : G \to H : g \to 2g$ is an isomorphism. First of all, φ is a one-to-one mapping of G onto H. Further, we have

$$\varphi(g + h) = 2(g + h) = 2g + 2h = \varphi(g) + \varphi(h) ,$$

thus, φ is an isomorphism.

2.25 Exercise. Prove that a commutative group and a noncommutative group cannot be isomorphic.

2.26 Example. Let g be a fixed chosen element of the group G. For arbitrary $h \in G$ we have $ghg^{-1} \in G$. We shall show that $\mathrm{Int}_g : G \to G : h \to ghg^{-1}$ for all $h \in G$ is an isomorphism of the group G upon itself. It is the so-called *inner automorphism* of the group G determined by the element $g \in G$. First we prove that Int_g is a one-to-one mapping of G on to G. Thus, let h be an arbitrary element from G and seek h_1 so that $\mathrm{Int}_g(h_1) = h$, i.e. that $gh_1g^{-1} = h$. This implies that $h_1 = g^{-1}hg$ so that h_1 exists and is determined uniquely. This proves that Int_g is a one-to-one mapping. Further, we have $\mathrm{Int}_g(h_1) \,.\, \mathrm{Int}_g(h_2) = gh_1g^{-1} \,.\, gh_2g^{-1} = gh_1h_2g^{-1} =$

$\mathrm{Int}_g(h_1h_2)$ for all h_1 and h_2 from G. This means that Int_g is an isomorphism of G on to G, for every $g \in G$.

2.27 Exercise. Show that the set $\mathbf{R}^+$ of all positive real numbers constitutes a group under the operation of multiplication. Further, show that this group is isomorphic with the group of all real numbers under the operation of addition. (The natural logarithm is an example of an isomorphism from $\mathbf{R}^+$ on to $\mathbf{R}$.)

2.28 Theorem. *The set G of all transformations of a set X constitutes a group under the operation of composition of mappings.*

Proof. By Definition 1.79, the composition of mappings is a binary operation on G which is associative by 1.80. Let $\mathrm{id}(x) = x$ for all $x \in X$. Then we have $(\varphi \circ \mathrm{id})(x) = (\mathrm{id} \circ \varphi)(x)$ for all $\varphi \in G$; so id is the unit element from G. For the inverse transformation φ^{-1} to the transformation φ we have $(\varphi \circ \varphi^{-1})(x) = \mathrm{id}(x)$, which proves the theorem.

2.29 Definition. Every subgroup of the group of all transformations of the set X is called a *group of transformations operating on the set X*.

2.30 Exercise. Show that the set of all inner automorphisms Int_g of group G is a group of transformations operating on the set G.

2.31 Definition. The group G of transformations operating on the set X is called *transitive* if to each pair of elements $x, y \in X$ there exists a transformation $g \in G$ such that

$$g(x) = y\,.$$

A group which is not transitive is called *intransitive*.

3. BASIC TOPOLOGICAL CONCEPTS

In the exposition which follows we discuss the continuity and differentiation of functions in a more general setting than in the case of functions of one real variable. For instance, we shall investigate continuity of functions on the set of straight lines of the Euclidean space, or a continuous or, rather, differentiable curve in the set of all congruences of the Euclidean space.

To avoid the necessity of introducing these concepts separately in each concrete case, we introduce the concept of differentiable manifold in Section 5. However, this concept is relatively complicated. For this reason we give the definition in two steps. First, we elucidate the concept of continuity, and then – in Section 5 – we discuss differentiability.

Should we wish to treat continuity only, we find sufficient the concepts of topology and topological space. We introduce these concepts now. For the sake of closer insight we present several simple properties and examples. The details related to this part of the book can be found e.g. in [VIII].

3.1 Definition. An arbitrary set E is called a *topological space* if a system $\mathscr{T}(E)$ of its subsets is selected with the following properties:

1. $\emptyset \in \mathscr{T}(E)$
2. $E \in \mathscr{T}(E)$
3. the union of an arbitrary collection of sets from $\mathscr{T}(E)$ belongs to $\mathscr{T}(E)$,
4. the intersection of a finite number of sets from $\mathscr{T}(E)$ belongs to $\mathscr{T}(E)$.

Remark: First of all note that the elements of the system $\mathscr{T}(E)$ are sets, they are subsets of the set E. $\emptyset$ denotes the empty set – it is a set which contains no element.

If a system $\mathscr{T}(E)$ which satisfies conditions 1 to 4 of Definition 3.1 is given in the set E, we also say that a *topology* is given in E. The elements from $\mathscr{T}(E)$, i.e. those subsets of the set E which belong to $\mathscr{T}(E)$, are called *open sets*. Thus to define a topology means, in essence, to declare some subsets of the set E to be open so as to satisfy properties 1 to 4. These four properties can also be expressed as follows: $\emptyset$ and the set E itself are open, the union of an arbitrary number of open subsets of the set E is an open subset of the set E, the intersection of a finite number of open subsets of the set E is an open subset of the set E.

3.2 Definition. A topological space is called a *Hausdorff space* if, in addition, the following holds:

5. To every two elements m and n from E, $m \neq n$, there exist open sets $U \subset E$, $V \subset E$, such that $m \in U$, $n \in V$, $U \cap V = \emptyset$.

In this case we also say that the open sets U and V *separate* the points m and n, i.e. every two distinct points are separable in a Hausdorff topological space. In the sequel, we deal only with Hausdorff topological spaces; the definition of the general topological space was given not to contrast with current topological literature.

3.3 Example. For illustration we present several examples of topological spaces:

a) Let E be an arbitrary set. As open sets take only the empty set $\emptyset$ and the entire set E. The obtained topology is called *trivial*.

b) Let E be an arbitrary set again. As open sets [i.e., elements of the system $\mathscr{T}(E)$] we take all the subsets of the set E. We obtain a topology in E. This topology is called *discrete*.

The above two examples are not very instructive, however. Therefore, further examples follow.

c) Let E be an arbitrary set again. In E we define a topology $\mathscr{T}(E)$ as follows: $U \subset E$ belongs to $\mathscr{T}(E)$ if and only if $E - U$ is a finite set or if $U = \emptyset$. (A set is called finite if it has a finite number of elements.) Let us show, first, that $\mathscr{T}(E)$ is actually a topology in E. For this we have to verify properties 1 to 4 from Definition 3.1. Here, properties 1 and 2 are evident ($\emptyset$ is considered to be a finite set). To prove property 3: Let $U_\alpha \in \mathscr{T}(E)$, $\alpha \in I$, where I is an arbitrary set of indices. This means that $E - U_\alpha$ is a finite set for every $\alpha \in I$. Further, $E - \bigcup_{\alpha \in I} U_\alpha = \bigcap_{\alpha \in I} (E - U_\alpha)$ is a finite set so that $\bigcup_{\alpha \in I} U_\alpha \in \mathscr{T}(E)$.

It remains to prove property 4: Let $U_i \in \mathscr{T}(E)$, where $i = 1, \ldots, n$. For every i, $E - U_i$ is a finite set and we obtain $E - \bigcap_{i=1}^{n} U_i = \bigcap_{i=1}^{n} (E - U_i)$ which is a finite set again. Thus, $\bigcap_{i=1}^{n} U_i \in \mathscr{T}(E)$.

Not to complicate the above consideration unnecessarily a slight inaccuracy was permitted – it was assumed that all the sets U_α and U_i are nonempty. However, if any of the sets U_α is empty we can omit it; if any of the sets U_i is empty the assertion is trivial. We leave the verification to the reader.

d) Let E be the set of all real numbers. A set $U \subset E$ is called open either if it is empty or if to every number $x \in U$ an open interval **I** exists such that it contains x and lies in U. Symbolically, $U \in \mathcal{T}(E)$ if and only if to every $x \in U$ there exist x_0 and x_1 such that $x_0 < x < x_1$ and $(x_0, x_1) \subset U$.

e) We imagine the set of real numbers as a straight line, thus as the Euclidean space of dimension one. The topology introduced in (d) can be immediately generalized to the case of higher dimensional Euclidean spaces as follows:

Let E_n be the set of n-tuples $[x^1, \ldots, x^n]$ of real numbers. We call it the *n-dimensional Euclidean space.* Its elements are n-tuples of real numbers – they are called points of the Euclidean space. By the distance $\varrho(A, B)$ of two n-tuples (i.e. points) from E_n, where $A = [x^1, \ldots, x^n]$, $B = [y^1, \ldots, y^n]$, we understand the number given by the expression

$$\varrho(A, B) = [(x^1 - y^1)^2 + \ldots + (x^n - y^n)^2]^{1/2} .$$

The considered expression actually yields the usual distance of two points in the Cartesian coordinate system in the case of E_1, E_2 and E_3. Now, let us introduce a topology in E_n. First, we define the concept of *spherical neighbourhood of a point* from E_n as follows: Let $A \in E_n$ be an arbitrary point. The spherical neighbourhood $S(A, \varepsilon)$ of the point A with radius $\varepsilon > 0$ is the set of all points X from E_n for which $\varrho(A, X) < \varepsilon$. (In the case of $n = 1$ the spherical neighbourhood is an open interval; for $n = 2$ it is the interior of a circle, for $n = 3$ it is the interior of a sphere.)

Now, let us give the definition of an open set in E_n: The set $U \subset E_n$ is called open if it is empty or if to its every point A there exists a neighbourhood $S(A, \varepsilon)$ of the point A which entirely lies in U. The requirements of Definition 3.1 are easily verified. Also we see that case (d) is obtained for $n = 1$. The resulting topology is called the *ordinary topology* of the Euclidean space. If we write E_n we have always in mind the set of all n-tuples of real numbers provided with this ordinary topology. The set of all real numbers with the ordinary topology will be denoted **R**.

f) Let F be an arbitrary nonempty set. We say that distance is defined in F if to every two elements m and n from F a real number $\varrho(m, n)$ is assigned with the following properties:

α) $\varrho(m, n) \geqq 0$,

β) $\varrho(m, n) = 0$ if and only if $m = n$,

γ) $\varrho(m, n) = \varrho(n, m)$,
δ) $\varrho(m, n) + \varrho(n, p) \geqq \varrho(m, p)$, where $p \in F$.

The elements of F are called *points*, the number $\varrho(m, n)$ is the *distance* of the points m and n and F is called *metric space*. The *spherical neighbourhood* $S(M, \varepsilon)$ of the point $m \in F$ with radius ε is defined as in case (e): It is the set of all $x \in F$ such that $\varrho(m, x) < \varepsilon$.

In the metric space F we then introduce a topology in the same way as in case (e): the set $U \in F$ is called open if it is empty or if it contains with every point m also some spherical neighbourhood of this point. We show easily that a topology on the set F is obtained in this way, i.e. that properties 1 to 4 from Definition 3.1 are satisfied. It is also easily seen that if the set F is finite we obtain, in the described way, the discrete topology, no matter how the distance ϱ is chosen.

3.4 Exercise. Verify in detail that the open sets of all the cases of Example 3.3 satisfy the properties 1 to 4 of Definition 3.1. In case (c) show that if E is a finite set we obtain the discrete topology, while if E is infinite a topology other than discrete is obtained. Ascertain which of the topological spaces of Example 3.3 are Hausdorff.

Examining the topological spaces of Example 3.3 we see that the definition of an open set may occasionally be quite complicated. Consequently, the obvious question arises whether it is not possible to introduce only the "simplest" open sets and obtain the other ones by taking their unions. In fact, this is possible as we shall see in the following definition.

3.5 Definition. Let E be a topological space. A system $\mathscr{B}$ of open sets from E [i.e., $\mathscr{B} \subset \mathscr{T}(E)$] is called a *base for topology* $\mathscr{T}(E)$ in E if every open set from E can be obtained as a union of some sets from $\mathscr{B}$.

We say that a topological space E has a *countable base* if there exists such a base of its topology whose sets can be enumerated by positive integers. All the topological spaces we shall need in the sequel have countable bases. On account of this, we assume that every topological space given here possesses a countable base without explicitly noting the fact.

However, it would be good to know which systems of subsets of a given set can be bases of some topology. The following theorem answers this question:

3.6 Theorem. *The system $\mathscr{B}$ of subsets of the set E is a base of some topology in E if and only if the following holds:*

a) *The union of all sets from $\mathscr{B}$ is E, $\emptyset \in \mathscr{B}$.*
b) *If U and V belong to $\mathscr{B}$ and if $x \in U \cap V$, then a $W \in \mathscr{B}$ exists such that $x \in W \subset U \cap V$.*

3.7 Example. Obviously the system of all open sets in a given topology is also a base of this topology. However, this base is by no means interesting. Thus, let us give some other examples. Take the set **R** of real numbers (with the ordinary topology from Example 3.3 d). Then the system of all open intervals, including $\emptyset$, is a base of this topology.

Similarly, the system of all spherical neighbourhoods, with the empty set added, is a base of (the ordinary) topology in E_n. Let us prove this proposition. It is obvious that the union of all spherical neighbourhoods of all points constitutes the entire space E_n; the empty set belongs to it by definition. Further, let $S(A, \varepsilon)$ and $S(B, \delta)$ $(A, B \in E_n, \varepsilon > 0, \delta > 0)$ be two spherical neighbourhoods and let $X \in S(A, \varepsilon) \cap S(B, \delta)$. Denote by γ the minimum of the numbers $\varepsilon - \varrho(X, A)$ and $\delta - \varrho(X, B)$. Then $\gamma > 0$. We shall show that

$$S(X, \gamma) \subset S(A, \varepsilon) \cap S(B, \delta) .$$

Let $Y \in S(X, \gamma)$. Then $\varrho(X, Y) < \gamma \leqq \varepsilon - \varrho(X, A)$. Consequently,

$$\varrho(Y, A) \leqq \varrho(Y, X) + \varrho(X, A) < \varepsilon ,$$

i.e. $\varrho(Y, A) < \varepsilon$, so that $Y \in S(A, \varepsilon)$. Similarly, $Y \in S(B, \delta)$ so that $Y \in \in S(A, \varepsilon) \cap S(B, \delta)$. This implies $S(X, \gamma) \subset S(A, \varepsilon) \cap S(B, \delta)$, which was to be proved.

3.8 Exercise. Show that in the case of the ordinary topology in **R** it suffices to consider all intervals of the form (p, q) where $p \leqq q$ and both p and q are rational. We obtain a countable base of this topology.

3.9 Exercise. Let E be the set of real numbers. Show that the system of intervals $(-\infty, a)$, where $a \in E$, together with the empty set constitutes a base of some topology in the set of real numbers. Show that this topology differs from the ordinary topology in E, i.e. the two topologies have different systems of open sets.

3.10 Exercise. Let E be an arbitrary set. Show that the system of all one-point subsets from E together with $\emptyset$ constitutes a base of the discrete topology in E and, conversely, that every base of the discrete topology in E contains this system.

3.11 Exercise. Let P be the set of all positive integers. Denote by $A_n = \{n, n+1, \ldots\}$ the set of all positive integers larger than or equal n. Write $A_0 = \emptyset$. Then the system of all sets A_n is a base of some topology in P. Prove this and show that P with this topology is not a Hausdorff space.

3.12 Exercise. Find all topologies in a three-point and four-point set. Ascertain which of them are Hausdorff.

Let $\varphi: X \to Y$ be a mapping of the set X into the set Y. By the counter-image $\varphi^{-1}(y)$ of the element $y \in Y$ we understand the set of all $x \in X$ such that $\varphi(x) = y$. By the counter-image $\varphi^{-1}(U)$ of the subset $U \subset Y$ we understand the union of all counter-images of all points from U.

Note that the counter-image of an element from Y is a subset of the set X and that $\varphi^{-1}(y)$ can be the empty set for some $y \in Y$. Obviously, if φ is one-to-one, then φ^{-1} defines the inverse mapping to the mapping φ.

3.13 Definition. Let X and Y be topological spaces. A mapping $\varphi: X \to Y$ is called *continuous* if the counter-image of every open set from Y is an open set from X.

3.14 Example. Let X be a topological space with the discrete topology. If Y is an arbitrary topological space, then every mapping $\varphi: X \to Y$ is continuous.

3.15 Example. We show that it suffices to verify the continuity of a mapping for a base of the topology. Thus, let $f: E \to F$ be a mapping of arbitrary topological spaces E and F with bases $\mathscr{B}$ and $\mathscr{B}'$. Further, let X_α, $\alpha \in I$, be arbitrary subsets in F. Then we have

$$f^{-1}\Big(\bigcup_{\alpha \in I} X_\alpha\Big) = \bigcup_{\alpha \in I} [f^{-1}(X_\alpha)] .$$

As a matter of fact, let $x \in f^{-1}\big(\bigcup_{\alpha \in I} X_\alpha\big)$. Then $f(x) \in \bigcup_{\alpha \in I} X_\alpha$, i.e. $f(x) \in X_\beta$ for some $\beta \in I$, so that $x \in f^{-1}(X_\beta)$, i.e. $x \in \bigcup_{\alpha \in I} [f^{-1}(X_\alpha)]$. Conversely, let $x \in$

$\bigcup_{\alpha \in I} [f^{-1}(X_\alpha)]$. Then $x \in f^{-1}(X_\beta)$ for some $\beta \in I$, i.e. $f(x) \in X_\beta$, i.e. $f(x) \in \bigcup_{\alpha \in I} X_\alpha$, i.e. $x \in f^{-1}[\bigcup_{\alpha \in I} X_\alpha]$.

We now have to prove the following proposition: A mapping f is continuous if and only if $f^{-1}(X) \in \mathscr{T}$ holds for every set $X \in \mathscr{B}'$. If f is continuous, then the counter-image of an open set must be an open set. The sets of the base are open, thus the assertion holds for them. Conversely, let U be an open subset in F and let $f^{-1}(X) \in \mathscr{T}$ for every set $X \in \mathscr{B}'$.

Then we have $U = \bigcup_{\alpha \in I} X_\alpha$ where $X_\alpha \in \mathscr{B}'$. Further, we have $f^{-1}(U) = f^{-1}(\bigcup_{\alpha \in I} X_\alpha) = \bigcup_{\alpha \in I} [f^{-1}(X_\alpha)]$, with $f^{-1}(X_\alpha) \in \mathscr{T}$ by assumption, so that $\bigcup_{\alpha \in I} [f^{-1}(X_\alpha)] = f^{-1}(U)$ is a union of open sets, thus an open set.

3.16 Definition. Let X be a topological space. A set $Z \subset X$ is called *closed* if and only if the set $X - Z$ is open in the space X.

For more details on closed sets see [VIII].

3.17 Definition. Let X and Y be topological spaces. The mapping $\varphi: X \to Y$ is called a *homeomorphism* of X on Y if

1. φ is one-to-one,
2. φ and φ^{-1} are continuous mappings.

The topological spaces X and Y are homeomorphic if there exists a homeomorphism of X on Y.

3.18 Example. The mapping $\varphi(x) = x^3$ is a homeomorphism of $\mathbf{R}$ on $\mathbf{R}$ in the ordinary topology.

3.19 Exercise. Find all the mutually non-homeomorphic topologies on a four-point set.

If we have two topological spaces and if there exists a homeomorphism $\varphi: X \to Y$, this means, essentially, that the topological spaces X and Y have the "same" open sets, i.e. that they have the "same" topology. From the position of topology there is thus no sense in distinguishing such spaces (continuous functions, limits, etc., will be defined on them in the same manner). Consequently, the concept of homeomorphism is of special significance in topology. Before exploiting this concept below, let us add some remarks:

1. Let X be a topological space. By a continuous function on X we understand a continuous mapping of the topological space X into the set $\mathbf{R}$ of real numbers with the ordinary topology. Prove that if $X = E_n$, then our definition of a continuous function conforms with the usual definition of this concept.

2. Let X be a topological space and let U be an open set in X. The set U can be considered to be a topological space if we define as open sets in U those open sets from X which lie in U.

3. If X, Y, Z are topological spaces and if $f: X \to Y$, $g: Y \to Z$, are mappings, we define the mapping $g \circ f: X \to Z$ by (see Definition 1.79)

$$(g \circ f)(x) = g(f(x))$$

for all $x \in X$. Further, if the mappings f and g are continuous, the mapping $g \circ f$ is also continuous. Indeed, let U be an open set in Z, then $g^{-1}(U)$ is an open set in the space Y so that $f^{-1}(g^{-1}(U))$ is open in the set X. It now suffices to show that $(g \circ f)^{-1}(A) = f^{-1}(g^{-1}(A))$ for every subset A from Z. And this is easy.

If now f and g are homeomorphisms, f and g are continuous so that $g \circ f$ is also continuous. Further, f^{-1} and g^{-1} are continuous so that $(g \circ f)^{-1} = f^{-1} \circ g^{-1}$ is also continuous. The fact that $g \circ f$ is one-to-one is obvious. Thus the mapping $g \circ f$ is a homeomorphism of X on to Z.

3.20 Exercise. Let U be an open set of the topological space X. Show that the topology in U, defined in 2 above, is actually a topology.

3.21 Exercise. Consider the same situation as in Exercise 3.20. Show that the open sets in U can be defined also as intersections of all the open sets from X with the set U.

Now, if X and Y are topological spaces let U be an open set in X, V an open set in Y. By a homeomorphism of U onto V we then understand a homeomorphism of U and V as topological spaces with the above-defined topology. In this sense the notion of homeomorphisms of sets will be used in what follows. Similarly, for instance, a continuous function in one variable defined in the interval (a, b) is a continuous mapping of the interval (a, b) as a topological space into the set of real numbers.

3.22 Definition. A Hausdorff topological space X will be called *locally Euclidean* if to its every point $x \in X$ there exists an open set $U \subset X$ such that $x \in U$ and U is homeomorphic with some open subset of the Euclidean space E_n, where n is the same for all $x \in X$. Such a topological space is also called a *topological manifold*, the number n is the dimension of this manifold.

Definition 3.22 means that in the space X there exists a system of open sets, say U_i, and a system of mappings φ_i such that

1. every $x \in X$ belongs to some set U_i,

2. $\varphi_i: U_i \to E_n$ is a homeomorphism of the open set $U_i \subset X$ onto some open set $\varphi_i(U_i)$ in E_n.

Every system of open subsets of the set X which satisfies 1 and 2 is called an atlas on X, the sets U_i are then the *charts* of this atlas. Definition 3.22 then intuitively means that the entire space X can be mapped and that

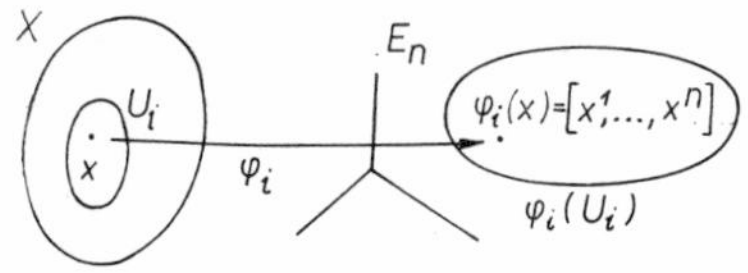

Fig. 2.

coordinates can be introduced into every chart. Let us discuss in more detail what this means (see Fig. 2). The pair (U_i, φ_i) will be called (*local*) *coordinate system* in the following exposition, the set U_i is called a *coordinate neighbourhood*, the mapping φ_i a *coordinate mapping*. If we choose an arbitrary point $x \in U_i$ then the mapping φ_i assigns to the point x a point $\varphi_i(x) \in E_n$. E_n is understood to be the set of n-tuples of real numbers. This means, as a matter of fact, that $\varphi_i(x)$ is such an n-tuple, denote it by $[x^1, \dots, x^n]$. We shall say that $x^1, \dots, x^n$ are the coordinates of the point x in the coordinate system (U_i, φ_i).

From what was just mentioned we see that a locally Euclidean space actually has, in the neighborhood of every point, topologically the same structure as the Euclidean space and that it can be locally described by coordinates. In fact, the spaces we will deal with in the sequel are of richer structure. Not only is it possible to describe them locally using coordinates, but it is even possible to transfer to them the calculus of functions of several variables – derivatives, etc. These topics will be discussed in Section 5 of this chapter.

3.23 Example. The Euclidean space E_n is a topological manifold. Here it suffices to consider the entire E_n as a chart and the identity mapping as φ. More generally – every open subset in E_n is also a topological manifold, which is shown in a similar way. In particular, the set of all regular $n \times n$ matrices, i.e. matrices a such that $\det a \neq 0$, is an n^2-dimensional topological manifold since it is an open subset of the Euclidean space E_{n^2} of dimension n^2.

The last assertion is easily proved: Consider the function det which assigns to every $n \times n$ matrix its determinant. The function det is a real function of n^2 variables. It is given as a polynomial of n-th degree which is a continuous function. Matrices whose determinants are nonzero are mapped by the function det into the intervals $(-\infty, 0)$ and $(0, \infty)$. These intervals are open sets; conversely, the counter-image of these two intervals are precisely all the regular matrices. Since the counter-image of an open set under a continuous mapping is an open set again, we see that the set of all regular $n \times n$ matrices is an open set in E_{n^2} (an $n \times n$ matrix can be viewed as an n^2-tuple of real numbers written in line). Consequently, it is a topological manifold.

Before giving further examples we present several remarks.

If a topological manifold is given and an atlas in it, we know the form of all open sets which are "small", i.e. which lie in some coordinate neighborhood. As a matter of fact, these sets are open sets from E_n mapped into our manifold by the inverse of a coordinate homeomorphism. Thus, we are well informed as far as these open sets are concerned. It would seem that the other open sets can be obtained by merely taking unions of the former. This is actually true as shown by the following theorem.

3.24 Theorem. *Let X be a topological manifold, and let an atlas $\{(U_i, \varphi_i)\}$ be given in it. Denote by $\mathscr{B}$ the system of all open sets which lie in some U_i. Then $\mathscr{B}$ is a base of the topology of the space X.*

Proof. By definition we have $U \in \mathscr{B}$ if and only if U is an open set and U_i exists such that $U \subset U_i$. We are to show that every open set in X is the union of open sets from $\mathscr{B}$. Thus, let Y be an open set in X. Choose

$y \in Y$. Then there exists a coordinate neighbourhood $U_i(y)$ such that $y \in U_i(y)$ since U_i covers the entire set X according to the definition of an atlas. Further, $U_i(y) \cap Y$ is an open set since it is an intersection of two open sets. Besides, $U_i(y) \cap Y$ is a part of $U_i(y)$ and thus belongs to $\mathscr{B}$. However, the union of all such $U_i(y) \cap Y$ for all $y \in Y$ yields precisely Y, which proves the theorem.

When defining the topological manifold we have proceeded in such a way that a topological space was considered and only then did we look for the possibility of introducing coordinates in this space. Imagine now that we would wish to proceed conversely: suppose that a set is given and that we want to introduce a topology in it so skilfully that it becomes a topological manifold of some dimension (however, we need not always succeed).

By Theorem 3.24 we expect that this can be managed piecewise, i.e. we introduce the topology only into certain parts of the given set and so obtain a topology on the entire set according to the preceding theorem. This means that we have to investigate what happens when some coordinate neighbourhoods intersect.

Consequently, consider the topological manifold X and on it the atlas $\{(U_i, \varphi_i)\}$. Let $m \in U_\alpha \cap U_\beta$. Then we have two homeomorphic mappings φ_α and φ_β of the open set $U_\alpha \cap U_\beta$ into E_n which originate by restriction of the domain of the mappings φ_α and φ_β on $U_\alpha \cap U_\beta$ (see Fig. 3). The point m has assigned coordinates with the aid of φ_α on the one hand – denote

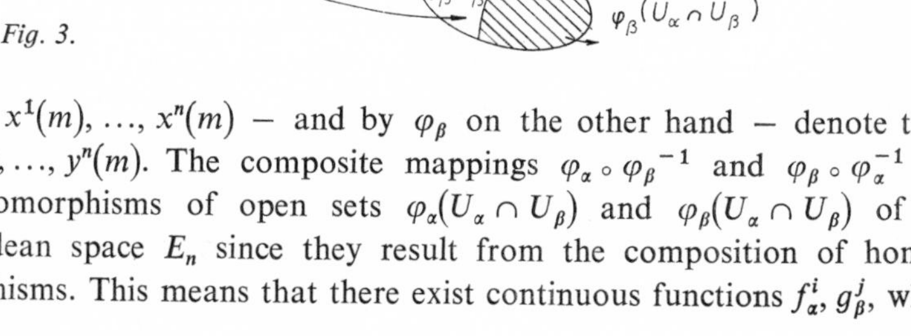

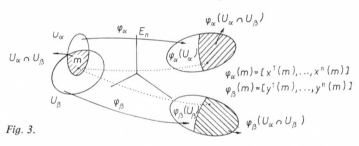

Fig. 3.

them $x^1(m), \ldots, x^n(m)$ – and by φ_β on the other hand – denote them $y^1(m), \ldots, y^n(m)$. The composite mappings $\varphi_\alpha \circ \varphi_\beta^{-1}$ and $\varphi_\beta \circ \varphi_\alpha^{-1}$ are homeomorphisms of open sets $\varphi_\alpha(U_\alpha \cap U_\beta)$ and $\varphi_\beta(U_\alpha \cap U_\beta)$ of the Euclidean space E_n since they result from the composition of homeomorphisms. This means that there exist continuous functions f^i_α, g^j_β, where

$i, j = 1, \ldots, n$, defined on $\varphi_\alpha(U_\alpha \cap U_\beta)$ and $\varphi_\beta(U_\alpha \cap U_\beta)$, respectively, such that $y^i = f_\alpha^i(x^1, \ldots, x^n)$, $x^j = g_\beta^j(y^1, \ldots, y^n)$.

Briefly speaking, every point from $U_\alpha \cap U_\beta$ on X possesses coordinates of two kinds: $x^1, \ldots, x^n$, and $y^1, \ldots, y^n$. Their relation is given by continuous functions. This relation between the coordinates is usually called *transformation of coordinates*. Thus, if we shall wish to introduce the structure of a topological manifold with the aid of an atlas we have to arrange it in such a way that transformations of coordinates be continuous on the intersections $U_\alpha \cap U_\beta$. Now, we are able to prove the following theorem:

3.25 Theorem. *Let a set X be given together with a system of pairs $\{(U_i, \varphi_i)\}$ such that*

1. *U_i is a subset in X and φ_i is a one-to-one mapping of U_i into E_n, with n being the same for all U_i,*
2. *every element $m \in X$ belongs to some set U_i,*
3. *all $\varphi_\alpha(U_\alpha \cap U_\beta)$ are open sets in E_n,*
4. *all mappings $\varphi_\alpha \circ \varphi_\beta^{-1}$ are continuous mappings of $\varphi_\beta(U_\alpha \cap U_\beta)$ on to $\varphi_\alpha(U_\alpha \cap U_\beta)$.*

Then the system of all sets from X, which are counter-images of open sets from E_n under the mappings φ_i^{-1}, is a base of a topology in X. If this topology is Hausdorff, then X is a topological manifold.

Proof. Let the assumptions of the theorem be satisfied and denote by $\mathscr{B}$ the system of all subsets of the set X which are counter-images of open sets from E_n for all mappings φ_i^{-1} (thus even $\emptyset \in \mathscr{B}$). We shall apply Theorem 3.6 and show that $\mathscr{B}$ is the base of some topology in X. For this it suffices to show that every element $m \in X$ lies in some set from $\mathscr{B}$ and that if $x \in U \cap V$, where U and V are from $\mathscr{B}$, then there exists W from $\mathscr{B}$ such that $x \in W \subset U \cap V$. The first assertion follows from assumption 2 since U_i are from $\mathscr{B}$. Now let U and V be from $\mathscr{B}$. We see that in this case even $U \cap V$ is from $\mathscr{B}$. This proves the second assertion since it suffices to put $W = U \cap V$. (See Fig. 4.) Consequently, if U and V are from $\mathscr{B}$, then there exist α and β such that $U \subset U_\alpha$, $V \subset U_\beta$, and $\varphi_\alpha(U)$ and $\varphi_\beta(V)$ are open sets in E_n. The sets $\varphi_\alpha(U_\alpha \cap U_\beta)$ and $\varphi_\beta(U_\alpha \cap U_\beta)$ are open in E_n by assumption 3. This means further that the sets $\varphi_\alpha(U) \cap \varphi_\alpha(U_\alpha \cap U_\beta) = \varphi_\alpha(U \cap U_\beta)$ and $\varphi_\beta(V) \cap \varphi_\beta(U_\alpha \cap U_\beta) = \varphi_\beta(U_\alpha \cap V)$ are open. Since $\varphi_\beta(U_\alpha \cap V) \subset \varphi_\beta(U_\alpha \cap U_\beta)$, it is possible to apply the homeomorphism $\varphi_\alpha \circ \varphi_\beta^{-1}$ to the set $\varphi_\beta(U_\alpha \cap V)$. Thus we obtain the open set $(\varphi_\alpha \circ \varphi_\beta^{-1})[\varphi_\beta(U_\alpha \cap V)] = \varphi_\alpha(U_\alpha \cap V)$.

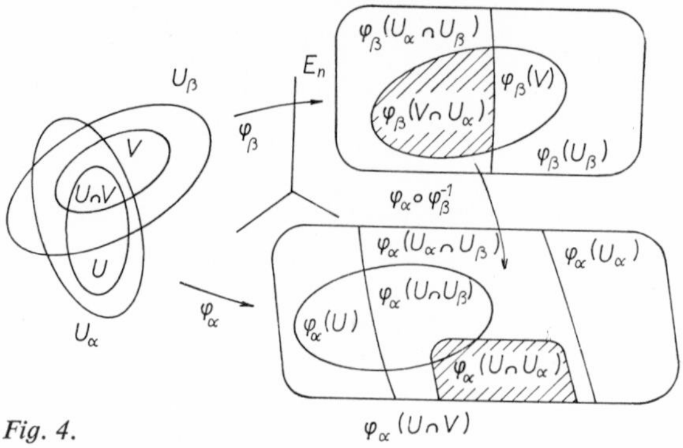

Fig. 4.

Consequently, the sets $\varphi_\alpha(U \cap U_\beta)$ and $\varphi_\alpha(U_\alpha \cap V)$ are open in E_n. Thus, even $\varphi_\alpha(U \cap U_\beta) \cap \varphi_\alpha(U_\alpha \cap V) = \varphi_\alpha(U \cap V)$ is an open set in E_n under the mapping φ_α^{-1}. Consequently, $U \cap V$ is from $\mathscr{B}$.

It remains to show that to every point $m \in X$ there exists a pair (U_i, φ_i) such that φ_i is a homeomorphism of U_i on to the open set $\varphi_i(U_i)$ in E_n and $m \in U_i$. We know that to every $m \in X$ there exists a set U_i and a mapping φ_i such that φ_i is a one-to-one mapping of U_i into E_n. We still have to show that φ_i is a homeomorphism. The mapping φ_i is continuous by definition – the counter-images of open sets from $\varphi_i(U_i)$ belong to the base $\mathscr{B}$ and are thus open in X. Thus, it suffices to show that the mappings φ_i^{-1} are continuous. To this aim we must show that the counter-image of every open set from U_i is an open set from E_n; it is sufficient to do this for the sets from the base. Thus, let $U \subset U_i$, $U \in \mathscr{B}$. Then $U = \varphi_\beta^{-1}(Y)$ for some β where Y is an open set in E_n. But then $Y \subset \varphi_\beta(U_\beta)$, i.e. $U \subset U_i \cap U_\beta$. Further, $\varphi_i(U) = \varphi_i \circ \varphi_\beta^{-1}(Y)$ is an open set in E_n since $\varphi_i \circ \varphi_\beta^{-1}$ is a homeomorphism, which was to be proved.

3.26 Example. In concluding let us show two examples of topological manifolds.

1. Consider the set of all spheres in E_3 with nonzero radius. Denote it S. Define a mapping $\varphi: S \to E_4$ which assigns, to every spherical surface with centre at the point with coordinates $[x^1, x^2, x^3]$ and radius $r > 0$,

a point from E_4 with coordinates $[x^1, x^2, x^3, r]$. $\varphi(S)$ is given by the condition $x^4 > 0$ so that $\varphi(S)$ is an open set in E_4. We see easily that this introduces in S the structure of a topological manifold of dimension 4 by Theorem 3.25.

2. Consider the sphere S^2 in E_3 (e.g., with unit radius and centre at the origin). Define mappings φ_1 and φ_2 as in Fig. 5: φ_1 is the projection of the sphere from the south pole J on to the tangent plane α to the sphere at the north pole S. φ_2 is the projection of the sphere from the north pole S

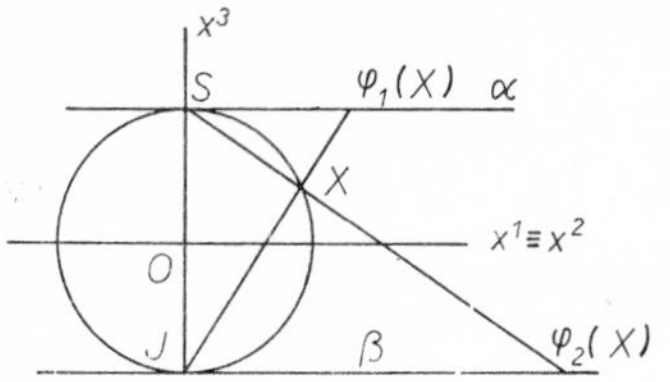

Fig. 5.

on to the tangent plane β at the south pole J. Here the domain of φ_1 is $U_1 = S^2 - \{J\}$; the domain of φ_2 is $U_2 = S^2 - \{S\}$. φ_1 is a one-to-one map of U_1 on to α, similarly φ_2 maps U_2 onto β one-to-one. We have $U_1 \cap U_2 = S^2 - \{S\} - \{J\}$ so that both $\varphi_1(U_1 \cap U_2)$ and $\varphi_2(U_1 \cap U_2)$ [as well as $\varphi_1(U_1)$ and $\varphi_2(U_2)$, naturally] are open in α or in β. The planes α and β serve as copies of the Euclidean space E_2. The transformations of coordinates $\varphi_2 \circ \varphi_1^{-1}$ and $\varphi_1 \circ \varphi_2^{-1}$ are continuous functions, which is easily verified. The concrete expression of these transformation functions is given in Section 5.

The assumptions of Theorem 3.25 are thus fulfilled and S^2 is a topological manifold of dimension 2 with the atlas $\{(U_1, \varphi_1), (U_2, \varphi_2)\}$.

4. DIFFERENTIAL GEOMETRY OF CURVES AND RULED SURFACES IN E_3

In this section we limit the exposition to the review of concepts and formulae of differential geometry of curves and ruled surfaces which will be necessary later. The reader who wishes to become more familiar with this material is referred to the respective literature [VII] and [IX].

In the considerations which follow we shall tacitly assume for all functions of one or more variables that they are continuous and have continuous partial derivatives of all necessary orders on some open set Ω.

Let a Cartesian coordinate system x^1, x^2, x^3, determined by an orthonormal frame $\mathcal{R} = \{0, \boldsymbol{f}_1, \boldsymbol{f}_2, \boldsymbol{f}_3\}$, be given in E_3.

4.1 Definition. By a *curve* in E_3 we understand the set of all points in E_3 whose coordinates are given by equations

$$x^i = x^i(\mathrm{t}), \quad i = 1, 2, 3, \tag{4.1}$$

where $x^i(\mathrm{t})$ are functions of the variable t on some open interval $\mathbf{I}$. The variable t is called the *parameter* of the curve. If at some point of the curve

$$[\dot{x}^1(\mathrm{t})]^2 + [\dot{x}^2(\mathrm{t})]^2 + [\dot{x}^3(\mathrm{t})]^2 = 0, \quad t \in \mathbf{I}, \tag{4.2}$$

where

$$\dot{x}^i = \frac{\mathrm{d}x^i}{\mathrm{dt}},$$

we call this point a *singular* point of the curve. The other points of the curve are called *regular*. If all points of a curve lie in one plane we speak of a *plane curve*.

If $\boldsymbol{r}$ is the position vector of a point (4.1) of the curve, then it is possible to express this curve by the vector equation

$$\boldsymbol{r} = x^1(\mathrm{t})\boldsymbol{f}_1 + x^2(\mathrm{t})\boldsymbol{f}_2 + x^3(\mathrm{t})\boldsymbol{f}_3, \quad \text{or rather} \quad \boldsymbol{r} = \boldsymbol{r}(\mathrm{t}).$$

If $\mathrm{t} = \mathrm{t}(\tau)$ is a function of one variable such that it maps an interval $\mathbf{I}_1$ on to $\mathbf{I}$ and $(\mathrm{dt}/\mathrm{d}\tau) \neq 0$ on $\mathbf{I}_1$, then this function is called a *transformation of the parameter* and the same curve can be expressed by the functions $x^i = x^i(\mathrm{t}(\tau))$. Curves which differ only by their parametric representation are considered equal.

Now let $\boldsymbol{r} = \boldsymbol{r}(\mathrm{t})$ be a curve with regular points only (a *regular curve*). Then the function

$$\mathrm{s}(\mathrm{t}) = \int_{t_0}^{t} [(\dot{x}^1)^2 + (\dot{x}^2)^2 + (\dot{x}^3)^2]^{1/2}\, \mathrm{d}\tau, \quad \mathrm{t}_0, \mathrm{t} \in \mathbf{I}, \tag{4.3}$$

where $\dot{x}^i = \mathrm{d}x^i/\mathrm{dt}$, is called the *arc length* of the curve $\boldsymbol{r} = \boldsymbol{r}(\mathrm{t})$; τ is the variable of integration. The geometrical significance of s(t) lies in the fact that, for $t > t_0$ or $t < t_0$, $s(t)$ or $-s(t)$, respectively, represents the length of the arc of the curve $\boldsymbol{r} = \boldsymbol{r}(t)$ where the endpoints $B(t_0)$, $B(t)$ of the arc have position vectors $\boldsymbol{r}(t_0)$, $\boldsymbol{r}(t)$.

Arc length can be chosen as a new parameter of the curve since $ds/dt = \|\dot{\boldsymbol{r}}\| \neq 0$.

4.2 Example. The parametric equations

$$x^1 = a \cos t \,,$$
$$x^2 = a \sin t \,, \quad a > 0 \,, \quad v_0 > 0 \,, \quad t \in (-\infty, \infty) \,,$$
$$x^3 = v_0 t$$

express a right-hand helix with reduced height of thread v_0 on a cylinder of revolution with axis x^3 and radius a. Let us find its vector equation in which the helix arc s is the parameter. By (4.3) we have

$$s(t) = \int_0^t [a^2 \sin^2 \tau + a^2 \cos^2 \tau + v_0^2]^{1/2} \, d\tau \,.$$

Hence

$$s = t(a^2 + v_0^2)^{1/2} \,, \quad t \in (-\infty, \infty) \,,$$

and further

$$t = s(a^2 + v_0^2)^{-1/2} \,, \quad s \in (-\infty, \infty) \,.$$

Thus the vector equation of the helix is

$$\boldsymbol{r} = a \cos [s(a^2 + v_0^2)^{-1/2}] \boldsymbol{f}_1$$
$$+ a \sin [s(a^2 + v_0^2)^{-1/2}] \boldsymbol{f}_2 + v_0 \, s(a^2 + v_0^2)^{-1/2} \boldsymbol{f}_3 \,, \quad s \in (-\infty, \infty) \,.$$

Derivatives with respect to the arc s will be denoted by a prime in what follows.

4.3 Definition. The vector

$$\dot{\boldsymbol{r}}(t_0) = \left(\frac{d\boldsymbol{r}}{dt}\right)_{t_0} = \dot{x}^1(t_0) \boldsymbol{f}_1 + \dot{x}^2(t_0) \boldsymbol{f}_2 + \dot{x}^3(t_0) \boldsymbol{f}_3$$

is called the *tangent vector at the point* $B(t_0)$ of the curve $\boldsymbol{r} = \boldsymbol{r}(t)$. The line parallel to the vector $\dot{\boldsymbol{r}}(t_0)$ and passing through the point $B(t_0)$ is called the *tangent of the curve at the point* $B(t_0)$. The normed tangent vector is denoted

$$\boldsymbol{t} = \frac{\boldsymbol{r}^{\cdot}}{\|\boldsymbol{r}^{\cdot}\|} .$$

Equation (4.3) implies

$$\frac{ds}{dt} = [(\dot{x}^1)^2 + (\dot{x}^2)^2 + (\dot{x}^3)^2]^{1/2} = \|\boldsymbol{r}^{\cdot}\| . \tag{4.4}$$

If the parameter t is identical with the arc s we have

$$\|\boldsymbol{r}'\| = 1$$

and the tangent vector is the unit vector

$$\boldsymbol{r}' = \boldsymbol{t} .$$

4.4 Theorem. *The parameter* s *is an arc if and only if*

$$[(\boldsymbol{r}', \boldsymbol{r}')]^{1/2} = \|\boldsymbol{r}'\| = 1 .$$

4.5 Definition. The vector

$$\boldsymbol{r}''(s) = \boldsymbol{t}'(s)$$

is called the *vector of* 1st *curvature* of the curve $\boldsymbol{r} = \boldsymbol{r}(s)$ at the point $B(s)$. Its length

$$\kappa_1(s) = \|\boldsymbol{r}''(s)\|$$

is called the 1*st curvature* of the curve at the point $B(s)$. The expression

$$\varrho_1(s) = \frac{1}{\kappa_1(s)} , \quad \text{if} \quad \kappa_1(s) > 0 ,$$

is called the *radius of the* 1*st curvature* of the curve at the point $B(s)$. The line parallel to the vector $\boldsymbol{r}''(s)$ and passing through the point $B(s)$ is called the *principal normal* of the curve at the point $B(s)$.

The normed vector of 1st curvature is denoted by

$$\boldsymbol{n} = \frac{\boldsymbol{r}''}{\|\boldsymbol{r}''\|}$$

and called the *unit vector of the principal normal.*

Since

$$(\boldsymbol{r}', \boldsymbol{r}') = 1\,,$$

we obtain by differentiation of both sides of this equation with respect to s

$$2(\boldsymbol{r}', \boldsymbol{r}'') = 0\,.$$

At all points of the curve $\boldsymbol{r} = \boldsymbol{r}(\mathrm{s})$ at which $\boldsymbol{r}'' \neq \boldsymbol{0}$, the vectors $\boldsymbol{r}'$, $\boldsymbol{r}''$ and thus also the vectors $\boldsymbol{t}$, $\boldsymbol{n}$ are perpendicular.

It is easily proved that a curve whose 1st curvature is equal to zero at its every point is a straight line.

4.6 Definition. The unit vector

$$\boldsymbol{b}(\mathrm{s}) = \boldsymbol{t}(\mathrm{s}) \times \boldsymbol{n}(\mathrm{s})\,, \quad \text{where} \quad \kappa_1(\mathrm{s}) > 0\,,$$

is called the *unit vector of the binormal.* The line parallel to the vector $\boldsymbol{b}(\mathrm{s})$ and passing through the point $B(\mathrm{s})$ is called the *binormal* of the curve at $B(\mathrm{s})$.

4.7 Definition. The orthonormal frame $\{B(\mathrm{s}), \boldsymbol{t}(\mathrm{s}), \boldsymbol{n}(\mathrm{s}), \boldsymbol{b}(\mathrm{s})\}$ is called the *Frenet frame* of the curve $\boldsymbol{r} = \boldsymbol{r}(\mathrm{s})$ at the point $B(\mathrm{s})$.

4.8 Definition. At a point $B(\mathrm{s})$ of a curve $\boldsymbol{r} = \boldsymbol{r}(\mathrm{s})$ the tangent and the principal normal determine the so-called *osculating plane,* the tangent and the binormal determine the *rectification plane,* the binormal and the principal normal determine the *normal plane* of the curve at the point $B(\mathrm{s})$.

For the derivatives of the Frenet frame the *Frenet formulae* hold:

4.9 Theorem. *For the curve* $B(\mathrm{s}) = \boldsymbol{r}(\mathrm{s})$ *we have* $B' = \boldsymbol{t}$

$$\begin{aligned} \boldsymbol{t}' &= \kappa_1 \boldsymbol{n} \\ \boldsymbol{n}' &= -\kappa_1 \boldsymbol{t} + \kappa_2 \boldsymbol{b} \\ \boldsymbol{b}' &= -\kappa_2 \boldsymbol{n}\,, \end{aligned}$$

where

$$\kappa_2(\mathrm{s}) = \frac{|\boldsymbol{r}', \boldsymbol{r}'', \boldsymbol{r}'''|}{(\boldsymbol{r}'', \boldsymbol{r}'')} \quad \textit{for} \quad \kappa_1 > 0\,.$$

4.10 Definition. The number $\kappa_2(\mathrm{s})$ is called the 2nd *curvature* or *torsion* of a curve at the point $B(\mathrm{s})$.

The expression

$$\varrho_2(s) = \frac{1}{\kappa_2(s)}, \quad \kappa_2(s) \neq 0,$$

is the radius of the 2nd curvature at the point $B(s)$.

4.11 Theorem. *A curve is a plane curve if and only if its 2nd curvature is equal to zero at all its points.*

Let a curve **k** be given by a vector equation with general parameter $\boldsymbol{r} = \boldsymbol{r}(t)$. Then its 1st curvature is

$$\kappa_1 = \frac{[(\boldsymbol{r}^\cdot, \boldsymbol{r}^\cdot)(\boldsymbol{r}^{\cdot\cdot}, \boldsymbol{r}^{\cdot\cdot}) - (\boldsymbol{r}^\cdot, \boldsymbol{r}^{\cdot\cdot})^2]^{1/2}}{(\boldsymbol{r}^\cdot, \boldsymbol{r}^\cdot)^{3/2}} = \frac{\|\boldsymbol{r}^\cdot \times \boldsymbol{r}^{\cdot\cdot}\|}{(\boldsymbol{r}^\cdot, \boldsymbol{r}^\cdot)^{3/2}} \tag{4.6}$$

while the 2nd curvature is

$$\kappa_2 = \frac{|\boldsymbol{r}^\cdot, \boldsymbol{r}^{\cdot\cdot}, \boldsymbol{r}^{\cdot\cdot\cdot}|}{(\boldsymbol{r}^\cdot, \boldsymbol{r}^\cdot)(\boldsymbol{r}^{\cdot\cdot}, \boldsymbol{r}^{\cdot\cdot}) - (\boldsymbol{r}^\cdot, \boldsymbol{r}^{\cdot\cdot})^2}. \tag{4.7}$$

At the point $B(t)$ of the curve **k** the vector equation

$$(\boldsymbol{p} - \boldsymbol{r}(t), \boldsymbol{r}^\cdot(t)) = 0 \tag{4.8}$$

determines the osculating plane. Here $\boldsymbol{p}$ is the position vector of a current point of the plane.

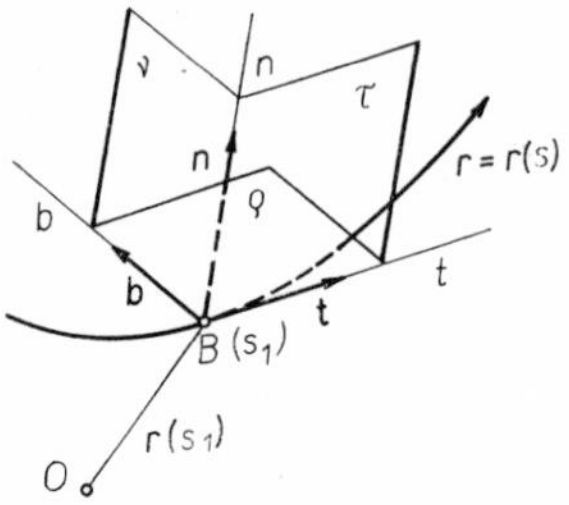

Fig. 6.

Fig. 6 shows the Frenet frame $\{B, \boldsymbol{t}, \boldsymbol{n}, \boldsymbol{b}\}$ of the curve $\boldsymbol{r} = \boldsymbol{r}(s)$ at the point $B(s_1)$. The osculating, normal, and rectification planes of the curve at the point $B(s_1)$ are denoted by τ, ν and ϱ, respectively.

4.12 Exercise. Compute the length of one thread of the helix $[t \in \langle 0, 2\pi \rangle]$ and its 1st and 2nd curvatures. Determine the equation of its osculating plane at the point $t = 0$.

A curve which lies on a sphere is called a *spherical curve.*

4.13 Definition. The set of all endpoints of unit tangent vectors (vectors of the binormal) of a curve **k** bound at one point is called the *spherical image of the tangents (binormals)* of the curve **k**.

The spherical images of the tangents and binormals of a curve lie on a sphere of radius 1, on the so-called *unit sphere.*

Let the curve $\boldsymbol{k}$ be expressed by the vector equation

$$\boldsymbol{r} = \boldsymbol{r}(s) ;$$

then the spherical images of its tangents k_1 and binormals k_3 are

$$\boldsymbol{r}_1 = \boldsymbol{t}(s) \quad \text{and} \quad \boldsymbol{r}_3 = \boldsymbol{b}(s) , \quad \text{respectively} ,$$

where s is the arc of the curve **k**. If s_1 and s_3 are arcs of the curves $\boldsymbol{r}_1$ and $\boldsymbol{r}_3$, respectively, then from relations (4.3) and (4.5) we obtain

$$\frac{ds_1}{ds} = \|\boldsymbol{t}'\| = \kappa_1 \quad \text{and} \quad \frac{ds_3}{ds} = \|\boldsymbol{b}'\| = \kappa_2 .$$

4.14 Example. Let us call K the terminal point of the position vector

$$\boldsymbol{p}(s) = \boldsymbol{r}(s) + \boldsymbol{n}(s)\,\varrho_1(s) ,$$

where $\varrho_1(s)$ is the radius of the 1st curvature of the curve $\boldsymbol{r}(s)$. K is the *centre of curvature* of the curve $\boldsymbol{r}(s)$ at the point $B(s)$. Let us show that the orthogonal projection of the centre of the sphere into the osculating plane of a spherical curve $\boldsymbol{r}(s)$ at the point $B(s)$ is the centre of curvature of the curve $\boldsymbol{r}(s)$. If we place the initial points of the position vectors to the centre of the sphere it suffices to prove that the vector $\boldsymbol{p}$ is perpendicular to the osculating plane, i.e. to the vectors $\boldsymbol{t}$ and $\boldsymbol{n}$ of the curve $\boldsymbol{r}$.

Obviously we have

$$(\boldsymbol{t}, \boldsymbol{p}) = (\boldsymbol{t}, \boldsymbol{r}) + (\boldsymbol{t}, \boldsymbol{n})\,\varrho_1 = (\boldsymbol{r}, \boldsymbol{r}') = 0 .$$

Differentiating this equation with respect to s we obtain

$$(\boldsymbol{r}', \boldsymbol{r}') + (\boldsymbol{r}'', \boldsymbol{r}) = 0 .$$

By (4.5) this equation yields

$$1 + (\boldsymbol{t}', \boldsymbol{r}) = 1 + \kappa_1(\boldsymbol{n}, \boldsymbol{r}) = 0$$

and upon modification

$$\varrho_1 + (\boldsymbol{n}, \boldsymbol{r}) = 0 .$$

Hence and from the equation for the position vector $\boldsymbol{p}$ it follows finally that

$$(\boldsymbol{n}, \boldsymbol{p}) = (\boldsymbol{n}, \boldsymbol{r}) + \varrho_1 = 0 ,$$

which was to be proved.

4.15 Exercise. Prove that the curve

$$\boldsymbol{r} = (a \cos^2 t,\ a \cos t \sin t,\ a \sin t), \quad \alpha \neq 0, \quad t \in \langle 0, 2\pi \rangle ,$$

is a spherical curve which lies on a cylinder of revolution whose directing circle in the plane $x_3 = 0$ has the centre $S = [\frac{1}{2}a, 0, 0]$ and radius $\frac{1}{2}a$.

4.16 Definition. A *surface* in E_3 is the set of all points in E_3 whose coordinates satisfy the equations

$$x^i = x^i(u, v), \quad i = 1, 2, 3, \quad [u, v] \in \Omega, \tag{4.10}$$

where Ω is an open set in the plane.

If the rank of the matrix

$$\begin{pmatrix} \dfrac{\partial x^1}{\partial u}, & \dfrac{\partial x^2}{\partial u}, & \dfrac{\partial x^3}{\partial u} \\ \dfrac{\partial x^1}{\partial v}, & \dfrac{\partial x^2}{\partial v}, & \dfrac{\partial x^3}{\partial v} \end{pmatrix} \tag{4.11}$$

is equal to two at some point of the surface, then this point is called a *regular point of the surface*. In the opposite case the point of the surface is called *singular*. Equations (4.10) are the *parametrical equations* of the surface.

If $\boldsymbol{r}$ is the position vector of a point of the surface (4.10), then the surface can be represented by the vector equation

$$\boldsymbol{r} = x^i(u, v)\,\boldsymbol{f}_i, \quad i = 1, 2, 3 \quad \text{or} \quad \boldsymbol{r} = \boldsymbol{r}(u, v). \tag{4.12}$$

Putting u = const. (or v = const.) in equations (4.10) we obtain a curve on the surface. This curve is called the parametric u-curve (or v-curve).

4.17 Example. Let us determine the vector equation of a surface of revolution whose generating curve $\boldsymbol{h}(v) = [h^1(v), 0, h^3(v)]$ in the plane $x^2 = 0$ revolves about the axis x^3. The trajectory of the revolving point $\boldsymbol{h}(v_1)$, the so-called *parallel*, has the parametric equation

$$\begin{aligned} x^1 &= h^1(v_1) \cos u \,, \\ x^2 &= h^1(v_1) \sin u \,, \\ x^3 &= h^3(v_1) \,, \end{aligned} \tag{4.13}$$

where $u \in \langle 0, 2\pi)$. The parallels are parametric v-curves while the positions of the revolving curve $\boldsymbol{h}(v)$ are parametric u-curves of the surface of revolution. Finally, its vector equation is

$$\boldsymbol{r}(u, v) = [h^1(v) \cos u, \; h^1(v) \sin u, \; h^3(v)]$$

for $u \in \langle 0, 2\pi)$ and $v \in \langle a, b \rangle$.

4.18 Exercise. Taking account of Example 4.17, determine the vector equation of the sphere with centre at the origin of coordinates O and with radius $\boldsymbol{r}$ and the vector equation of the cone with vertex O, vertex angle δ, and axis x^3. Determine the eventual singular points of the surface.

Similarly as in the case of curves we can use different parametric equations to represent a single surface. Such a transition is called a *transformation of parameters* u and v. A transformation of parameters is a mapping $\varphi: \bar{\Omega} \to \Omega$ such that φ is a one-to-one mapping of the open set $\bar{\Omega}$ in the plane on to Ω given by equations $u = u(\bar{u}, \bar{v})$, $v = v(\bar{u}, \bar{v})$, with

$$\begin{vmatrix} \dfrac{\partial u}{\delta \bar{u}}, & \dfrac{\partial u}{\partial \bar{v}} \\ \dfrac{\partial v}{\partial \bar{u}}, & \dfrac{\partial v}{\partial \bar{v}} \end{vmatrix} \neq 0 \,.$$

If the functions $u = u(\mathrm{t})$, $v = v(\mathrm{t})$ are given, a curve on the surface is determined. This curve is $\boldsymbol{r} = \boldsymbol{r}(\mathrm{t}) = \boldsymbol{r}(u(\mathrm{t}), v(\mathrm{t}))$. A tangent of a curve on a surface is called a *tangent of the surface*.

4.19 Theorem. *All tangents of any surface at its nonsingular point $B(u, v)$ lie in a plane. Its vector equation is*

$$|\boldsymbol{p} - \boldsymbol{r}(u, v),\ \boldsymbol{r}_u(u, v),\ \boldsymbol{r}_v(u, v)| = 0\,,$$

where $\boldsymbol{p}$ *is the position vector of a current point of any tangent plane and*

$$\boldsymbol{r}_u = \frac{\partial \boldsymbol{r}}{\partial u}\,, \quad \boldsymbol{r}_v = \frac{\partial \boldsymbol{r}}{\partial v}\,.$$

The vectors $\boldsymbol{r}_u$ and $\boldsymbol{r}_v$ are the tangent vectors of the parametric $\boldsymbol{v}$-curve and $\boldsymbol{u}$-curve of the surface, respectively.

4.20 Definition. The plane from Theorem 4.19 is called the *tangent plane* of the surface. Its point B is called the point of contact. The vector perpendicular to the tangent plane at the point B of the surface is called the *normal vector* of the surface at point B. The line parallel to the normal vector and passing through point B is called the *normal* of the surface at point B.

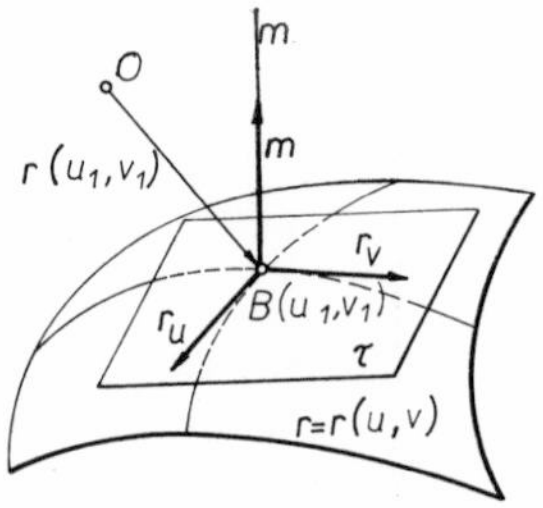

Fig. 7.

In Fig. 7 we have the tangent plane τ of the surface $\boldsymbol{r} = \boldsymbol{r}(u, v)$ at the point $B(u_1, v_1)$.

For the unit normal vector $\boldsymbol{m}$ of the surface we have

$$\boldsymbol{m} = \frac{\boldsymbol{r}_u \times \boldsymbol{r}_v}{\|\boldsymbol{r}_u \times \boldsymbol{r}_v\|} = \frac{\boldsymbol{r}_u \times \boldsymbol{r}_v}{\left[\boldsymbol{r}_u^2 \boldsymbol{r}_v^2 - (\boldsymbol{r}_u, \boldsymbol{r}_v)^2\right]^{1/2}}\,. \tag{4.14}$$

A special type of surfaces are *ruled surfaces* which can be generated by the movement of a line in space. Let the moving line be determined by its point $\boldsymbol{k}(u)$ and by the unit vector $\boldsymbol{q}(u)$ of its direction, $u \in \mathbf{I}$. Then the equation of the ruled surface $\boldsymbol{r}(u, v)$ is

$$\boldsymbol{r}(u, v) = \boldsymbol{k}(u) + v\,\boldsymbol{q}(u)\,, \quad v \in (-\infty, \infty)\,. \tag{4.15}$$

The parametric u-curve of this surface is a straight line of the surface which is called *ruling*. The parametric v-curve for $v = 0$ is the curve $\boldsymbol{k}(u)$ which is called *generating* curve of the surface (see Fig. 8).

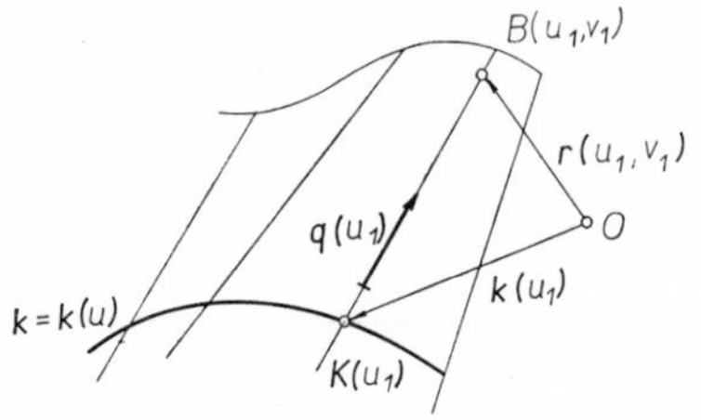

Fig. 8.

The unit normal vector $\boldsymbol{m}$ of a ruled surface is, by (4.14) and (4.15),

$$\boldsymbol{m} = \frac{(\boldsymbol{k}^{\cdot} + v\boldsymbol{q}^{\cdot}) + \boldsymbol{q}}{[(\boldsymbol{k}^{\cdot} + v\boldsymbol{q}^{\cdot})^2 - (\boldsymbol{k}, \boldsymbol{q})^2]^{1/2}}\,,$$

$$\text{where} \quad \boldsymbol{k}^{\cdot} = \frac{d\boldsymbol{k}}{du}\,, \quad \boldsymbol{q}^{\cdot} = \frac{d\boldsymbol{q}}{du}\,. \tag{4.16}$$

Let us investigate normal vectors at distinct points $v_1 \neq v_2$ of the same ruling of a ruled surface. For their cross-product we have

$$\begin{aligned}
&[(\boldsymbol{k}^{\cdot} + v_1\boldsymbol{q}^{\cdot}) \times [(\boldsymbol{k}^{\cdot} + v_2\boldsymbol{q}^{\cdot}) \times \boldsymbol{q}]\\
&= |\boldsymbol{k}^{\cdot} + v_1\boldsymbol{q}^{\cdot}, \boldsymbol{q}, \boldsymbol{q}|(\boldsymbol{k}^{\cdot} + v_2\boldsymbol{q}^{\cdot}) - |\boldsymbol{k}^{\cdot} + v_1\boldsymbol{q}^{\cdot}, \boldsymbol{q}, \boldsymbol{k}^{\cdot} + v_2\boldsymbol{q}^{\cdot}|\,\boldsymbol{q}\\
&= (v_1 - v_2)\,|\boldsymbol{k}^{\cdot}, \boldsymbol{q}, \boldsymbol{q}^{\cdot}|\,\boldsymbol{q}\,.
\end{aligned}$$

If $|\boldsymbol{k}^{\cdot}, \boldsymbol{q}, \boldsymbol{q}^{\cdot}| = 0$, then the normal vectors are colinear at all points of the same ruling and at the nonsingular points of the surface its tangent planes are identical. We then say that the tangent plane contacts the surface along a ruling. Such a ruling is called a *torsal* ruling. If $|\boldsymbol{k}^{\cdot}, \boldsymbol{q}, \boldsymbol{q}^{\cdot}| \neq 0$, then the tangent planes of a ruled surface are distinct at all points of the same ruling. It can be shown that they constitute a pencil of planes.

4.21 Definition. A ruled surface whose all rulings are torsal is called a *developable surface*. The remaining ruled surfaces are called *skew surfaces*.

4.22 Theorem. *A ruled surface is developable if and only if at all its points*

$$|\boldsymbol{k}^{\cdot}, \boldsymbol{q}, \boldsymbol{q}^{\cdot}| = 0\,. \tag{4.17}$$

4.23 Example. Let a ruled surface be constituted by all tangents of a space curve $\boldsymbol{k} = \boldsymbol{k}(u)$. This surface is called the *surface of tangents of a space curve.* By (4.15) its equation is

$$\boldsymbol{r}(u, v) = \boldsymbol{k}(u) + v\,\boldsymbol{k}^{\cdot}(u)\,, \quad \text{where} \quad \boldsymbol{k}^{\cdot}(u) = \frac{\mathrm{d}\boldsymbol{k}}{\mathrm{d}u}\,. \tag{4.18}$$

Then we have

$$r_u = \boldsymbol{k}^{\cdot} + v\boldsymbol{k}^{\cdot\cdot}\,, \quad \boldsymbol{r}_v = \boldsymbol{k}^{\cdot} \tag{4.19}$$

and

$$\boldsymbol{r}_u \times \boldsymbol{r}_v = (\boldsymbol{k}^{\cdot} + v\boldsymbol{k}^{\cdot\cdot}) \times \boldsymbol{k}^{\cdot} = v\boldsymbol{k}^{\cdot\cdot} \times \boldsymbol{k}^{\cdot}\,. \tag{4.20}$$

Hence it follows that all points with the parameter $v = 0$, i.e. all points of the curve $\boldsymbol{k}$, are singular. The curve $\boldsymbol{k}$ is called the *edge of regression* of the surface of tangents of a space curve.

Since for the unit vector $\boldsymbol{q}$ it is

$$\boldsymbol{q} = \frac{\boldsymbol{k}^{\cdot}}{\|\boldsymbol{k}^{\cdot}\|}\,, \tag{4.21}$$

we have

$$|\boldsymbol{k}^{\cdot}, \boldsymbol{q}, \boldsymbol{q}^{\cdot}| = \frac{1}{\|\boldsymbol{k}^{\cdot}\|}\,|\boldsymbol{k}^{\cdot}, \boldsymbol{k}^{\cdot}, \boldsymbol{q}^{\cdot}| = 0\,.$$

By Theorem 4.22 the surface of tangents of a space curve is a developable surface.

4.24 Exercise. Prove that the tangent plane of the surface of the tangents of a space curve along a ruling p is identical with the osculating plane of its edge of regression at the point of contact of the line p.

We find out easily that the conical surface ($\boldsymbol{k}^{\cdot} = 0$), the cylindrical surface ($\boldsymbol{q}^{\cdot} = 0$), and the plane ($\boldsymbol{k}^{\cdot}, \boldsymbol{q}, \boldsymbol{q}^{\cdot}$ are coplanar) belong to developable surfaces. However, we shall not consider them in what follows.

Let the unit normal vector $\boldsymbol{m}(u, v)$ of a skew surface converge to the vector $\boldsymbol{a}$ for $v \to \pm\infty$ at the points of its nontorsal ruling $u = u_1$. Relation (4.16) implies

$$\boldsymbol{a} = \lim_{v \to \pm\infty} \boldsymbol{m}(u_1, v) = \frac{(\boldsymbol{q} \times \boldsymbol{q}^{\cdot})}{\|\boldsymbol{q}^{\cdot}\|}. \tag{4.22}$$

4.25 Definition. The plane of a skew surface which passes through its ruling u_1 and is perpendicular to the vector $\boldsymbol{a}$ from (4.22) is called the *asymptotic plane* α. The tangent plane γ passing through the ruling u_1 which is perpendicular to the asymptotic plane α is called the *central plane*. Its point of contact C is called the *central* point of the ruling u_1. The straight lines which pass through point C and are perpendicular to the planes α and γ are called the *central tangent* and *central normal*, respectively.

The perpendicularity of the vectors $\boldsymbol{q}$, $\boldsymbol{q}^{\cdot}$ and relation (4.22) imply the representation of the unit vector $\boldsymbol{h}$ of the central normal in the form

$$\boldsymbol{h} = \frac{\boldsymbol{q}^{\cdot}}{\|\boldsymbol{q}^{\cdot}\|}. \tag{4.23}$$

Substituting the parameter v of the central point C into equation (4.16) we get

$$\boldsymbol{h} \times \boldsymbol{m} = \boldsymbol{0},$$

and thus

$$\boldsymbol{q}^{\cdot} \times [(\boldsymbol{k}^{\cdot} + v\boldsymbol{q}^{\cdot}) \times \boldsymbol{q}] = (\boldsymbol{q}^{\cdot}, \boldsymbol{q})(\boldsymbol{k}^{\cdot} + v\boldsymbol{q}^{\cdot}) - (\boldsymbol{q}^{\cdot}, \boldsymbol{k}^{\cdot} + v\boldsymbol{q}^{\cdot})\,\boldsymbol{q} = \boldsymbol{0}.$$

Hence we obtain, after some manipulation,

$$v = -\frac{(\boldsymbol{k}^{\cdot}, \boldsymbol{q}^{\cdot})}{\boldsymbol{q}^{\cdot 2}}. \tag{4.24}$$

4.26 Definition. The set of central points of all the rulings of a skew surface is called the *line of striction* of the surface.

By (4.15) and (4.24) the equation of the line of striction $\boldsymbol{c} = \boldsymbol{c}(u)$ is

$$\boldsymbol{c}(u) = \boldsymbol{k} - \boldsymbol{q}\,\frac{(\boldsymbol{k}^{\cdot}, \boldsymbol{q}^{\cdot})}{\boldsymbol{q}^{\cdot 2}}. \tag{4.25}$$

Hence it follows that the generating curve $\boldsymbol{k}$ of a skew surface will be its line of striction if and only if

$$(\boldsymbol{k}^{\cdot}, \boldsymbol{q}^{\cdot}) = 0\,. \tag{4.26}$$

It can be shown that the limit position of the shortest transversal of two rulings $u = u_1$, $u = u_2$ of a skew surface for $u_2 \to u_1$ is the central tangent of the ruling u_1.

4.27 Definition. Let w be the length of the shortest transversal of two rulings $u = u_1$, $u = u_2$ of a ruled surface. Let φ be the angle between u_1 and u_2. If the limit

$$\lim_{u_2 \to u_1} \frac{w}{\varphi}$$

exists we call it the *parameter of distribution* of the ruling u_1 and denote it d.

For the parameter of distribution d of a ruling u we have

$$d = \frac{|\boldsymbol{k}^{\cdot}, \boldsymbol{q}, \boldsymbol{q}^{\cdot}|}{\boldsymbol{q}^{\cdot 2}}\,. \tag{4.27}$$

By (4.17) all rulings of developable surfaces for which $\boldsymbol{q}^{\cdot 2} \neq 0$, i.e. with the exception of cylindrical surfaces, have distribution parameter $d = 0$. For rulings of skew surfaces, with the exception of their torsal rulings, the distribution parameter is $d \neq 0$.

Now, let us pay attention to the geometrical interpretation of the distribution parameter. Let the generating curve of a skew surface be its line of striction. Then

$$\boldsymbol{q}^{\cdot} \times (\boldsymbol{k}^{\cdot} \times \boldsymbol{q}) = (\boldsymbol{q}^{\cdot}, \boldsymbol{q})\,\boldsymbol{k}^{\cdot} - (\boldsymbol{q}^{\cdot}, \boldsymbol{k}^{\cdot})\,\boldsymbol{q} = \boldsymbol{0}$$

and the nonzero vectors $\boldsymbol{q}^{\cdot}$, $\boldsymbol{k}^{\cdot} \times \boldsymbol{q}$ are colinear, i.e. $\boldsymbol{k} \times \boldsymbol{q} = \lambda \boldsymbol{q}^{\cdot}$. This implies

$$(\boldsymbol{k}^{\cdot} \times \boldsymbol{q}, \boldsymbol{q}^{\cdot}) = \lambda \boldsymbol{q}^{\cdot 2}\,.$$

Hence equation (4.27) yields $\lambda = d$ and, finally, $\boldsymbol{k}^{\cdot} \times \boldsymbol{q} = \mathrm{d}\boldsymbol{q}^{\cdot}$. By (4.16) the unnormed normal vector of the ruled surface is

$$\overline{\boldsymbol{m}} = \mathrm{d}\boldsymbol{q}^{\cdot} + v(\boldsymbol{q}^{\cdot} \times \boldsymbol{q}) \tag{4.28}$$

and depends along the ruling u on the parameter v only.

The vectors $q^{\cdot}$ and $q^{\cdot} \times q$ are of the same length, mutually perpendicular and constant for every ruling u of the surface. In the base $\{q^{\cdot}, q^{\cdot} \times q\}$ the vector $\overline{m}$ has the coordinates $[d, v]$ which satisfy the equation

$$\frac{v}{d} = \tan \lambda\,, \quad d \neq 0\,.$$

Here λ is the angle between the vectors $\overline{m}$ and $q^{\cdot}$.

Consequently we have

4.28 Theorem. *For the angle λ between the tangent plane of a skew surface at the point $[u, v]$ of a nontorsal ruling u and the central plane we have*

$$\frac{v}{d} = \operatorname{tg} \lambda\,, \quad d \neq 0\,. \tag{4.29}$$

Here d is the distribution parameter of the ruling u and the central point has the coordinates $[u, 0]$.

Let us now define the Frenet frame of a skew surface:

4.29 Definition. The orthonormal frame $\{C, q, h, a\}$, where C is the central point of the ruling of a skew surface and q, h, a are the unit vectors of the ruling, the central normal and the central tangent, respectively, is called the *Frenet frame of the skew surface.*

The set of all bound vectors $q(u)$ at the point O constitutes the *directing cone* of the ruled surface. The end-points of these vectors circumscribe a spherical curve k_1, called the spherical image of the ruled surface whose arc is denoted by s_1.

Let us define the *Frenet frame of the directing cone* as the orthonormal frame $\{O, q, n, z\}$ where

$$n = \frac{dq}{ds_1} = q'\,.$$

Since we have

$$q' = \frac{q^{\cdot}}{\|q^{\cdot}\|} = h$$

by (4.23), the tangent planes of the directing cone are parallel to the asymptotic planes of the skew surface. Finally, we have

$$z = q \times h = a$$

and the Frenet frame of the directing cone has the same base as the Frenet frame of the skew surface, i.e. $\{O, q, h, a\}$.

Compute the derivatives of the vectors h, a, with respect to the arc s_1 of the generating curve k_1. We have $h^2 = 1$, thus $(h, h') = 0$. Consequently, it is possible to write

$$h' = \alpha q + \beta a .$$

From $(h, q) = 0$ it follows that

$$(h', q) + (q', h) = \alpha + 1 = 0 ,$$

and if we put $\beta = k$, where k is the *conical curvature* of a cone, we have

$$h' = -q + ka .$$

From $(h, a) = 0$ we obtain

$$(h', a) + (h, a') = k + (h, a') = 0 ,$$

and thus

$$(h, a') = -k . \tag{4.30}$$

Further, $(a, a) = 1$ and $(a, q) = 0$ imply the relations

$$(a, a') = 0 \quad \text{and} \quad (a', q) + (a, q') = (a', q) = 0 .$$

This implies that the vector a' is colinear with the vector h, i.e. $a' = \gamma h$. By (4.30) we get

$$(h, a') = \gamma = -k$$

and thus

$$a' = -kh . \tag{4.31}$$

For the spherical curve k_3 with arc s_3 circumscribed by the bound vector a at the point O we have

$$\frac{ds_3}{ds_1} = \|a'\| = k \tag{4.32}$$

by (4.4).

4.30 Theorem. *For the derivatives of the vectors of the Frenet frame of a skew surface and of its directing cone* $\{O, \boldsymbol{q}, \boldsymbol{h}, \boldsymbol{a}\}$ *with respect to the arc* s_1 *of its generating curve* k_1 *we have the Frenet formulae*

$$\begin{aligned} \boldsymbol{q}' &= \quad\ \boldsymbol{h}\,, \\ \boldsymbol{h}' &= -\boldsymbol{q} \qquad + k\boldsymbol{a}\,, \\ \boldsymbol{a}' &= \quad -k\boldsymbol{h}\,, \end{aligned} \tag{4.33}$$

where $k = \mathrm{d}s_3/\mathrm{d}s_1$ *by* (4.32).

Let s be the arc of the line of striction of a skew surface. Furthermore, we call $\mathrm{d}s_1/\mathrm{d}s = \kappa_1$ the 1st curvature, $\mathrm{d}s_3/\mathrm{d}s = \kappa_2$ the 2nd curvature of the skew surface or rather of its directing cone. Then we have

$$\kappa_2 = \kappa_1 k\,. \tag{4.34}$$

Multiplying equations (4.33) by the 1st curvature $\kappa_1 = \mathrm{d}s_1/\mathrm{d}s$ we obtain the following

4.31 Theorem. *For the derivatives of the vectors of the Frenet frame with base* $\{\boldsymbol{q}, \boldsymbol{h}, \boldsymbol{a}\}$ *of a skew surface and of its directing cone with respect to the arc s of the line of striction of the skew surface we have the Frenet formulae*

$$\begin{aligned} \frac{\mathrm{d}\boldsymbol{q}}{\mathrm{d}s} &= \qquad \kappa_1\boldsymbol{h}\,, \\ \frac{\mathrm{d}\boldsymbol{h}}{\mathrm{d}s} &= -\kappa_1\boldsymbol{q} \qquad +\kappa_2\boldsymbol{a}\,, \\ \frac{\mathrm{d}\boldsymbol{a}}{\mathrm{d}s} &= \qquad -\kappa_2\boldsymbol{h}\,, \end{aligned} \tag{4.35}$$

where

$$\kappa_1 = \frac{\mathrm{d}s_1}{\mathrm{d}s}, \quad \kappa_2 = \frac{\mathrm{d}s_3}{\mathrm{d}s}$$

and s_1, s_3 *are the arcs of the spherical curves* k_1, k_3 *circumscribed by the bound vectors* $\boldsymbol{q}$ *and* $\boldsymbol{a}$.

Furthermore, note that in the deliberations which follow we will consider the edge of regression of the tangent surface of a space curve to be the line of striction of this surface. By definition (4.7) the Frenet frame of the edge of regression $\{C, \boldsymbol{t}, \boldsymbol{n}, \boldsymbol{b}\}$ will here be the Frenet frame of the surface $\{C, \boldsymbol{q}, \boldsymbol{b}, \boldsymbol{a}\}$. For the arc u of the edge of regression $\boldsymbol{k} = \boldsymbol{k}(u)$ we have

$$\left(\frac{\mathrm{d}\boldsymbol{k}}{\mathrm{d}u}, \frac{\mathrm{d}^2\boldsymbol{k}}{\mathrm{d}u^2}\right) = (\boldsymbol{t}, \boldsymbol{t}') = 0 .$$

This is the familiar condition (4.26) which shows that the generating curve of the surface is its line of striction.

We will show that the tangent of the line of striction of the ruled surface at the point C is perpendicular to the vector $\boldsymbol{h}$ of the Frenet frame $\{C, \boldsymbol{q}, \boldsymbol{h}, \boldsymbol{a}\}$. Equation (4.25) implies

$$\frac{\mathrm{d}\boldsymbol{c}}{\mathrm{d}u} = \boldsymbol{k}^{\boldsymbol{\cdot}} + v\boldsymbol{q}^{\boldsymbol{\cdot}} + \dot{v}\boldsymbol{q} ,$$

and further we have, by (4.23) and (4.24),

$$\left(\boldsymbol{h}, \frac{\mathrm{d}\boldsymbol{c}}{\mathrm{d}u}\right) = (\boldsymbol{k}^{\boldsymbol{\cdot}}, \boldsymbol{h}) + (v\boldsymbol{q}^{\boldsymbol{\cdot}}, \boldsymbol{h}) = \frac{(\boldsymbol{k}^{\boldsymbol{\cdot}}, \boldsymbol{q}^{\boldsymbol{\cdot}})}{\|\boldsymbol{q}^{\boldsymbol{\cdot}}\|} - \frac{(\boldsymbol{k}^{\boldsymbol{\cdot}}, \boldsymbol{q}^{\boldsymbol{\cdot}})(\boldsymbol{q}^{\boldsymbol{\cdot}}, \boldsymbol{q}^{\boldsymbol{\cdot}})}{\boldsymbol{q}^{\boldsymbol{\cdot} 2}\|\boldsymbol{q}^{\boldsymbol{\cdot}}\|} = 0 .$$

Let σ be the angle between the line of striction and the ruling, i.e. $\sigma = \sphericalangle \boldsymbol{q}\boldsymbol{t}$ where $\boldsymbol{t}$ is the tangent vector of the line of striction. The positive sense of rotation in the central plane is determined by the sense of rotation of $\boldsymbol{q}$ into $\boldsymbol{a}$ through a right angle. The orientation of the vector $\boldsymbol{t}$ is the same as the orientation of the line of striction C. Assume further that

$$-\tfrac{1}{2}\pi \leqq \sigma \leqq \tfrac{1}{2}\pi .$$

Then we can write

$$\boldsymbol{t} = \frac{\mathrm{d}\boldsymbol{c}}{\mathrm{d}s} = \boldsymbol{q} \cos \sigma + \boldsymbol{a} \sin \sigma . \tag{4.36}$$

The equation of the line of striction is thus

$$\boldsymbol{c}(s) = \int [\boldsymbol{q}(s) \cos \sigma(s) + \boldsymbol{a}(s) \sin \sigma(s)] \, \mathrm{d}s$$

while the equation of the ruled surface is

$$\boldsymbol{r}(s, v) = \boldsymbol{c}(s) + v\, \boldsymbol{q}(s) .$$

For the parameter of distribution we have

$$d = \frac{\left| q, \dfrac{dq}{ds}, \dfrac{dc}{ds} \right|}{\left(\dfrac{dq}{ds}\right)^2} = \frac{|q, h, t|}{\kappa_1} \tag{4.37}$$

by (4.27) and (4.35). Hence and from equations (4.36) and (4.37) it follows that

$$\sin \sigma = (a, t) = (q \times h, t) = |q, h, t| = \kappa_1 d\,. \tag{4.38}$$

4.32 Example. The ruled surfaces for which $\kappa_1 \kappa_2 \neq 0$ and $\kappa_2/\kappa_1 = \text{const.}$ have a cone of revolution as their directing cone. If $\kappa_1 \neq 0$, $\kappa_2 = 0$ is true for all rulings of a ruled surface we obtain a directing plane instead of a directing cone. These ruled surfaces are the so-called *conoids*. If $\sigma = \frac{1}{2}\pi$ for all rulings, the corresponding ruled surfaces are constituted by the binormals of the line of striction.

Let s_1 be the arc of the spherical curve k_1 of the directing cone of a ruled surface. By (4.38), equation (4.36) and the Frenet formulae of a surface are

$$\begin{aligned}
\frac{dc}{ds_1} &= \frac{\cos \sigma}{\kappa_1} q + \frac{\sin \sigma}{\kappa_1} a = fq + da\,,\\
\frac{dq}{ds_1} &= h\,,\\
\frac{dh}{ds_1} &= -q + ka\,,\\
\frac{da}{ds_1} &= -kh\,.
\end{aligned} \tag{4.39}$$

The functions $f(s_1)$, $d(s_1)$, $k(s_1)$ are the *invariants* of the ruled surface. They determine the ruled surface uniquely up to its position in the space.

In conclusion we introduce the *Plücker coordinates* of an oriented line in E_3. Such a line can be given by the position vector b of its arbitrary point B and by its unit vector q. The vector equation of the line is

$$p = b + vq\,, \quad v \in (-\infty, \infty)\,.$$

Construct the vector

$$\bar{\boldsymbol{q}} = \boldsymbol{b} \times \boldsymbol{q} .$$

We immediately see that $(\boldsymbol{q}, \bar{\boldsymbol{q}}) = 0$ and that the vector $\bar{\boldsymbol{q}}$ is independent of the choice of the point B since for an arbitrary point P with position vector $\boldsymbol{p}$ it is

$$\boldsymbol{p} \times \boldsymbol{q} = (\boldsymbol{b} + v\boldsymbol{q}) \times \boldsymbol{q} = \boldsymbol{b} \times \boldsymbol{q} = \bar{\boldsymbol{q}} .$$

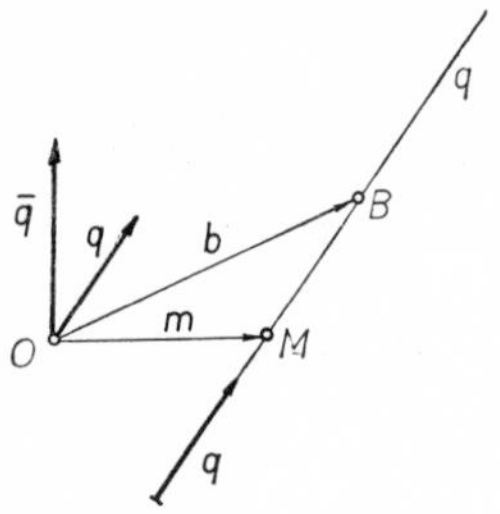

Fig. 9.

Thus, we can choose the foot M of the perpendicular $\boldsymbol{m}$ dropped from the point O to the given line (Fig. 9). To the oriented line we then assign an ordered pair of vectors $(\boldsymbol{q}; \bar{\boldsymbol{q}})$ which satisfy the conditions

$$\boldsymbol{q}^2 = 1 , \quad (\boldsymbol{q}, \bar{\boldsymbol{q}}) = 0 . \tag{4.40}$$

Since $\boldsymbol{m} \times \boldsymbol{q} = \bar{\boldsymbol{q}}$ and in addition

$$\boldsymbol{q} \times \bar{\boldsymbol{q}} = \boldsymbol{q} \times (\boldsymbol{m} \times \boldsymbol{q}) = \boldsymbol{q}^2\boldsymbol{m} - (\boldsymbol{q}, \boldsymbol{m})\, \boldsymbol{q} = \boldsymbol{m} ,$$

it is possible to construct a unique oriented line $\boldsymbol{p} = \boldsymbol{q} \times \bar{\boldsymbol{q}} + v\boldsymbol{q}$ to the given pair of vectors $(\boldsymbol{q}; \bar{\boldsymbol{q}})$ with conditions (4.40).

4.33 Theorem. *There is one-to-one correspondence between oriented lines of* E_3 *and ordered pairs of vectors* $(\boldsymbol{q}; \bar{\boldsymbol{q}})$ *such that* $\boldsymbol{q}^2 = 1$, $(\boldsymbol{q}, \bar{\boldsymbol{q}}) = 0$.

4.34 Definition. The coordinates of the pair of vectors $(\boldsymbol{q}; \bar{\boldsymbol{q}})$ from Theorem 4.33 are called the *normed Plücker coordinates* of a line. They are denoted $p_1 = q^1$, $p_2 = q^2$, $p_3 = q^3$, $p_4 = \bar{q}^1$, $p_5 = \bar{q}^2$, $p_6 = \bar{q}^3$.

Four of these six Plücker coordinates are independent. The set of all lines in E_3 constitutes thus a four-dimensional space.

4.35 Definition. The set of all lines $(\boldsymbol{q}; \overline{\boldsymbol{q}})$ in E_3 which satisfy the vector equation

$$(\boldsymbol{a}, \overline{\boldsymbol{q}}) + (\overline{\boldsymbol{a}}, \boldsymbol{q}) = 0, \tag{4.41}$$

where $\boldsymbol{a}, \overline{\boldsymbol{a}}$ are arbitrary vectors, is called a *general linear complex* for $(\boldsymbol{a}, \overline{\boldsymbol{a}}) \neq 0$ and a *special linear complex* for $(\boldsymbol{a}, \overline{\boldsymbol{a}}) = 0$, $\boldsymbol{a}^2 = 1$.

A linear complex is thus the set of all lines in the space which depend on three independent parameters. A special linear complex can be geometrically interpreted as the set of all lines $(\boldsymbol{q}; \overline{\boldsymbol{q}})$ which intersect the line $(\boldsymbol{a}, \overline{\boldsymbol{a}})$.

5. DIFFERENTIABLE MANIFOLDS, LIE GROUPS

In this section the reader will find a brief overview of the topics from the theory of differentiable manifolds and Lie groups necessary for the exposition which follows. The details can be found, e.g. in [III] to [VI] and [XII].

5.1 Definition. Let X be a topological manifold with an atlas $\{(U_i, \varphi_i)\}$. X is called a *differentiable manifold* (*of class* C^∞) if transformations of coordinates (i.e. the functions $\varphi_j \circ \varphi_i^{-1}$) have all partial derivatives of all orders. X is called an *analytic manifold* if transformations of coordinates are real analytic functions at every point.

5.2 Remark. A function $f(x^1, \ldots, x^n)$ is real analytic at a point $A = [x_0^1, \ldots, x_0^n] \in E_n$ if there exists a spherical neighbourhood of the point A such that the Taylor expansion of the function f at the point A converges on this neighbourhood to the values of the function. An example of a function which has all derivatives but is nowhere analytic is given in [III].

5.3 Exercise. Show that the topological manifolds given in Examples 3.23 and 3.26 with the atlases given there are differentiable manifolds.

5.4 Theorem. *Let $f(x^1, x^2, x^3)$ be a function of three variables in E_3 which has all partial derivatives. Let the equation $f(x^1, x^2, x^3) = 0$ have at least one solution, and let no point of the solution of this equation be singular. Then the set of all points from E_3 which satisfy equation $f(x^1, x^2, x^3) = 0$ is a differentiable manifold of dimension* 2.

Proof: see [IV]. A point from E_3 is called singular if at this point

$$\frac{\partial f}{\partial x^1} = \frac{\partial f}{\partial x^2} = \frac{\partial f}{\partial x^3} = 0 .$$

In [IV] it is also described in detail how to find the respective charts. As the local coordinates in the neighbourhood of each point a suitable pair of the coordinates x^1, x^2, x^3 can be taken.

5.5 Example. Let us consider the ellipsoid given by the equation

$$f \equiv \left(\frac{x^1}{a}\right)^2 + \left(\frac{x^2}{b}\right)^2 + \left(\frac{x^3}{c}\right)^2 - 1 = 0 .$$

Only the point $[0, 0, 0]$ which is not a point of the surface can be singular. Thus, the ellipsoid is a differentiable manifold.

5.6 Exercise. Show that the equation $f(x, y, z) = x^2 + y^2 - z^2 - 1 = 0$ determines a differentiable manifold. Construct a differentiable atlas on this manifold. Show that the equation $f(x, y, z) = x^2 - y^2 - z^2 = 0$ does not yield a manifold but that a differentiable manifold is obtained after removing the singular points.

5.7 Remark. Let f be a function on the set M (i.e. a mapping of M into $\mathbf{R}$). If $N \subset M$ we denote by $f|_N$ the *restriction* of the function f to the subset N. i.e., $f|_N(x) = f(x)$, for $x \in N$ while $f|_N(x)$ is not defined for $x \in M - N$.

5.8 Definition. Let M be a differentiable manifold, and let $m \in M$. If f is a function defined on M and if (U_i, φ_i) is a coordinate neighbourhood which includes the point m, then it is possible to define the function

$$f \circ \varphi_i^{-1}(x^1, \ldots, x^n) = f|_{U_i} \circ \varphi_i^{-1}(x^1, \ldots, x^n) = f(\varphi_i^{-1}(x^1, \ldots, x^n)) .$$

This function is defined on the set $\varphi_i(U_i) \subset E_n$. We shall say that f is *differentiable at the point* m if and only if the function $f \circ \varphi_i^{-1}$ has all partial derivatives at the point $\varphi_i(m)$. A function differentiable at all points of the manifold is called *differentiable*. (See [III], [IV], [XII].)

5.9 Example. Consider the unit sphere S^2 in E_3 with the coordinates $\zeta^1, \zeta^2, \zeta^3$. Denote $S = [0, 0, 1]$, $J = [0, 0, -1]$, $\alpha \equiv \zeta^3 + 1 = 0$, $\beta \equiv$

$\zeta^3 - 1 = 0$. Denote the Cartesian coordinates in the plane α, with origin at J and axes parallel to ζ^1 and ζ^2, by x^1 and x^2. Similarly, let us define y^1 and y^2 in the plane β. Further, choose a point X on the sphere with coordinates $\zeta^1, \zeta^2, \zeta^3$ and project it from the point S into the plane α (this will be the coordinate mapping φ_1); similarly we proceed with φ_2, see Section 2. For the sake of brevity we write $\varepsilon = \pm 1$, the point $Z = [0, 0, \varepsilon]$ is either S or J, and the projection is performed into the plane $\zeta^3 + \varepsilon = 0$. The line $p \equiv ZX$ has the equation

$$p \equiv [0, 0, \varepsilon] + \lambda[\xi^1, \xi^2, \xi^3 - \varepsilon] .$$

The line p is the projection line of the point X. The coordinates of its point of intersection with the plane $\zeta^3 + \varepsilon = 0$ are

$$\left[\frac{2\varepsilon\zeta^1}{\varepsilon - \xi^3}, \frac{2\varepsilon\xi^2}{\varepsilon - \xi^3}, -\varepsilon\right].$$

The coordinate mappings φ_1 and φ_2 are thus given by the relations

$$\begin{aligned} &\varphi_1 : [\xi^1, \xi^2, \xi^3] \to [x^1, x^2] , \\ &\text{where} \quad x^1 = \frac{2\xi^1}{1 - \xi^3} , \quad x^2 = \frac{2\xi^2}{1 - \xi^3} , \end{aligned} \tag{5.1}$$

and

$$\begin{aligned} &\varphi^2 : [\xi^1, \xi^2, \xi^3] \to [y^1, y^2] , \\ &\text{where} \quad y^1 = \frac{2\xi^1}{1 + \xi^3} , \quad y^2 = \frac{2\xi^2}{1 + \xi^3} . \end{aligned} \tag{5.2}$$

They are defined on $S^2 - \{S\}$ and $S^2 - \{J\}$, respectively. (5.1) and (5.2) yield the transformation equations

$$\begin{aligned} y^1 &= \frac{4x^1}{(x^1)^2 + (x^2)^2} & x^1 &= \frac{4y^1}{(y^1)^2 + (y^2)^2} , \\ y^2 &= \frac{4x^2}{(x^1)^2 + (x^2)^2} & x^2 &= \frac{4y^2}{(y^1)^2 + (y^2)^2} , \end{aligned} \tag{5.3}$$

which hold on $(S^2 - \{S\}) \cap (S^2 - \{J\})$, i.e. for $[x^1, x^2] \neq [0, 0]$ and for $[y^1, y^2] \neq [0, 0]$.

In the mentioned domain the transformation equations are differentiable, and the structure of a differentiable manifold of dimension 2 is thus given on S^2.

If $f(\zeta^1, \zeta^2, \zeta^3)$ is a function of three variables given at all points of the sphere S^2, a function on the sphere is given. From this function we obtain the function $f_1(x^1, x^2) = f \circ \varphi^1$ and $f_2(y^1, y^2) = f \circ \varphi^2$. These are functions of two variables defined in the entire planes α and β, respectively.

For the sake of simplicity, denote $[(x^1)^2 + (x^2)^2]^{1/2} = r$, $[(y^1)^2 + (y^2)^2]^{1/2} = s$. For the functions f_1 and f_2 we naturally have

$$f_1(4y^1s^{-2}, 4y^2s^{-2}) \equiv f_2(y^1, y^2), \quad \text{if} \quad [y^1, y^2] \neq [0, 0], \tag{5.4}$$

$$f_2(4x^1r^{-2}, 4x^2r^{-2}) \equiv f_1(x^1, x^2), \quad \text{if} \quad [x^1, x^2] \neq [0, 0]. \tag{5.5}$$

Consider, e.g., the function $f(\xi^1, \xi^2, \xi^3) = \xi^1\xi^2 + \xi^2\xi^3 + \xi^3\xi^1$. Then

$$f_1 = f \circ \varphi = 4\frac{4x^1x^2 + (x^1 + x^2)(r^2 - 4)}{(r^2 + 4)^2},$$

$$f_2 = f \circ \varphi = 4\frac{4y^1y^2 + (y^1 + y^2)(4 - s^2)}{(s^2 + 4)^2}.$$

Substituting the expressions for x^1 and x^2 from (5.3) into the function f_1 we obtain f_2.

Conversely, if we wish to define a function on the sphere with the aid of its coordinate representation it suffices to give two functions $f_1(x^1, x^2)$ and $f_2(y^1, y^2)$ in two variables for which (5.4) and (5.5) hold.

5.10 Definition. Denote by $\mathscr{F}(M)$ the set of all differentiable functions on a differentiable manifold M. A *tangent vector* ζ_m of the manifold M at the point $m \in M$ is a mapping (or functional) $\mathscr{F}(M) \to \mathbf{R}$ with the following properties:

1. $\xi_m(\lambda f + \mu g) = \lambda\, \xi_m(f) + \mu\, \xi_m(g)$,
2. $\xi_m(fg) = f(m)\, \xi_m(g) + \xi_m(f)\, g(m)$,

where $f, g \in \mathscr{F}(M)$ and $\lambda, \mu \in \mathbf{R}$. (See [III], [V], [VI].)

5.11 Theorem. *Let* $m \in M$. *Let* ζ_m: $\mathscr{F}(M) \to \mathbf{R}$ *be a functional with the properties* 1 *and* 2 *from Definition* 5.10. *Further, let* $(U(x^1, \ldots, x^n), \varphi)$

be a coordinate system in the neighbourhood of the point m. Then

$$\xi_m = \lambda^i \left. \frac{\partial}{\partial x^i} \right|_m ,$$

where $\lambda^i \in \mathbf{R}$.

For the proof see [VI].

Remark. Let the vector $\boldsymbol{v}$ be a vector from E_3 at the point $A \in E_3$. If f is a differentiable function on E_3, then it is possible to define the functional $\boldsymbol{v}_A$ as follows:

$$\boldsymbol{v}_A(f) = \frac{\mathrm{d}}{\mathrm{d}t} [f(A + t\boldsymbol{v})]|_{t=0} .$$

Let us find the coordinate representation of the functional $\boldsymbol{v}_A$:

Let

$$A = [a^1, a^2, a^3] , \quad \boldsymbol{v} = [v^1, v^2, v^3] , \quad f = f(x^1, x^2, x^3) .$$

Then

$$\boldsymbol{v}_A(f) = \frac{\mathrm{d}}{\mathrm{d}t} f(a^1 + tv^1, a^2 + tv^2, a^3 + tv^3)|_{t=0}$$

$$= \left. \frac{\partial f}{\partial x^1} \right|_A v^1 + \left. \frac{\partial f}{\partial x^2} \right|_A v^2 + \left. \frac{\partial f}{\partial x^3} \right|_A v^3 ,$$

which can be written as

$$\boldsymbol{v}_A(f) = \left. \left(v^1 \frac{\partial}{\partial x^1} + v^2 \frac{\partial}{\partial x^2} + v^3 \frac{\partial}{\partial x^3} \right) \right|_A f .$$

Consequently, we see that the functional $\boldsymbol{v}_A$ can be written as a linear combination of the functionals

$$\left. \frac{\partial}{\partial x^1} \right|_A , \quad \left. \frac{\partial}{\partial x^2} \right|_A , \quad \left. \frac{\partial}{\partial x^3} \right|_A$$

with coefficients equal to the coordinates of the vector $\boldsymbol{v}$. These functionals thus constitute a base of the tangent space of E_3 at the point A which can be identified with the original base for vectors in E_3.

5.12 Definition. Let ζ_m and η_m be tangent vectors of the manifold M at the point m, $\lambda \in \mathbf{R}$. Define

$$(\xi_m + \eta_m)(f) = \xi_m(f) + \eta_m(f), \quad (\lambda \xi_m)(f) = \lambda\, \xi_m(f)$$

for all $f \in \mathscr{F}(M)$.

The set of all tangent vectors of the manifold at the point m constitutes a vector space of the same dimension as is the dimension of the manifold. This follows directly from Theorem 5.11. This vector space is called the *tangent space of the manifold M* at the point m. It is denoted by $T(M)_m$, or only T_m.

Remark. Whenever a manifold will be mentioned, here as well as in the sequel, a differentiable manifold will be meant. The similar is true for functions.

5.13 Definition. By a *vector field* on a manifold we mean an operator $\boldsymbol{X}: \mathscr{F}(M) \to \mathscr{F}(M)$ with the following properties:

1. $\boldsymbol{X}(\lambda f + \mu g) = \lambda\, \boldsymbol{X}(f) + \mu\, \boldsymbol{X}(g)$,
2. $\boldsymbol{X}(fg) = \boldsymbol{X}(f)\, g + f\, \boldsymbol{X}(g)$,

where $f, g \in \mathscr{F}(M)$, $\lambda, \mu \in \mathbf{R}$.

5.14 Theorem. *Let $\boldsymbol{X}$ be a vector field on M, let $U(x^1, \ldots, x^n)$ be a system of local coordinates on M. Then there exist differentiable functions $\lambda^1(x^1, \ldots, x^n), \ldots, \lambda^n(x^1, \ldots, x^n)$ on U such that on U we have*

$$\boldsymbol{X} = \lambda^1 \frac{\partial}{\partial x^1} + \ldots + \lambda^n \frac{\partial}{\partial x^n}.$$

For the proof see [VI].

Remark. Note that the differentiability of a vector field is included in its definition – this is not always done.

Usually by the notion of a vector field we understand a rule which assigns to every point a tangent vector at this point. Obviously, such a rule can be easily obtained from our definition of vector field: let $\boldsymbol{X}$ be a vector field on the manifold M. Then the functional $\boldsymbol{X}_m$ defined by the formula $\boldsymbol{X}_m(f) = (\boldsymbol{X}f)_m$ is a tangent vector of the manifold M at the point m.

5.15 Definition. Let X and Y be vector fields on $M, f, g \in \mathscr{F}(M)$. Define

$$(X + Y)(f) = X(f) + Y(f), \quad (gX)(f) = g\,X(f).$$

It can be verified easily that the vector fields on M constitute a vector space over **R**.

If a mapping φ of the interval (a, b) into M is given, then to every function $f \in \mathscr{F}(M)$ a function $f \circ \varphi$ of one variable t is assigned on the interval (a, b) by the formula $(f \circ \varphi)(\mathrm{t}) = f(\varphi(\mathrm{t}))$.

5.16 Definition. A mapping $\varphi: (a, b) \to M$ is called a *differentiable curve* if the function $f \circ \varphi$ is differentiable for all $f \in \mathscr{F}(M)$.

If $U = U(x^1, \dots, x^n)$ is a coordinate system on M, then it is possible to describe the part of the curve which lies in U, i.e. $\varphi[(a, b)] \cap U$, parametrically with the aid of coordinates: $\varphi(\mathrm{t}) = [x^1(\mathrm{t}), \dots, x^n(\mathrm{t})]$. A curve is differentiable if and only if all its parametric representations are given by differentiable functions. As a matter of fact, if φ is a differentiable curve and if $f \in \mathscr{F}(M)$, then $f \circ \varphi$ is differentiable as well. Choosing $f = x^i$ $(i = 1, \dots, n)$, we see that $f \circ \varphi = x^i(\mathrm{t})$ is differentiable. Conversely, if x^i and $f(x^1, \dots, x^n)$ are differentiable, then $f \circ \varphi = f(x^1(\mathrm{t}), \dots, x^n(\mathrm{t}))$ is differentiable as well.

5.17 Definition. Let $\varphi: (a, b) \to M$ be a differentiable curve. By the *tangent vector of* the curve φ at the point $\varphi(\mathrm{t}_0)$ we understand the functional

$$\varphi'(t_0): f \to \frac{\mathrm{d}}{\mathrm{d}t}(f \circ \varphi)\Big|_{t_0}.$$

5.18 Exercise. Show that the tangent vector of a curve φ at the point $\varphi(\mathrm{t}_0)$ is a tangent vector of the manifold M at the point $\varphi(\mathrm{t}_0)$.

5.19 Theorem. *Let* $[x^1(\mathrm{t}), \dots, x^n(\mathrm{t})]$ *be a representation of a differentiable curve on the manifold M in local coordinates. For the tangent vector of the curve* φ *at the point* $\varphi(\mathrm{t}_0)$ *we then have*

$$\varphi'(\mathrm{t}_0) = \frac{\mathrm{d}x^1}{\mathrm{d}t}\bigg|_{t_0} \frac{\partial}{\partial x^1}\bigg|_{\varphi(t_0)} + \dots + \frac{\mathrm{d}x^n}{\mathrm{d}t}\bigg|_{t_0} \frac{\partial}{\partial x^n}\bigg|_{\varphi(t_0)}.$$

Proof. The theorem is a direct consequence of the formula for the derivative of a composite function.

5.20 Definition. Let X be a vector field on the manifold M. By an *integral curve of the vector field* X we understand a curve $\varphi: (a, b) \to M$ such that its every tangent vector belongs to the vector field X.

5.21 Theorem. *If* $\lambda^i(x^1, \ldots, x^n)(\partial/\partial x^i)$ *is a vector field on the coordinate neighbourhood* $U(x^1, \ldots, x^n)$, *then the integral curves of this field are solutions of the system of differential equations of first order*

$$\frac{dx^i(t)}{dt} = \lambda^i(x^1(t), \ldots, x^n(t)), \quad i = 1, \ldots, n,$$

for the unknown functions $x^1(t), \ldots, x^n(t)$.

Proof. The theorem follows directly from Definition 5.20 and Theorem 5.19.

5.22 Example. In the plane E_2 let us consider the vector field

$$X = x^1 \frac{\partial}{\partial x^1} + x^1 x^2 \frac{\partial}{\partial x^2}.$$

The differential equations for the integral curves are

$$\frac{dx^1}{dt} = x^1, \quad \frac{dx^2}{dt} = x^1 x^2,$$

where $x^1(t)$ and $x^2(t)$ are the unknown functions. The integral curves are

$$x^1 = C_1 \exp t, \quad x^2 = C_2 \exp(C_1 \exp t),$$

where C_1 and C_2 are integration constants.

5.23 Exercise. Determine an arbitrary vector field on the sphere S^2 with the aid of the coordinate systems (U_1, φ_1) and (U_2, φ_2) and show which relations the coordinates of the vector field must satisfy on the intersection $U_1 \cap U_2$ of the coordinate neighbourhoods.

5.24 Definition. Let V be a vector space over $\mathbf{R}$ of dimension n. A vector field X on V is called *linear* if $X_v = A(v)$ for all $v \in V$, where A is a linear mapping from V into V (see 1.86).

5.25 Theorem. *Let* E_3 *be a three-dimensional Euclidean vector space with the unit sphere* S^2. *Let an orthonormal base* $\{u_1, u_2, u_3\}$ *be given* *in E_3. A linear vector field $\boldsymbol{X}_{\boldsymbol{v}} = \boldsymbol{A}(\boldsymbol{v})$ determines a vector field of tangent vectors on the sphere S^2 if and only if the representation of the mapping $\boldsymbol{A}$ in the base $\{\boldsymbol{u}_1, \boldsymbol{u}_2, \boldsymbol{u}_3\}$ is given by a skew-symmetric matrix.*

The proof is accomplished by direct computation and it is easy.

5.26 Theorem. *Let $\boldsymbol{A}$ be a linear mapping in E_3 given by a skew-symmetric matrix in an orthonormal base. Then it is possible to find an orthonormal base in E_3 such that the matrix of the mapping $\boldsymbol{A}$ assumes the form*

$$\begin{pmatrix} 0, & \lambda, & 0 \\ -\lambda, & 0, & 0 \\ 0, & 0, & 0 \end{pmatrix},$$

where $\lambda \in \mathbf{R}$.

The proof is simple.

5.27 Theorem. *The integral curves of a linear vector field on a sphere are circles lying in parallel planes.*

Proof. Applying Theorem 5.26, the differential equations for the integral curves are easily solved.

Let an orthonormal frame $\mathcal{R} = \{O, \boldsymbol{u}_1, \boldsymbol{u}_2, \boldsymbol{u}_3\}$ in E_3 be given. By a linear vector field in E_3 we shall understand any vector field $\boldsymbol{X}$ given by the formula

$$\begin{pmatrix} 0 \\ \boldsymbol{X}_A \end{pmatrix} = \begin{pmatrix} 0, & 0 \\ t, & r \end{pmatrix} \begin{pmatrix} 1 \\ A \end{pmatrix}.$$

Here $\boldsymbol{X}_A$ is the value of the vector field at the point $A \in E_3$, t is a column of three numbers, r is a 3×3 matrix, A appearing on the right-hand side stands for the column of coordinates of the point A in the frame $\mathcal{R}$. We see easily that a vector field which is linear in one coordinate system is linear in any coordinate system (which need not be orthonormal).

5.28 Theorem. *Let $\boldsymbol{X}$ be a linear vector field in E_3 determined in the orthonormal frame $\mathcal{R} = \{O, \boldsymbol{u}_1, \boldsymbol{u}_2, \boldsymbol{u}_3\}$ by the matrix $a = \begin{pmatrix} 0, & 0 \\ t, & r \end{pmatrix}$, where*

r is a skew-symmetric matrix. The integral curves of this vector field X are the following:

1. *If the rank of the matrix a is equal to three, then the integral curves are helices with common axis and with the same parameter.*
2. *If the rank of the matrix a is equal to two, then the integral curves are circles in parallel planes whose centres lie on a straight line perpendicular to those planes.*
3. *If the rank of the matrix a is equal to one, then the integral curves are parallel straight lines.*

The proof is accomplished by direct computation, applying Theorem 5.26.

5.29 Definition. Let M and N be differentiable manifolds, let $\varphi: M \to N$ be a mapping of M into N. Then it is possible to assign to every function f on N a function on M (denote it by $f \circ \varphi$) by the formula

$$(f \circ \varphi)(m) = f(\varphi(m)).$$

The mapping φ is called a *differentiable mapping* if $f \circ \varphi \in \mathscr{F}(M)$ for all $f \in \mathscr{F}(N)$. (See [V], [VI].)

5.30 Example. Let $M = \mathbf{R} \times \mathbf{R}$, $N = \mathbf{R}$. Let φ be the mapping defined by the formula $\varphi(u, v) = u^2 + v^2$, where $(u, v) \in \mathbf{R} \times \mathbf{R}$. Let f be a differentiable function on $\mathbf{R}$. Then the function $(f \circ \varphi)(u, v) = f(u^2 + v^2)$ is an everywhere differentiable function of two variables. This means that the mapping φ is differentiable.

5.31 Remark. Let $U(x^1, \ldots, x^n)$ be a coordinate system in the neighbourhood of the point $m \in M$. Let φ be a mapping of M into N. Let $V(y^1, \ldots, y^r)$ be a coordinate system in the neighbourhood of the point $\varphi(m) \in N$. Then it is possible to write, on $\varphi(U) \cap V$,

$$\varphi[x^1, \ldots, x^n] = [y^1(x^1, \ldots, x^n), \ldots, y^r(x^1, \ldots, x^n)],$$

i.e. $y^1 = y^1(x^1, \ldots, x^n), \ldots, y^r = y^r(x^1, \ldots, x^n)$ is the coordinate representation of the mapping φ in the neighbourhood of the point m. If $f(y^1, \ldots, y^r)$ is a function on V it is obvious that $f \circ \varphi = f(y^1(x^1, \ldots, x^n), \ldots, y^r(x^1, \ldots \ldots, x^n))$. Hence we see that φ is a differentiable mapping if and only if the functions $y^1(x^1, \ldots, x^n), \ldots, y^r(x^1, \ldots, x^n)$ are differentiable functions for all coordinate systems on M and on N.

5.32 Example. A congruence of the Euclidean space E_3 is a differentiable mapping since it is given by linear, and thus differentiable, functions.

5.33 Definition. Let $\varphi: M \to N$ be a differentiable mapping. By the *differential* φ'_m of the mapping φ at the point $m \in M$ we understand the mapping $\varphi'_m: T(M)_m \to T(N)_{\varphi(m)}$ defined as follows. Let $\zeta_m \in T(M)_m$. Then $(\varphi'_m\zeta_m)(f) = \zeta_m(f \circ \varphi)$ for all $f \in \mathscr{F}(N)$. (See [III], [V], [VI].)

5.34 Exercise. Show that $\varphi'_m(\zeta_m)$ is a tangent vector of the manifold N at the point $\varphi(m)$.

Hint: Verify the properties from Definition 5.10.

5.35 Exercise. Show that to every tangent vector ζ_m of the manifold M at the point m there exists a curve on M such that ζ_m is one of its tangent vectors.

5.36 Exercise. Show that φ'_m is a linear mapping (see 1.86).

5.37 Theorem. *Let $\varphi: M \to N$ be a differentiable mapping, where M, N are differentiable manifolds. Further, let $m \in M$ and $\zeta_m \in T(M)_m$. Let $u(t)$ be a curve on M such that $u'(0) = \zeta_m$. Then $\varphi'_m\zeta_m$ is the tangent vector of the curve $\varphi(t)$ at the point $\varphi(u(0)) = \varphi(m)$, i.e. $\varphi'_m\zeta_m = [\varphi(u(0))]'$.*

Proof. The theorem follows directly from the respective definitions.

5.38 Example. Let $\varphi: \mathbf{R} \times \mathbf{R} \to \mathbf{R} \times \mathbf{R}$ be a mapping given by the equation

$$\varphi[x, y] = [x^2, xy],$$

where $x, y \in \mathbf{R}$. The mapping φ is differentiable. Choose a point $p = [x_0, y_0] \in \mathbf{R} \times \mathbf{R}$. Then $\varphi(p) = [x_0^2, x_0 y_0]$. Further, let $\zeta_p = \lambda_1(\partial/\partial x)|_p + \lambda_2(\partial/\partial y)|_p$ be an arbitrary tangent vector at p. The curve $u(t)$: $x = x_0 + \lambda_1 t$, $y = y_0 + \lambda_2 t$ has ζ_p as its tangent vector at p. The image of $u(t)$ under φ is

$$\varphi(u(t)): x = (x_0 + \lambda_1 t)^2, \quad y = (x_0 + \lambda_1 t)(y_0 + \lambda_2 t).$$

For $t = 0$ its tangent vector has the coordinates

$$[\varphi(u(0))]' : [2x_0\lambda_1, \lambda_1 y_0 + \lambda_2 x_0] .$$

Thus we have

$$(\varphi'\xi_p)|_{\varphi(p)} = 2x_0\lambda_1 \left(\frac{\partial}{\partial x}\right)\bigg|_{\varphi(p)} + (\lambda_1 y_0 + \lambda_2 x_0)\left(\frac{\partial}{\partial y}\right)\bigg|_{\varphi(p)} .$$

Further, let us investigate how the differential φ' maps the individual tangent spaces of the manifold $M = \mathbf{R} \times \mathbf{R}$. We have

$$\dim \varphi' T_{[0,0]} = 0 , \quad \dim \varphi' T_{[0,y_0]} = 1 \quad \text{for} \quad y_0 \neq 0 ,$$
$$\dim \varphi' T_{[x_0,y_0]} = 2 \quad \text{for} \quad x_0 \neq 0 .$$

5.39 Definition. Let M and N be manifolds. The mapping $\varphi: M \to N$ is called a *diffeomorphism* if φ is one-to-one and if φ as well as φ^{-1} are differentiable mappings.

5.40 Theorem. *Let $\varphi: M \to N$ be a diffeomorphism. Then to every vector field $\boldsymbol{X}$ on M there exists a unique vector field* (denoted by $\varphi'\boldsymbol{X}$) *on N such that $\varphi_p'(\boldsymbol{X}_p) = (\varphi'\boldsymbol{X})_{\varphi(p)}$ for all $p \in M$. It is such a vector field $\boldsymbol{Y}$ on N for which $(\boldsymbol{Y}f) \circ \varphi = \boldsymbol{X}(f \circ \varphi)$ for all $f \in \mathscr{F}(N)$.* (See [VI].)

5.41 Example. A congruence of the Euclidean space E_3 is a diffeomorphism.

5.42 Definition. An algebra V is called a *Lie algebra* if the operation of multiplication in V, which is denoted by $[\boldsymbol{u}, \boldsymbol{v}]$ for $\boldsymbol{u}, \boldsymbol{v}$ from V, satisfies

1. $[\boldsymbol{u}, \boldsymbol{v}] = -[\boldsymbol{v}, \boldsymbol{u}]$
2. $[\boldsymbol{u}, [\boldsymbol{v}, \boldsymbol{w}]] + [\boldsymbol{v}, [\boldsymbol{w}, \boldsymbol{u}]] + [\boldsymbol{w}, [\boldsymbol{u}, \boldsymbol{v}]] = 0 .$

The operation $[\boldsymbol{u}, \boldsymbol{v}]$ is called the *bracket* of vectors $\boldsymbol{u}$ and $\boldsymbol{v}$. Condition 1. immediately implies $[\boldsymbol{u}, \boldsymbol{u}] = 0$.

5.43 Exercise. Let V_3 be a vector space of dimension three over $\mathbf{R}$ with the ordinary cross product $\boldsymbol{u} \times \boldsymbol{v}$. Show that if we define $[\boldsymbol{u}, \boldsymbol{v}] = \boldsymbol{u} \times \boldsymbol{v}$, then V_3 is a Lie algebra of dimension three.

5.44 Example. $\mathscr{D}_3$ is a Lie algebra of dimension 6 over $\mathbf{R}$ with the bracket $[\boldsymbol{u}, \boldsymbol{v}] = \boldsymbol{u} \times \boldsymbol{v}$. (For the definition of $\mathscr{D}_3$ see Chapter III, Section 10.)

5.45 Example. Let M be a manifold, let $\mathscr{X}(M)$ be the vector space of all vector fields on M. For $\boldsymbol{X}, \boldsymbol{Y} \in \mathscr{X}(M)$ we define

$$[\boldsymbol{X}, \boldsymbol{Y}](f) = \boldsymbol{X}(\boldsymbol{Y}(f)) - \boldsymbol{Y}(\boldsymbol{X}(f))$$

for all $f \in \mathscr{F}(M)$. Then $\mathscr{X}(M)$ is a Lie algebra over $\mathbf{R}$. This can be easily verified by direct calculations.

5.46 Example. Let $M = E_2$,

$$\boldsymbol{X} = x^2 \frac{\partial}{\partial x^1} - x^1 \frac{\partial}{\partial x^2}, \quad Y = \frac{\partial}{\partial x^1},$$

let f be a differentiable function on E_2. Then

$$[\boldsymbol{X}, \boldsymbol{Y}](f) = \boldsymbol{X}(\boldsymbol{Y}(f)) - \boldsymbol{Y}(\boldsymbol{X}(f)) =$$

$$= \boldsymbol{X}\left(\frac{\partial f}{\partial x^1}\right) - \boldsymbol{Y}\left(x^2 \frac{\partial f}{\partial x^1} - x^1 \frac{\partial f}{\partial x^2}\right) =$$

$$= \left(x^2 \frac{\partial}{\partial x^1} - x^1 \frac{\partial}{\partial x^2}\right)\left(\frac{\partial f}{\partial x^1}\right) - \frac{\partial}{\partial x^1}\left(x^2 \frac{\partial f}{\partial x^1} - x^1 \frac{\partial f}{\partial x^2}\right) =$$

$$= \frac{\partial}{\partial x^2}(f),$$

i.e.

$$[\boldsymbol{X}, \boldsymbol{Y}] = \frac{\partial}{\partial x^2}.$$

5.47 Exercise. Denote by $\mathfrak{GL}(n, \mathbf{R})$ the set of all $n \times n$ matrices. Define $[a, b] = ab - ba$ for $a, b \in \mathfrak{GL}(n, \mathbf{R})$. Show that $\mathfrak{GL}(n, \mathbf{R})$ is a Lie algebra of dimension n^2.

5.48 Definition. A set G is called a *Lie group* if the following holds:

1. G is a group,
2. G is an analytic manifold,
3. the mapping $G \times G \to G: (g_1, g_2) \to g_1 g_2$ is analytic.

5.49 Definition. A subset H of a Lie group G is called a *closed subgroup* of the Lie group G if H is a subgroup of the group G and the set H is a closed subset in G.

5.50 Theorem. *Let H be a closed subgroup of a Lie group G. Then there exists a unique analytic structure on H such that H is a Lie group and its topology is the topology induced on H by the topology of G.*

For the proof see [VI].

Remark. Let E be a topological space with the topology $\mathscr{T}(E$ – see) Section 3. Let $F \subset E$. Then it is possible to define a topology $\mathscr{T}(F)$ in $F : X \in \mathscr{T}(F)$ if and only if there exists an $Y \in \mathscr{T}(E)$ such that $X = Y \cap F$. The topology $\mathscr{T}(F)$ is called the topology induced by the topology $\mathscr{T}(E)$ in F.

Consider the set of all $n \times n$ matrices with real coefficients. According to Section 3 this set is a topological manifold. From 2.16 and 2.17 we know that $GL(n)$ is a group with respect to matrix multiplication. It also is an analytic manifold since it is an open subset of E_{n^2} and is thus described by a single local coordinate system. The product of two matrices is represented in the coordinates by linear functions. These functions are certainly analytic so that $GL(n)$ is a Lie group.

5.51 Theorem. *Let G be a subgroup of the group $GL(n)$ defined by some system of polynomial equations. Then G is a Lie group in the induced topology* (see the remark which follows Theorem 5.50).

For the proof see [V].

From Theorem 5.51 it follows, for instance, that the group $O(3) \subset \subset GL(3)$ is a Lie group. In fact, this assertion follows from Theorem 5.50 as well.

Consider now a Lie group G. The mapping L_g is obviously a diffeomorphism of the group G for all $g \in G$ (the definition of L_g is given in Section 2). A vector field $\boldsymbol{X}$ on G is called left invariant if $L'_g(\boldsymbol{X}) = \boldsymbol{X}$ for all g from G, L'_g is the differential of the mapping L_g. The relation $L'_g(\boldsymbol{X}) = \boldsymbol{X}$ means explicitly that $(L'_g)_h(\boldsymbol{X}_h) = \boldsymbol{X}_{gh}$ for all $h \in G$. Now, let us show that left invariant vector fields exist on every Lie group.

5.52 Theorem. *Choose $\boldsymbol{X}_0 \in T_e = T(G)_e$ where e is the unit element of the Lie group G. Then there exists one and only one left invariant vector field $\boldsymbol{X}$ on G such that $\boldsymbol{X}_0 = \boldsymbol{X}_e$. This vector field is analytic, i.e. it is given by analytic functions in every analytic coordinate system on G.*

For the proof see [V], [VI].

Theorem 5.52 indicates that left invariant vector fields on G are in one-to-one correspondence with their values at the point e. Denote by $\mathscr{X}^{\mathrm{L}}(G)$, or by $\mathscr{X}^{\mathrm{L}}$ only, the set of all left invariant vector fields on G.

5.53 Theorem. *Let $X, Y \in \mathscr{X}^{\mathrm{L}}(G)$ where G is a Lie group, and let $\lambda \in \mathbf{R}$. Then we have:*

$$X + Y \in \mathscr{X}^{\mathrm{L}}(G)\,, \quad \lambda X \in \mathscr{X}^{\mathrm{L}}(G), \quad [X, Y] \in \mathscr{X}^{\mathrm{L}}(G)\,.$$

For the proof see [V], [VI].

5.54 Lemma. $\mathscr{X}^{\mathrm{L}}(G)$ is a Lie algebra of the same dimension as the Lie group G.

It is convenient to introduce the following definition:

Let e be the unit element of the Lie group G. Let $\zeta, \eta \in T_e(G)$. Denote by X, Y the left invariant vector fields determined by ζ, η. Then we define

$$[\xi, \eta] = [X, Y]_e\,.$$

Thus the tangent space T_e at the unit element of the group becomes a Lie algebra [which is the same as $\mathscr{X}^{\mathrm{L}}(G)$, of course; we limit ourselves to the unit of the group for the sake of simplification only]. This Lie algebra is called the *Lie algebra of the group* G and it is denoted by the corresponding fraktur type $\mathfrak{G}$.

5.55 Example. For illustration let us repeat some of the considerations for a concrete example. Consider $E_2 = \mathbf{R} \times \mathbf{R}$, let the coordinates be x^1 and x^2. For $x = [x^1, x^2]$, $y = [y^1, y^2]$ we define the product $z = x \,.\, y$, where $z = [z^1, z^2]$, by the relations

$$z^1 = x^1 + y^1 + x^1 y^1\,, \quad z^2 = y^2 + x^1 y^2 + x^2(1 + y^1)^{-1}\,,$$

as far as $y^1 \neq -1$. The reader can verify easily that $G = E_2 - \{[-1, x^2]\}$ is a group. Obviously, it also is a Lie group since both z^1 and z^2 are analytic functions, and G is an analytic manifold obtained from E_2 by the omission of the line $x^1 = -1$. The unit element of the group is the point $[0, 0]$. First, let us find the left invariant vector fields on G to the vectors

$$u_1 = \left.\frac{\partial}{\partial x^1}\right|_e\,, \quad u_2 = \left.\frac{\partial}{\partial x^2}\right|_e\,.$$

For u_1 we obtain:

$u(t) = [t, 0]$ is a curve with tangent vector u_1. Its image under $L_{[x^1,x^2]}$ is the curve

$$z^1(t) = (x^1 + y^1 + x^1 y^1)|_{y^1=t, y^2=0} = t + x^1 + x^1 t\,,$$
$$z^2(t) = [y^2 + x^1 y^2 + x^2(1 + y^1)^{-1}]|_{y^1=t, y^2=0} = x^2(1 + t)^{-1}\,.$$

Its tangent vector at $t = 0$ is

$$(1 + x^1)\left(\frac{\partial}{\partial x^1}\right)\Big|_x - x^2\left(\frac{\partial}{\partial x^2}\right)\Big|_x\,.$$

For the desired left invariant vector field X_1 we then have

$$X_1 = (1 + x^1)\frac{\partial}{\partial x^1} - x^2\frac{\partial}{\partial x^2}\,.$$

Analogously,

$$X_2 = (1 + x^1)\frac{\partial}{\partial x^2}\,.$$

Thus, if $v = \lambda^1 u_1 + \lambda^2 u_2$ is an arbitrary vector from T_e, then

$$\lambda_1\left[(1 + x^1)\frac{\partial}{\partial x^1} - x^2\frac{\partial}{\partial x^2}\right] + \lambda^2\left[(1 + x^1)\frac{\partial}{\partial x^2}\right] = X$$

is a left invariant vector field on G for which $X_e = v$ holds.

If we wish to find the structure of the Lie algebra of the group G we compute $[X_1, X_2] = 2X_2$.

The structure of the Lie algebra on T_e is thus determined by the relation $[u_1, u_2] = 2u_2$.

5.56 Example. Let us find the Lie algebra of the group $GL(n)$ – see [V]. Let $x, y, z \in GL(n)$, and let $z = xy$. For the sake of simplicity, we shall write the two matrix indices as subscripts in this example, i.e. let $z_{ik} = x_{ij}y_{jk}$. Further, let

$$A = a_{ij}\left(\frac{\partial}{\partial x_{ij}}\right)\Big|_e$$

be a tangent vector at the unit matrix e, i.e. $e_{ij} = \delta_{ij}$. Let us find the respective left invariant vector field X, i.e. a field such that $X_e = A$. The curve

$u(t)_{ij} = a_{ij}t + \delta_{ij}$ has the tangent vector $\boldsymbol{A}$ for $t = 0$. The image of this curve under left translation by the matrix (x_{ij}) is the curve

$$(x\, u(t))_{ik} = x_{ij}(a_{jk}t + \delta_{jk}) .$$

Its tangent vector at $t = 0$ is $x_{ij}a_{jk}(\partial/\partial x_{ik})|_x$ so that

$$\boldsymbol{X} = x_{ij}a_{jk}\frac{\partial}{\partial x_{ik}}$$

is the desired left invariant vector field on $G\,L(n)$. If

$$\boldsymbol{Y} = x_{ij}b_{jk}\frac{\partial}{\partial x_{ik}}$$

is another left invariant vector field on $G\,L(n)$, then

$$[\boldsymbol{X}, \boldsymbol{Y}] = \left[x_{ij}a_{jk}\frac{\partial}{\partial x_{ik}},\ x_{\alpha\beta}b_{\beta\gamma}\frac{\partial}{\partial x_{\alpha\gamma}}\right] = x_{ij}(a_{jk}b_{km} - b_{jk}a_{km})\frac{\partial}{\partial x_{im}} .$$

Writing symbolically $a = (a_{ij})$, $b = (b_{ij})$, $x = (x_{ij})$, we obtain for the coordinates of the left invariant vector fields $\boldsymbol{X}$, $\boldsymbol{Y}$, and $[\boldsymbol{X}, \boldsymbol{Y}]$ the following relations:

$$\boldsymbol{X} = xa\,, \quad \boldsymbol{Y} = xb\,, \quad [\boldsymbol{X}, \boldsymbol{Y}] = x(ab - ba)\,.$$

Thus, if we identify the vector $\boldsymbol{A} = a_{ij}(\partial/\partial x_{ij})|_e$ with the matrix $a = (a_{ij})$ we see that the Lie algebra of the group $G\,L(n)$ is the Lie algebra $\mathfrak{GL}(n)$ defined in 5.47.

5.57 Remark. If H is a Lie subgroup of the Lie group G, then the Lie algebra $\mathfrak{H}$ is a subalgebra of the algebra $\mathfrak{G}$. Thus, the bracket of two vectors from $\mathfrak{H}$ does not depend on whether it is computed in $\mathfrak{H}$ or in $\mathfrak{G}$.

5.58 Example. Consider the Lie group $O(n)$. Let us find its Lie algebra, i.e. the tangent space at the unit element. Let $a_{ij}(t)$ be a curve on the group $O(n)$ such that $a_{ij}(0) = \delta_{ij}$. Denote by $a'_{ij}(0) = A_{ij}$ the coordinates of its tangent vector at the unit of the group. For the curve $a_{ij}(t)$

$$a_{ij}(t)\,a_{jk}(t) = \delta_{ik} \quad \text{must hold for all} \quad t \in (a, b)\,.$$

By differentiation we obtain

$$a'_{ij}(0)\,a_{kj}(0) + a_{ij}(0)\,a'_{kj}(0) = 0\,, \quad \text{i.e.} \quad A_{ik} + A_{ki} = 0\,.$$

Hence we see already that $\mathfrak{O}(n)$ is the algebra of all skew-symmetric matrices [$\mathfrak{O}(n)$ cannot be smaller since, for instance, it is possible to find three independent vectors in $T_e(O(3))$ – e.g. the tangent vectors of rotations about the coordinate axes].

5.59 Example. Consider the group J of all unit quaternions $\boldsymbol{\alpha} = a_0 + a_1\mathbf{i} + a_2\mathbf{j} + a_3\mathbf{k}$ with the group operation of quaternion multiplication. J is a Lie group of dimension three (it is a three-dimensional sphere in E_4). Let us find its Lie algebra $\mathfrak{J}$. Thus, let $\boldsymbol{\alpha}(t) = a_0(t) + a_1(t)\,\mathbf{i} + a_2(t)\,\mathbf{j} + a_3(t)\,\mathbf{k}$ be a curve on J, and let $\boldsymbol{\alpha}(0) = 1$, i.e. $a_0(0) = 1$, $a_m(0) = 0$ for $m = 1, 2, 3$. By differentiation the equation $\sum_{m=0}^{3} [a_m(t)]^2 = 1$ yields the equation $a_m(t)\, a'_m(t) = 0$. Substituting $t = 0$ we obtain $a'_0(0) = 0$.

The Lie algebra $\mathfrak{J}$ is thus constituted by vectors of the form $\zeta = \zeta^m(\partial/\partial a_m)|_{\alpha=1}$ where $m = 1, 2, 3$. The vector ζ is formally written in the form $\zeta = \zeta^1\mathbf{i} + \zeta^2\mathbf{j} + \zeta^3\mathbf{k}$. Let us find the left invariant vector field $\boldsymbol{X}$ on J for which $X_{\alpha=1} = \zeta$. Let $\boldsymbol{\beta}(t)$ be a curve on J such that $\boldsymbol{\beta}(0) = 1$, $\boldsymbol{\beta}'(0) = \zeta$. Then $L_\alpha(\boldsymbol{\beta}(t)) = \boldsymbol{\alpha}\,\boldsymbol{\beta}(t)$ is the left translation of the curve $\boldsymbol{\beta}(t)$ by the quaternion $\boldsymbol{\alpha}$, its tangent vector is $\boldsymbol{\alpha}\,\boldsymbol{\beta}'(0) = \boldsymbol{\alpha}\zeta$. In particular, denote by $\boldsymbol{X}_m$ those left invariant vector fields on J for which

$$\boldsymbol{X}_m\Big|_{\alpha=1} = \frac{\partial}{\partial a_m}\Big|_{\alpha=1},$$

where $m = 1, 2, 3$. These three vector fields are represented at the point $\boldsymbol{\alpha} = 1$, in quaternion notation, by the quaternions $\mathbf{i}$, $\mathbf{j}$, and $\mathbf{k}$. For the components of these vector fields at the point $\boldsymbol{\alpha} = a_0 + a_1\mathbf{i} + a_2\mathbf{j} + a_3\mathbf{k}$ we have $(\boldsymbol{X}_1)_\alpha = \boldsymbol{\alpha}\mathbf{i}$, $(\boldsymbol{X}_2)_\alpha = \boldsymbol{\alpha}\mathbf{j}$, $(\boldsymbol{X}_3)_\alpha = \boldsymbol{\alpha}\mathbf{k}$.

The computations yield

$$\boldsymbol{X}_1 = -a_1\frac{\partial}{\partial a_0} + a_0\frac{\partial}{\partial a_1} + a_3\frac{\partial}{\partial a_2} - a_2\frac{\partial}{\partial a_3},$$

$$\boldsymbol{X}_2 = -a_2\frac{\partial}{\partial a_0} - a_3\frac{\partial}{\partial a_1} + a_0\frac{\partial}{\partial a_2} + a_1\frac{\partial}{\partial a_3},$$

$$\boldsymbol{X}_3 = -a_3\frac{\partial}{\partial a_0} + a_2\frac{\partial}{\partial a_1} - a_1\frac{\partial}{\partial a_2} + a_0\frac{\partial}{\partial a_3},$$

where all the partial derivatives are at the point $\boldsymbol{\alpha}$. Further, we obtain

$$[X_1, X_2] = 2X_3\,, \quad [X_2, X_3] = 2X_1\,, \quad [X_3, X_1] = 2X_2\,.$$

If we limit ourselves to the values at the point $\boldsymbol{\alpha} = 1$ we obtain, in quaternion notation,

$$[\mathbf{i}, \mathbf{j}] = 2\mathbf{k}\,, \quad [\mathbf{j}, \mathbf{k}] = 2\mathbf{i}\,, \quad [\mathbf{k}, \mathbf{i}] = 2\mathbf{j}\,.$$

If we desire to replace $[\ ,\ ]$ by quaternion multiplication, we have to omit the real part since, e.g., $[\mathbf{i}, \mathbf{i}] = 0$ while $\mathbf{i}^2 = -1$. For $\boldsymbol{\zeta}, \boldsymbol{\eta} \in \mathfrak{J}$ we thus have $[\boldsymbol{\zeta}, \boldsymbol{\eta}] = 2\boldsymbol{\zeta\eta}|_{\mathrm{mod}\,\mathbf{R}}$, i.e. the quaternion $2\boldsymbol{\zeta\eta}$ without its real part.

Let a Lie group G be given. To its every element g the mapping $\mathrm{Int}\, g\colon G \to G\colon x \to gxg^{-1}$ for all $x \in G$ is assigned (see Section 2). The mapping $\mathrm{Int}\, g$ is a differentiable isomorphism of the group G, and we have $\mathrm{Int}\, g(e) = g\, e\, g^{-1} = e$. This means that the differential of the mapping $\mathrm{Int}\, g$ at the point e, i.e. $(\mathrm{Int}\, g)'_e$, maps $T_e(G)$ into $T_e(G)$.

5.60 Definition. Denote $(\mathrm{Int}\, g)'_e = \mathrm{ad} g$. Then $\mathrm{ad} g\colon \mathfrak{G} \to \mathfrak{G}$. The mapping ad: $g \to \mathrm{ad}\, g$ which assigns to every $g \in G$ the mapping $\mathrm{ad} g$ is called the *adjoint representation of the group G.*

We have $\mathrm{ad}(gh) = \mathrm{ad}g.\ \mathrm{ad}h$, see [VI]. This formula says that the mapping ad is a mapping of the group G into the group of all linear transformations of the vector space T_e such that to the product of elements in the group G there corresponds the composition of corresponding transformations.

5.61 Theorem. *Let* $g \in G$, $\zeta, \eta \in \mathfrak{G}$. *Then* $\mathrm{ad}g[\zeta, \eta] = [\mathrm{ad}g\, \zeta, \mathrm{ad}g\, \eta]$.

For the proof see [VI].

5.62 Example. Let us find the adjoint representation of the group G from Example 5.55.

The inverse element x^{-1} to the element $x = [x^1, x^2]$ is $x^{-1} = [-x^1(1 + x^1)^{-1}, -x^2]$. Now choose the curve $k(t) = [t, 0]$ with tangent vector $\boldsymbol{u}_1$ for $t = 0$ and find the curve $(gkg^{-1})(t)$ where $g = [x^1, x^2]$. We obtain

$$\begin{aligned} gkg^{-1} = [&x^1 + t + x^1t - x^1(1 + x^1)^{-1} \\ &- x^1(1 + x^1)^{-1}(x^1 + t + x^1t)\,, \\ &- x^2 + (x^1 + t + x^1t)(-x^2) + x^2(1 + t)^{-1}(1 + x^1)]\,. \end{aligned}$$

The tangent vector of this curve for $t = 0$ is $[1, -2x^2(1 + x^1)]$. Analogously, choose $k(t) = [0, t]$ with tangent vector $\boldsymbol{u}_2$ for $t = 0$. Now the tangent vector of the curve $gkg^{-1}(t)$ for $t = 0$ is $[0, (1 + x^1)^2]$. Consequently, we have

$$\mathrm{ad}g\{\boldsymbol{u}_1, \boldsymbol{u}_2\} = \{\boldsymbol{u}_1 - 2x^2(1 + x^1)\,\boldsymbol{u}_2,\ (1 + x^1)^2\,\boldsymbol{u}_2\},$$

which means that ad$g$ is given in the frame $\mathscr{R} = \{\boldsymbol{u}_1, \boldsymbol{u}_2\}$ by the matrix

$$\mathrm{ad}g = \begin{pmatrix} 1, & 0 \\ -2x^2(1 + x^1), & (1 + x^1)^2 \end{pmatrix}.$$

Let G be some group of linear transformations of a vector space V, i.e. let a regular linear mapping $V \to V$ be assigned to every element g from G. We say that a symmetrical bilinear form $\mathbf{B}$ on the vector space V is invariant with respect to G if $\mathbf{B}(\boldsymbol{u}, \boldsymbol{v}) = \mathbf{B}(g(\boldsymbol{u}), g(\boldsymbol{v}))$ for all $\boldsymbol{u}$ and $\boldsymbol{v}$ from V and g from G. If G is a Lie group with the Lie algebra $\mathfrak{G}$, we say that $\mathbf{B}$ is invariant on $\mathfrak{G}$ if it is invariant under ad$G$, i.e. if

$$\mathbf{B}(\mathrm{ad}g\ \xi,\ \mathrm{ad}g\ \eta) = \mathbf{B}(\xi, \eta)$$

for all $g \in G$, $\zeta, \eta \in \mathfrak{G}$. Below we give an important example of an invariant bilinear form.

5.63 Definition. Let $\mathfrak{G}$ be a Lie algebra. For $\boldsymbol{X} \in \mathfrak{G}$ we denote $\mathrm{Ad}\boldsymbol{X}\colon \mathfrak{G} \to \mathfrak{G}\colon \boldsymbol{Y} \to [\boldsymbol{Y}, \boldsymbol{X}]$ for all $\boldsymbol{Y} \in \mathfrak{G}$. Let us define $\mathbf{K}(\boldsymbol{X}, \boldsymbol{Y}) = \mathrm{Tr}(\mathrm{Ad}\boldsymbol{X}\,\mathrm{Ad}\boldsymbol{Y})$ for all $\boldsymbol{X}, \boldsymbol{Y} \in \mathfrak{G}$. The form $\mathbf{K}(\boldsymbol{X}, \boldsymbol{Y})$ is called the *Killing bilinear form on* $\mathfrak{G}$.

Remark. $\mathrm{Tr}(\mathrm{Ad}\boldsymbol{X}\,\mathrm{Ad}\boldsymbol{Y})$ stands for the trace of the mapping $(\mathrm{Ad}\boldsymbol{X}).(\mathrm{Ad}\boldsymbol{Y})\colon \mathfrak{G} \to \mathfrak{G}\colon \boldsymbol{Z} \to [\boldsymbol{X}, [\boldsymbol{Y}, \boldsymbol{Z}]]$, i.e. the trace of the matrix of this mapping in any arbitrary base. $\mathbf{K}(\boldsymbol{X}, \boldsymbol{Y})$ is a symmetrical bilinear form. This follows immediately from the properties of Tr given in 1.2.

5.64 Theorem. *The Killing bilinear form is invariant.*

Proof. The theorem is a direct consequence of Theorem 5.61.

5.65 Example. Let us compute the Killing form for the group G from Example 5.55. Let $\boldsymbol{X} = \alpha^1\boldsymbol{u}_1 + \alpha^2\boldsymbol{u}_2$ and $\boldsymbol{Y} = \beta^1\boldsymbol{u}_1 + \beta^2\boldsymbol{u}_2$ be two vectors from $\mathfrak{G}$. Then we have

$$\mathrm{Ad}\boldsymbol{X}(\boldsymbol{u}_1) = [\alpha^1\boldsymbol{u}_1 + \alpha^2\boldsymbol{u}_2, \boldsymbol{u}_1] = -2\alpha^2\boldsymbol{u}_2\,,\quad \mathrm{Ad}\boldsymbol{X}(\boldsymbol{u}_2) = 2\alpha^1\boldsymbol{u}_2\,,$$

so that the matrices of the mappings AdX, AdY, and AdX AdY are:

$$\mathrm{Ad}X = \begin{pmatrix} 0, & 0 \\ -2\alpha^2, & 2\alpha^1 \end{pmatrix}, \quad \mathrm{Ad}Y = \begin{pmatrix} 0, & 0 \\ -2\beta^2, & 2\beta^1 \end{pmatrix},$$

$$\mathrm{Ad}X\,\mathrm{Ad}Y = 4\begin{pmatrix} 0, & 0 \\ -\alpha^1\beta^2, & \alpha^1\beta^1 \end{pmatrix}.$$

Consequently, we obtain $\mathbf{K}(X, Y) = \mathrm{Tr}(\mathrm{Ad}X\,\mathrm{Ad}Y) = 4\alpha^1\beta^1$.

Verify that $\mathbf{K}(X, Y)$ is actually invariant. The verification is simple.

5.66 Example. Let us construct the adjoint representation of the group J. If $\boldsymbol{\alpha} \in J$ we obviously have ad $\boldsymbol{\alpha}(X) = \boldsymbol{\alpha} X \bar{\boldsymbol{\alpha}}$, where $X = a_1\mathbf{i} + a_2\mathbf{j} + a_3\mathbf{k} \in J$ and $\bar{\boldsymbol{\alpha}}$ is the conjugate quaternion to $\boldsymbol{\alpha}$. Indeed we have $\boldsymbol{\alpha}\bar{\boldsymbol{\alpha}} = 1$, i.e. $\bar{\boldsymbol{\alpha}} = \boldsymbol{\alpha}^{-1}$.

Let us find the Killing form now. Consider the vector space of all pure imaginary quaternions to be V_3 with the cross product and inner product under identification $\mathbf{i} \equiv \boldsymbol{u}_1$, $\mathbf{j} \equiv \boldsymbol{u}_2$, $\mathbf{k} \equiv \boldsymbol{u}_3$, where $\{\boldsymbol{u}_1, \boldsymbol{u}_2, \boldsymbol{u}_3\}$ is an orthonormal base in V_3. For $X, Y \in \mathfrak{J}$ it is then possible to write

$$X = a^i\boldsymbol{u}_i\,, \quad Y = b^i\boldsymbol{u}_i\,, \quad [X, Y] = 2X \times Y\,, \quad (X, Y) = a^i b^i\,,$$

where $i = 1, 2, 3$. Further, we obtain

$$\tfrac{1}{4}(\mathrm{Ad}X)(\mathrm{Ad}Y)(Z) = X \times (Y \times Z) = (X, Z)Y - (X, Y)Z\,.$$

From this relation we easily find the matrix (a_i^j) of the mapping $\tfrac{1}{4}\mathrm{Ad}X\,\mathrm{Ad}Y$:

$$\tfrac{1}{4}\mathrm{Ad}X\,\mathrm{Ad}Y(\boldsymbol{u}_i) = (X, \boldsymbol{u}_i)\,Y - (X, Y)\,\boldsymbol{u}_i = a^i Y - (X, Y)\,\boldsymbol{u}_i\,.$$

Thus, for a_i^j we obtain

$$a_i^j = a^i b^j - (X, Y)\,\delta_i^j\,,$$

so that

$$\tfrac{1}{4}\mathrm{Tr}(\mathrm{Ad}X\,\mathrm{Ad}Y) = a_i^i = a^i b^i - 3(X, Y) = (X, Y) - 3(X, Y)$$
$$= -2(X, Y).$$

Hence, $\mathbf{K}(X, Y) = -8(X, Y)$. As a matter of fact, the Killing form is the inner product in $\mathfrak{J} = V_3$. This means that the group ad J preserves the inner product. Consequently, it is a subgroup of the group $O(3)$. If $\boldsymbol{\alpha} \in J$, we shall show that the mapping ad$\boldsymbol{\alpha}$ is a rotation of the vector space V_3 about a certain axis through a certain angle, and we will find them.

Every $\alpha \in J$ can be written in the form $\alpha = \cos\varphi + \varepsilon_1 \sin\varphi$, where ε_1 is a normed quaternion. We obtain

$$\begin{aligned}\mathrm{ad}\alpha(\varepsilon_1) &= \mathrm{ad}(\cos\varphi + \varepsilon_1 \sin\varphi)(\varepsilon_1) \\ &= (\cos\varphi + \varepsilon_1 \sin\varphi)(\varepsilon_1)(\cos\varphi - \varepsilon_1 \sin\varphi) = \varepsilon_1 .\end{aligned}$$

This means that to the quaternion α corresponds a rotation about the axis determined by ε_1.

Further, let us complete ε_1 to a right-hand orthonormal base by the quaternions ε_2 and ε_3. For purely imaginary quaternions ε, δ we have the relation

$$\varepsilon . \delta = -(\varepsilon, \delta) + \varepsilon \times \delta$$

so that

$$\varepsilon_1 . \varepsilon_2 = \varepsilon_3 , \quad \varepsilon_2 . \varepsilon_3 = \varepsilon_1 , \quad \varepsilon_3 . \varepsilon_1 = \varepsilon_2 .$$

Now, let us find the image of the quaternion ε_2 under the mapping

$$\begin{aligned}\mathrm{ad}\alpha\colon \mathrm{ad}\alpha(\varepsilon_2) &= \alpha\varepsilon_2\bar{\alpha} = (\cos\varphi + \varepsilon_1 \sin\varphi)\,\varepsilon_2(\cos\varphi - \varepsilon_1 \sin\varphi) \\ &= \varepsilon_3 \sin(2\varphi) + \varepsilon_2 \cos(2\varphi) .\end{aligned}$$

Thus, we see that the mapping $\mathrm{ad}\alpha$ belonging to the quaternion $\alpha = \cos\varphi + \varepsilon_1 \sin\varphi$ is the rotation about the axis determined by the vector ε_1 through the angle 2φ. For $0 < \varphi < \frac{1}{2}\pi$ the quaternions ε_1, ε_2, $\mathrm{ad}\,\alpha\varepsilon_2$ define a right-hand coordinate system. We have found out also that the mapping $\mathrm{ad}\alpha$ is an orientation preserving congruence since it maps a right-hand frame into a right-hand frame so that – by 1.85 – the determinant of this transformation is equal to one. All the elements of $O(3)$ whose determinant is equal to one constitute a subgroup of the group $O(3)$. It is a Lie group of dimension three and has the same Lie algebra as the group $O(3)$. We call it the special orthogonal group of degree three and denote it by $S\,O(3)$. The above facts imply that the groups $\mathrm{ad}J$ and $S\,O(3)$ are isomorphic. Also, we see that the mapping $\mathrm{ad}\colon J \to \mathrm{ad}J$ is not one-to-one: to the quaternions α and $-\alpha$ the same rotation corresponds.

The relations just derived for $\mathrm{ad}\alpha$ will be applied to represent the elements of the matrix of rotation about the given axis through a given angle. Let the axis of rotation be given by the normed quaternion ε, and let φ be an angle. If we consider the quaternion

$$\alpha(\varepsilon, \varphi) = \cos\left(\tfrac{1}{2}\varphi\right) + \varepsilon \sin\left(\tfrac{1}{2}\varphi\right),$$

$\mathrm{ad}\alpha$ is obviously the rotation about the axis o through the angle φ in the positive sense. Denote by (a_i^j) the matrix of this rotation. If a_1', a_2', a_3' are the columns of this matrix, then the colum a_1' is the column of the coordinates of the vector $\alpha\mathbf{i}\bar{\alpha}$. The similar holds for the other two columns. Further, we have

$$\begin{aligned} \alpha\mathbf{i}\bar{\alpha} &= \left[\cos\left(\tfrac{1}{2}\varphi\right) + \varepsilon \sin\left(\tfrac{1}{2}\varphi\right)\right] \mathbf{i}\left[\cos\left(\tfrac{1}{2}\varphi\right) - \varepsilon \sin\left(\tfrac{1}{2}\varphi\right)\right] \\ &= \mathbf{i} \cos\varphi + (1 - \cos\varphi)(\varepsilon, \mathbf{i})\,\varepsilon + (\varepsilon \times \mathbf{i}) \sin\varphi , \end{aligned}$$

where we use the relations

$$\varepsilon \,.\, \mathbf{i} = -(\varepsilon, \mathbf{i}) + \varepsilon \times \mathbf{i}\,, \quad \varepsilon\mathbf{i}\varepsilon = \left[-(\varepsilon, \mathbf{i}) + \varepsilon \times \mathbf{i}\right] \varepsilon = \mathbf{i} - 2(\varepsilon, \mathbf{i})\,\varepsilon\,.$$

Thus, the column a_i' is constituted by the coordinates of the vector

$$\boldsymbol{u}_i \cos\varphi + (1 - \cos\varphi)(\varepsilon, \boldsymbol{u}_i)\,\varepsilon + (\varepsilon \times \boldsymbol{u}_i) \sin\varphi\,.$$

Then we obtain

$$a_i^j = (\boldsymbol{u}_i, \boldsymbol{u}_j) \cos\varphi + (1 - \cos\varphi)(\varepsilon, \boldsymbol{u}_i)(\varepsilon, \boldsymbol{u}_j) + (\varepsilon \times \boldsymbol{u}_i, \boldsymbol{u}_j) \sin\varphi\,,$$

so that

$$a_i^j = \delta_j^i \cos\varphi + (1 - \cos\varphi)(\varepsilon, \boldsymbol{u}_i)(\varepsilon, \boldsymbol{u}_j) + (\varepsilon, \varepsilon_{ijk}\boldsymbol{u}_k) \sin\varphi$$

since $(\varepsilon \times \boldsymbol{u}_i, \boldsymbol{u}_j) = (\varepsilon, \boldsymbol{u}_i \times \boldsymbol{u}_j)$. Writing in coordinate form $\varepsilon = x^1\boldsymbol{u}_1 + x^2\boldsymbol{u}_2 + x^3\boldsymbol{u}_3$, we obtain

$$a_i^j = \delta_i^j \cos\varphi + (1 - \cos\varphi)\, x^i x^j + \varepsilon_{ijk} x^k \sin\varphi\,.$$

Let us also find the image of an arbitrary vector $\boldsymbol{Y} = y^i\boldsymbol{u}_i$. We obtain the vector

$$\boldsymbol{Y} \cos\varphi + (1 - \cos\varphi)(\varepsilon, \boldsymbol{Y})\,\varepsilon + (\varepsilon \times \boldsymbol{Y}) \sin\varphi\,.$$

5.67 Example. We present here a different representation of the group J than the one found in Example 5.66. The quaternions $1, \mathbf{i}, \mathbf{j}, \mathbf{k}$ constitute a base of the vector space of all quaternions. Let us define a mapping for them as follows:

$$\sigma(1) = \begin{pmatrix} 1, & 0 \\ 0, & 1 \end{pmatrix}, \quad \sigma(\mathbf{i}) = \begin{pmatrix} \mathbf{i}, & 0 \\ 0, & -\mathbf{i} \end{pmatrix}, \quad \sigma(\mathbf{j}) = \begin{pmatrix} 0, & 1 \\ -1, & 0 \end{pmatrix}, \quad \sigma(\mathbf{k}) = \begin{pmatrix} 0, & \mathbf{i} \\ \mathbf{i}, & 0 \end{pmatrix}.$$

The mapping σ is extended linearly to all quaternions, i.e. if

$$\boldsymbol{\alpha} = a_0 + a_1\mathbf{i} + a_2\mathbf{j} + a_3\mathbf{k}$$

is an arbitrary quaternion we put

$$\sigma(\boldsymbol{\alpha}) = a_0\,\sigma(1) + a_1\,\sigma(\mathbf{i}) + a_2\,\sigma(\mathbf{j}) + a_3\,\sigma(\mathbf{k}).$$

Hence, we obtain a mapping of the set of all quaternions into the set of all complex 2×2 matrices. By computation we verify easily that

$$\sigma(\boldsymbol{\alpha}_1 + \boldsymbol{\alpha}_2) = \sigma(\boldsymbol{\alpha}_1) + \sigma(\boldsymbol{\alpha}_2), \quad \sigma(\boldsymbol{\alpha}_1\boldsymbol{\alpha}_2) = \sigma(\boldsymbol{\alpha}_1)\,\sigma(\boldsymbol{\alpha}_2)$$

for arbitrary quaternions $\boldsymbol{\alpha}_1$ and $\boldsymbol{\alpha}_2$. Writing $a_0 + a_1\mathbf{i} = z_1$, $a_2 + a_3\mathbf{i} = z_2$, we see that

$$\sigma(\boldsymbol{\alpha}) = \begin{pmatrix} z_1, & z_2 \\ -\bar{z}_2, & \bar{z}_1 \end{pmatrix}.$$

We have $\boldsymbol{\alpha}\bar{\boldsymbol{\alpha}} = 1$ if and only if $\det \sigma(\boldsymbol{\alpha}) = z_1\bar{z}_1 + z_2\bar{z}_2 = 1$.

The group of all matrices of the form

$$\begin{pmatrix} z_1, & z_2 \\ -\bar{z}_2, & \bar{z}_1 \end{pmatrix}$$

with determinant equal to 1 is called the special unitary group and is denoted by $S\,U(2)$. The mapping σ is one-to-one, and it preserves multiplication. Consequently, it is an isomorphism of the group J onto the group $S\,U(2)$. Applying σ we can now represent each element from $S\,O(3)$ by a matrix from $S\,U(2)$. This relationship is illustrated by the diagram

$$\begin{array}{ccc} J & \overset{\sigma}{\longleftrightarrow} & S\,U(2) \\ & \searrow \quad O(3) \quad \swarrow & \end{array}$$

which means explicitly

$$\sigma(\boldsymbol{\alpha}) = \begin{pmatrix} \cos\left(\tfrac{1}{2}\varphi\right) + \mathbf{i}x^1 \sin\left(\tfrac{1}{2}\varphi\right), & x^2 \sin\left(\tfrac{1}{2}\varphi\right) + \mathbf{i}x^2 \sin\left(\tfrac{1}{2}\varphi\right) \\ -x^2 \sin\left(\tfrac{1}{2}\varphi\right) + \mathbf{i}x^3 \sin\left(\tfrac{1}{2}\varphi\right), & \cos\left(\tfrac{1}{2}\varphi\right) - \mathbf{i}x^1 \sin\left(\tfrac{1}{2}\varphi\right) \end{pmatrix}$$

$$\boldsymbol{\alpha} = \cos\left(\tfrac{1}{2}\varphi\right) + \boldsymbol{\varepsilon} \sin\left(\tfrac{1}{2}\varphi\right)$$

$$a_i^j = \delta_i^j \cos\varphi + (1 - \cos\varphi)\, x^i x^j + \varepsilon_{ijk} x^k \sin\varphi\,.$$

Furthermore we note that the matrix from $S\,U(2)$ corresponding to a given rotation is determined uniquely up to its sign.

5.68 Exercise. Demonstrate the relations just discussed for the concrete case of the rotation about the axis $\boldsymbol{u} = 3^{-1/2}(1, 1, 1)$ through the angle $\frac{1}{3}\pi$ in the positive sense. Find the two corresponding quaternions, the two corresponding matrices from $S\,U(2)$, and the corresponding matrix from $S\,O(3)$.

Chapter II

Motion on the unit sphere

1. DEFINITION OF SPHERICAL MOTION

In the first chapter we have built up an apparatus sufficient for the investigation of general motion in space. Since some computations are rather complicated for general space motion we accomplish all the considerations for the simpler case of space motion with one fixed point first. Denote this point by O. Every space motion which preserves the point O at its place preserves also all the spheres whose centre lies at the point O. Besides, it suffices to consider only the sphere with radius equal to one. A motion which takes place on this sphere is called *spherical motion*; this chapter is devoted to the investigation of such motions.

For the reader who is interested in applications only and who has omitted reading the first chapter we review briefly the concepts and notation which will be used in what follows. Should the reader encounter difficulties in this section he should return to the respective section of Chapter I.

Consequently, let us consider the Euclidean space E_3 and fix one of its points. We denote this point O and call it the *origin*. Further, choose a sphere with centre at the point O and with radius equal to one. Denote this sphere by S^2 and call it *the unit sphere*. Points on this unit sphere will be represented by unit vectors (i.e. vectors of length one).

Choose a fixed orthonormal frame $\mathscr{R} = \{O, \boldsymbol{u}_1, \boldsymbol{u}_2, \boldsymbol{u}_3\}$. The point O is not mentioned in what follows since it does not change throughout this chapter. The points from E_3 will simply be identified with their position vectors. Every point X of the sphere S^2 is thus represented by a unit vector with coordinates x^1, x^2, x^3 where

$$X = x^1\boldsymbol{u}_1 + x^2\boldsymbol{u}_2 + x^3\boldsymbol{u}_3\,, \quad (x^1)^2 + (x^2)^2 + (x^3)^2 = 1\,.$$

(To be accurate, we should write $X - O = x^1\boldsymbol{u}_1 + x^2\boldsymbol{u}_2 + x^3\boldsymbol{u}_3$, naturally.)

See Fig. 10. If $\mathscr{R}' = \{\boldsymbol{u}_1', \boldsymbol{u}_2', \boldsymbol{u}_3'\}$ is another orthonormal frame, it is possible to decompose each of the vectors $\boldsymbol{u}_1', \boldsymbol{u}_2', \boldsymbol{u}_3'$ into its components in the frame $\boldsymbol{u}_1, \boldsymbol{u}_2, \boldsymbol{u}_3$. Thus we obtain

$$\begin{aligned} \boldsymbol{u}_1' &= a_1^1\boldsymbol{u}_1 + a_1^2\boldsymbol{u}_2 + a_1^3\boldsymbol{u}_3 \,, \\ \boldsymbol{u}_2' &= a_2^1\boldsymbol{u}_1 + a_2^2\boldsymbol{u}_2 + a_2^3\boldsymbol{u}_3 \,, \\ \boldsymbol{u}_3' &= a_3^1\boldsymbol{u}_1 + a_3^2\boldsymbol{u}_2 + a_3^3\boldsymbol{u}_3 \,, \end{aligned}$$

where a_i^j are coordinates of vector $\boldsymbol{u}_i'$. The relation between frames $\mathscr{R}'$ and $\mathscr{R}$ can then be formally written as follows: write the coordinates of

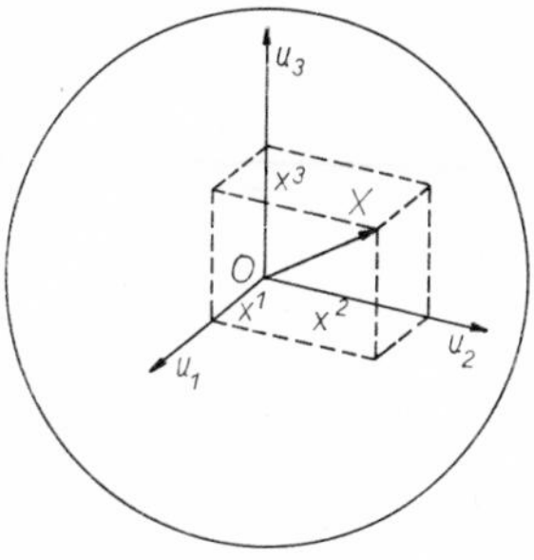

Fig. 10.

the vectors $\boldsymbol{u}_1', \boldsymbol{u}_2', \boldsymbol{u}_3'$ in the frame $\mathscr{R}$ into three columns, one beside the other. A 3×3 matrix is obtained, and we denote it by g:

$$g = \begin{pmatrix} a_1^1, & a_2^1, & a_3^1 \\ a_1^2, & a_2^2, & a_3^2 \\ a_1^3, & a_2^3, & a_3^3 \end{pmatrix}.$$

(Recall that the superscripts denote rows while the subscripts denote columns.) For the frame $\mathscr{R}$ and $\mathscr{R}'$ we may now formally write $\mathscr{R}' = \mathscr{R}g$. Here the right-hand side is the formal product of row $\mathscr{R}$ and matrix g. Let $\mathscr{S}$ be a congruence of the Euclidean space E_3 which preserves the origin, i.e. $\mathscr{S}(O) = O$. The congruence $\mathscr{S}$ maps every point $X \in E_3$ into some point $X' \in E_3$; we write $\mathscr{S}(X) = X'$. In particular, we denote by

$$\boldsymbol{u}_1' = \mathscr{S}(\boldsymbol{u}_1)\,, \quad \boldsymbol{u}_2' = \mathscr{S}(\boldsymbol{u}_2)\,, \quad \boldsymbol{u}_3' = \mathscr{S}(\boldsymbol{u}_3)$$

the images of the vectors of the frame $\mathscr{R}$ under $\mathscr{S}$. The vectors $\{\boldsymbol{u}_1', \boldsymbol{u}_2', \boldsymbol{u}_3'\}$ constitute an orthonormal frame again; let us denote it by $\mathscr{R}'$. Thus, to every congruence $\mathscr{S}$ a certain orthonormal frame $\mathscr{R}'$ is assigned. Conversely, if the orthonormal frame $\mathscr{R}'$ is given, then the congruence $\mathscr{S}$ is determined as well since to every point $X = x^1\boldsymbol{u}_1 + x^2\boldsymbol{u}_2 + x^3\boldsymbol{u}_3 \in E_3$ we know how to find its image $X' = \mathscr{S}(X)$. In fact, every congruence is a linear mapping, i.e.

$$\mathscr{S}(X) = x^1\,\mathscr{S}(\boldsymbol{u}_1) + x^2\,\mathscr{S}(\boldsymbol{u}_2) + x^3\,\mathscr{S}(\boldsymbol{u}_3)\,,$$

so that

$$X' = x^1\boldsymbol{u}_1' + x^2\boldsymbol{u}_2' + x^3\boldsymbol{u}_3'\,.$$

Consequently, every congruence is uniquely determined by the frame $\mathscr{R}'$, and this frame is determined by the matrix g. The congruence $\mathscr{S}$ is thus determined by the relation $\mathscr{R}' = \mathscr{R}g$.

The congruence $\mathscr{S}$ maps point X into the point $X' = \mathscr{S}(X)$, with the coordinates of the point X' in frame $\mathscr{R}'$ being the same as the coordinates of the point X in frame $\mathscr{R}$. Now, let us find the coordinates of the point X' in the original frame $\mathscr{R}$. We write the coordinates into columns. If it will not be obvious from the context with respect to which frame are the coordinates computed, we add to the column of coordinates the notation of the respective frame as a subscript. Thus, if we have

$$X = x^1\boldsymbol{u}_1 + x^2\boldsymbol{u}_2 + x^3\boldsymbol{u}_3 \quad \text{and} \quad \mathscr{R} = \{\boldsymbol{u}_1, \boldsymbol{u}_2, \boldsymbol{u}_3\}\,,$$

we write

$$X_{\mathscr{R}} = \begin{pmatrix} x^1 \\ x^2 \\ x^3 \end{pmatrix}.$$

Formally it is possible to write $X = \mathscr{R}X_{\mathscr{R}}$; on the right-hand side we have the product of the row $\mathscr{R} = \{\boldsymbol{u}_1, \boldsymbol{u}_2, \boldsymbol{u}_3\}$ and of the column $X_{\mathscr{R}}$.

Further, consider the congruence $\mathscr{S}$ given by the matrix g, i.e. let $\mathscr{R}' = \mathscr{R}g$. The condition that the point X is mapped by the congruence $\mathscr{S}$ into the point X' is given in coordinate form by the relation $X'_{\mathscr{R}'} = X_{\mathscr{R}}$. The coordinates of the point X' in the frame $\mathscr{R}$ are found easily now. As a matter of fact, we have:

$$\mathscr{R}' = \mathscr{R}g\,, \quad X' = \mathscr{R}'X'_{\mathscr{R}'}\,, \quad X = \mathscr{R}X_{\mathscr{R}}\,.$$

Upon substitution we obtain

$$X' = \mathscr{R}'X'_{\mathscr{R}'} = \mathscr{R}gX_{\mathscr{R}} = \mathscr{R}X'_{\mathscr{R}},$$

so that $X'_{\mathscr{R}} = gX_{\mathscr{R}}$. The basic formulae for the change of frames and points under congruence are the following:

$$\mathscr{R}' = \mathscr{R}g\,, \quad X'_{\mathscr{R}} = gX_{\mathscr{R}}\,.$$

Remark. As we know already, the numbers a^i_j in the matrix g cannot be arbitrary. In fact, we have $\boldsymbol{u}'_i = a^i_j\boldsymbol{u}_j$, the $\boldsymbol{u}'_i$ constitute an orthonormal frame,

$$\delta_{\alpha\beta} = (\boldsymbol{u}'_\alpha, \boldsymbol{u}'_\beta) = (a^i_\alpha\boldsymbol{u}_i, a^i_\beta\boldsymbol{u}_j) = a^i_\alpha a^j_\beta(\boldsymbol{u}_i, \boldsymbol{u}_j) = a^i_\alpha a^j_\beta\delta_{ij} = a^i_\alpha a^i_\beta\,,$$

so that

$$a^i_\alpha a^i_\beta = \delta_{\alpha\beta}\,.$$

Such matrices are called orthogonal and they constitute the group $O(3)$. (See I,5.)

1.1 Example. Let us consider the rotation of the Euclidean space E_3 about the axis o, determined by the vector $\boldsymbol{o} = \frac{1}{2}\sqrt{(2)}\,(\boldsymbol{u}_1 + \boldsymbol{u}_2)$, through the angle 30° in the positive sense. Positive sense of rotation is defined as

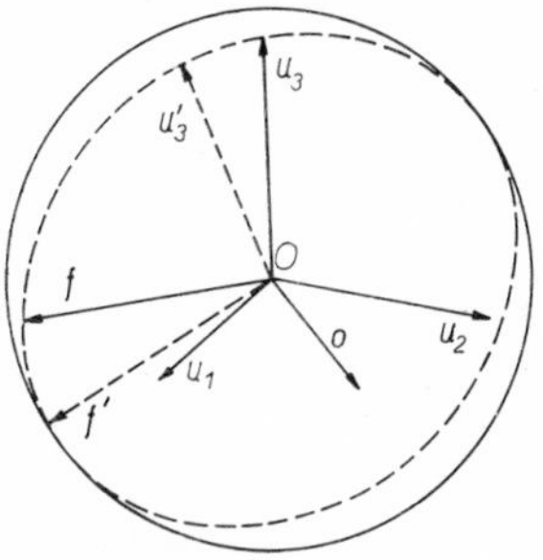

Fig. 11.

follows: looking from the terminal point of vector $\boldsymbol{o}$ in the direction of its initial point the rotation is realized counterclockwise. The rotation just described is a congruence which preserves the origin; let us denote it by $\mathscr{S}$. For $\mathscr{S}$, let us find the frame $\mathscr{R}'$, the matrix g and the image of an arbitrary point X. $\mathscr{S}$ preserves every plane perpendicular to the axis of rotation. In particular, the plane which passes through the origin of coordinates. This is the plane with equation $x^1 + x^2 = 0$. Thus, let us choose an auxiliary coordinate system in the plane $x^1 + x^2 = 0$ (see Fig. 11) composed of the vectors $\boldsymbol{u}_3, \boldsymbol{f}$ so that the vectors $\{\boldsymbol{O}, \boldsymbol{u}_3, \boldsymbol{f}\}$ constitute a right-handed frame, i.e. let us put $\boldsymbol{f} = \boldsymbol{o} \times \boldsymbol{u}_3$. The computations yield

$$\boldsymbol{f} = \begin{vmatrix} \boldsymbol{u}_1, & \boldsymbol{u}_2, & \boldsymbol{u}_3 \\ \frac{\sqrt{2}}{2}, & \frac{\sqrt{2}}{2}, & 0 \\ 0, & 0, & 1 \end{vmatrix} {}^{*} = \frac{\sqrt{2}}{2}(\boldsymbol{u}_1 - \boldsymbol{u}_2)\,,$$

which is also obvious from the figure. The frame $\{\boldsymbol{u}_3, \boldsymbol{f}\}$ rotates now in the plane $x^1 + x^2 = 0$ through the 30° angle in the positive sense. For rotation in a plane through the 30° angle we have the formulae

$$\boldsymbol{u}_3' = \boldsymbol{u}_3 \cos\varphi + \boldsymbol{f}\sin\varphi\,, \quad \boldsymbol{f}' = -\boldsymbol{u}_3\sin\varphi + \boldsymbol{f}\cos\varphi\,.$$

Here, $\boldsymbol{u}_3', \boldsymbol{f}'$ are the vectors $\boldsymbol{u}_3$ and $\boldsymbol{f}$ after rotation. In our case we have $\sin\varphi = \frac{1}{2}$, $\cos\varphi = \frac{1}{2}\sqrt{3}$, so that

$$\mathscr{S}(\boldsymbol{u}_3) = \frac{\sqrt{3}}{2}\boldsymbol{u}_3 + \frac{\boldsymbol{f}}{2}, \quad \mathscr{S}(\boldsymbol{f}) = -\frac{\boldsymbol{u}_3}{2} + \frac{\sqrt{3}}{2}\boldsymbol{f}, \quad \mathscr{S}(\boldsymbol{o}) = \boldsymbol{o}\,.$$

Now, let us write out these three equations in coordinate form:

$$\mathscr{S}(\boldsymbol{u}_3) = \frac{\sqrt{3}}{2}\boldsymbol{u}_3 + \frac{1}{2}\left(\frac{\sqrt{2}}{2}\boldsymbol{u}_1 - \frac{\sqrt{2}}{2}\boldsymbol{u}_2\right)$$

$$= \tfrac{1}{4}\left[\sqrt{(2)}\,\boldsymbol{u}_1 - \sqrt{(2)}\,\boldsymbol{u}_2 + 2\sqrt{(3)}\,\boldsymbol{u}_3\right],$$

$$\mathscr{S}\left(\frac{\sqrt{2}}{2}\boldsymbol{u}_1 - \frac{\sqrt{2}}{2}\boldsymbol{u}_2\right) = -\frac{1}{2}\boldsymbol{u}_3 + \frac{\sqrt{3}}{2}\left(\frac{\sqrt{2}}{2}\boldsymbol{u}_1 - \frac{\sqrt{2}}{2}\boldsymbol{u}_2\right)$$

$$= \tfrac{1}{4}\left[\sqrt{(6)}\,\boldsymbol{u}_1 - \sqrt{(6)}\,\boldsymbol{u}_2 - 2\boldsymbol{u}_3\right],$$

$$\mathscr{S}\left(\frac{\sqrt{2}}{2}\boldsymbol{u}_1 + \frac{\sqrt{2}}{2}\boldsymbol{u}_2\right) = \frac{\sqrt{2}}{2}\boldsymbol{u}_1 + \frac{\sqrt{2}}{2}\boldsymbol{u}_2\,.$$

* The expression is formally viewed as a determinant.

Adding the last two equations and taking into account that congruence is a linear mapping we obtain

$$\mathcal{S}(\sqrt{[2]}\,\boldsymbol{u}_1) = \tfrac{1}{4}[(\sqrt{6} + 2\sqrt{2})\,\boldsymbol{u}_1 + (-\sqrt{6} + 2\sqrt{2})\,\boldsymbol{u}_2 - 2\boldsymbol{u}_3]\,.$$

Subtracting the second equation from the third we have

$$\mathcal{S}(\sqrt{[_{\beta}]}\boldsymbol{u}_2) = \tfrac{1}{4}[(-\sqrt{6} + 2\sqrt{2})\,\boldsymbol{u}_1 + (\sqrt{6} + 2\sqrt{2})\,\boldsymbol{u}_2 + 2\boldsymbol{u}_3]\,.$$

Upon modification there is

$$\mathcal{S}(\boldsymbol{u}_1) = \tfrac{1}{4}[(\sqrt{3} + 2)\,\boldsymbol{u}_1 + (-\sqrt{3} + 2)\,\boldsymbol{u}_2 - \sqrt{(2)}\,\boldsymbol{u}_3]\,,$$
$$\mathcal{S}(\boldsymbol{u}_2) = \tfrac{1}{4}[(-\sqrt{3} + 2)\,\boldsymbol{u}_1 + (\sqrt{3} + 2)\,\boldsymbol{u}_2 + \sqrt{(2)}\,\boldsymbol{u}_3]\,,$$
$$\mathcal{S}(\boldsymbol{u}_3) = \tfrac{1}{4}[\sqrt{(2)}\,\boldsymbol{u}_1 - \sqrt{(2)}\,\boldsymbol{u}_2 + 2\sqrt{(3)}\,\boldsymbol{u}_3]\,.$$

For the image $\mathcal{R}' = \{\boldsymbol{u}_1', \boldsymbol{u}_2', \boldsymbol{u}_3'\}$ of the frame $\mathcal{R} = \{\boldsymbol{u}_1, \boldsymbol{u}_2, \boldsymbol{u}_3\}$ under $\mathcal{S}$ we have

$$\boldsymbol{u}_1' = \tfrac{1}{4}[(2 + \sqrt{3})\,\boldsymbol{u}_1 + (2 - \sqrt{3})\,\boldsymbol{u}_2 - \sqrt{(2)}\,\boldsymbol{u}_3]\,,$$
$$\boldsymbol{u}_2' = \tfrac{1}{4}[(2 - \sqrt{3})\,\boldsymbol{u}_1 + (2 + \sqrt{3})\,\boldsymbol{u}_2 + \sqrt{(2)}\,\boldsymbol{u}_3]\,,$$
$$\boldsymbol{u}_3' = \tfrac{1}{4}[\sqrt{(2)}\,\boldsymbol{u}_1 - \sqrt{(2)}\,\boldsymbol{u}_2 + 2\sqrt{(3)}\,\boldsymbol{u}_3]\,.$$

The matrix g in the equation $\mathcal{R}' = \mathcal{R}g$ is:

$$g = \tfrac{1}{4}\begin{pmatrix} 2 + \sqrt{3}, & 2 - \sqrt{3}, & \sqrt{2} \\ 2 - \sqrt{3}, & 2 + \sqrt{3}, & -\sqrt{2} \\ -\sqrt{2}, & \sqrt{2}, & 2\sqrt{3} \end{pmatrix}.$$

The equation $a_\alpha^i a_\beta^i = \delta_{\alpha\beta}$ is verified easily by direct computation. If we wish to find the image of the point X under $\mathcal{S}$, i.e. the point $X' = \mathcal{S}(X)$, where

$$X' = (x^1)'\,\boldsymbol{u}_1 + (x^2)'\,\boldsymbol{u}_2 + (x^3)'\,\boldsymbol{u}_3\,,$$

we have

$$\begin{pmatrix} (x^1)' \\ (x^2)' \\ (x^3)' \end{pmatrix} = \tfrac{1}{4}\begin{pmatrix} 2 + \sqrt{3}, & 2 - \sqrt{3}, & \sqrt{2} \\ 2 - \sqrt{3}, & 2 + \sqrt{3}, & -\sqrt{2} \\ -\sqrt{2}, & \sqrt{2}, & 2\sqrt{3} \end{pmatrix}\begin{pmatrix} x^1 \\ x^2 \\ x^3 \end{pmatrix}$$
$$= \tfrac{1}{4}\begin{pmatrix} (2 + \sqrt{3})\,x^1 + (2 - \sqrt{3})\,x^2 + \sqrt{(2)}\,x^3 \\ (2 - \sqrt{3})\,x^1 + (2 + \sqrt{3})\,x^2 - \sqrt{(2)}\,x^3 \\ -\sqrt{(2)}\,x^1 + \sqrt{(2)}\,x^2 + 2\sqrt{(3)}\,x^3 \end{pmatrix}.$$

For instance, for $x^1 = 1$, $x^2 = 1$, $x^3 = 0$ we obtain

$$(x^1)' = \tfrac{1}{4}[(2 + \sqrt{3}) + (2 - \sqrt{3})] = 1\,,$$
$$(x^2)' = \tfrac{1}{4}[(2 - \sqrt{3}) + (1 + \sqrt{3})] = 1\,,$$
$$(x^3)' = \tfrac{1}{4}(-\sqrt{2} + \sqrt{2}) = 0\,.$$

This is correct since the point with these coordinates lies on the axis of rotation.

1.2 Example. In the course of computation of Example 1.1 we could note that the representation of congruence $\mathscr{S}$ would be very simple in the frame $\{\boldsymbol{o}, \boldsymbol{u}_3, \boldsymbol{f}\}$. Let us try to find out how the representation of a congruence changes if we pass over from the given basic frame $\mathscr{R}$ to another basic frame. Thus, let us assume that in place of the original frame $\mathscr{R}$ some other orthonormal frame was chosen, say $\mathscr{R}_1$, and let $\mathscr{R} = \mathscr{R}_1\gamma$, i.e. let $\mathscr{R}_1 = \mathscr{R}\gamma^{-1}$. Then the following equations hold:

$$X = \mathscr{R}X_{\mathscr{R}} = \mathscr{R}_1X_{\mathscr{R}_1} = \mathscr{R}\gamma^{-1}X_{\mathscr{R}_1}$$

so that

$$X_{\mathscr{R}} = \gamma^{-1}X_{\mathscr{R}_1}\,;$$
$$X' = \mathscr{R}'X'_{\mathscr{R}'} = \mathscr{R}'X_{\mathscr{R}} = \mathscr{R}gX_{\mathscr{R}} = \mathscr{R}_1\gamma gX_{\mathscr{R}} = \mathscr{R}_1X'_{\mathscr{R}_1}$$

so that

$$X'_{\mathscr{R}_1} = \gamma gX_{\mathscr{R}}\,.$$

Upon substitution for $X_{\mathscr{R}}$ we have $X'_{\mathscr{R}_1} = \gamma g\gamma^{-1}X_{\mathscr{R}_1}$. Consequently, our original congruence is determined by the matrix $\gamma g\gamma^{-1}$. We also have $\mathscr{R}'_1 = \mathscr{R}_1\gamma g\gamma^{-1}$.

1.3 Example. For illustration, let us use the results of Example 1.2 for the computation of Example 1.1. In Example 1.3 the notation of Example 1.1 is used without definition, to avoid unnecessary repetitions. As the frame $\mathscr{R}_1$ let us choose the frame

$$\mathscr{R}_1 = \{\boldsymbol{o}, \boldsymbol{u}_3, \boldsymbol{f}\} = \left\{\frac{\sqrt{2}}{2}\boldsymbol{u}_1 + \frac{\sqrt{2}}{2}\boldsymbol{u}_2, \boldsymbol{u}_3, \frac{\sqrt{2}}{2}\boldsymbol{u}_1 - \frac{\sqrt{2}}{2}\boldsymbol{u}_2\right\},$$

where $\mathscr{R} = \mathscr{R}_1\gamma$, or $\mathscr{R}_1 = \mathscr{R}\gamma^{-1}$. Denote by g the matrix of the congruence $\mathscr{S}$ in the original frame $\mathscr{R}$; let h be its matrix in the frame $\mathscr{R}_1$. Then $h =$

$\gamma g \gamma^{-1}$ by Example 1.2, i.e. $g = \gamma^{-1} h \gamma$. Here we have

$$\gamma^{-1} = \tfrac{1}{2}\begin{pmatrix} \sqrt{2}, & 0, & \sqrt{2} \\ \sqrt{2}, & 0, & -\sqrt{2} \\ 0, & 2, & 0 \end{pmatrix}, \quad h = \tfrac{1}{2}\begin{pmatrix} 2, & 0, & 0 \\ 0, & \sqrt{3}, & -1 \\ 0, & 1, & \sqrt{3} \end{pmatrix},$$

since h is the matrix of rotation through 30° about the first coordinate axis. Further, we have $\gamma^{-1} = \gamma^T$ since γ is an orthogonal matrix. Thus, we obtain

$$g = \tfrac{1}{8}\begin{pmatrix} \sqrt{2}, & 0, & \sqrt{2} \\ \sqrt{2}, & 0, & -\sqrt{2} \\ 0, & 2, & 0 \end{pmatrix}\begin{pmatrix} 2, & 0, & 0 \\ 0, & \sqrt{3}, & -1 \\ 0, & 1, & \sqrt{3} \end{pmatrix}\begin{pmatrix} \sqrt{2}, & \sqrt{2}, & 0 \\ 0, & 0, & 2 \\ \sqrt{2}, & -\sqrt{2}, & 0 \end{pmatrix}$$
$$= \tfrac{1}{4}\begin{pmatrix} 2+\sqrt{3}, & 2-\sqrt{3}, & \sqrt{2} \\ 2-\sqrt{3}, & 2+\sqrt{3}, & -\sqrt{2} \\ -\sqrt{2}, & \sqrt{2}, & 2\sqrt{3} \end{pmatrix}.$$

This is in agreement with the result of Example 1.1.

1.4 Exercise. Find the matrix of rotation about the axis determined by the vector $\boldsymbol{o} = \boldsymbol{u}_1 + \boldsymbol{u}_2 - \boldsymbol{u}_3$ through the angles 45° and 270° in the positive sense.

Now, let us assume that we have two coinciding (or co-local) Euclidean spaces E_3 and $\bar{E}_3$ (i.e., each point of one space is at the same time a point of the other space) with same origin O (or two coinciding unit spheres S^2, $\bar{S}^2$ with the same centre, which is almost the same). If $\bar{E}_3$, or $\bar{S}^2$, moves in E_3, or S^2, in such a manner that their origins, or centres, respectively, coincide, we speak of *spherical motion*. $\bar{E}_3$ is called the *moving space* (*moving system*), E_3 the *fixed space* (*fixed system*). Analogously, $\bar{S}^2$ is called the *moving sphere*, S^2 the *fixed sphere*. Let us consider now the problem how to determine such a spherical motion. First of all, let us adopt the convention that the concept of "frame" will in this chapter mean an orthonormal frame since no other frames will be used. Thus, let us choose the frame $\mathscr{R} = \{\boldsymbol{u}_1, \boldsymbol{u}_2, \boldsymbol{u}_3\}$ in the fixed space E_3 and the frame $\bar{\mathscr{R}} = (\bar{\boldsymbol{u}}_1, \bar{\boldsymbol{u}}_2, \bar{\boldsymbol{u}}_3)$ in the moving space $\bar{E}_3$ (the bar will here as well as in the sequel denote membership to the moving space). From what was said above it follows that a motion is given if the frame $\bar{\mathscr{R}}$ is given as a function of some parameter; if we denote it by t we must have

$$\bar{\mathscr{R}}(\mathrm{t}) = \{\bar{\boldsymbol{u}}_1(\mathrm{t}), \bar{\boldsymbol{u}}_2(\mathrm{t}), \bar{\boldsymbol{u}}_3(\mathrm{t})\}.$$

The frame $\bar{\mathcal{R}}(t)$ can be represented, in accordance with the above considerations, with the aid of frame $\mathcal{R}$, i.e. $\bar{\mathcal{R}}(t) = \mathcal{R}\, g(t)$. Here, $g(t)$ is an orthogonal matrix which depends on the parameter t. As is familiar, all orthogonal matrices constitute the Lie group $O(3)$. This means that spherical motion determines a curve on the orthogonal group. These considerations lead to the following definition:

1.5 Definition. Let E_3, $\bar{E}_3$ be two coinciding Euclidean spaces with the same origin O. Let in each of these spaces an orthonormal frame be given. By *spherical motion* we mean a regular curve on the group $O(3)$, i.e. a differentiable mapping g of an interval $(a, b) \subset \mathbf{R}$ into $O(3)$ which has nonzero differential at every point, i.e. $g: (a, b) \to O(3): t \to g(t)$, where $g'_t(\partial/\partial t) \neq 0$ for every $t \in (a, b)$.

1.6 Remark. Consequently, spherical motion is given if the coefficients a^i_j of the matrix g are given as functions of the parameter t. The condition of regularity means that for every t at least one of the numbers $(a^i_j(t))'$ is different from zero. In the exposition which follows we will not discuss how many derivatives do the functions $a^i_j(t)$ have. However, we shall always assume that they possess as many derivatives as necessary for our further considerations, so as not to burden the exposition by unnecessary details.

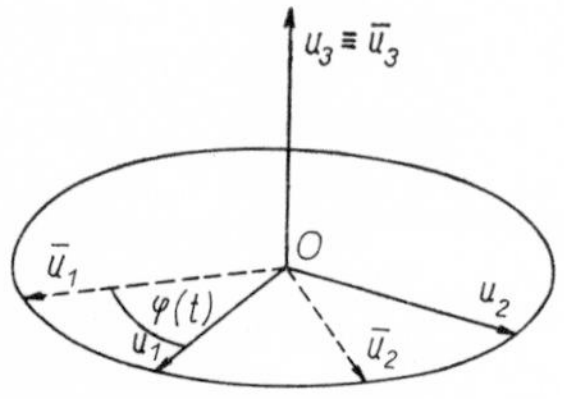

Fig. 12.

1.7 Example. Let us consider rotation through the angle $\varphi(t)$, $t \in \mathbf{R}$, of the space E_3 in the positive sense about the axis $\boldsymbol{u}_3$ of the frame $\mathcal{R}$. For a suitable choice of the frame $\bar{\mathcal{R}}$ we have

$$\bar{\mathcal{R}}(t) = \{\bar{\boldsymbol{u}}_1(t), \bar{\boldsymbol{u}}_2(t), \bar{\boldsymbol{u}}_3(t)\},$$

where

$$\bar{\boldsymbol{u}}_1(t) = \boldsymbol{u}_1 \cos\varphi(t) + \boldsymbol{u}_2 \sin\varphi(t),$$
$$\bar{\boldsymbol{u}}_2(t) = -\boldsymbol{u}_1 \sin\varphi(t) + \boldsymbol{u}_2 \cos\varphi(t),$$

$$\bar{\boldsymbol{u}}_3(\mathrm{t}) = \quad \boldsymbol{u}_3 ,$$

so that

$$\bar{\mathscr{R}}(\mathrm{t}) = \{\boldsymbol{u}_1 \cos \varphi(\mathrm{t}) + \boldsymbol{u}_2 \sin \varphi(\mathrm{t}), \; -\boldsymbol{u}_1 \sin \varphi(\mathrm{t}) + \boldsymbol{u}_2 \cos \varphi(\mathrm{t}), \boldsymbol{u}_3\} \; ;$$

see Fig. 12. The curve $g(\mathrm{t})$ on the group $O(3)$ which corresponds to this motion is

$$g(\mathrm{t}) = \begin{pmatrix} \cos \varphi(\mathrm{t}), & -\sin \varphi(\mathrm{t}), & 0 \\ \sin \varphi(\mathrm{t}), & \cos \varphi(\mathrm{t}), & 0 \\ 0, & 0, & 1 \end{pmatrix} .$$

1.8 Definition. Let $g(\mathrm{t})$ be a spherical motion. The curve $X(\mathrm{t}) = g(\mathrm{t})\,\bar{X}$ in E_3 is called the *trajectory* of the point $\bar{X} \in E_3$. It is the path along which the point $\bar{X} \in \bar{E}_3$ travels in the space E_3 under motion $g(\mathrm{t})$.

1.9 Remark. Note that in Definition 1.5 of spherical motion it was not assumed that $\bar{\mathscr{R}}(0) = \mathscr{R}$ necessarily (as is the case in Example 1.7), i.e. that at the start of the motion the frame in the moving space is identical with the frame in the fixed space. Consequently, this means that the trajectory of the point $\bar{X}$ of the moving system need not pass at all through the point of the fixed system which is identical with $\bar{X}$ at the start of motion. However, in the examples below we shall mostly assume that $\bar{\mathscr{R}}(0) = \mathscr{R}$, for the sake of simplicity. As a consequence, for $\mathrm{t} = 0$ the trajectory of a point passes always through this point. Theoretically, however, this assumption is in no way inevitable. Rather, it would complicate our considerations. For illustration let us present a simple example:

Consider the motion $g(\varphi)$ given by the matrix

$$g(\varphi) = \begin{pmatrix} \cos \varphi, & 0, & -\sin \varphi \\ \sin \varphi, & 0, & \cos \varphi \\ 0, & -1, & 0 \end{pmatrix} .$$

The trajectory of the point $\bar{X} = \begin{pmatrix} 0 \\ 0 \\ 1 \end{pmatrix}$ is the curve

$$X(\varphi) = g(\varphi)\,\bar{X} = \begin{pmatrix} \cos \varphi, & 0, & -\sin \varphi \\ \sin \varphi, & 0, & \cos \varphi \\ 0, & -1, & 0 \end{pmatrix} \begin{pmatrix} 0 \\ 0 \\ 1 \end{pmatrix} = \begin{pmatrix} -\sin \varphi \\ \cos \varphi \\ 0 \end{pmatrix},$$

which does not pass through the point $X = \begin{pmatrix} 0 \\ 0 \\ 1 \end{pmatrix}$ which has the same coordinates as the point $\bar{X}$.

1.10 Example. Let us find the trajectory of the point $\bar{X} = \begin{pmatrix} 1 \\ -1 \\ 2 \end{pmatrix}$ under the motion from Example 1.7. We obtain

$$X(t) = \begin{pmatrix} \cos\varphi(t), & -\sin\varphi(t), & 0 \\ \sin\varphi(t), & \cos\varphi(t), & 0 \\ 0, & 0, & 1 \end{pmatrix} \begin{pmatrix} 1 \\ -1 \\ 2 \end{pmatrix} = \begin{pmatrix} \cos\varphi(t) + \sin\varphi(t) \\ \sin\varphi(t) - \cos\varphi(t) \\ 2 \end{pmatrix}.$$

The parametric representation of the trajectory $X(t)$ is

$$x^1(t) = \cos\varphi(t) + \sin\varphi(t),$$
$$x^2(t) = \sin\varphi(t) - \cos\varphi(t),$$
$$x^3 = 2.$$

If we exclude the parameter t, we obtain the equations of the trajectory of the point $\bar{X}$ in the form

$$(x^1)^2 + (x^2)^2 = 2, \quad x^3 = 2.$$

Naturally, these are the equations of a circle.

1.11 Exercise. Recall the formula of I,5.66: The matrix $g = (a_i^j)$ of rotation about the axis determined by the unit vector $\boldsymbol{o} = x^1\boldsymbol{u}_1 + x^2\boldsymbol{u}_2 + x^3\boldsymbol{u}_3$ through the angle φ in the positive sense is

$$a_i^j = \delta_i^j \cos\varphi + (1 - \cos\varphi)\, x^i x^j + \varepsilon_{ijk} x^k \sin\varphi.$$

Find the matrix of rotation about the axis determined by the vector with coordinates $x^1 = \frac{1}{2}\sqrt{2}$, $x^2 = \frac{1}{2}\sqrt{2}$, $x^3 = 0$ through the angle φ in the positive sense and compare the result with Examples 1.1 and 1.3.

1.12 Example. Find the matrix of the motion given by rotation about the axis o determined by the vector $\boldsymbol{v} = 2\boldsymbol{u}_1 + 3\boldsymbol{u}_2 + 6\boldsymbol{u}_3$ through the angle t. According to the formula from Exercise 1.11 we again have $a_i^j = \delta_i^j \cos t + (1 - \cos t)\, x^i x^j + \varepsilon_{ijk} x^k \sin t$, where x^1, x^2, x^3 are coordinates of the unit vector of the axis of rotation. Consequently, we have to norm the given vector $\boldsymbol{v}$ of the axis of rotation. We have

$$\|\boldsymbol{v}\|^2 = 4 + 9 + 36 = 49, \quad \|\boldsymbol{v}\| = 7, \quad x^1 = \tfrac{2}{7}, \quad x^2 = \tfrac{3}{7}, \quad x^3 = \tfrac{6}{7}.$$

Substituting we obtain

$$g(t) = \cos t \begin{pmatrix} 1, & 0, & 0 \\ 0, & 1, & 0 \\ 0, & 0, & 1 \end{pmatrix} + \frac{1 - \cos t}{49} \begin{pmatrix} 4, & 6, & 12 \\ 6, & 9, & 18 \\ 12, & 18, & 36 \end{pmatrix} + \frac{\sin t}{7} \begin{pmatrix} 0, & -6, & 3 \\ 6, & 0, & -2 \\ -3, & 2, & 0 \end{pmatrix},$$

i.e.,

$$g(t) = \tfrac{1}{49} \begin{pmatrix} 4 + 45 \cos t & 6 - 6 \cos t - 42 \sin t, & 12 - 12 \cos t + 21 \sin t \\ 6 - 6 \cos t + 42 \sin t, & 9 + 40 \cos t, & 18 - 18 \cos t - 14 \sin t \\ 12 - 12 \cos t - 21 \sin t, & 18 - 18 \cos t + 14 \sin t, & 36 + 13 \cos t \end{pmatrix}.$$

1.13 Exercise. Find the matrix of the motion which is given as the rotation about the axis determined by the vector $\boldsymbol{v} = 2\boldsymbol{u}_1 - \boldsymbol{u}_2 + \boldsymbol{u}_3$ through the angle t by the methods of Examples 1.1, 1.3 and 1.12, and compare the results.

Now, let us proceed further with the considerations initiated at the beginning of this section. It was assumed that in the moving space $\bar{E}_3$ the frame $\bar{\mathscr{R}}$ is given, in the fixed space E_3 the frame $\mathscr{R}$. A spherical motion is then determined by a curve on the group $O(3)$. Let us now investigate how this curve will change if we choose some other frames in $\bar{E}_3$ and E_3 for reference. So let the frames $\bar{\mathscr{R}}_1$ in $\bar{E}_3$ and $\mathscr{R}_1$ in E_3 be given, where $\bar{\mathscr{R}} = \bar{\mathscr{R}}_1\bar{\gamma}$, $\mathscr{R} = \mathscr{R}_1\gamma$ with $\gamma, \bar{\gamma} \in O(3)$. Let the investigated motion be given by the matrix $g(t)$ with respect to frames $\bar{\mathscr{R}}$ and $\mathscr{R}$, i.e. let $\bar{\mathscr{R}}(t) = \mathscr{R}\, g(t)$. Substitution gives $\bar{\mathscr{R}}_1(t)\,\bar{\gamma} = \mathscr{R}_1\gamma\, g(t)$. Right multiplication by the matrix $\bar{\gamma}^{-1}$ yields $\bar{\mathscr{R}}_1(t) = \mathscr{R}_1\gamma\, g(t)\, \bar{\gamma}^{-1}$; the result is as follows:

1.14 Lemma. The change of frame in the moving or fixed space effects the curve in $O(3)$ describing the motion by right or left multiplication by a constant matrix from $O(3)$.

It is obvious that properties of motion are only such properties of the curve $g(t)$ on the group $O(3)$ which do not depend on the arbitrary choice of a frame in $\bar{E}_3$ or in E_3, i.e. only those properties of curves on the group $O(3)$ which do not change under right or left translations. Besides, kinematic geometry considers only geometric properties of spherical motion, i.e. it does not consider the behaviour of motion in dependence on time (that is on the parameter t). It treats only curves generated by motion. Thus, we can say: kinematic geometry of spherical motion treats those properties of curves on the group $O(3)$ which do not change under either right or left translations on the group $O(3)$, and not even under the change of the parameter of a curve. (A change of parameter is any function of one variable which has its first derivative everywhere different from zero.)

Kinematics of spherical motion considers those properties of curves on the group $O(3)$ which do not change under translations from the group $O(3)$. This means that every kinematic–geometrical property of spherical motion is a kinematic property as well. The converse is not true, however. We would like to illustrate the above considerations by a simple example:

1.15 Example. Let us consider motion given by the matrix

$$g(t) = \begin{pmatrix} \cos(t^2+4), & -\sin(t^2+4), & 0 \\ \sin(t^2+4), & \cos(t^2+4), & 0 \\ 0, & 0, & 1 \end{pmatrix}, \quad t \in (-\infty, +\infty).$$

Putting $t_1 = t^2 + 4$ we obtain a new motion

$$g_1(t_1) = \begin{pmatrix} \cos t_1, & -\sin t_1, & 0 \\ \sin t_1, & \cos t_1, & 0 \\ 0, & 0, & 1 \end{pmatrix}, \quad t_1 \in (4, +\infty).$$

The function $t_1 = t^2 + 4$ is not a change of parameter, naturally, since $dt_1/dt = 2t$ so that $dt_1/dt = 0$ for $t = 0$. This belongs to the domains of both motions. Thus, let us limit the domains to $t \in (0, +\infty)$, $t_1 \in (4, \infty)$. Then the function $t_1 = t^2 + 4$ is a change of parameter. The difference between kinematic geometry and kinematics of motion now means that from the position of kinematics the motions $g(t)$ and $g_1(t_1)$ are two distinct motions while from the point-of-view of kinematic geometry they are one and the same motion – namely rotation about the axis $\boldsymbol{u}_3$. Trajectories of this motion are circles in planes which are perpendicular to the axis $\boldsymbol{u}_3$. From the position of kinematic geometry it is not interesting whether they were generated by a "faster" motion $g(t)$ or a "slower" motion $g_1(t_1)$. However, this fact is of interest when studying motion from the kinematic point of view. The instant which was troublesome for the change of parameter (i.e., $t = 0$) was the instant at which the sense of rotation changes. Such instants were excluded already when defining motion.

In the text which follows we will treat, as a rule, the geometrical properties of motion first, i.e. kinematic geometry first, and only then its kinematic properties.

2. VARIOUS WAYS OF REPRESENTING MOTION, EULER ANGLES

In the preceding section it was shown that motion is given as a regular curve on the group $O(3)$, i.e. as an orthogonal matrix depending on a single parameter. In Chapter I, in Example 5.66 and 5.67, an isomorphism was established between the groups $O(3)$, ad J and $S\,U(2)$ (even though not one-to-one). This yields the possibility of representing spherical motion in three ways: by an orthogonal matrix, a quaternion, or a 2×2 complex matrix. Let us now introduce relations which make it possible to pass from one representation to another.

A. *Relations Between the Groups* $O(3)$, ad J, $S\,U(2)$

a) There exists a one-to-one isomorphism σ between the group J and the group $S\,U(2)$. It was described in I,5.67: Let

$$\boldsymbol{\alpha} = \cos\frac{\varphi}{2} + \boldsymbol{\varepsilon}\sin\frac{\varphi}{2}$$

be a normed quaternion,

$$\boldsymbol{\varepsilon} = x^1\mathbf{i} + x^2\mathbf{j} + x^3\mathbf{k}\,.$$

Then

$$\sigma(\boldsymbol{\alpha}) = \begin{bmatrix} \cos\frac{\varphi}{2} + \mathrm{i}x^1\sin\frac{\varphi}{2}, & x^2\sin\frac{\varphi}{2} + \mathrm{i}x^3\sin\frac{\varphi}{2} \\ -x^2\sin\frac{\varphi}{2} + \mathrm{i}x^3\sin\frac{\varphi}{2}, & \cos\frac{\varphi}{2} - \mathrm{i}x^1\sin\frac{\varphi}{2} \end{bmatrix} \in S\,U(2)\,.$$

b) Conversely, let $u \in S\,U(2)$ be given. Let us find $\sigma^{-1}(u)$. We write

$$u = \begin{pmatrix} z_1, & z_2 \\ -\bar{z}_2, & \bar{z}_1 \end{pmatrix}.$$

By comparison we obtain the following relations for the quaternion $\boldsymbol{\alpha} = \cos(\frac{1}{2}\varphi) + \boldsymbol{\varepsilon}\sin(\frac{1}{2}\varphi)$:

$$\cos\frac{\varphi}{2} = \tfrac{1}{2}(z_1 + \bar{z}_1)\,, \quad \sin\frac{\varphi}{2} = \tfrac{1}{2}[4 - (z_1 + \bar{z}_1)^2]^{1/2}\,,$$

$$x^1 \sin\frac{\varphi}{2} = \frac{z_1 - \bar{z}_1}{2i}\,, \quad x^2 \sin\frac{\varphi}{2} = \frac{z_2 + \bar{z}_2}{2}\,, \quad x^3 \sin\frac{\varphi}{2} = \frac{z_2 - \bar{z}_2}{2i}\,.$$

[Formally we obtain two solutions: $\cos(\frac{1}{2}\varphi)$, $\sin(\frac{1}{2}\varphi)$, $\boldsymbol{\varepsilon}$, and $\cos(\frac{1}{2}\varphi)$, $-\sin(\frac{1}{2}\varphi)$, $-\boldsymbol{\varepsilon}$. However, both yield the same quaternion.]

c) Now, let

$$\boldsymbol{\alpha} = \cos\frac{\varphi}{2} + \boldsymbol{\varepsilon}\sin\frac{\varphi}{2}$$

be a normed quaternion again. By I,5.66 we know that the corresponding matrix is $g = (a_i^j) \in O(3)$ where

$$a_i^j = \delta_i^j \cos\varphi + (1 - \cos\varphi)\, x^i x^j + \varepsilon_{ijk} x^k \sin\varphi\,. \tag{2.1}$$

d) Conversely, let the matrix $(a_i^j) \in O(3)$ be given, and let us find the normed quaternion corresponding to this matrix. This means that equations (2.1) have to be solved for the unknown φ, x^1, x^2, x^3; recall that

$$\operatorname{Tr} a = a_i^i = a_1^1 + a_2^2 + a_3^3\,.$$

(2.1) yields

$$\begin{aligned}\operatorname{Tr} a = a_i^i &= \delta_i^i \cos\varphi + (1 - \cos\varphi)\, x^i x^i + \varepsilon_{iik} x^k \sin\varphi \\ &= \delta_i^i \cos\varphi + (1 - \cos\varphi) \sum_{i=1}^{3} (x^i)^2 = 3\cos\varphi + (1 - \cos\varphi) \\ &= 1 + 2\cos\varphi\,,\end{aligned}$$

so that

$$\operatorname{Tr} a = 1 + 2\cos\varphi\,, \quad \text{i.e.} \quad \cos\varphi = \tfrac{1}{2}(\operatorname{Tr} a - 1)\,.$$

For $\sin\varphi$ we now apply the relation $\sin\varphi = \pm(1 - \cos^2\varphi)^{1/2}$ where we substitute for $\cos\varphi$. For the angle φ we then have the following possibilities: φ, $\varphi + 2\pi$, $-\varphi$, $-\varphi - 2\pi$. (The angles $\varphi + 2\pi$, $-\varphi - 2\pi$ must also be considered since we will pass to half the angle.) Further, we have

$$\begin{aligned} a_i^j - a_j^i &= \\ &= \delta_i^j \cos\varphi - \delta_j^i \cos\varphi + (1 - \cos\varphi)(x^i x^j - x^j x^i) \\ &\quad + (\varepsilon_{ijk} x^k - \varepsilon_{jik} x^k) \sin\varphi \\ &= (\varepsilon_{ijk} x^k + \varepsilon_{ijk} x^k) \sin\varphi = 2\varepsilon_{ijk} x^k \sin\varphi , \end{aligned}$$

i. e. we obtain the equations

$$a_i^j - a_j^i = 2\varepsilon_{ijk} x^k \sin\varphi ,$$

from which we easily compute x^k since we obtain, after writing the equations out, e.g. $a_1^2 - a_1^2 = 2x^3 \sin\varphi$. Symbolically it is possible to write $2x^k \sin\varphi = \varepsilon_{ijk} a_i^j$. Finally we have to ascertain the number of solutions. Note first that to the angles φ and $-\varphi$ the same quaternion corresponds. Indeed, we then have $\cos\varphi = \cos(-\varphi)$, $\sin\varphi = -\sin(-\varphi)$, and the vectors $\boldsymbol{\varepsilon}$ differ only in their signs so the resulting quaternions are the same. It remains to consider the angles φ and $\varphi + 2\pi$. We have $\sin(\varphi + 2\pi) = \sin\varphi$ so that the vectors $\boldsymbol{\varepsilon}$ are equal. But $\cos(\frac{1}{2}\varphi + \pi) = -\cos(\frac{1}{2}\varphi)$, $\sin(\frac{1}{2}\varphi + \pi) = -\sin(\frac{1}{2}\varphi)$ so that two distinct quaternions are obtained as the result: $\boldsymbol{\alpha} = \cos(\frac{1}{2}\varphi) + \boldsymbol{\varepsilon}\sin(\frac{1}{2}\varphi)$, and $-\boldsymbol{\alpha} = -\cos(\frac{1}{2}\varphi) - \boldsymbol{\varepsilon}\sin(\frac{1}{2}\varphi)$. Thus, the correspondence is ambiguous, in fact, as we already know from I,5.66. Finally, let us write the two formulae obtained:

$$\cos\varphi = \tfrac{1}{2}(\operatorname{Tr} a - 1) , \quad x^k \sin\varphi = \tfrac{1}{2} a_i^j \varepsilon_{ijk} .$$

B. *Introducing Euler Angles*

We know that the group $O(3)$ is of dimension 3 as a Lie group since its Lie algebra is constituted by all skew-symmetric matrices of type 3×3. The vector space of all such matrices has dimension 3 (as will be seen below). Every matrix from $O(3)$ has nine elements, however. This means that the members are not independent, and it is difficult to choose from them three independent ones which would constitute a system of local coordinates. If we wish to find a system of local coordinates on the group $O(3)$ we usually proceed otherwise. The most frequently used system of local coordinates – the Euler angles – will be discussed here. They are denoted ψ_1, ψ_2, ψ_3. In introducing them we proceed as follows: Let $\mathscr{R} = \{\boldsymbol{u}_1, \boldsymbol{u}_2, \boldsymbol{u}_3\}$, $\bar{\mathscr{R}} = \{\bar{\boldsymbol{u}}_1, \bar{\boldsymbol{u}}_2, \bar{\boldsymbol{u}}_3\}$ be two arbitrary right-handed frames, i.e. let an arbitrary element $g \in O(3)$ be given such that $\bar{\mathcal{R}} = \mathcal{R}g$. Denote by p the line of intersection of the planes determined by the pairs of vectors $\boldsymbol{e}_1, \boldsymbol{e}_2$ and $\bar{\boldsymbol{e}}_1, \bar{\boldsymbol{e}}_2$ (the case when these planes coincide is easy, we leave its analysis to the reader). Let us rotate the frame $\mathcal{R}$ in the positive sense through the angle ψ_1 about the axis $\boldsymbol{u}_3$ so that the vector $\boldsymbol{u}_1$ lies after rotation on the line p_1. Here we have $0 \leqq \psi_1 < \pi$. In this way we obtain a new frame $\mathcal{R}_1 = \{\boldsymbol{f}_1, \boldsymbol{f}_2, \boldsymbol{f}_3\}$, where

$$\boldsymbol{f}_3 = \boldsymbol{u}_3\,, \quad \boldsymbol{f}_1 = \boldsymbol{u}_1 \cos \psi_1 + \boldsymbol{u}_2 \sin \psi_1\,,$$

$$\boldsymbol{f}_2 = -\boldsymbol{u}_1 \sin \psi_1 + \boldsymbol{u}_2 \cos \psi_1$$

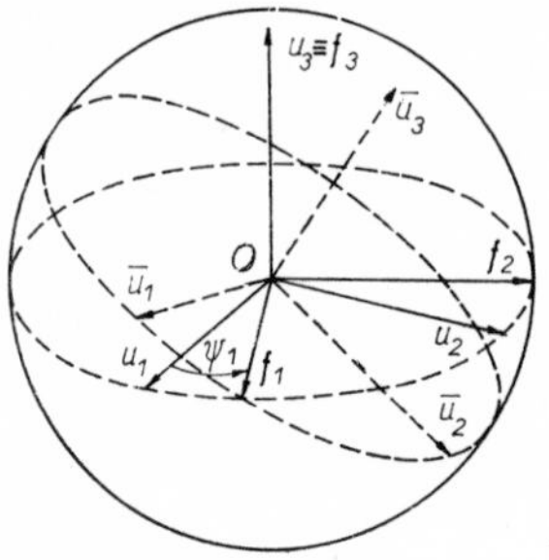

Fig. 13.

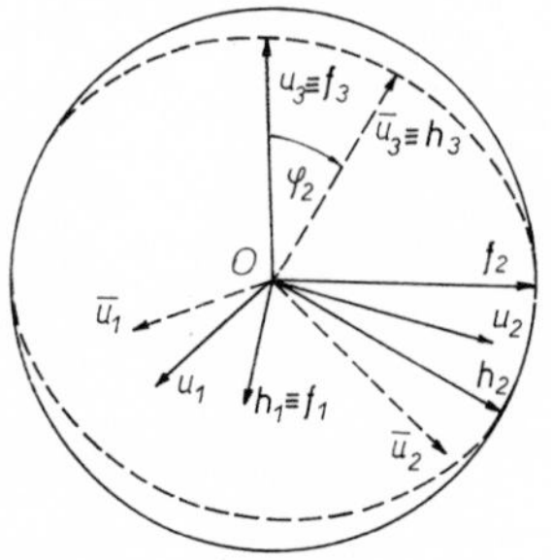

Fig. 14.

(see Fig. 13). On account of the fact that $\boldsymbol{f}_1$ is perpendicular to $\boldsymbol{f}_3$ and to $\bar{\boldsymbol{u}}_3$ as well (since it lies on the straight line p_1, i.e. in the plane determined by the vectors $\bar{\boldsymbol{u}}_1$ and $\bar{\boldsymbol{u}}_2$), it is possible to rotate the frame $\mathcal{R}_1$ about the axis $\boldsymbol{f}_1$ through the angle ψ_2 so that $\boldsymbol{f}_3$ coincides with $\bar{\boldsymbol{u}}_3$. The resulting frame is denoted by $\mathcal{R}_2 = \{\boldsymbol{h}_1, \boldsymbol{h}_2, \boldsymbol{h}_3\}$ where

$$\boldsymbol{h}_1 = \boldsymbol{f}_1\,,\ \boldsymbol{h}_2 = \boldsymbol{f}_2 \cos\psi_2 + \boldsymbol{f}_3 \sin\psi_2\,,\ \boldsymbol{h}_3 = -\boldsymbol{f}_2 \sin\psi_2 + \boldsymbol{f}_3 \cos\psi_2\,;$$

$$\boldsymbol{h}_3 = \bar{\boldsymbol{u}}_3 \quad \text{(see Fig. 14).}$$

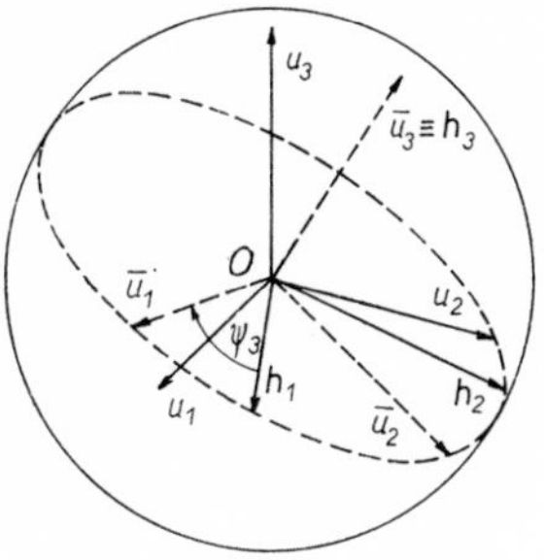

Fig. 15.

Finally, we rotate the frame $\mathscr{R}_2$ about the axis $\boldsymbol{h}_3 = \bar{\boldsymbol{u}}_3$ through the angle ψ_3 so that it coincides with the frame $\bar{\mathscr{R}}$, i.e. (see Fig. 15)

$$\bar{\boldsymbol{u}}_1 = \boldsymbol{h}_1 \cos\psi_3 + \boldsymbol{h}_2 \sin\psi_3\,,$$

$$\bar{\boldsymbol{u}}_3 = -\boldsymbol{h}_1 \sin\psi_3 + \boldsymbol{h}_2 \cos\psi_3\,, \quad \bar{\boldsymbol{u}}_3 = \boldsymbol{h}_3\,.$$

Denote $\mathscr{R}_1 = \mathscr{R} g_1$, $\mathscr{R}_2 = \mathscr{R}_1 g_2$, $\bar{\mathscr{R}} = \mathscr{R}_2 g_3$. Then

$$\bar{\mathscr{R}} = \mathscr{R} g = \mathscr{R} g_1 g_2 g_3\,, \quad \text{so that} \quad g = g_1 g_2 g_3\,,$$

where

$$g_1 = \begin{pmatrix} \cos\psi_1, & -\sin\psi_1, & 0 \\ \sin\psi_1, & \cos\psi_1, & 0 \\ 0, & 0, & 1 \end{pmatrix}, \quad g_2 = \begin{pmatrix} 1, & 0, & 0 \\ 0, & \cos\psi_2, & -\sin\psi_2 \\ 0, & \sin\psi_2, & \cos\psi_2 \end{pmatrix}, \tag{2.2}$$

$$g_3 = \begin{pmatrix} \cos\psi_3, & -\sin\psi_3, & 0 \\ \sin\psi_3, & \cos\psi_3, & 0 \\ 0, & 0, & 1 \end{pmatrix}.$$

If the reader will have sufficient patience to multiply these three matrices he will probably obtain the following result:

$$g = \begin{pmatrix} \cos\psi_1\cos\psi_3 - \sin\psi_1\cos\psi_2\sin\psi_3, & -\cos\psi_1\sin\psi_3 - \sin\psi_1\cos\psi_2\cos\psi_3, & \sin\psi_1\sin\psi_2 \\ \sin\psi_1\cos\psi_3 + \cos\psi_1\cos\psi_2\sin\psi_3, & -\sin\psi_1\sin\psi_3 + \cos\psi_1\cos\psi_2\cos\psi_3, & -\cos\psi_1\sin\psi_2 \\ \sin\psi_2\sin\psi_3, & \sin\psi_2\cos\psi_3, & \cos\psi_2 \end{pmatrix}. \quad (2.3)$$

Furthermore, let us note that the inverse element to the element $g(\psi_1, \psi_2, \psi_3) \in O(3)$ is $g(-\psi_3, -\psi_2, -\psi_1)$, as is obvious from the decomposition $g = g_1 g_2 g_3$, according to (2.2).

2.1 Exercise. Find the Euler angles of the rotation from Example 1.1.

2.2 Example. The orthogonal matrix $g = (a^i_j)$ was represented with the aid of Euler angles ψ_1, ψ_2, ψ_3, where $0 \leqq \psi_1 < \pi$. Let us solve the converse problem now: the elements a^j_i of the orthogonal matrix g are given, we are to find the corresponding Euler angles. From (2.3) we see: $\cos\psi_2 = a^3_3$. This relation determines the angle ψ_2 up to its sign. From the equation $\sin\psi_1 \sin\psi_2 = a^1_3$ we determine the sign of the angle ψ_2 since $0 \leqq \psi_1 < \pi$, i.e. $\sin\psi_1 \geqq 0$. Hence, $\sin\psi_2$ must have the same sign as a^1_3. From the equations

$$\sin\psi_1 \sin\psi_2 = a^1_3\,, \quad -\cos\psi_1 \sin\psi_2 = a^2_3$$

we now uniquely determine the angle ψ_1 (the first equation has two solutions, we select the solution which satisfies the other equation as well). Similarly we find ψ_3 with the aid of the elements a^1_3, a^2_3.

2.3 Example. Motion given by Euler angles can be easily represented with the aid of unitary matrices. In fact, let a matrix $g = g(\psi_1, \psi_2, \psi_3) \in O(3)$ be given by the three Euler angles ψ_1, ψ_2, ψ_3. Then the decomposition (2.2) is: $g = g_1 g_2 g_3$. Thus, if we wish to find for the matrix g its unitary representation it suffices to find the unitary representations for the matrices g_1, g_2, g_3 and to multiply these representations, since the correspondence between the group $O(3)$ and $SU(2)$ discussed in part A is an isomorphism. As we know from (2.2), g_1 and g_3 are rotations about the third axis through the angles ψ_1 and ψ_3, respectively. I.e. to g_1 and g_3 correspond the quaternions

$$\boldsymbol{\alpha}_1 = \cos\frac{\psi_1}{2} + \mathbf{k}\sin\frac{\psi_1}{2} \quad \text{and} \quad \boldsymbol{\alpha}_3 = \cos\frac{\psi_3}{2} + \mathbf{k}\sin\frac{\psi_3}{2},$$

respectively. Analogously, we obtain

$$\boldsymbol{\alpha}_2 = \cos\frac{\psi_2}{2} + \mathbf{i}\sin\frac{\psi_2}{2}.$$

Then the corresponding unitary matrices u_1, u_2, u_3 are, by Aa),

$$u_1 = \begin{pmatrix} \cos\frac{\psi_1}{2}, & \mathbf{i}\sin\frac{\psi_1}{2} \\ \mathbf{i}\sin\frac{\psi_1}{2}, & \cos\frac{\psi_1}{2} \end{pmatrix},$$

$$u_2 = \begin{pmatrix} \cos\frac{\psi_2}{2} + \mathbf{i}\sin\frac{\psi_2}{2}, & 0 \\ 0, & \cos\frac{\psi_2}{2} - \mathbf{i}\sin\frac{\psi_2}{2} \end{pmatrix} = \begin{pmatrix} e^{\mathrm{i}\psi_2/2}, & 0 \\ 0, & e^{-\mathrm{i}\psi_2/2} \end{pmatrix},$$

$$u_3 = \begin{pmatrix} \cos\frac{\psi_3}{2}, & \mathbf{i}\sin\frac{\psi_3}{2} \\ \mathbf{i}\sin\frac{\psi_3}{2}, & \cos\frac{\psi_3}{2} \end{pmatrix}.$$

Consequently, for $u = u_1u_2u_3$ we have

$$u = \begin{pmatrix} \cos\frac{\psi_1}{2}, & \mathbf{i}\sin\frac{\psi_1}{2} \\ \mathbf{i}\sin\frac{\psi_1}{2}, & \cos\frac{\psi_1}{2} \end{pmatrix} \begin{pmatrix} e^{\mathrm{i}\psi_2/2}, & 0 \\ 0, & e^{-\mathrm{i}\psi_2/2} \end{pmatrix} \begin{pmatrix} \cos\frac{\psi_3}{2}, & \mathbf{i}\sin\frac{\psi_3}{2} \\ \mathbf{i}\sin\frac{\psi_3}{2}, & \cos\frac{\psi_3}{2} \end{pmatrix}.$$

After multiplication

$$u = e^{\mathrm{i}\psi_2/2} \begin{pmatrix} \cos\frac{\psi_1}{2}\cos\frac{\psi_3}{2}, & -\mathbf{i}\cos\frac{\psi_1}{2}\sin\frac{\psi_3}{2} \\ \mathbf{i}\sin\frac{\psi_1}{2}\cos\frac{\psi_3}{2}, & -\sin\frac{\psi_1}{2}\sin\frac{\psi_3}{2} \end{pmatrix}$$

$$+ e^{-i\psi_2/2} \begin{pmatrix} -\sin\frac{\psi_1}{2}\sin\frac{\psi_3}{2}, & \mathbf{i}\sin\frac{\psi_1}{2}\cos\frac{\psi_3}{2} \\ \mathbf{i}\cos\frac{\psi_1}{2}\sin\frac{\psi_3}{2}, & \cos\frac{\psi_1}{2}\cos\frac{\psi_3}{2} \end{pmatrix}.$$

2.4 Exercise. Find the Euler angles of the product of two matrices, i.e. determine ψ_1, ψ_2, ψ_3 from the equation

$$g(\psi_1, \psi_2, \psi_3) = g_1(\psi_1^1, \psi_2^1, \psi_3^1)\, g_2(\psi_1^2, \psi_2^2, \psi_3^2)\,.$$

2.5 Example. Now, let us show on a concrete example how the different means of representing motion look like. For this purpose we choose spherical analogy of the so-called *elliptic motion in the plane.* The elliptic motion in the plane is given as follows (see Fig. 16):

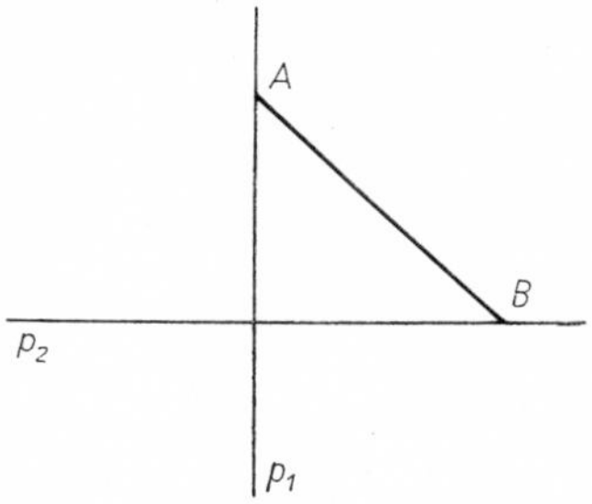

Fig. 16.

The segment $\overline{AB}$ moves so that its endpoints lie all the time on concurrent straight lines p_1, p_2, i.e. $A \in p_1$, $B \in p_2$. For the sake of simplicity, choose the lines p_1 and p_2 perpendicular. Now, consider the spherical analogy of this motion (see Fig. 17). On the unit sphere we choose two perpendicular great circles. Denote them by k_1 and k_2. The fixed frame $\mathscr{R} = \{\boldsymbol{u}_1, \boldsymbol{u}_2, \boldsymbol{u}_3\}$ is chosen by Fig. 17 ($\boldsymbol{u}_3$ lies on the line of intersection of the planes of the two circles while $\boldsymbol{u}_1$ and $\boldsymbol{u}_2$ lie in the planes of the circles k_1 and k_2, respectively). Further, let us choose a spherical segment of length d determined by the vectors $\boldsymbol{u}_3$ and $\boldsymbol{f} = \boldsymbol{u}_3 \cos d + \boldsymbol{u}_1 \sin d$, $0 \neq d \neq \frac{1}{2}\pi$. (By a spherical segment we understand a part of a great circle limited by its two points.) Now, the motion takes place so that the endpoints of the segment d move along the circles k_1 and k_2 (or at least along their parts). This means that $\bar{\boldsymbol{u}}_3$ moves in the plane determined by the vectors $\boldsymbol{u}_2$ and $\boldsymbol{u}_3$ while $\boldsymbol{f}$ moves in the plane determined by the vectors $\boldsymbol{u}_1$ and $\boldsymbol{u}_2$. The angle t of the vectors $\boldsymbol{u}_3$ and $\bar{\boldsymbol{u}}_3$ is chosen as the parameter of motion.

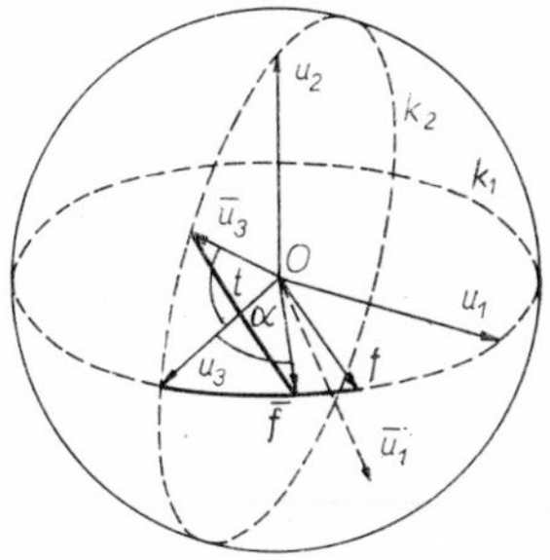

Fig. 17.

For the moving frame $\bar{\mathcal{R}} = \{\bar{\boldsymbol{u}}_1, \bar{\boldsymbol{u}}_2, \bar{\boldsymbol{u}}_3\}$ which is obtained by the movement of the basic frame $\mathcal{R}$ we thus have the following conditions:

$$\bar{\boldsymbol{u}}_3 = \boldsymbol{u}_3 \cos \mathrm{t} + \boldsymbol{u}_2 \sin \mathrm{t}\,, \quad \boldsymbol{f} = \bar{\boldsymbol{u}}_3 \cos d + \bar{\boldsymbol{u}}_1 \sin d\,, \quad (\bar{\boldsymbol{f}}, \boldsymbol{u}_2) = 0\,.$$

These are the fundamental conditions determining our motion. We use them to represent the vectors $\bar{\boldsymbol{u}}_1$ and $\bar{\boldsymbol{u}}_2$:

Write $\bar{\boldsymbol{u}}_1 = \alpha_1 \boldsymbol{u}_1 + \alpha_2 \boldsymbol{u}_2 + \alpha_3 \boldsymbol{u}_3$. We must have $(\bar{\boldsymbol{u}}_1, \bar{\boldsymbol{u}}_1) = 1$, $(\bar{\boldsymbol{u}}_1, \bar{\boldsymbol{u}}_3) = 0$, i.e.

$$\alpha_1^2 + \alpha_2^2 + \alpha_3^2 = 1\,, \quad \alpha_2 \sin \mathrm{t} + \alpha_3 \cos \mathrm{t} = 0\,. \tag{2.4}$$

Into the equation $(\bar{\boldsymbol{f}}, \boldsymbol{u}_2) = 0$ we substitute the representation of the vector $\bar{\boldsymbol{f}}$:

$$\begin{aligned}\bar{\boldsymbol{f}} &= \bar{\boldsymbol{u}}_3 \cos d + \bar{\boldsymbol{u}}_1 \sin d \\ &= (\boldsymbol{u}_3 \cos \mathrm{t} + \boldsymbol{u}_2 \sin \mathrm{t}) \cos d + (\alpha_1 \boldsymbol{u}_1 + \alpha_2 \boldsymbol{u}_2 + \alpha_3 \boldsymbol{u}_3) \sin d \\ &= A \boldsymbol{u}_1 + (\cos d \sin \mathrm{t} + \alpha_2 \sin d)\, \boldsymbol{u}_2 + B \boldsymbol{u}_3\,,\end{aligned}$$

($A, B \in \mathbf{R}$, merely the second coordinate is needed.) Then

$$(\bar{\boldsymbol{f}}, \boldsymbol{u}_2) = \cos d \sin \mathrm{t} + \alpha_2 \sin d = 0\,, \quad \text{i.e.} \quad \alpha_2 = -\sin \mathrm{t} \cot d\,.$$

To simplify further computations we put $\cot d = \lambda$. Then $\alpha_2 = -\lambda \sin \mathrm{t}$.

Equations (2.4) then yield

$$-\lambda \sin^2 \mathrm{t} + \alpha_3 \cos \mathrm{t} = 0\,, \quad \text{i.e.} \quad \alpha_3 = \lambda \frac{\sin^2 \mathrm{t}}{\cos \mathrm{t}}\,,$$

where $\cos t \neq 0$. Further,

$$\alpha_1^2 = 1 - \alpha_2^2 - \alpha_3^2 = 1 - \lambda^2 \sin^2 t - \lambda^2 \frac{\sin^4 t}{\cos^2 t}$$

$$= 1 - \lambda^2 \tan^2 t(\sin^2 t + \cos^2 t) = 1 - \lambda^2 \tan^2 t$$

so that

$$\alpha_1 = (1 - \lambda^2 \tan^2 t)^{1/2}$$

(in the root the plus sign has to be considered since $\alpha_1 = 1$ for $t = 0$). Finally we have

$$\bar{u}_1 = u_1(1 - \lambda^2 \tan^2 t)^{1/2} - u_2\lambda \sin t + u_3\lambda \frac{\sin^2 t}{\cos t},$$

$$\bar{u}_3 = u_2 \sin t + u_3 \cos t .$$

The relation $\bar{u}_2 = \bar{u}_3 \times \bar{u}_1$ yields

$$\bar{u}_2 = \begin{vmatrix} u_1, & u_2, & u_3 \\ 0, & \sin t, & \cos t \\ (1 - \lambda^2 \tan^2 t)^{1/2}, & -\lambda \sin t, & \lambda \dfrac{\sin^2 t}{\cos t} \end{vmatrix}, *$$

so that

$$\bar{u}_2 = u_1\lambda \tan t + u_2(1 - \lambda^2 \tan^2 t)^{1/2} \cos t$$
$$- u_3(1 - \lambda^2 \tan^2 t)^{1/2} \sin t .$$

For the frame $\bar{\mathscr{R}} = \{\bar{u}_1, \bar{u}_2, \bar{u}_3\}$ we have thus obtained the following representation:

$$\bar{u}_1 = u_1(1 - \lambda^2 \tan^2 t)^{1/2} - u_2\lambda \sin t + u_3\lambda \frac{\sin^2 t}{\cos t},$$

$$\bar{u}_2 = u_1\lambda \tan t + u_2(1 - \lambda^2 \tan^2 t)^{1/2} \cos t$$
$$- u_3(1 - \lambda^2 \tan^2 t)^{1/2} \sin t ,$$

$$\bar{u}_3 = u_2 \sin t + u_3 \cos t .$$

* The right-hand side expression is formally considered as a determinant.

Consequently, the matrix representation of our motion is

$$g(t) = \begin{pmatrix} (1 - \lambda^2 \tan^2 t)^{1/2}, & \lambda \tan t, & 0 \\ -\lambda \sin t, & (1 - \lambda^2 \tan^2 t)^{1/2} \cos t, & \sin t \\ \lambda \sin t \tan t, & -(1 - \lambda^2 \tan^2 t)^{1/2} \sin t, & \cos t \end{pmatrix}.$$

Now, we pass to other representations of the motion $g(t)$. First, let us try to find the quaternion function corresponding to the matrix $g(t)$. Write this function in the form

$$\boldsymbol{\alpha}(t) = \cos\frac{\varphi}{2} + \boldsymbol{\varepsilon}\sin\frac{\varphi}{2}, \quad \text{where} \quad \boldsymbol{\varepsilon} = \boldsymbol{\varepsilon}(t), \quad \varphi = \varphi(t)$$

are functions of the angle t. If we wish to find the quaternion function $\boldsymbol{\alpha}(t)$ we have to solve the equations from 2.A.d):

$$\cos\varphi = \tfrac{1}{2}(\operatorname{Tr} g - 1), \quad a_i^j - a_j^i = 2\varepsilon_{ijk}x^k \sin\varphi,$$

where $g = (a_i^j)$, $\boldsymbol{\varepsilon} = x^1\mathbf{i} + x^2\mathbf{j} + x^3\mathbf{k}$.

The solution of these equations yields

$$\begin{aligned}\cos\varphi(t) &= \tfrac{1}{2}[\cos t + (1 - \lambda^2\tan^2 t)^{1/2}\cos t + (1 - \lambda^2 \tan^2 t)^{1/2} - 1] \\ &= \tfrac{1}{2}[(1 - \lambda^2\tan^2 t)^{1/2}(\cos t + 1) + \cos t - 1].\end{aligned}$$

From the equation $\sin^2\varphi = 1 - \cos^2\varphi$ we get

$$\sin^2\varphi = \frac{\sin^2 t}{4}[2(1 - \lambda^2\tan^2 t)^{1/2} + \lambda^2(\cos^{-1} t + 1)^2 + 2].$$

From the familiar relations

$$\cos\frac{\varphi}{2} = \left(\frac{1 + \cos\varphi}{2}\right)^{1/2}, \quad \sin\frac{\varphi}{2} = \left(\frac{1 - \cos\varphi}{2}\right)^{1/2},$$

we represent $\cos(\frac{1}{2}\varphi)$ and $\sin(\frac{1}{2}\varphi)$. Further, we have equations for the normed quaternion $\boldsymbol{\varepsilon}$:

$$\begin{aligned} a_1^2 - a_2^1 &= 2x^3 \sin\varphi, \\ a_1^3 - a_3^1 &= -2x^2\sin\varphi, \\ a_2^3 - a_3^2 &= 2x^1\sin\varphi, \end{aligned}$$

i.e. the equations

$$-\lambda \sin t - \lambda \tan t = 2x^3 \sin \varphi ,$$
$$-\lambda \sin t \tan t = 2x^2 \sin \varphi ,$$
$$-(1 - \lambda^2 \tan^2 t)^{1/2} \sin t - \sin t = 2x^1 \sin \varphi .$$

After substitution for $\sin \varphi$ we obtain the desired coordinates x^1, x^2, x^3. The final expression is somewhat more complicated. We leave to the reader the decision to carry out the computations till the end. Nevertheless, from the presented computation it is obvious that the passage from the quaternion to the unitary representation would be complicated. It is considerably more advantageous to use Euler angles for this case. This is immediately understood upon noting that in the matrix representation of motion we have $a_3^3 = \cos t$ while in the matrix of motion represented by the Euler angles we have $\cos \psi_2$ in this place. Consequently, let us try to represent our motion with the aid of Euler angles. The matrix (2.3) yields the following equations:

$$\cos \psi_2 = \cos t , \quad \sin \psi_1 \sin \psi_2 = 0 , \quad \cos \psi_1 \sin \psi_2 = -\sin t ,$$
$$\sin \psi_2 \sin \psi_3 = \lambda \sin t \tan t ,$$
$$\sin \psi_2 \cos \psi_3 = -(1 - \lambda^2 \tan^2 t)^{1/2} \sin t .$$

From the first equation we have $\psi_2 = \pm t$. The second equation implies $\sin \psi_1 = 0$, i.e. $\psi_1 = 0$, since $0 \leqq \psi_1 < \pi$. The third equation yields $\psi_2 = -t$. The fourth equation will assume the form

$$\sin \psi_3 = -\lambda \tan t ;$$

the last equation is

$$\cos \psi_3 = (1 - \lambda^2 \tan^2 t)^{1/2} ,$$

so that $\cos \psi_3 \geqq 0$, i.e. $-\frac{1}{2}\pi \leqq \psi_3 \leqq \frac{1}{2}\pi$, so that

$$\psi_3 = \arcsin (-\lambda \tan t) .$$

Consequently, it is possible to represent our motion by Euler angles as follows:

$$\psi_1 = 0 , \quad \psi_2 = -t , \quad \psi_3 = \arcsin (-\lambda \tan t) .$$

As a matter of fact, the motion $g(t)$ is thus defined as a curve on the group $O(3)$ given by parametric equations for the coordinates ψ_1, ψ_2, ψ_3.

Now it is easy to find the unitary matrix $u(\mathrm{t})$ which determines our motion. By Example 2.3 we have

$$u = \begin{pmatrix} e^{-it/2}\cos\dfrac{\psi_3}{2}, & \mathrm{i}e^{-it/2}\sin\dfrac{\psi_3}{2} \\ \\ \mathrm{i}e^{it/2}\sin\dfrac{\psi_3}{2}, & e^{it/2}\cos\dfrac{\psi_3}{2} \end{pmatrix};$$

taking account of the fact that $\cos\psi_3 = (1 - \lambda^2\tan^2\mathrm{t})^{1/2}$ we obtain

$$\cos\frac{\psi_3}{2} = \left[\frac{1 + (1 - \lambda^2\tan^2\mathrm{t})^{1/2}}{2}\right]^{1/2},$$

$$\sin\frac{\psi_3}{2} = \pm\left[\frac{1 - (1 - \lambda^2\tan^2\mathrm{t})^{1/2}}{2}\right]^{1/2},$$

since $-\frac{1}{4}\pi \leqq \frac{1}{2}\psi_3 \leqq \frac{1}{4}\pi$.

From the theoretical point-of-view we have at our disposal three equivalent approaches to the representation of motion: orthogonal matrices, normed quaternions, and unitary matrices. Example 2.5 shows that for practical calculations it can happen that the representation of motion with the aid of one approach is simpler than with the aid of another. However, theoretical considerations will be performed for one approach only, namely for the representation of motion by orthogonal matrices. The reader who studied Chapter I carefully will easily find the eventual transfer relations from one approach to another. Nevertheless, we shall explicitly derive all formulae for the representation of motion with the aid of Euler angles since this representation is advantageous for the computation of some examples.

3. DIRECTING CONES, CENTRODES, AND INVARIANTS OF SPHERICAL MOTION

Now, let us proceed with the theory of spherical motion. From Theorem I,5.51 we know that $O(3)$ is a Lie group. Its Lie algebra is the algebra of all skew-symmetric matrices as ascertained earlier. This algebra was denoted by $\mathfrak{O}(3)$. In $\mathfrak{O}(3)$ we shall first find the bracket operation and compute the Killing invariant bilinear form. The procedure is similar to that of Example I, 5.66 – we repeat it in brief. Let $X \in \mathfrak{O}(3)$ be an arbitrary vector. Then X is a matrix of the form

$$X = \begin{pmatrix} 0, & x_2^1, & x_3^1 \\ -x_2^1, & 0, & x_3^2 \\ -x_3^1, & -x_3^2, & 0 \end{pmatrix}.$$

It is advantageous to write the vectors from $\mathfrak{O}(3)$ in coordinate form as follows:

Choose the base X_1, X_2, X_3 of algebra $\mathfrak{O}(3)$ by the matrices

$$X_1 = \begin{pmatrix} 0, & 0, & 0 \\ 0, & 0, & -1 \\ 0, & 1, & 0 \end{pmatrix}, \quad X_2 = \begin{pmatrix} 0, & 0, & 1 \\ 0, & 0, & 0 \\ -1, & 0, & 0 \end{pmatrix}, \quad X_3 = \begin{pmatrix} 0, & -1, & 0 \\ 1, & 0, & 0 \\ 0, & 0, & 0 \end{pmatrix}.$$

Then a vector $X \in \mathfrak{O}(3)$ can be written in the form

$$X = \begin{pmatrix} 0, & -x^3, & x^2 \\ x^3, & 0, & -x^1 \\ -x^2, & x^1, & 0 \end{pmatrix}.$$

The two expressions are connected by the relation $x_j^i = -\varepsilon_{ijk} x^k$, in coordinate form we can write $X = x^1 X_1 + x^2 X_2 + x^3 X_3$. Further, let us compute the bracket $[X, Y]$ of two vectors $X, Y \in \mathfrak{O}(3)$ where

$$Y = \begin{pmatrix} 0, & -y^3, & y^2 \\ y^3, & 0, & -y^1 \\ -y^2, & y^1, & 0 \end{pmatrix}.$$

By I, 5.56 we know that

$$[X, Y] = XY - YX = \begin{pmatrix} 0, & x^2y^1 - x^1y^2, & x^3y^1 - x^1y^3 \\ x^1y^2 - x^2y^1, & 0, & x^3y^2 - x^2y^3 \\ x^1y^3 - x^3y^1, & x^2y^3 - x^3y^2, & 0 \end{pmatrix}.$$

Investigating the resulting matrix we see that its element are the coordinates of the cross product of the vectors $X = x^i X_i$ and $Y = y^i X_i$ in the orthonormal base X_i *, i.e. it is possible to write $[X, Y] = X \times Y$. Consequently, we may view the vectors from $\mathfrak{O}(3)$ in two ways – as skew-symmetric

* Unless explicitly stated, i runs over the subscripts 1, 2, 3. The same is also true in the further exposition.

matrices or as vectors of the usual three-dimensional vector space. In what follows we will use both of these possibilities according to which of the two will be more advantageous in the given case.

Further, let us compute the Killing bilinear form $\mathbf{K}(X, Y)$. By definition we have $\mathbf{K}(X, Y) = \mathrm{Tr}\,(\mathrm{Ad}X\,\mathrm{Ad}Y)$, where $\mathrm{Ad}X(Z) = (X, Z) = X \times Z$. This means that if we wish to compute $\mathbf{K}(X, Y)$ we have to use the operator $\mathrm{Ad}X\,\mathrm{Ad}Y$. This is a linear operator which assigns to every vector $Z \in \mathfrak{O}(3)$ the vector $\mathrm{Ad}X\,\mathrm{Ad}Y(Z) = [X, [Y, Z]] = X \times (Y \times Z)$. If we are to find the matrix of the operator in some base, we have to let this operator act upon the vectors of this base. The coordinates of the images of the vectors of the base then yield the elements of the matrix of the operator, as we know from Section 1. Thus, if (a_i^j) is the matrix of the operator $\mathrm{Ad}X\,\mathrm{Ad}Y$ in the base X_i, then $\mathrm{Ad}X\,\mathrm{Ad}Y(X_i) = a_i^j X_j$. Consequently, let X_i be the base of $\mathfrak{O}(3)$ defined above, $X = x^i X_i$, $Y = y^i X_i$; let us denote by $(X, Y) = x^i y^i$ the usual inner product. Then similarly as in Example I, 5.66 we get

$$\mathrm{Ad}X\,\mathrm{Ad}Y(X_i) = [x^i y^j - (X, Y)\,\delta_i^j]\,X_j$$

and

$$\mathrm{Tr}\,(\mathrm{Ad}X\,\mathrm{Ad}Y) = -2(X, Y)\,.$$

This means that the usual inner product can be considered as the Killing invariant bilinear form of $\mathfrak{O}(3)$ (the factor -2 can be neglected as unimportant for our considerations). Let us summarize the results into the following theorem:

3.1 Theorem. *The Lie algebra $\mathfrak{O}(3)$ of the group of rotations of the Euclidean space with fixed origin is constituted by the usual three-dimensional vector space, the bracket is the cross product. The Killing bilinear form is the inner product. The group* $\mathrm{Ad}\,O(3)$ *is the group of all mappings of the three-dimensional vector space $\mathfrak{O}(3)$ which preserve the inner product.*

Remark. Theorem 3.1 holds for dimension three only, for higher dimensions no similar theorem exists!

3.2 Definition. Let $g(t)$ be a spherical motion. Denote by $g\mathrm{d}/\mathrm{d}t$ the tangent vector of the curve $g(t)$ at the point t. The conical surface determined by the vector function $(L_{g^{-1}})'\,\mathrm{d}g/\mathrm{d}t$ is called the *moving directing conical surface of the spherical motion* $g(\mathrm{t})$, and we denote it by $\bar{\boldsymbol{r}}(\mathrm{t})$. Similarly we define the *fixed directing conical surface of the spherical motion* $g(\mathrm{t})$ as the conical surface determined by the vector function $\boldsymbol{r}(\mathrm{t}) = (R_{g^{-1}})' \cdot (\mathrm{d}g/\mathrm{dt})$. The conical surfaces $\boldsymbol{r}(\mathrm{t})$ and $\bar{\boldsymbol{r}}(\mathrm{t})$ thus lie in the vector space $\mathfrak{O}(3)$. We recall that L'_g and R'_g stand for the differentials of the left and right translations, respectively.

Remark. We find out easily that in every group G of matrices [and thus also in $O(3)$] we have $(L_{g-1})' \boldsymbol{X}_g = g^{-1}\boldsymbol{X}_g$, $(R_{g^{-1}})' \boldsymbol{X}_g = \boldsymbol{X}_g g^{-1}$, where $\boldsymbol{X}_g \in T_g(G)$ (see I, 5.12). In particular, in $O(3)$ we have $g^{-1} = g^{\mathrm{T}}$ so that for the vector functions $\bar{\boldsymbol{r}}(\mathrm{t})$ and $\boldsymbol{r}(\mathrm{t})$ concrete expressions are obtained in the form

$$\bar{\boldsymbol{r}}(\mathrm{t}) = g^{\mathrm{T}} \frac{\mathrm{d}g}{\mathrm{dt}}, \quad \boldsymbol{r}(\mathrm{t}) = \frac{\mathrm{d}g}{\mathrm{dt}} g^{\mathrm{T}}.$$

Further, let us investigate how the directing conical surfaces change if the frame in the fixed and moving systems changes. Thus, let

$$g_1(\mathrm{t}) = \gamma_1 \, g(\mathrm{t}) \, \gamma_2 \, .$$

Then

$$\boldsymbol{r}(\mathrm{t})_1 = [\gamma_1 \, g(\mathrm{t}) \, \gamma_2]^{\mathrm{T}} \frac{\mathrm{d}}{\mathrm{dt}} [\gamma_1 \, g(\mathrm{t}) \, \gamma_2] = \gamma_2^{\mathrm{T}} \, g(\mathrm{t})^{\mathrm{T}} \, \gamma_1^{\mathrm{T}} \gamma_1 \left(\frac{\mathrm{d}g}{\mathrm{dt}}\right) \gamma_2$$

$$= \gamma_2^{\mathrm{T}} \, g(\mathrm{t})^{\mathrm{T}} \left(\frac{\mathrm{d}g}{\mathrm{dt}}\right) \gamma_2 = \gamma_2^{\mathrm{T}} \, \bar{\boldsymbol{r}}(\mathrm{t}) \, \gamma_2 = \mathrm{ad}\gamma_2^{-1} \, \bar{\boldsymbol{r}}(\mathrm{t}) \, ,$$

i.e.

$$\bar{\boldsymbol{r}}(\mathrm{t})_1 = \mathrm{ad}\gamma_2^{-1} \, \bar{\boldsymbol{r}}(\mathrm{t}) \, .$$

Analogously we obtain

$$\boldsymbol{r}(\mathrm{t})_1 = \mathrm{ad}\gamma_1 \, \boldsymbol{r}(\mathrm{t}) \, .$$

Consequently, the directing conical surfaces change by way of the mapping $\mathrm{ad}\,O(3)$. This means that only those properties of the directing conical surfaces or of the directing vector functions $\bar{\boldsymbol{r}}(\mathrm{t})$ and $\boldsymbol{r}(\mathrm{t})$ which do not depend on the mapping $\mathrm{ad}\,O(3)$ can stand out as kinematic properties of motion.

Further, let us investigate what happens if the parameter of motion changes. Thus, let t = t(τ) be a change of the parameter. Then we have

$$\bar{\boldsymbol{r}}(\tau) = g^{\mathrm{T}}(t(\tau))\,\frac{\mathrm{d}g(t(\tau))}{\mathrm{d}\tau} = g^{\mathrm{T}}(t(\tau))\,\frac{\mathrm{d}g(t)}{\mathrm{d}t}\,\frac{\mathrm{d}t}{\mathrm{d}\tau} = \bar{\boldsymbol{r}}(t(\tau))\,\frac{\mathrm{d}t}{\mathrm{d}\tau}$$

or, more precisely,

$$\bar{\boldsymbol{r}}(\tau) = \bar{\boldsymbol{r}}(t)\big|_{t=t(\tau)}\,\frac{\mathrm{d}t}{\mathrm{d}\tau}\,.$$

Similarly we obtain

$$\boldsymbol{r}(\tau) = \boldsymbol{r}(t)\big|_{t=t(\tau)}\,\frac{\mathrm{d}t}{\mathrm{d}\tau}\,.$$

This means that with a change of the parameter both vector functions $\bar{\boldsymbol{r}}(t)$ and $\boldsymbol{r}(t)$ are multiplied by the same factor. This implies that the unit vector of the same direction as $\boldsymbol{r}(t)$ or $\bar{\boldsymbol{r}}(t)$ yields some property of motion since the magnitude of the vector is independent of the mappings from $\mathrm{ad}\,O(3)$, as was shown in Theorem 3.1. This leads to the following definition. Recall that $\|\boldsymbol{X}\| = (\boldsymbol{X}, \boldsymbol{X})^{1/2}$ denotes the length of the vector $\boldsymbol{X}$.

3.3 Definition. The *moving directing cone* $\bar{\boldsymbol{R}}(t)$ of the motion $g(t)$ is the vector function

$$\bar{\boldsymbol{R}}(t) = \frac{\bar{\boldsymbol{r}}(t)}{\|\bar{\boldsymbol{r}}(t)\|}\,,$$

the *fixed directing cone* of the motion $g(t)$ is the vector function

$$\boldsymbol{R}(t) = \frac{\boldsymbol{r}(t)}{\|\boldsymbol{r}(t)\|}\,.$$

Remark. Do not forget that motion was defined as a regular mapping, which means that $\mathrm{d}g/\mathrm{d}t \neq 0$, so that $\bar{\boldsymbol{r}}(t) = (L_{g^{-1}})'\,(\mathrm{d}g/\mathrm{d}t)$ and $\boldsymbol{r}(t) = (R_{g^{-1}})'\,(\mathrm{d}g/\mathrm{d}t)$ are non-zero. Consequently, we have $\|\bar{\boldsymbol{r}}(t)\| \neq 0$ and $\|\boldsymbol{r}(t)\| \neq 0$ as well.

3.4 Assertion.

a) $\boldsymbol{r}(t) \;= \mathrm{ad}\,g(t)\,\bar{\boldsymbol{r}}(t)\,,$

b) $\boldsymbol{R}(\mathrm{t}) = \operatorname{ad}g(\mathrm{t})\,\bar{\boldsymbol{R}}(\mathrm{t})$,
c) $\|\bar{\boldsymbol{r}}(\mathrm{t})\| = \|\boldsymbol{r}(\mathrm{t})\|$.

Proof. The inner product in $\mathfrak{O}(3)$ is a multiple of the Killing form and is thus invariant by $\operatorname{ad}O(3)$, as mentioned above. Equation (a) is obvious since $\boldsymbol{r}(\mathrm{t}) = g'g^{-1} = g(g^{-1}g')\,g^{-1} = \operatorname{ad}g(\mathrm{t})\,\bar{\boldsymbol{r}}(\mathrm{t})$. This equation implies

$$(\boldsymbol{r}(\mathrm{t}), \boldsymbol{r}(\mathrm{t})) = (\operatorname{ad}g(\mathrm{t})\,\bar{\boldsymbol{r}}(\mathrm{t}), \operatorname{ad}g(\mathrm{t})\,\bar{\boldsymbol{r}}(\mathrm{t})) = (\bar{\boldsymbol{r}}(\mathrm{t}), \bar{\boldsymbol{r}}(\mathrm{t}))\,,$$

which yields equation (c). (b) follows from (c) and from Definition 3.3.

3.5 Definition. The motion $\operatorname{ad}g(\mathrm{t})$ in $\mathfrak{O}(3)$ is called the *adjoint motion* of the given motion.

Assertion 3.4 (a) can then be interpreted so that during adjoint motion the rulings of the directing conical surfaces identify themselves successively.

3.6 Example. Let us find the geometrical meaning of adjoint motion. If $\boldsymbol{X}$ is a skew-symmetric matrix, let us denote for the moment by $\boldsymbol{X}^v$ the column of coordinates of the vector which corresponds to this matrix (see the consideration at the beginning of Section 3). Let us show that

$$gXg^{-1} = gX^v$$

for all $g \in O(3)$: express g with the aid of Euler angles. Then $g = g_1 g_2 g_3$, where g_1, g_2, g_3 are the matrices described in Section 2, B. Note also that, e.g.,

$$g_2 g_3 X g_3^{-1} g_2^{-1} = g_2(g_3 X g_3^{-1})\,g_2^{-1}\,, \quad g_2 g_3 X^v = g_2(g_3 X^v)\,,$$

so that it suffices to prove our equality for the components g_1, g_2, g_3 only and for an arbitrary skew-symmetric matrix $\boldsymbol{X}$. For g_1 we have

$$g_1 X^v = \begin{pmatrix} \cos\psi_1, & -\sin\psi_1, & 0 \\ \sin\psi_1, & \cos\psi_1, & 0 \\ 0, & 0, & 1 \end{pmatrix} \begin{pmatrix} x^1 \\ x^2 \\ x^3 \end{pmatrix} = \begin{pmatrix} x^1\cos\psi_1 - x^2\sin\psi_1 \\ x^1\sin\psi_1 + x^2\cos\psi_1 \\ x^3 \end{pmatrix},$$

and

$$g_1 X g_1^{-1} =$$

$$= \begin{pmatrix} \cos\psi_1, & -\sin\psi_1, & 0 \\ \sin\psi_1, & \cos\psi_1, & 0 \\ 0, & 0, & 1 \end{pmatrix} \begin{pmatrix} 0, & -x^3, & x^2 \\ x^3, & 0, & -x^1 \\ -x^2, & x^1, & 0 \end{pmatrix} \begin{pmatrix} \cos\psi_1, & \sin\psi_1, & 0 \\ -\sin\psi_1, & \cos\psi_1, & 0 \\ 0, & 0, & 1 \end{pmatrix}$$

$$= \begin{pmatrix} 0, & -x^3, & x^2 \cos \psi_1 + x^1 \sin \psi_1 \\ x^3, & 0, & x^2 \sin \psi_1 - x^1 \cos \psi_1 \\ -x^2 \cos \psi_1 - x^1 \sin \psi_1, & x^1 \cos \psi_1 - x^2 \sin \psi_1, & 0 \end{pmatrix}.$$

The proof is similar for g_2, the same for g_3.

Example 3.6 shows that ad$g\boldsymbol{X}$ is merely a different notation for $g\boldsymbol{X}^v$, so that ad$g$(t) is only a different notation for the original motion (this circumstance is unique in its sense; later we shall see that already in the case of general space motion it is not true). As far as it will be advantageous we will use adjoint motion instead of the original one, since adjoint motion has the property that it takes place directly in the Lie algebra $\mathfrak{O}(3)$ of the group $O(3)$. Thus we need not look for the relations between the space in which the motion takes place and the Lie algebra of the group $O(3)$.

Now the question arises whether it is not possible to choose directly the parameter of motion so that $\bar{\boldsymbol{r}} = \bar{\boldsymbol{R}}$ and $\boldsymbol{r} = \boldsymbol{R}$, i.e. so that $\|\boldsymbol{r}\| = 1$. Let us try: introduce a change of the parameter $\mathrm{t} = \mathrm{t}(\varphi)$. Then

$$\|\boldsymbol{r}(\varphi)\| = \|\boldsymbol{r}(\mathrm{t})\| \,\big|_{t=t(\varphi)} \left|\frac{\mathrm{dt}}{\mathrm{d}\varphi}\right| .$$

For the parameter φ we thus obtain the condition

$$\left|\frac{\mathrm{dt}}{\mathrm{d}\varphi}\right| = \left(\frac{1}{\|\boldsymbol{r}(\mathrm{t})\|}\right)\bigg|_{t=t(\varphi)},$$

so that – by the theorem on the derivative of the inverse function – we have

$$\left|\frac{\mathrm{d}\varphi}{\mathrm{dt}}\right| = \|\boldsymbol{r}(\mathrm{t})\| , \quad \text{i.e.} \quad |\varphi| = \int \|\boldsymbol{r}(\mathrm{t})\| \,\mathrm{dt} .$$

This leads naturally to the following definition and theorem:

3.7 Definition. The parameter φ for which $\|\boldsymbol{r}(\varphi)\| = 1$ is called the *canonical* parameter.

3.8 Theorem. *The canonical parameter always exists. If φ_1, φ_2 are two canonical parameters, then $\varphi_1 = \pm\varphi_2 + C$, $C = const.$*

Proof. The existence of the canonical parameter is obvious from the preceding exposition. The assumption $\boldsymbol{r}(\mathrm{t}) \neq \boldsymbol{0}$ ensures that $\mathrm{t} = \mathrm{t}(\varphi)$ is a change of the parameter. If φ_1, φ_2 are two canonical parameters, then

$$\left|\frac{d\varphi_1}{dt}\right| = \left|\frac{d\varphi_2}{dt}\right|, \quad \text{i.e.} \quad \varphi_1 = \pm\varphi_2 + C\,.$$

3.9 Remark. From this moment on it is always possible to assume that motion is given as a function of the canonical parameter. This we shall do whenever advantageous. The canonical parameter will always be denoted by φ while an arbitrary parameter will be denoted by t.

3.10 Definition. Let $\bar{\mathscr{R}}(\mathrm{t}) = \mathscr{R}\, g(\mathrm{t})$ be a motion of $\bar{E}_3$ in E_3. The motion originating in the interchange of the moving and fixed systems and in the change of the sense of running through the domain of parameter t is called the *inverse motion* to the given motion.

3.11 Theorem. *The inverse motion to the given motion $g(\mathrm{t})$ is the motion given by the curve $g^{-1}(-\mathrm{t})$. Denote by $\bar{\boldsymbol{R}}^{in}$ and $\boldsymbol{R}^{in}$, respectively, the moving and fixed directing cones of the inverse motion. Then $\bar{\boldsymbol{R}}^{in} = \boldsymbol{R}$ and $\boldsymbol{R}^{in} = \bar{\boldsymbol{R}}$.*

Proof. The first part of the assertion of the theorem is obvious. Further, let us introduce the canonical parameter φ. Differentiation of equation $g(\varphi)\, g^{-1}(\varphi) = e$ with respect to φ yields

$$g'(\varphi)\, g^{-1}(\varphi) + g(\varphi)\,[g^{-1}(\varphi)]' = 0\,.$$

Writing $\tau = -\varphi$ we have

$$\bar{\boldsymbol{R}}^{in}(\varphi) = \bar{\boldsymbol{R}}^{in}(-\tau) = g(-\tau)\frac{d}{d\tau}[g^{-1}(-\tau)]$$

$$= -g(\varphi)\frac{d}{d\varphi}[g^{-1}(\varphi)] = \boldsymbol{R}(\varphi)\,.$$

Similarly, the second assertion is obtained by differentiation of the equation $g^{-1}(\varphi)\, g(\varphi) = e$.

Now we pass to the investigation of the properties of motion. It may be expected that the most simple will be those motions whose directing cones are constant vectors. Therefore, we deal with these first.

3.12 Theorem. $\mathrm{ad}g(\mathrm{t})\,\bar{\boldsymbol{R}}'(\mathrm{t}) = \boldsymbol{R}'(\mathrm{t})$.

Proof. Again, let us choose the canonical parameter as the parameter of motion. Then

$$\boldsymbol{R} = g'g^{-1}\,, \quad \bar{\boldsymbol{R}} = g^{-1}g'\,.$$

Further, we have

$$(\bar{\boldsymbol{R}})' = (g^{-1})'\,g' + g^{-1}g''\,, \quad \mathrm{ad}g(\bar{\boldsymbol{R}})' = g(g^{-1})'\,g'g^{-1} + g''g^{-1}\,,$$

$$\boldsymbol{R}' = g''g^{-1} + g'(g^{-1})'\,.$$

Differentiating again the relation $gg^{-1} = e$ we obtain $g'g^{-1} + g(g^{-1})' = 0$. Substituting into the expression for $\mathrm{ad}g(\bar{\boldsymbol{R}})'$ we obtain

$$\begin{aligned}\mathrm{ad}g(\bar{\boldsymbol{R}})' &= -g(g^{-1})'\,g(g^{-1})' + g''g^{-1} = g'g^{-1}\,g(g^{-1})' + g''g^{-1} \\ &= g'(g^{-1})' + g''g^{-1} = \boldsymbol{R}'\,,\end{aligned}$$

which was to be proved.

Theorem 3.12 expresses the fact that not only the elements of the directing cones but also their tangent planes coincide during adjoint motion. Later we shall prove that these cones revolve. Theorem 3.12 was introduced already at this stage because it is needed for some other purpose.

3.13 Definition. The motion whose fixed directing cone is a constant nonzero vector is called *elementary* motion.

3.14 Remark. From Theorem 3.12 we find out immediately that $\boldsymbol{R}(\mathrm{t}) =$ *const.* implies $\bar{\boldsymbol{R}}(\mathrm{t}) = $ *const.*, since if $\boldsymbol{R}(\mathrm{t}) =$ *const.*, then $\boldsymbol{R}'(\mathrm{t}) = 0$, i.e. $\mathrm{ad}g(\mathrm{t})\,\bar{\boldsymbol{R}}'(\mathrm{t}) = 0$, i.e. $\bar{\boldsymbol{R}}'(\mathrm{t}) = 0$, i.e. $\bar{\boldsymbol{R}}(\mathrm{t}) =$ *const.*

3.15 Theorem. *Elementary motions are all motions of the form* $\gamma_1\,g(\varphi)\,\gamma_2$, *where* $\gamma_1, \gamma_2 \in O(3)$ *are constant matrices and*

$$g(\varphi_1 + \varphi_2) = g(\varphi_1)\,g(\varphi_2)\,.$$

Proof. Let $\boldsymbol{X}$ be a skew-symmetric matrix from $\mathfrak{O}(3)$ which corresponds to a unit vector. We look for all motions whose directing cone is $\boldsymbol{X}$. This means that we have to solve the equation $\boldsymbol{X} = g'g^{-1}$, or $g' = \boldsymbol{X}g$. This is a system of linear differential equations with constant coefficients. Such a system has exactly one solution which satisfies the initial condition $g(0) = \gamma$, where γ is a given matrix. From this it follows that there exists precisely one solution $g(\varphi)$ for which $g(0) = e$. Further, let $h(\varphi)$ be an arbitrary solution; denote $h(0) = \gamma$. Upon substitution we see that $g(\varphi)\,\gamma$ is a solution as well, and $h(0) = g(0)\,\gamma = \gamma$ so that $h(\varphi) = g(\varphi)\,\gamma$ because of the uniqueness.

Thus it suffices to show that $g(\varphi_1 + \varphi_2) = g(\varphi_1)\,g(\varphi_2)$. Consider the functions $g_1(\varphi) = g(\varphi + \varphi_2)$, $g_2(\varphi) = g(\varphi)\,g(\varphi_2)$ for arbitrary φ_2. Then $g_1(\varphi)$ and $g_2(\varphi)$ are also solutions of the equation $g' = \boldsymbol{X}g$ and $g_1(0) = g_2(0) = g(\varphi_2)$. Hence, $g(\varphi_1 + \varphi_2) = g(\varphi_1).\ g(\varphi_2)$ for arbitrary φ_1 and φ_2, again as a consequence of the uniqueness of the solution.

It remains to prove the second part of the theorem, namely that all motions of the form $\gamma_1\,g(\varphi)\,\gamma_2$, where $g(\varphi_1 + \varphi_2) = g(\varphi_1)\,g(\varphi_2)$, are elementary. Obviously, it suffices to show this for $g(\varphi)$. Thus, we have to show that $g'(\varphi)\,g^{-1}(\varphi) = const.$ To this it is sufficient to show that $g'(\varphi) = g'(0)\,g(\varphi)$. First of all, we have $g(0) = g(0 + 0) = g(0)\,g(0)$ so that $g(0) = e$. Further, we have

$$g'(\varphi) = \lim_{t\to 0} \frac{g(\varphi + \mathrm{t}) - g(\varphi)}{\mathrm{t}} = \lim_{t\to 0} \frac{g(\varphi)\,g(\mathrm{t}) - g(\varphi)\,g(0)}{\mathrm{t}}$$

$$= \lim_{t\to 0} \left[g(\varphi)\,\frac{g(\mathrm{t}) - g(0)}{\mathrm{t}} \right] = g(\varphi) \lim_{t\to 0} \frac{g(\mathrm{t}) - g(0)}{\mathrm{t}} = g(\varphi)\,g'(0)\,,$$

which was to be shown.

Since elementary motions will play a significant role in the further exposition let us devote more attention to them. In the proof of Theorem 3.15 we have seen that to find the elementary motion we had to solve a matrix differential equation of the form $g' = Xg$, where X was a constant matrix. The solution of an ordinary differential equation of the same form, i.e. the equation $y'(\varphi) = k\,y(\varphi)$, $k \in \mathbf{R}$, is the function $y = Ce^{k\varphi}$, $C \in \mathbf{R}$. The situation is entirely analogous for the matrix equation. We formally introduce the exponential function for matrices in exactly the same way as in the case of the ordinary exponential function e^{φ}: let X be a square matrix. Then we define

$$\exp X = e + \frac{X}{1!} + \frac{X^2}{2!} + \dots .$$

It can be proved that the series on the right-hand side converges always to some regular matrix.

Further, it can be shown that for the matrix equation $g' = Xg$ with a constant matrix X the matrix function $g(\varphi) = \exp(\varphi X)$ is its solution for which $g(0) = e$.

Further still, $g(\exp X)\, g^{-1} = \exp(gXg^{-1})$ holds for regular g. We show now that any elementary motion in spherical kinematical geometry is a rotation about a fixed axis up to a choice of frame. As a matter of fact, we will show that $\exp(\varphi X)$ for a skew-symmetric matrix X means rotation about the axis determined by the vector X through the angle φ. Since the group ad $O(3)$ acts in the Lie algebra $\mathfrak{O}(3)$ in the same way as $O(3)$ does in E_3 it can be assumed, by I, 5.26, that

$$X = \begin{pmatrix} 0, & -1, & 0 \\ 1, & 0, & 0 \\ 0, & 0, & 0 \end{pmatrix}. \quad \text{Denote } Y = \begin{pmatrix} 1, & 0, & 0 \\ 0, & 1, & 0 \\ 0, & 0, & 0 \end{pmatrix}.$$

Then $X^{2n+1} = (-1)^n X, X^{2n} = (-1)^n Y$, so that $\exp(\varphi X) = \cos\varphi Y + \sin\varphi X$, where the Taylor expansion of the functions cos and sin was used.

This also gives the geometrical interpretation of the canonical parameter as the angle of rotation of an elementary motion.

3.16 Remark. Curves of the form $\exp(\varphi X)$ on the group $O(3)$ are called one-parametric subgroups. They are subgroups, in fact, since for them $g(\varphi_1)\, g(\varphi_2) = g(\varphi_1 + \varphi_2)$, i.e. they are closed with respect to the group operation of matrix multiplication. For the geometry of the group $O(3)$ their significance is similar to the significance of straight lines in the Euclidean space. In this sense, one-parametric subgroups are merely all the straight lines passing through the unit element of the group. All other lines can be obtained from them by displacements, using right translations (it can be shown that there is no difference in applying right or left translations – both cases yield the same resulting set of lines). Thus, all curves on a group which are of the form $g(\varphi)\,\gamma$, $\gamma \in O(3)$, can be viewed as straight lines. The analogue just mentioned will be exploited for the following consideration: If a curve is given e.g. in the plane E_2, then we know how to construct not only the tangent vector of this curve but also its tangent. Here the difference seems to be unimportant, but take into account that in the case of the plane E_2 we identify E_2 with its tangent space $T_p(E_2)$ at arbitrary points $p \in E_2$. This fact starts to be interesting only when we realize that for the group $O(3)$ it is not possible to identify $O(3)$ with its tangent space [the group $O(3)$ is "curved"]. Therefore, we know how to construct the tangent vector to a curve on a group and we call the line constituted by its multiples the tangent. But this tangent lies in the tangent space of the group $O(3)$ at the point of contact and not on the group $O(3)$; thus, it is not a motion. The existence of one-parametric subgroups gives us the possibility of constructing the "tangent" of motion which lies on the group, i.e. the "tangent motion". This will be accomplished in the following definition.

3.17 Definition. Let $g(\mathrm{t})$ be a motion with fixed directing cone $\boldsymbol{R}(\mathrm{t})$. The motion $\{\exp(\varphi \boldsymbol{R}(\mathrm{t}_0))\}\, g(\mathrm{t}_0)$ is called the *instantaneous motion of the motion* $g(\mathrm{t})$ at the time instant $\mathrm{t} = \mathrm{t}_0$.

3.18 Theorem. *Let $g(\varphi)$ be a motion with fixed directing cone $\boldsymbol{R}(\mathrm{t})$, and let $\bar{X}$ be a point from E_3. For the tangent vector $X'(\varphi)$ of the trajectory of the point $\bar{X}$ at the point $X(\varphi)$ we have*

$$X'(\varphi) = \boldsymbol{R}(\varphi)\, X(\varphi)\,.$$

Proof. We have $X(\varphi) = g(\varphi)\,\bar{X}$. This implies

$$X'(\varphi) = g'(\varphi)\,\bar{X} = g'(\varphi)\, g^{-1}(\varphi)\, g(\varphi)\,\bar{X} = \boldsymbol{R}(\varphi)\, X(\varphi)\,.$$

3.19 Corollary of Theorem 3.18. The vector field of tangent vectors of trajectories of points is the Killing vector field determined by the vector $\boldsymbol{R}$ of the fixed directing cone.

3.20 Remark. Thus, the integral curves of the field of the tangent vectors are circles, as familiar from Chapter I. It is also obvious that they are the trajectories of instantaneous motion for a given instant φ_0. Instantaneous motion can also be defined as the motion whose trajectories are the integral curves of the vector field of the tangent vectors of the trajectories at the given moment φ_0 (but this can be done only at this stage, when we know that such a motion exists).

Now we can devote our attention to the properties of instantaneous motion.

3.21 Theorem. *Let $g(t)$ be a spherical motion with directing cones $\bar{R}(t)$ and $R(t)$. Further, let $\gamma(\varphi) = \{\exp(\varphi R(t_0)\} g(t_0)$ be the instantaneous motion of motion $g(t)$ at the moment t_0. Denote by $\bar{S}(\varphi)$, $S(\varphi)$ the directing cones of the instantaneous motion $\gamma(\varphi)$. Then we have $\bar{S}(\varphi) = \bar{R}(t_0)$, $S(\varphi) = R(t_0)$.*

Proof.

$$\begin{aligned}\bar{S}(\varphi) &= [\gamma(\varphi)]^{-1}\gamma'(\varphi) = [g(t_0)]^{-1}\{\exp(\varphi R(t_0))\}^{-1} \\ &\quad . \{\exp(\varphi R(t_0))\}' g(t_0) \\ &= [g(t_0)]^{-1} R(t_0)\, g(t_0) = [\mathrm{ad} g(t_0)]^{-1} R(t_0) = \bar{R}(t_0),\end{aligned}$$

$$S(\varphi) = \{\exp(\varphi R(t_0))\}' g(t_0)[g(t_0)]^{-1}\{\exp(\varphi R(t_0))\}^{-1} = R(t_0).$$

3.22 Assertion. Let the same situation as in Theorem 3.21 be given. Then the following is true: the given motion at the moment t_0 and the instantaneous motion for all φ have the same tangents of their trajectories.

Proof. $X' = S(\varphi) X = R(t_0) X$ by Theorem 3.21.

Let us note that for the tangent vector of the trajectory we now have three equivalent means of notation:

a) $X' = RX$ – the product of the matrix R and the column X,
b) $X' = R \times X$ – the cross product of the vector R and the vector X,
c) $X' = [R, X]$ – the bracket of two matrices R and X, if we use adjoint motion. (3.1)

Here, the result in cases (a) and (b) is a column or a vector in case (c) the result is a skew-symmetric matrix.

3.23 Exercise. Make sure that in all the three cases of formula (3.1) the same result is obtained. First concretely for

$$R = \tfrac{1}{4}\begin{pmatrix} 2 \\ 3 \\ \sqrt{3} \end{pmatrix}, \quad X = \begin{pmatrix} 1 \\ -1 \\ 3 \end{pmatrix}$$

and then generally.

Before proceeding with the theory of spherical motion we first compute several examples.

3.24 Example. Let motion $g(t)$ be given by means of Euler angles, i.e. let $g(t) = g(\psi_1(t), \psi_2(t), \psi_3(t))$. Let us find the directing vector functions of this motion. Write

$$g(t) = g_1(\psi_1(t))\, g_2(\psi_2(t))\, g_3(\psi_3(t)),$$

where g_1, g_2, g_3 are given in (2.2).

For the directing vector function $\boldsymbol{r}(t)$ we obtain

$$\begin{aligned}\boldsymbol{r}(t) &= \{g_1(t)\, g_2(t)\, g_3(t)\}' \{g_1(t)\, g_2(t)\, g_3(t)\}^{-1} \\ &= (g_1 g_2 g_3)'\, g_3^{-1} g_2^{-1} g_1^{-1} \\ &= (g_1' g_2 g_3 + g_1 g_2' g_3 + g_1 g_2 g_3')\, g_3^{-1} g_2^{-1} g_1^{-1} \\ &= g_1' g_1^{-1} + g_1 (g_2' g_2^{-1})\, g_1^{-1} + g_1 g_2 (g_3' g_3^{-1})\, g_2^{-1} g_1^{-1} \\ &= \boldsymbol{r}_1 + \operatorname{ad} g_1 \boldsymbol{r}_2 + \operatorname{ad}(g_1 g_2)\, \boldsymbol{r}_3 = \boldsymbol{r}_1 + \operatorname{ad} g_1 (\boldsymbol{r}_2 + \operatorname{ad} g_2 \boldsymbol{r}_3),\ ^*\end{aligned}$$

where $\boldsymbol{r}_1, \boldsymbol{r}_2, \boldsymbol{r}_3$ are directing functions for the motions g_1, g_2, g_3. We compute easily that in vector notation we have

$$\boldsymbol{r}_1 = \begin{pmatrix} 0 \\ 0 \\ \psi_1' \end{pmatrix}, \quad \boldsymbol{r}_2 = \begin{pmatrix} \psi_2' \\ 0 \\ 0 \end{pmatrix}, \quad \boldsymbol{r}_3 = \begin{pmatrix} 0 \\ 0 \\ \psi_3' \end{pmatrix}.$$

The formula for $\boldsymbol{r}(t)$ assumes in vector notation the form $\boldsymbol{r} = \boldsymbol{r}_1 + g_1(\boldsymbol{r}_2 + g_2 \boldsymbol{r}_3)$, so that we obtain successively

$$g_2 \boldsymbol{r}_3 = \begin{pmatrix} 1, & 0, & 0 \\ 0, & \cos\psi_2, & -\sin\psi_2 \\ 0, & \sin\psi_2, & \cos\psi_2 \end{pmatrix} \begin{pmatrix} 0 \\ 0 \\ \psi_3' \end{pmatrix} = \begin{pmatrix} 0 \\ -\psi_3' \sin\psi_2 \\ \psi_3' \cos\psi_2 \end{pmatrix},$$

$$g_1(\boldsymbol{r}_2 + g_2 \boldsymbol{r}_3) = \begin{pmatrix} \cos\psi_1, & -\sin\psi_1, & 0 \\ \sin\psi_1, & \cos\psi_1, & 0 \\ 0, & 0, & 1 \end{pmatrix} \begin{pmatrix} \psi_2' \\ -\psi_3' \sin\psi_2 \\ \psi_3' \cos\psi_2 \end{pmatrix},$$

$$\boldsymbol{r} = \boldsymbol{r}_1 + g_1(\boldsymbol{r}_2 + g_2 \boldsymbol{r}_3) = \begin{pmatrix} \psi_2' \cos\psi_1 + \psi_3' \sin\psi_1 \sin\psi_2 \\ \psi_2' \sin\psi_1 - \psi_3' \cos\psi_1 \sin\psi_2 \\ \psi_1' + \psi_3' \cos\psi_2 \end{pmatrix}. \tag{3.2}$$

* Here and in the further exposition we will omit the argument in case of more complicated expressions.

For the computation of the moving directing function it suffices to take into account that the element $g = g(\psi_1, \psi_2, \psi_3)$ has the inverse element $g^{-1} = g(-\psi_3, -\psi_2, -\psi_1)$. Thus, we obtain the moving directing function easily as the fixed directing function of the inverse motion, i.e. we substitute in the formula of $\boldsymbol{r}(t)$: $\psi_1 \to -\psi_3, \psi_2 \to -\psi_2, \psi_3 \to -\psi_1$ and change signs so that

$$\bar{\boldsymbol{r}} = \begin{pmatrix} \psi_2' \cos \psi_3 + \psi_1' \sin \psi_2 \sin \psi_3 \\ -\psi_2' \sin \psi_3 + \psi_1' \sin \psi_2 \cos \psi_3 \\ \psi_3' + \psi_1' \cos \psi_2 \end{pmatrix} . \tag{3.3}$$

3.25 Example. Let us find the directing functions and directing cones for the motion of Example 2.7. The motion $g(t)$ was represented with the aid of Euler angles as follows:

$$\psi_1 = 0, \ \psi_2 = -t, \ \psi_3 = \arcsin(-\lambda \tan t)$$

where $\cos \psi_3 = (1 - \lambda^2 \tan^2 t)^{1/2}$. Further, we have

$$\psi_1' = 0, \ \psi_2' = -1, \ \psi_3' = -\frac{\lambda}{(1 - \lambda^2 \tan^2 t)^{1/2} \cos^2 t},$$

so that substitution into formulae (3.2) and (3.3) yields in the frame $\mathcal{R} = \{\boldsymbol{u}_1, \boldsymbol{u}_2, \boldsymbol{u}_3\}$ the following expressions:

$$\boldsymbol{r}(t) = -\boldsymbol{u}_1 - \frac{\lambda}{(1 - \lambda^2 \tan^2 t)^{1/2} \cos^2 t} (\boldsymbol{u}_2 \sin t + \boldsymbol{u}_3 \cos t),$$

$$\boldsymbol{r}(t) = -\boldsymbol{u}_1 (1 - \lambda^2 \tan^2 t)^{1/2} - \boldsymbol{u}_2 \lambda \tan t - \boldsymbol{u}_3 \frac{\lambda}{(1 - \lambda^2 \tan^2 t)^{1/2} \cos^2 t}.$$

From (3.2) we compute also the length of the vector $\boldsymbol{r}(t)$:

$$(\boldsymbol{r}(t), \boldsymbol{r}(t)) = (\psi_1')^2 + (\psi_2')^2 + (\psi_3')^2 + 2\psi_1'\psi_3' \cos \psi_2 .$$

In our case we have

$$(\boldsymbol{r}(t), \boldsymbol{r}(t)) = 1 + \frac{\lambda^2}{(1 - \lambda^2 \tan^2 t) \cos^4 t},$$

i.e.

$$\|\boldsymbol{r}(t)\| = \frac{[\lambda^2 + (1 - \lambda^2 \tan^2 t) \cos^4 t]^{1/2}}{(1 - \lambda^2 \tan^2 t)^{1/2} \cos^2 t}.$$

Now, we obtain the directing cones in the form

$$\boldsymbol{R}(\mathrm{t}) = \frac{-\boldsymbol{u}_1(1 - \lambda^2 \tan^2 \mathrm{t})^{1/2} \cos^2 \mathrm{t} - \boldsymbol{u}_2 \lambda \sin \mathrm{t} - \boldsymbol{u}_3 \lambda \cos \mathrm{t}}{[(1 - \lambda^2 \tan^2 \mathrm{t}) \cos^4 \mathrm{t} + \lambda^2]^{1/2}},$$

$$\bar{\boldsymbol{R}}(\mathrm{t}) = \frac{-\boldsymbol{u}_1(1-\lambda^2 \tan^2 \mathrm{t}) \cos^2 \mathrm{t} - \boldsymbol{u}_2 \lambda (1-\lambda^2 \tan^2 \mathrm{t})^{1/2} \tan \mathrm{t} \cos^2 t - \boldsymbol{u}_3 \lambda}{[(1 - \lambda^2 \tan^2 \mathrm{t}) \cos^4 \mathrm{t} + \lambda^2]^{1/2}}.$$

Note that for the canonical parameter φ we obtain the relation

$$\varphi = \int \left[\frac{(1 - \lambda^2 \tan^2 \mathrm{t}) \cos^4 \mathrm{t} + \lambda^2}{(1 - \lambda^2 \tan^2 \mathrm{t}) \cos^4 \mathrm{t}} \right]^{1/2} \mathrm{dt}$$

$$= \int \left[1 + \frac{\lambda^2}{(1 - \lambda^2 \tan^2 \mathrm{t}) \cos^4 t} \right]^{1/2} \mathrm{dt},$$

which signals that in concrete cases of motions we will probably have to do without the canonical parameter. This fact we shall have in mind when deriving formulae for the computation of curvatures, etc.

3.26 Example. Let us calculate the directing functions for the spherical analogy of *conchoidal plane motion*. Conchoidal plane motion is given as follows: The straight line $\bar{a}$, with a point $\bar{A}$ lying on it, moves in such a way

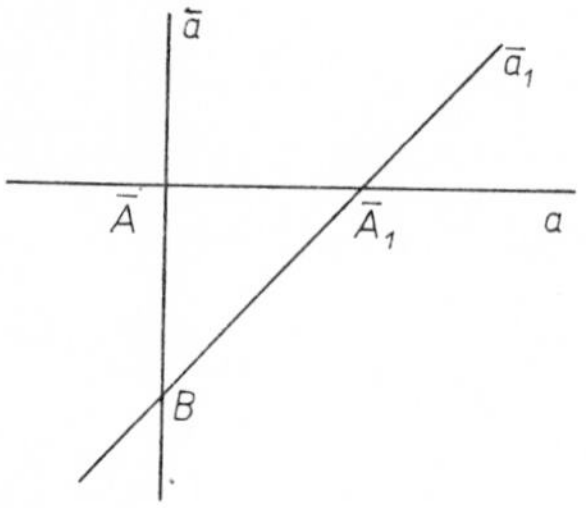

Fig. 18.

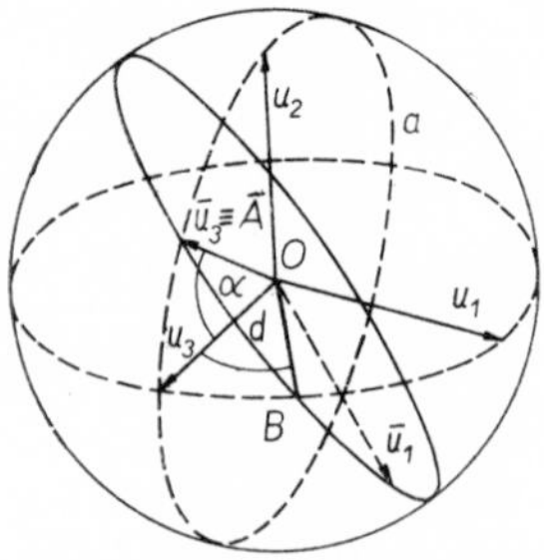

Fig. 19.

that the point $\bar{A}$ lies all the time on a fixed line a while the line $\bar{a}$ passes through a point B all the time (see Fig. 18). The figure shows two positions of the line $\bar{a}$ and the point $\bar{A}$: $\bar{a}$, $\bar{A}$ and $\bar{a}_1$, $\bar{A}_1$. Further, consider the spherical analogy of this motion (see Fig. 19). On the unit sphere the great circle a is given and a point B not lying on a. The basic frame $\mathscr{R} = \{\boldsymbol{u}_1, \boldsymbol{u}_2, \boldsymbol{u}_3\}$

is chosen as shown in Fig. 19 [$\boldsymbol{u}_2$ and $\boldsymbol{u}_3$ lie in the plane of circle a, the plane $(\boldsymbol{u}_1\boldsymbol{u}_3)$ contains point B]. The frame $\mathcal{R}$ moves now in such a way that the vector $\bar{\boldsymbol{u}}_3$ rotates in great circle a while the plane $(\bar{\boldsymbol{u}}_1\bar{\boldsymbol{u}}_3)$ passes through the point B all the time. Let us choose the angle α of the vectors $\boldsymbol{u}_3$ and $\bar{\boldsymbol{u}}_3$ as the parameter of motion. Then

$$\bar{\boldsymbol{u}}_3 = \boldsymbol{u}_3 \cos\alpha + \boldsymbol{u}_2 \sin\alpha .$$

Further, denote

$$\bar{\boldsymbol{u}}_1 = x^1\boldsymbol{u}_1 + x^2\boldsymbol{u}_2 + x^3\boldsymbol{u}_3 .$$

Then $(\bar{\boldsymbol{u}}_1, \bar{\boldsymbol{u}}_3) = 0$, i.e.

$$(\boldsymbol{u}_3 \cos\alpha + \boldsymbol{u}_2 \sin\alpha,\ x^1\boldsymbol{u}_1 + x^2\boldsymbol{u}_2 + x^3\boldsymbol{u}_3) = 0 ,$$

i.e.

$$x^3 \cos\alpha + x^2 \sin\alpha = 0 .$$

Further, let us write $B = \boldsymbol{u}_3 \cos d + \boldsymbol{u}_1 \sin d$. Since the plane $(\bar{\boldsymbol{u}}_1\bar{\boldsymbol{u}}_3)$ has to pass through the point B all the time, the vectors $\bar{\boldsymbol{u}}_1$, $\bar{\boldsymbol{u}}_3$, B must be coplanar, $|\bar{\boldsymbol{u}}_1, \bar{\boldsymbol{u}}_3 B| = 0$. Thus, we obtain the equation

$$\begin{vmatrix} x^1 & x^2, & x^3 \\ 0, & \sin\alpha, & \cos\alpha \\ \sin d, & 0, & \cos d \end{vmatrix} = x^3 \sin\alpha \sin d - x^1 \sin\alpha \cos d - x^2 \cos\alpha \sin d = 0 .$$

Consequently, we have three equations altogether for the unknown x^1, x^2, x^3:

$$\begin{aligned} &(x^1)^2 + (x^2)^2 + (x^3)^2 = 1 , \\ &x^2 \sin\alpha + x^3 \cos\alpha = 0 , \\ &(x^3 \sin\alpha - x^2 \cos\alpha) \sin d - x^1 \sin\alpha \cos d = 0 . \end{aligned}$$

From the second equation we obtain $x^3 = \lambda \sin\alpha$, $x^2 = -\lambda \cos\alpha$ for some $\lambda \in \mathbf{R}$. Let us substitute into the third equation: We obtain

$$\lambda \sin d - x^1 \sin\alpha \cos d = 0 , \quad \text{i.e.} \quad x^1 \sin\alpha = \lambda \tan d .$$

Multiply the first equation by $\sin^2\alpha$. Upon substitution we then have

$$\sin^2\alpha = \lambda^2 \tan^2 d + \lambda^2 \sin^2\alpha = \lambda^2(\sin^2\alpha + \tan^2 d) ,$$

i.e.

$$\lambda = \frac{\pm \sin \alpha}{(\sin^2 \alpha + \tan^2 d)^{1/2}} .$$

Choose the plus sign and denote $\tan d = \mu$, $(\sin^2 \alpha + \tan^2 d)^{-1/2} = M$. Then $\lambda = M \sin \alpha$ and we have the solution

$$x^1 = \mu M , \quad x^2 = -M \sin \alpha \cos \alpha , \quad x^3 = M \sin^2 \alpha ,$$

so that

$$\bar{\boldsymbol{u}}_1 = M(\boldsymbol{u}_1 \mu - \boldsymbol{u}_2 \sin \alpha \cos \alpha + \boldsymbol{u}_3 \sin^2 \alpha) .$$

Further, we have $\bar{\boldsymbol{u}}_2 = \bar{\boldsymbol{u}}_3 \times \bar{\boldsymbol{u}}_1$, i.e.

$$\begin{aligned} \bar{\boldsymbol{u}}_2 &= M \begin{vmatrix} \boldsymbol{u}_1, & \boldsymbol{u}_2, & \boldsymbol{u}_3 \\ 0, & \sin \alpha, & \cos \alpha \\ \mu, & \sin \alpha \cos \alpha, & \sin^2 \alpha \end{vmatrix} \\ &= M(\boldsymbol{u}_1 \sin \alpha + \boldsymbol{u}_2 \mu \cos \alpha - \boldsymbol{u}_3 \mu \sin \alpha) . \end{aligned}$$

The moving frame $\bar{\mathscr{R}} = \{\bar{\boldsymbol{u}}_1, \bar{\boldsymbol{u}}_2, \bar{\boldsymbol{u}}_3\}$ is thus given by the relation

$$\begin{aligned} \bar{\mathscr{R}} = \{ & M(\boldsymbol{u}_1 \mu - \boldsymbol{u}_2 \sin \alpha \cos \alpha + \boldsymbol{u}_3 \sin^2 \alpha) , \\ & M(\boldsymbol{u}_1 \sin \alpha + \boldsymbol{u}_2 \mu \cos \alpha - \boldsymbol{u}_3 \mu \sin \alpha) , \\ & \boldsymbol{u}_2 \sin \alpha + \boldsymbol{u}_3 \cos \alpha \} . \end{aligned}$$

The matrix of motion is then

$$g(\alpha) = \begin{pmatrix} \mu M, & M \sin \alpha, & 0 \\ -M \sin \alpha \cos \alpha, & \mu M \cos \alpha, & \sin \alpha \\ M \sin^2 \alpha, & -\mu M \sin \alpha, & \cos \alpha \end{pmatrix} .$$

The element a_3^3 in the matrix $g(\alpha)$ calls our attention again to the fact that it will be advantageous to exploit Euler angles. From (2.3) we obtain the equations

$$\cos \psi_2 = \cos \alpha , \quad \sin \psi_1 \sin \psi_2 = 0 , \quad -\cos \psi_1 \sin \psi_2 = \sin \alpha ,$$
$$\sin \psi_2 \sin \psi_3 = M \sin^2 \alpha , \quad \sin \psi_2 \cos \psi_3 = -\mu M \sin \alpha ,$$

so that the second equation yields $\psi_1 = 0$, from the first and third we have $\psi_2 = -\alpha$, from the last two we obtain

$$\cos \psi_3 = \mu M , \quad \sin \psi_3 = -M \sin \alpha , \quad \text{so that} \quad \psi_3 = \arccos (\mu M) .$$

Thus, the considered motion is represented with the aid of Euler angles by

$$g(\alpha) = g(\psi_1(\alpha), \psi_2(\alpha), \psi_3(\alpha)),$$

where

$$\psi_1 = 0, \quad \psi_2 = -\alpha, \quad \psi_3 = \arccos(\mu M).$$

The derivatives of the Euler angles are

$$\psi_1' = 0, \quad \psi_2' = -1, \quad \psi_3' = \frac{-\mu M'}{(1 - \mu^2 M^2)^{1/2}},$$

where

$$M' = -\tfrac{1}{2}(\sin^2 \alpha + \mu^2)^{-3/2} \,.\, 2 \sin \alpha \cos \alpha = -M^3 \sin \alpha \cos \alpha ,$$

so that

$$\psi_3' = \frac{-\mu}{M \sin \alpha}(-M^3 \sin \alpha \cos \alpha) = \mu M^2 \cos \alpha ,$$

since

$$1 - \mu^2 M^2 = 1 - \cos^2 \psi_3 = \sin^2 \psi_3 = M^2 \sin^2 \alpha .$$

Upon substituting into formulae (3.2) and (3.3) we obtain

$$\begin{aligned} \boldsymbol{r}(\alpha) &= -\boldsymbol{u}_1 + \mu M^2 \cos \alpha(\boldsymbol{u}_2 \sin \alpha + \boldsymbol{u}_3 \cos \alpha), \\ \bar{\boldsymbol{r}}(\alpha) &= M(-\mu \boldsymbol{u}_1 - \boldsymbol{u}_2 \sin \alpha + \boldsymbol{u}_3 \mu M \cos \alpha). \end{aligned}$$

3.27 Exercise. Find the matrix of the spherical analogue of the following plane motion: The right angle with sides $\bar{a}$ and $\bar{b}$ upon whose side $\bar{b}$ a fixed point $\bar{B}$ is given moves so that side $\bar{a}$ passes through a fixed point M permanently while the point B lies on a fixed given straight line m all the time. Prove, further, that the motion originating in this way is the elliptic motion (i.e. the motion from Example 3.25) and find a point on the sphere, different from $\bar{B}$, which moves along a great circle as well. This point is found easily by comparing the matrix of our motion with the matrix of elliptic motion.

Remark. In plane kinematics the two considered motions differ from each other. (See Fig. 20.)

3.28 Exercise. For the plane motion of Exercise 3.27 the trajectory of point $\bar{A}$ is called the Kappa curve. Find the equation of its spherical analogue.

Result.

$$(x^2x^3)^2 + (x^1)^2\, m^2 = (x^2)^2\,,$$
$$(x^1)^2 + (x^2)^2 + (x^3)^2 = 1\,, \quad m \in \mathbf{R}\,.$$

3.29 Exercise. For the motion of Example 2.5 prove that the only points whose trajectory is a great circle are the end-points of the moving straight line segment and the points lying opposite to them.

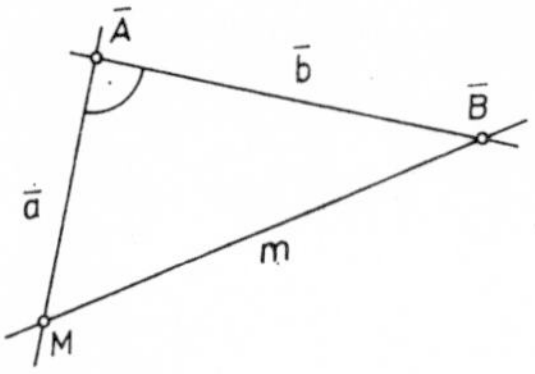

Fig. 20.

Hint. For suitable values of the parameter find three noncolinear points of the trajectory of a point.

3.30 Exercise. Find the representation of the motion which is obtained similarly as the motion of Example 2.5, but with k_1 and k_2 chosen in a general position (thus not necessarily mutually perpendicular).

Till now we have dealt only with trajectories of points on the unit sphere. But all that was said holds for arbitrary points of the space as well. From now on we will deal with motion of the unit sphere $\bar{S}^2$ only (we will not take account of the fact that together with the sphere $\bar{S}^2$ the entire space $\bar{E}_3$ moves). Let us recall that we have the group $O(3)$ with its Lie algebra $\mathfrak{O}(3)$. The curve $g(t)$ on $O(3)$ determines a motion of the unit sphere $\bar{S}^2$, the adjoint motion $\mathrm{ad}g(t)$ is a motion in $\mathfrak{O}(3)$ and thus also a motion on the unit sphere S^2 in $\mathfrak{O}(3)$. Taking into account that $\mathrm{ad}g(t)$ is merely a different representation of motion $g(t)$, it is possible to identify these two spheres (for this reason we have also used the same letter for their notation). Vectors from $\mathfrak{O}(3)$ can then be conceived either as matrices or as vectors. A bracket of two matrices thus yields the same result as the cross product of the corresponding vectors. And if $\boldsymbol{R}$, $\bar{\boldsymbol{R}}$ are the directing cones of motion $g(t)$, then the tangent vector X' of the trajectory $X(t)$ of the point $\bar{X} \in \bar{S}^2$

is given either by the formula $X' = [\boldsymbol{R}, X]$ or by the formula $X' = \boldsymbol{R} \times X$, according to the applied notation. Further, by Assertion 3.5 we have

$$\boldsymbol{R}(\mathrm{t}) = \mathrm{ad}g(\mathrm{t})\,\bar{\boldsymbol{R}}(\mathrm{t})$$

and, by Theorem 3.12,

$$\boldsymbol{R}'(\mathrm{t}) = \mathrm{ad}g(\mathrm{t})\,\bar{\boldsymbol{R}}'(\mathrm{t}),$$

or, for the other type of notation,

$$\boldsymbol{R}(\mathrm{t}) = g(\mathrm{t})\,\bar{\boldsymbol{R}}(\mathrm{t}), \quad \boldsymbol{R}'(\mathrm{t}) = g(\mathrm{t})\,\bar{\boldsymbol{R}}'(\mathrm{t}).$$

3.31 Definition. The curves $\boldsymbol{R}(\mathrm{t})$ on S^2 and $\bar{\boldsymbol{R}}(\mathrm{t})$ on $\bar{S}^2$ are called the *fixed* and the *moving centrode* of motion, respectively.

3.32 Assertion. The fixed centrode is the set of centres of rotation of instantaneous motions, the moving centrode is the set of points which will become these centres successively during the motion.

Proof. The vector field of tangent vectors of the trajectories is simultaneously the field of tangent vectors for instantaneous motion. For the centre of rotation $\boldsymbol{S}$ we must have $\boldsymbol{S}' = 0$, i.e. $[\boldsymbol{R}, \boldsymbol{S}] = 0$, i.e. $\boldsymbol{S} = \pm\boldsymbol{R}$.

3.33 Remark. The reader has certainly noted a certain inconsistency in Assertion 3.32. Naturally, simultaneously with $\boldsymbol{R}(\mathrm{t})$, $-\boldsymbol{R}(\mathrm{t})$ is also an instantaneous centre of rotation for which the same holds as for $\boldsymbol{R}(\mathrm{t})$. Namely, $-\boldsymbol{R}(\mathrm{t})$ is the directing cone of motion $g(-\mathrm{t})$. As centrodes we consider only the curves $\bar{\boldsymbol{R}}(\mathrm{t})$ and $\boldsymbol{R}(\mathrm{t})$ on $\bar{S}^2$ and S^2, respectively. However, we have to keep in mind that together with the curves $\bar{\boldsymbol{R}}(\mathrm{t}), \boldsymbol{R}(\mathrm{t})$ we also have on $\bar{S}^2$, S^2 the curves $-\bar{\boldsymbol{R}}(\mathrm{t})$, $-\boldsymbol{R}(\mathrm{t})$, equivalent with the former curves. Roughly speaking, we can say that the fixed centrode is the set of points on S^2 which have zero "velocity" at some instant, i.e. a null tangent vector. Now, we introduce the following convention: the derivative with respect to the canonical parameter will be denoted by a dash, the derivative with respect to an arbitrary parameter by a dot.

Let now $g(\varphi)$ be a motion with fixed directing cone $\boldsymbol{R}(\varphi)$. If $\boldsymbol{R}(\varphi) \neq 0$, we denote by $\boldsymbol{T}$ the unit vector in the direction of $\boldsymbol{R}'(\varphi)$, i.e. $(\boldsymbol{T}, \boldsymbol{T}) = 1$, $\boldsymbol{T} = \mu(\mathrm{d}\boldsymbol{R}/\mathrm{d}\varphi)$ where $\mu > 0$. Further, denote $\boldsymbol{N} = \boldsymbol{R} \times \boldsymbol{T}$. The frame $\mathscr{F} = \{\boldsymbol{R}, \boldsymbol{T}, \boldsymbol{N}\}$ is called the *Frenet frame* of the fixed directing cone, $\boldsymbol{T}$ is the *unit vector of the tangent*, $\boldsymbol{N}$ is the *unit vector of the spherical normal.* (See Fig. 21.) Analogous notation is introduced for the moving directing cone as well.

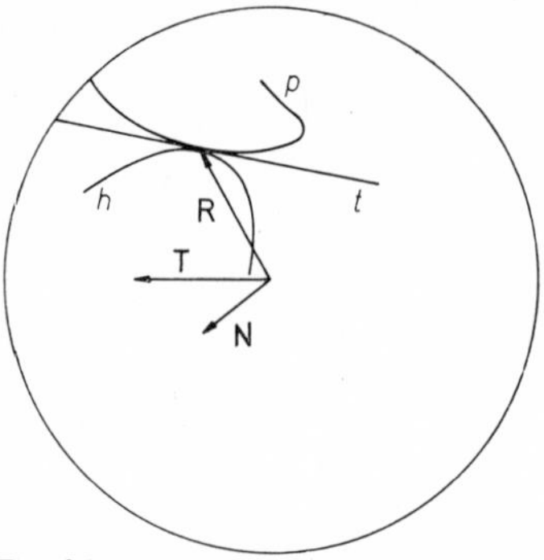

Fig. 21.

3.34 Theorem. *Let $g(\varphi)$ be a motion with fixed directing cone $\boldsymbol{R}(\varphi)$. Let $\boldsymbol{R}(\varphi) \neq 0$. Then the following Frenet formulae hold:*

$$\begin{aligned} \boldsymbol{R}' &= \kappa_1 \boldsymbol{T}, & & [\boldsymbol{R}, \boldsymbol{T}] = \boldsymbol{N}, \\ \boldsymbol{T}' &= -\kappa_1 \boldsymbol{R} + \kappa_2 \boldsymbol{N}, & & [\boldsymbol{T}, \boldsymbol{N}] = \boldsymbol{R}, \\ \boldsymbol{N}' &= -\kappa_2 \boldsymbol{T}, \qquad \kappa_1 > 0, & & [\boldsymbol{N}, \boldsymbol{R}] = \boldsymbol{T}. \end{aligned}$$

The functions κ_1 and κ_2 are called the first and second curvatures of the fixed directing cone, respectively.

Proof. First of all, we have $(\boldsymbol{R}, \boldsymbol{R}) = 1$, i.e. $(\boldsymbol{R}, \boldsymbol{R}') = 0$, i.e. $(\boldsymbol{R}, \boldsymbol{T}) = 0$. Since $(\boldsymbol{T}, \boldsymbol{T}) = 1$, $\{\boldsymbol{R}, \boldsymbol{T}, \boldsymbol{N}\}$ constitute an orthonormal frame. This proves the second column of the Frenet formulae. Further, we have $(\boldsymbol{T}, \boldsymbol{T}') = 0$. Thus, denote $\boldsymbol{T}' = \alpha \boldsymbol{R} + \beta \boldsymbol{N}$, $\alpha, \beta \in \mathbf{R}$. Then $(\boldsymbol{R}, \boldsymbol{T}) = 0$ implies

$$(\boldsymbol{R}', \boldsymbol{T}) + (\boldsymbol{R}, \boldsymbol{T}') = 0, \quad \text{i.e.} \quad \alpha + \kappa_1 = 0, \quad \text{i.e.} \quad \alpha = -\kappa_1.$$

Denote $\beta = \kappa_2$. Consequently, $\boldsymbol{T}' = -\kappa_1 \boldsymbol{R} + \kappa_2 \boldsymbol{N}$. Now

$$\begin{aligned} &(\boldsymbol{T}, \boldsymbol{N}) = 0, \quad \text{i.e.} \quad (\boldsymbol{T}', \boldsymbol{N}) + (\boldsymbol{T}, \boldsymbol{N}') = 0, \\ &(\boldsymbol{N}, \boldsymbol{N}) = 1, \quad \text{i.e.} \quad (\boldsymbol{N}', \boldsymbol{N}) = 0, \\ &(\boldsymbol{R}, \boldsymbol{N}) = 0, \quad \text{i.e.} \quad (\boldsymbol{R}', \boldsymbol{N}) + (\boldsymbol{R}, \boldsymbol{N}') = 0, \quad \text{i.e.} \quad (\boldsymbol{R}, \boldsymbol{N}') = 0. \end{aligned}$$

Denote $\boldsymbol{N}' = \gamma \boldsymbol{T}$, $\gamma \in \mathbf{R}$. Then $\kappa_2 + \gamma = 0$, i.e. $\gamma = -\kappa_2$, i.e. $\boldsymbol{N}' = -\kappa_2 \boldsymbol{T}$. This accomplishes the proof of the Frenet formulae.

3.35 Theorem. *Let the assumptions of Theorem 3.34 be satisfied. For the moving directing cone we then have*

$$\begin{aligned} \bar{\boldsymbol{R}}' &= \bar{\kappa}_1 \bar{\boldsymbol{T}}, & \bar{\kappa}_1 > 0, \quad [\bar{\boldsymbol{R}}, \bar{\boldsymbol{T}}] &= \bar{\boldsymbol{N}}, \\ \bar{\boldsymbol{T}}' &= -\bar{\kappa}_1 \bar{\boldsymbol{R}} + \bar{\kappa}_2 \bar{\boldsymbol{N}}, & [\bar{\boldsymbol{T}}, \bar{\boldsymbol{N}}] &= \bar{\boldsymbol{R}}, \\ \bar{\boldsymbol{N}}' &= -\bar{\kappa}_2 \bar{\boldsymbol{T}}, & [\bar{\boldsymbol{N}}, \bar{\boldsymbol{R}}] &= \bar{\boldsymbol{T}}. \end{aligned}$$

$\bar{\kappa}_1$ and $\bar{\kappa}_2$ are called the first and second curvature of the moving directing cone, respectively.

Proof. Since $\boldsymbol{R}' \neq 0$, $\boldsymbol{R}' = \operatorname{ad}g\bar{\boldsymbol{R}}'$, then $\bar{\boldsymbol{R}}' \neq 0$ also. We apply Theorem 3.34 to the motion $g(-\varphi)^{\mathrm{T}}$.

3.36 Theorem. *For the curvatures of directing cones we have*

$$\bar{\kappa}_1 = \kappa_1, \quad \kappa_2 - \bar{\kappa}_2 = 1.$$

Proof. We have $\bar{\boldsymbol{R}}' = \bar{\kappa}_1 \bar{\boldsymbol{T}}$, $\boldsymbol{R}' = \kappa_1 \boldsymbol{T}$, $\boldsymbol{R}' = \operatorname{ad}g\bar{\boldsymbol{R}}'$. This implies

$$\kappa_1^2 = (\boldsymbol{R}', \boldsymbol{R}') = (\operatorname{ad}g\bar{\boldsymbol{R}}', \operatorname{ad}g\bar{\boldsymbol{R}}') = (\bar{\boldsymbol{R}}', \bar{\boldsymbol{R}}') = \bar{\kappa}_1^2$$

since the inner product is invariant under ad$g$. Since $\kappa_1 > 0$, $\bar{\kappa}_1 > 0$, we have $\kappa_1 = \bar{\kappa}_1$. Equation $\boldsymbol{R}' = \operatorname{ad}g\bar{\boldsymbol{R}}'$ yields $\kappa_1 \boldsymbol{T} = \operatorname{ad}g(\kappa_1 \bar{\boldsymbol{T}})$, i.e. $\boldsymbol{T} = \operatorname{ad}g\bar{\boldsymbol{T}}$. Differentiate the last equation:

$$\begin{aligned} \boldsymbol{T}' &= (g\bar{\boldsymbol{T}}g^{-1})' = g'\bar{\boldsymbol{T}}g^{-1} + g\bar{\boldsymbol{T}}(g^{-1})' + g\bar{\boldsymbol{T}}'g^{-1} \\ &= g'g^{-1}g\bar{\boldsymbol{T}}g^{-1} + g\bar{\boldsymbol{T}}g^{-1}\, g(g^{-1})' + \operatorname{ad}g\bar{\boldsymbol{T}}' \\ &= \boldsymbol{R}\operatorname{ad}g\bar{\boldsymbol{T}} - \operatorname{ad}g\bar{\boldsymbol{T}}\boldsymbol{R} + \operatorname{ad}g\bar{\boldsymbol{T}}' = [\boldsymbol{R}, \boldsymbol{T}] + \operatorname{ad}g\bar{\boldsymbol{T}}' = \boldsymbol{N} + \operatorname{ad}g\bar{\boldsymbol{T}}', \end{aligned}$$

where use was made of the equations

$$g'g^{-1} + g(g^{-1})' = 0, \quad \boldsymbol{RT} - \boldsymbol{TR} = [\boldsymbol{R}, \boldsymbol{T}] = \boldsymbol{N}.$$

Thus, we have the relation $\boldsymbol{T}' = \boldsymbol{N} + \operatorname{ad}g\bar{\boldsymbol{T}}'$. Substitute now from the Frenet formulae. Then

$$-\kappa_1 \boldsymbol{R} + \kappa_2 \boldsymbol{N} = \boldsymbol{N} + \operatorname{ad}g(-\kappa_1 \bar{\boldsymbol{R}} + \bar{\kappa}_2 \bar{\boldsymbol{N}}) = \boldsymbol{N} - \kappa_1 \boldsymbol{R} + \bar{\kappa}_2 \boldsymbol{N}$$

since

$$\operatorname{ad}g\bar{\boldsymbol{N}} = \operatorname{ad}g[\bar{\boldsymbol{R}}, \bar{\boldsymbol{T}}] = [\operatorname{ad}g\bar{\boldsymbol{R}}, \operatorname{ad}g\bar{\boldsymbol{T}}] = [\boldsymbol{R}, \boldsymbol{T}] = \boldsymbol{N},$$

so that, in all, $\kappa_2 \boldsymbol{N} = \boldsymbol{N} + \bar{\kappa}_2 \boldsymbol{N}$, i.e. $\kappa_2 - \bar{\kappa}_2 = 1$. This concludes the proof.

3.37 Theorem. $\kappa_1 = (\boldsymbol{R}', \boldsymbol{R}')^{1/2}$, $\kappa_2 = \kappa_1^{-2}|\boldsymbol{R}, \boldsymbol{R}', \boldsymbol{R}''|$.

Proof.

$$\boldsymbol{R}' = \kappa_1 \boldsymbol{T}, \quad \text{i.e.} \quad (\boldsymbol{R}', \boldsymbol{R}') = \kappa_1^2 .$$

Further,

$$\begin{aligned} \boldsymbol{R}'' &= \kappa_1' \boldsymbol{T} + \kappa_1 \boldsymbol{T}' = \kappa_1' \boldsymbol{T} + \kappa_1(-\kappa_1 \boldsymbol{R} + \kappa_2 \boldsymbol{N}) = \\ &= -\kappa_1^2 \boldsymbol{R} + \kappa_1' \boldsymbol{T} + \kappa_1 \kappa_2 \boldsymbol{N}, \end{aligned}$$

so that

$$\begin{aligned} |\boldsymbol{R}, \boldsymbol{R}', \boldsymbol{R}''| &= |\boldsymbol{R}, \kappa_1 \boldsymbol{T}, -\kappa_1^2 \boldsymbol{R} + \kappa_1' \boldsymbol{T} + \kappa_1 \kappa_2 \boldsymbol{N}| = \kappa_1^2 \kappa_2 |\boldsymbol{R}, \boldsymbol{T}, \boldsymbol{N}| = \\ &= \kappa_1^2 \kappa_2 , \quad \text{i.e.} \quad \kappa_1 = (\boldsymbol{R}', \boldsymbol{R}')^{1/2} , \quad \kappa_2 = \kappa_1^{-2} |\boldsymbol{R}, \boldsymbol{R}', \boldsymbol{R}''| , \end{aligned}$$

which was to be proved.

As seen in Example 3.25, the introduction of the canonical parameter can be complicated. Therefore, let us find formulae for the computation of the curvatures κ_1 and κ_2 under the assumption that motion is given as a function of an arbitrary parameter t. Then

$$\boldsymbol{r}(\mathrm{t}) = \frac{\mathrm{d}g}{\mathrm{dt}} g^{-1}(\mathrm{t}) .$$

Now, let $\mathrm{t} = \mathrm{t}(\varphi)$ where φ is the canonical parameter. Then

$$\begin{aligned} \boldsymbol{R}(\varphi) &= \frac{\mathrm{d}g(\mathrm{t}(\varphi))}{\mathrm{d}\varphi} g^{-1}(\mathrm{t}(\varphi)) \\ &= \left.\frac{\mathrm{d}g(\mathrm{t})}{\mathrm{dt}}\right|_{t=t(\varphi)} \frac{\mathrm{dt}}{\mathrm{d}\varphi} g^{-1}(\mathrm{t}(\varphi)) = \boldsymbol{r}(\mathrm{t}(\varphi)) \frac{\mathrm{dt}(\varphi)}{\mathrm{d}\varphi} . \end{aligned}$$

Thus,

$$\boldsymbol{R}(\varphi) = \boldsymbol{r}(\mathrm{t}(\varphi)) \frac{\mathrm{dt}}{\mathrm{d}\varphi}, \quad \text{i.e.} \quad \boldsymbol{r}(\mathrm{t}) = \boldsymbol{R}(\varphi(\mathrm{t})) \frac{\mathrm{d}\varphi}{\mathrm{dt}},$$

where $\varphi = \varphi(\mathrm{t})$ is the inverse function to the function $\mathrm{t} = \mathrm{t}(\varphi)$, and where we apply the theorem on the derivative of the inverse function. Thus, we have $\boldsymbol{r} = \boldsymbol{R}\dot{\varphi}$. This implies first of all (as we already know)

$$(\boldsymbol{r}, \boldsymbol{r}) = (\dot{\varphi})^2 , \quad \text{i.e.} \quad \dot{\varphi} = (\boldsymbol{r}, \boldsymbol{r})^{1/2} .$$

3.38 Theorem. *Let $g(t)$ be a motion with the fixed directing function $\boldsymbol{r}(t)$ and fixed directing cone $\boldsymbol{R}(t)$. Then*

$$\boldsymbol{r} = \boldsymbol{R}\dot{\varphi}\,, \quad \dot{\varphi} = (\boldsymbol{r}, \boldsymbol{r})^{1/2}\,,$$
$$\kappa_1 = (\dot{\varphi})^{-1}(\boldsymbol{R}^{\bullet}, \boldsymbol{R}^{\bullet})^{1/2}\,, \quad \kappa_2 = (\dot{\varphi})^{-1}(\boldsymbol{R}^{\bullet}, \boldsymbol{R}^{\bullet})^{-1}\left|\boldsymbol{R}, \boldsymbol{R}^{\bullet}, \boldsymbol{R}^{\bullet\bullet}\right|.$$

Proof. The first relation was derived above. Further, we have

$$\boldsymbol{R}^{\bullet} = \frac{\mathrm{d}\boldsymbol{R}}{\mathrm{d}t} = \frac{\mathrm{d}\boldsymbol{R}}{\mathrm{d}\varphi}\frac{\mathrm{d}\varphi}{\mathrm{d}t} = \boldsymbol{R}'\dot{\varphi}\,,$$

so that

$$(\boldsymbol{R}^{\bullet}, \boldsymbol{R}^{\bullet}) = (\boldsymbol{R}', \boldsymbol{R}')\,\dot{\varphi}^2 = \kappa_1^2(\dot{\varphi})^2\,, \quad \text{i.e.} \quad \kappa_1 = (\dot{\varphi})^{-1}(\boldsymbol{R}^{\bullet}, \boldsymbol{R}^{\bullet})^{1/2}\,.$$

Further,

$$\boldsymbol{R}^{\bullet\bullet} = \left(\frac{\mathrm{d}\boldsymbol{R}}{\mathrm{d}\varphi}\frac{\mathrm{d}\varphi}{\mathrm{d}t}\right)^{\bullet} = \boldsymbol{R}''\frac{\mathrm{d}\varphi}{\mathrm{d}t}\frac{\mathrm{d}\varphi}{\mathrm{d}t} + \frac{\mathrm{d}\boldsymbol{R}}{\mathrm{d}\varphi}\frac{\mathrm{d}^2\varphi}{\mathrm{d}t^2} = \boldsymbol{R}''(\dot{\varphi})^2 + \boldsymbol{R}'\ddot{\varphi}\,,$$

so that

$$\left|\boldsymbol{R}, \boldsymbol{R}^{\bullet}, \boldsymbol{R}^{\bullet\bullet}\right| = \left|\boldsymbol{R}, \boldsymbol{R}'\dot{\varphi}, \boldsymbol{R}''(\dot{\varphi})^2 + \boldsymbol{R}'\ddot{\varphi}\right| = (\dot{\varphi})^3\left|\boldsymbol{R}, \boldsymbol{R}', \boldsymbol{R}''\right|.$$

Consequently, we have

$$\kappa_2 = \kappa_1^{-2}\left|\boldsymbol{R}, \boldsymbol{R}', \boldsymbol{R}''\right| = (\dot{\varphi})^2(\boldsymbol{R}^{\bullet}, \boldsymbol{R}^{\bullet})^{-1}\left|\boldsymbol{R}, \boldsymbol{R}^{\bullet}, \boldsymbol{R}^{\bullet\bullet}\right|(\dot{\varphi})^{-3}$$
$$= (\dot{\varphi})^{-1}(\boldsymbol{R}^{\bullet}, \boldsymbol{R}^{\bullet})^{-1}\left|\boldsymbol{R}, \boldsymbol{R}^{\bullet}, \boldsymbol{R}^{\bullet\bullet}\right|.$$

Theorem 3.38 is still not very advantageous for practical calculations. To obtain the directing cone the directing function $\boldsymbol{r}(t)$ is to be divided by the expression $\dot{\varphi}$. This expression is a square root and so the derivatives of the vector $\boldsymbol{R}$ are very unpleasant for computations. For this reason, let us try to compute the curvatures κ_1 and κ_2 directly from the directing function $\boldsymbol{r}(t)$. The result is formulated in the following theorem:

3.39 Theorem. *Let $g(t)$ be a motion with the directing function $\boldsymbol{r}(t)$. Then*

$$(\dot{\varphi})^2 = \boldsymbol{r}^2\,,$$
$$\kappa_1 = (\dot{\varphi})^{-3}\left\{(\boldsymbol{r}^{\bullet})^2(\dot{\varphi})^2 - (\boldsymbol{r}, \boldsymbol{r}^{\bullet})^2\right\}^{1/2},$$
$$\kappa_2 = \left\{(\boldsymbol{r}^{\bullet})^2(\dot{\varphi})^2 - (\boldsymbol{r}, \boldsymbol{r}^{\bullet})^2\right\}^{-1}\left|\boldsymbol{r}, \boldsymbol{r}^{\bullet}, \boldsymbol{r}^{\bullet\bullet}\right|.$$

Proof. We have $\boldsymbol{R} = \boldsymbol{r}(\dot{\varphi})^{-1}$ so that $\boldsymbol{r}^2 = (\dot{\varphi})^2$. Differentiating this relation we obtain $2(\boldsymbol{r}, \boldsymbol{r}^{\bullet}) = 2\dot{\varphi}\ddot{\varphi}$, i.e. $\ddot{\varphi} = (\dot{\varphi})^{-1}(\boldsymbol{r}, \boldsymbol{r}^{\bullet})$. Differentiation of the vector $\boldsymbol{R}$ yields

$$\boldsymbol{R}^{\bullet} = \boldsymbol{r}^{\bullet}(\dot{\varphi})^{-1} - \boldsymbol{r}(\dot{\varphi})^{-2}\,\ddot{\varphi}\,,$$

$$\boldsymbol{R}^{\bullet\bullet} = \boldsymbol{r}^{\bullet\bullet}(\dot{\varphi})^{-1} - 2\boldsymbol{r}^{\bullet}(\dot{\varphi})^{-2}\,\ddot{\varphi} + 2\boldsymbol{r}(\dot{\varphi})^{-3}\,(\ddot{\varphi})^2 - \boldsymbol{r}(\dot{\varphi})^{-2}\,\dddot{\varphi}\,,$$

$$\begin{aligned}(\boldsymbol{R}^{\bullet}, \boldsymbol{R}^{\bullet}) &= (\boldsymbol{r}^{\bullet})^2\,(\dot{\varphi})^{-2} - 2(\boldsymbol{r}, \boldsymbol{r}^{\bullet})\,(\dot{\varphi})^{-3}\,\ddot{\varphi} + \boldsymbol{r}^2(\dot{\varphi})^{-4}\,(\ddot{\varphi})^2\\
&= (\boldsymbol{r}^{\bullet})^2\,(\dot{\varphi})^{-2} - 2(\boldsymbol{r}, \boldsymbol{r}^{\bullet})\,(\dot{\varphi})^{-3}\,(\dot{\varphi})^{-1}\,(\boldsymbol{r}, \boldsymbol{r}^{\bullet})\\
&\quad + \boldsymbol{r}^2(\dot{\varphi})^{-4}\,(\dot{\varphi})^{-2}\,(\boldsymbol{r}, \boldsymbol{r}^{\bullet})^2\\
&= (\dot{\varphi})^{-2}\,\{(\boldsymbol{r}^{\bullet})^2 - (\boldsymbol{r}, \boldsymbol{r}^{\bullet})^2\,(\dot{\varphi})^{-2}\}\\
&= (\dot{\varphi})^{-4}\,\{(\boldsymbol{r}^{\bullet})^2\,(\dot{\varphi})^2 - (\boldsymbol{r}, \boldsymbol{r}^{\bullet})^2\}\,.\end{aligned}$$

$$\begin{aligned}|\boldsymbol{R}, \boldsymbol{R}^{\bullet}, \boldsymbol{R}^{\bullet\bullet}| &=\\
&= |\boldsymbol{r}(\dot{\varphi})^{-1}, \boldsymbol{r}^{\bullet}(\dot{\varphi})^{-1} - \boldsymbol{r}(\dot{\varphi})^{-2}\,\ddot{\varphi}, \boldsymbol{r}^{\bullet\bullet}(\dot{\varphi})^{-1} - 2\boldsymbol{r}^{\bullet}(\dot{\varphi})^{-2}\,\ddot{\varphi}\\
&\quad + 2\boldsymbol{r}(\dot{\varphi})^{-3}\,(\ddot{\varphi})^2 - \boldsymbol{r}(\dot{\varphi})^{-2}\,\dddot{\varphi}| = (\dot{\varphi})^{-3}\,|\boldsymbol{r}, \boldsymbol{r}^{\bullet}, \boldsymbol{r}^{\bullet\bullet}|\,.\end{aligned}$$

Finally, we thus obtain

$$\begin{aligned}\kappa_1 &= (\dot{\varphi})^{-1}\,(\boldsymbol{R}^{\bullet}, \boldsymbol{R}^{\bullet})^{1/2} = (\dot{\varphi})^{-1}\,(\dot{\varphi})^{-2}\,\{(\boldsymbol{r}^{\bullet})^2\,(\dot{\varphi})^2 - (\boldsymbol{r}, \boldsymbol{r}^{\bullet})^2\}^{1/2}\\
&= (\dot{\varphi})^{-3}\,\{(\boldsymbol{r}^{\bullet})^2\,(\dot{\varphi})^2 - (\boldsymbol{r}, \boldsymbol{r}^{\bullet})^2\}^{1/2}\,,\\
\kappa_2 &= (\dot{\varphi})^{-1}\,(\boldsymbol{R}^{\bullet}, \boldsymbol{R}^{\bullet})^{-1}\,|\boldsymbol{R}, \boldsymbol{R}^{\bullet}, \boldsymbol{R}^{\bullet\bullet}|\\
&= (\dot{\varphi})^{-1}\,(\dot{\varphi})^4\,\{(\boldsymbol{r}^{\bullet})^2\,(\dot{\varphi})^2 - (\boldsymbol{r}, \boldsymbol{r}^{\bullet})^2\}^{-1}\,(\dot{\varphi})^{-3}\,|\boldsymbol{r}, \boldsymbol{r}^{\bullet}, \boldsymbol{r}^{\bullet\bullet}|\\
&= \{(\boldsymbol{r}^{\bullet})^2\,(\dot{\varphi})^2 - (\boldsymbol{r}, \boldsymbol{r}^{\bullet})^2\}^{-1}\,|\boldsymbol{r}, \boldsymbol{r}^{\bullet}, \boldsymbol{r}^{\bullet\bullet}|\,,\end{aligned}$$

which accomplishes the proof.

3.40 Theorem. *Let $\boldsymbol{R}(\varphi)$ be a differentiable vector function of the parameter φ in $O(3)$, and let $(\boldsymbol{R}, \boldsymbol{R}) = 1$. Then there exists a unique motion $g(\varphi)$ with the canonical parameter φ whose fixed directing cone is $\boldsymbol{R}(\varphi)$ and for which $g(0) = e$.*

Proof. We have to solve the differential equation $g'g^{-1} = \boldsymbol{R}(\varphi)$ for the unknown matrix g, i.e. the equation $g'(\varphi) = \boldsymbol{R}(\varphi)\,g(\varphi)$. We require that the matrix $g(\varphi)$ be regular. A system of nine linear differential equations is obtained which always has a solution satisfying the initial condition $g(0) = e$, and this solution is unique. It remains to show that this solution belongs to the group $O(3)$. Thus, we have $g' = \boldsymbol{R}g$ so that $g'g^{-1} = \boldsymbol{R}$, i.e. $-g(g^{-1})' = \boldsymbol{R}$, i.e. $-(g^{-1})' = g^{-1}\boldsymbol{R}$ where the equation $g'g^{-1} +$

$g(g^{-1})' = 0$ was applied. Further, since $\boldsymbol{R}$ is a skew-symmetric matrix we have $\boldsymbol{R}^{\mathrm{T}} = -\boldsymbol{R}$. The transposition of the matrix $g' = \boldsymbol{R}g$ yields $(g')^{\mathrm{T}} = -g^{\mathrm{T}}\boldsymbol{R}$. Thus, if we consider the equation $y' = -y\boldsymbol{R}$ for the unknown matrix y, it is obvious that this equation has a unique solution such that $y(0) = e$. Since this equation has the solutions g^{T} and g^{-1} while $g^{\mathrm{T}}(0) = g^{-1}(0) = e$, we necessarily have $g^{\mathrm{T}} = g^{-1}$, which was to be proved.

It is now possible to prove the theorem which states that motion is uniquely determined (up to the choice of base) by its curvature $\kappa_1(\varphi)$ and $\kappa_2(\varphi)$.

3.41 Theorem. *Let differentiable functions $\kappa_1(\varphi) > 0$, $\kappa_2(\varphi)$ be given. Then there exists a motion $g(\varphi)$ such that the functions $\kappa_1(\varphi)$ and $\kappa_2(\varphi)$ are its curvatures and φ is its canonical parameter. Conversely, two motions $g_1(\varphi)$ and $g_2(\varphi)$ have the same curvatures if and only if there exist $\gamma_1, \gamma_2 \in O(3)$ such that $g_1(\varphi) = \gamma_1\, g_2(\varphi)\, \gamma_2$.*

Proof. We have to solve the system of differential equations

$$\boldsymbol{R}' = \kappa_1 \boldsymbol{T}, \quad \boldsymbol{T}' = -\kappa_1 \boldsymbol{R} + \kappa_2 \boldsymbol{N}, \quad \boldsymbol{N}' = -\kappa_2 \boldsymbol{T}.$$

This is a system of linear differential equations and as such always possesses a solution. We easily see that if $\boldsymbol{R}_1(\varphi)$ is a solution then $\boldsymbol{R}_2(\varphi) = \mathrm{ad}\gamma_1\, \boldsymbol{R}_1(\varphi)$ is a solution as well since then $\boldsymbol{T}_2 = \mathrm{ad}\gamma_1 \boldsymbol{T}_1$, $\boldsymbol{N}_2 = \mathrm{ad}\gamma_1 \boldsymbol{N}_1$, and what remains is easy. Conversely, let us show that any two solutions differ by a constant matrix. Thus, let $\boldsymbol{R}_2(\varphi)$ and $\boldsymbol{R}_1(\varphi)$ be arbitrary two solutions of the given system. Then the vectors $\boldsymbol{T}_2$, $\boldsymbol{N}_2$, and $\boldsymbol{T}_1$, $\boldsymbol{N}_1$ are defined, i.e. the frames $\mathscr{R}_2 = \{\boldsymbol{R}_2, \boldsymbol{T}_2, \boldsymbol{N}_2\}$, $\mathscr{R}_1 = \{\boldsymbol{R}_1, \boldsymbol{T}_1, \boldsymbol{N}_1\}$. This means that there exists a uniquely given curve $\gamma_1(\varphi)$ such that

$$\begin{aligned}
\boldsymbol{R}_2(\varphi) &= \mathrm{ad}\gamma_1(\varphi)\, \boldsymbol{R}_1(\varphi), \\
\boldsymbol{T}_2(\varphi) &= \mathrm{ad}\gamma_1(\varphi)\, \boldsymbol{T}_1(\varphi), \\
\boldsymbol{N}_2(\varphi) &= \mathrm{ad}\gamma_1(\varphi)\, \boldsymbol{N}_1(\varphi).
\end{aligned}$$

Denote $S(\varphi) = \gamma_1'\gamma_1^{-1}$. Now we need to compute the derivatives of the function $\boldsymbol{R}_2(\varphi)$ by way of the derivatives of the function $\boldsymbol{R}_1(\varphi)$. Let $\boldsymbol{X}(\varphi)$ be a vector function, $\boldsymbol{X}(\varphi) \in \mathfrak{D}(3)$. Then

$$\begin{aligned}
[\mathrm{ad}\gamma_1(\varphi)\, \boldsymbol{X}(\varphi)]' &= (\gamma_1 \boldsymbol{X} \gamma_1^{-1})' = \gamma_1 \boldsymbol{X}' \gamma_1^{-1} + \gamma_1' \boldsymbol{X} \gamma_1^{-1} + \gamma_1 \boldsymbol{X} (\gamma_1^{-1})' \\
&= \mathrm{ad}\gamma_1 \boldsymbol{X}' + \gamma_1'\gamma_1^{-1}\gamma_1 \boldsymbol{X}\gamma_1^{-1} - \gamma_1 \boldsymbol{X} \gamma_1^{-1}\gamma_1'\gamma_1^{-1} \\
&= \mathrm{ad}\gamma_1 \boldsymbol{X}' + [S, \mathrm{ad}\gamma_1 \boldsymbol{X}].
\end{aligned}$$

For the function $\boldsymbol{R}_2(\varphi)$ we have

$$\begin{aligned}\boldsymbol{R}_2'(\varphi) &= \kappa_1 \boldsymbol{T}_2(\varphi) = [S, \mathrm{ad}\gamma_1 \boldsymbol{R}_1(\varphi)] + \mathrm{ad}\gamma_1 \boldsymbol{R}_1'(\varphi) \\ &= [S, \boldsymbol{R}_2] + \kappa_1 \mathrm{ad}\gamma_1 \boldsymbol{T}_1 = [S, \boldsymbol{R}_2] + \kappa_1 \boldsymbol{T}_2 ,\end{aligned}$$

so that $[S, \boldsymbol{R}_2] = 0$. Analogously, we obtain $[S, \boldsymbol{T}_2] = [S, \boldsymbol{N}_2] = 0$. These relations imply $S = 0$, i.e. $\gamma_1'\gamma_1^{-1} = 0$, i.e. $\gamma_1' = 0$ since γ_1 is a regular matrix, i.e. $\gamma_1 = const.$ Till now we have proved the following: if $\boldsymbol{R}_1(\varphi)$ is a solution, then all the other solutions are $\boldsymbol{R}_2(\varphi) = \mathrm{ad}\gamma_1 \boldsymbol{R}_1(\varphi)$, where $\gamma_1 \in O(3)$. Choose now a fixed solution $\boldsymbol{R}_1(\varphi)$. According to Theorem 3.39 there exists one and only one solution $g(\varphi)$ with the directing cone $\boldsymbol{R}_1(\varphi)$ for which $g(0) = e$. From the proof of Theorem 3.39 we know that all the other solutions are of the form $g(\varphi)\gamma_2$, $\gamma_2 \in O(3)$, $\gamma_2 = \text{const}$. It is now easy to show that the solutions corresponding to $\boldsymbol{R}_2(\varphi)$ are $\gamma_1 g(\varphi)\gamma_2$. This accomplishes the proof of the theorem.

3.42 Theorem. *Let differentiable functions $\kappa_1(\varphi) > 0$ and $\kappa_2(\varphi)$ be given. Further, let $\boldsymbol{X}, \boldsymbol{Y} \in \mathfrak{O}(3)$ be an arbitrary pair of perpendicular vectors. Then there exists a unique motion $g(\varphi)$ such that κ_1 and κ_2 are the curvatures of the fixed directing cone, φ is the canonical parameter, $g(0) = e$, $\boldsymbol{R}(0) = \boldsymbol{X}$, $\boldsymbol{T}(0) = \boldsymbol{Y}$.*

Proof. Let $g(\varphi)$ be a fixed chosen motion such that κ_1 and κ_2 are the curvatures of the fixed directing cone. All the other solutions will then be $\gamma_1 g(\varphi)\gamma_2$, as we know from Theorem 3.41. Denote the desired solution $g_1(\varphi)$. Should $g_1(0) = e$, then $\gamma_1 g(0)\gamma_2 = e$ must hold, i.e. $\gamma_2 = [g(0)]^{-1} \cdot \gamma_1^{-1}$, i.e. $g_1(\varphi) = \gamma_1 g(\varphi)[g(0)]^{-1}\gamma_1^{-1}$. Thus, let $g_2(\varphi)$ be a solution such that $g_2(0) = e$. Then all other solutions $g_3(\varphi)$ with the property $g_3(0) = e$ are of the form $g_3(\varphi) = \gamma_1 g_2(\varphi)\gamma_1^{-1}$, where $\gamma_1 \in O(3)$. If $\boldsymbol{R}_3(0) = \boldsymbol{X}$, $\boldsymbol{T}_3(0) = \boldsymbol{Y}$ is to hold for the fixed directing cone of the motion $g_3(\varphi)$, we have to solve the equations $\mathrm{ad}\gamma_1 \boldsymbol{R}_2(0) = \boldsymbol{X}$, $\mathrm{ad}\gamma_1 \boldsymbol{T}_2(0) = \boldsymbol{Y}$ for the unknown matrix $\gamma_1 \in O(3)$. These equations are uniquely solvable according to the assumption of the perpendicularity of $\boldsymbol{X}$ and $\boldsymbol{Y}$ [in fact, the equations $\gamma_1 \boldsymbol{R}_2(0) = \boldsymbol{X}$, $\gamma_1 \boldsymbol{T}_2(0) = \boldsymbol{Y}$ are in question here]. By a single rotation γ_1 we are to map $\boldsymbol{R}_2$ into $\boldsymbol{X}$ and $\boldsymbol{T}_2$ into $\boldsymbol{Y}$, and this is possible in precisely one manner]. This proves the theorem. Note further that the assumptions $g(0) = e$, $\boldsymbol{R}(0) = \boldsymbol{X}$, $\boldsymbol{T}(0) = \boldsymbol{Y}$, imply also $\bar{\boldsymbol{R}}(0) = \boldsymbol{X}$, $\bar{\boldsymbol{T}}(0) = \boldsymbol{Y}$, so that the directing cones contact along the ruling $\boldsymbol{R}(0)$.

3.43 Example. Let us demonstrate the presented considerations on a simple concrete example. Choose the most simple possible case $\kappa_1 =$ const., $\kappa_2 = 0$, and find the corresponding motion. Let us write $\kappa_1 = k > 0$, and choose the initial conditions $g(0) = e$, $\boldsymbol{R}(0) = \boldsymbol{u}_1$, $\boldsymbol{T}(0) = \boldsymbol{u}_2$, where $\mathscr{R} = \{\boldsymbol{u}_1, \boldsymbol{u}_2, \boldsymbol{u}_3\}$ is a given fixed chosen frame in E_3. According to Theorem 3.42 there exists precisely one motion $g(\varphi)$ which satisfies the given conditions. We shall find this motion. First, we find the directing cone $\boldsymbol{R}(t)$ with the curvature $\kappa_1 = k$, $\kappa_2 = 0$, and with the initial condition $\boldsymbol{R}(0) = \boldsymbol{u}_1$, $\boldsymbol{T}(0) = \boldsymbol{u}_2$. Thus, we have to solve the system of differential equations

$$\boldsymbol{R}' = k\boldsymbol{T}, \quad \boldsymbol{T}' = -k\boldsymbol{R}, \quad \boldsymbol{N}' = \boldsymbol{0}.$$

Differentiate the first equation, and substitute from the second, i.e. $\boldsymbol{R}'' = k\boldsymbol{T}'$, i.e. $\boldsymbol{R}'' = -k^2\boldsymbol{R}$. As a consequence, we have to solve the differential equation $\boldsymbol{R}'' + k^2\boldsymbol{R} = 0$. Its solution is found easily: We have $\boldsymbol{R} = \boldsymbol{A}\cos .(k\varphi) + \boldsymbol{B}\sin(k\varphi)$, where $\boldsymbol{A}$ and $\boldsymbol{B}$ are constant vectors chosen arbitrarily. The initial conditions yield

$$\boldsymbol{R}(0) = \boldsymbol{A} = \boldsymbol{u}_1, \quad \boldsymbol{R}'(0) = k\boldsymbol{B}, \quad \text{so that} \quad \boldsymbol{B} = \boldsymbol{u}_2.$$

Consequently, the directing cone is

$$\boldsymbol{R} = \boldsymbol{u}_1 \cos(k\varphi) + \boldsymbol{u}_2 \sin(k\varphi).$$

Now we have to solve the equation $g'g^{-1} = \boldsymbol{R}$, i.e. the system $g'(\varphi) = \boldsymbol{R}(\varphi)\, g(\varphi)$, where $\boldsymbol{R}$ has to be written in matrix form, i.e.

$$\boldsymbol{R} = \begin{pmatrix} 0, & 0, & \sin(k\varphi) \\ 0, & 0, & -\cos(k\varphi) \\ -\sin(k\varphi), & \cos(k\varphi), & 0 \end{pmatrix}.$$

The system $g' = \boldsymbol{R}g$ is a system of nine equations. Instead of solving this system we solve the system $X' = \boldsymbol{R}X$ for the trajectory of $\bar{X}$ and note that, e.g., the trajectory of point $\boldsymbol{u}_1$ yields

$$\boldsymbol{u}_1(t) = g(t)\,\boldsymbol{u}_1 = g_1^1\boldsymbol{u}_1 + g_1^2\boldsymbol{u}_2 + g_1^3\boldsymbol{u}_3.$$

The initial condition is, e.g., $\boldsymbol{u}_1(0) = \boldsymbol{u}_1$; and similarly for the other ones.

Thus, we shall solve the system of equations $X' = \boldsymbol{R}X$ for the unknown column X. Let us introduce a new variable matrix Y by the relation

$X = A(\varphi)\,Y$ where we choose

$$A(\varphi) = \begin{pmatrix} \cos k\varphi, & -\sin k\varphi, & 0 \\ \sin k\varphi, & \cos k\varphi, & 0 \\ 0, & 0 & 1 \end{pmatrix}.$$

$$X' = A'Y + AY',$$

substitution yields

$$A'Y + AY' = \boldsymbol{R}AY, \quad \text{i.e.} \quad AY' = \boldsymbol{R}AY - A'Y = (\boldsymbol{R}A - A')\,Y,$$

i.e.

$$Y' = A^{-1}(\boldsymbol{R}A - A')\,Y.$$

It is easily computed that

$$A^{-1}\boldsymbol{R}A = \begin{pmatrix} 0, & 0, & 0 \\ 0, & 0, & 1 \\ 0, & -1, & 0 \end{pmatrix}, \quad A^{-1}A' = \begin{pmatrix} 0, & -k, & 0 \\ k, & 0, & 0 \\ 0, & 0, & 0 \end{pmatrix}.$$

Thus, we have to solve the system of equations

$$\begin{pmatrix} y^1 \\ y^2 \\ y^3 \end{pmatrix}' = \begin{pmatrix} 0, & k, & 0 \\ -k, & 0, & 1 \\ 0, & -1, & 0 \end{pmatrix} \begin{pmatrix} y^1 \\ y^2 \\ y^3 \end{pmatrix}, \quad \text{i.e.}$$

$$(y^1)' = ky^2, \quad (y^2)' = -ky^1 + y^3, \quad (y^3)' = y^2.$$

Differentiate the second equation, and substitute into it from the other two. We obtain

$$(y^2)'' = -k(y^1)' + (y^3)' = -(k^2 + 1)\,y^2,$$

i.e. $(y^2)'' + (k^2 + 1)\,y^2 = 0$. Denote $\mu = \sqrt{(1 + k^2)}$. The solution of the equation in general form is

$$y^2 = A \cos \mu\varphi + B \sin \mu\varphi.$$

Integration of the first and third equations yields

$$y^3 = -\mu^{-1}[A \sin \mu\varphi - B \cos \mu\varphi + C],$$

$$y^1 = k\mu^{-1}[A \sin \mu\varphi - B \cos \mu\varphi + D],$$

where A, B, C, D are integration constants. The obtained solution need not satisfy the second equation but only its derivative. So let us substitute the obtained solution into the second equation. The result is $C = -k^2 D$.

Consequently, the solution is

$$y^1 = k\mu^{-1}[A \sin \mu\varphi - B \cos \mu\varphi + D],$$
$$y^2 = A \cos \mu\varphi + B \sin \mu\varphi,$$
$$y^3 = \mu^{-1}[-A \sin \mu\varphi + B \cos \mu\varphi + k^2 D].$$

Further, we have $X = A(\varphi) Y$, i.e.

$$x^1 = y^1 \cos k\varphi - y^2 \sin k\varphi,$$
$$x^2 = y^1 \sin k\varphi + y^2 \cos k\varphi,$$
$$x^3 = y^3.$$

Now, if we wish to find the coefficients $g_1^1(\varphi), g_1^2(\varphi), g_1^3(\varphi)$, of the matrix of the motion $g(\varphi)$, it suffices to find the trajectory $\boldsymbol{u}_1(\varphi)$ of the vector $\boldsymbol{u}_1$ with initial condition $\boldsymbol{u}_1(0) = \boldsymbol{u}_1$, the similar being true for the other ones. Thus, we find the values of the solution x^1, x^2, x^3 for $\varphi = 0$, and determine the integration constants from the initial conditions. For $\varphi = 0$ we obtain

$$x^1(0) = y^1(0) = k\mu^{-1}(D - B),$$
$$x^2(0) = y^2(0) = A,$$
$$x^3(0) = y^3(0) = \mu^{-1}(B + k^2 D).$$

Finally, for the three columns of matrix g we obtain the following three systems of equations, each consisting of three equations:

1. $x^1(0) = 1, \quad x^2(0) = x^3(0) = 0, \qquad$ i.e. $\boldsymbol{u}_1(0) = \boldsymbol{u}_1,$
2. $x^1(0) = 0, \quad x^2(0) = 1, \quad x^3(0) = 0, \quad$ i.e. $\boldsymbol{u}_2(0) = \boldsymbol{u}_2,$
3. $x^1(0) = x^2(0) = 0, \quad x^3(0) = 1, \qquad$ i.e. $\boldsymbol{u}_3(0) = \boldsymbol{u}_3,$

so that

1. $k\mu^{-1}(D - B) = 1, \quad A = 0, \quad \mu^{-1}(B + k^2 D) = 0,$ i.e.
   $A = 0, \quad B = -k\mu^{-1}, \quad D = k^{-1}\mu^{-1}.$

2. $k\mu^{-1}(D - B) = 0\,, \quad A = 1\,, \quad \mu^{-1}(B + k^2 D) = 0\,,$ i.e.
$A = 1\,, \quad B = D = 0\,.$

3. $k\mu^{-1}(D - B) = 0\,, \quad A = 0\,, \quad \mu^{-1}(B + k^2 D) = 1\,,$ i.e.
$A = 0\,, \quad B = D = \mu^{-1}\,.$

The final solutions are the following:

1.

$$x^1 = \mu^{-2}[k^2 \cos k\varphi \cos \mu\varphi + k\mu \sin k\varphi \sin \mu\varphi + \cos k\varphi]\,,$$
$$x^2 = \mu^{-2}[k^2 \sin k\varphi \cos \mu\varphi - k\mu \cos k\varphi \sin \mu\varphi + \sin k\varphi]\,,$$
$$x^3 = \mu^{-2}k[1 - \cos \mu\varphi]\,.$$

2.

$$x^1 = \mu^{-2}[k \cos k\varphi \sin \mu\varphi - \mu^2 \sin k\varphi \cos \mu\varphi]\,,$$
$$x^2 = \mu^{-2}[k \sin k\varphi \sin \mu\varphi + \mu^2 \cos k\varphi \cos \mu\varphi]\,,$$
$$x^3 = -\mu^{-1} \sin \mu\varphi\,.$$

3.

$$x^1 = \mu^{-2}[-k \cos k\varphi \cos \mu\varphi - \mu \sin k\varphi \sin \mu\varphi + k \cos k\varphi]\,,$$
$$x^2 = \mu^{-2}[-k \sin k\varphi \cos \mu\varphi + \mu \cos k\varphi \sin \mu\varphi + k \sin k\varphi]\,,$$
$$x^3 = \mu^{-2}[\cos \mu\varphi + k^2]\,.$$

The resulting matrix of motion is $g(\varphi) = \mu^{-2}(g^i_j(\varphi))$, where

$$g^1_1 = \cos k\varphi[1 + k^2 \cos \mu\varphi] + k\mu \sin k\varphi \sin \mu\varphi\,,$$
$$g^2_1 = \sin k\varphi[1 + k^2 \cos \mu\varphi] - k\mu \cos k\varphi \sin \mu\varphi\,,$$
$$g^3_1 = k[1 - \cos \mu\varphi]\,,$$
$$g^1_2 = k\mu \cos k\varphi \sin \mu\varphi - \mu^2 \sin k\varphi \cos \mu\varphi\,,$$
$$g^2_2 = k\mu \sin k\varphi \sin \mu\varphi + \mu^2 \cos k\varphi \cos \mu\varphi\,,$$
$$g^3_2 = -\mu \sin(\mu\varphi)\,,$$
$$g^1_3 = k \cos k\varphi[1 - \cos \mu\varphi] - \mu \sin k\varphi \sin \mu\varphi\,,$$
$$g^2_3 = k \sin k\varphi[1 - \cos \mu\varphi] + \mu \cos k\varphi \sin \mu\varphi\,,$$
$$g^3_3 = k^2 + \cos \mu\varphi\,.$$

3.44 Example. Let us present an example on the computation of the curvature of motion. Consider the spherical motion of a spherical segment of length $\frac{1}{2}\pi$ such that its endpoints move along circles, one of them being a great circle. A special case of a spherical hinged quadrangle is thus obtained

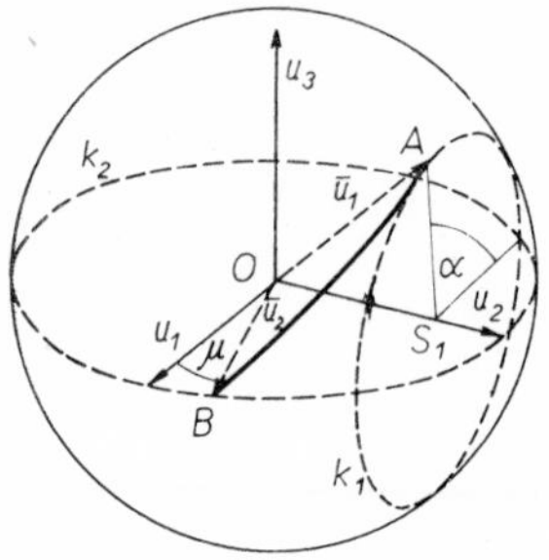

Fig. 22.

which will be discussed in more detail later. Choose a frame $\mathcal{R} = \{\boldsymbol{u}_1, \boldsymbol{u}_2, \boldsymbol{u}_3\}$. Further, let a circle $k_1 \equiv x^2 = \frac{1}{2}\sqrt{2}$ with radius $r_1 = \frac{1}{2}\sqrt{2}$ be given on the sphere S^2, also let a great circle $k_2 \equiv x^3 = 0$ be given. The segment AB of length $\frac{1}{2}\pi$ moves so that $A \in k_1$, $B \in k_2$ (see Fig. 22). The position vectors of the endpoints of segment AB are chosen as the vectors $\bar{\boldsymbol{u}}_1$ and $\bar{\boldsymbol{u}}_2$ of the moving frame $\bar{\mathcal{R}}$. The vector $\bar{\boldsymbol{u}}_3$ is then defined by the relation $\bar{\boldsymbol{u}}_3 = \bar{\boldsymbol{u}}_1 \times \bar{\boldsymbol{u}}_2$. Let S_1 be the center of circle k_1. As the parameter of motion we choose the angle μ of the vectors $\boldsymbol{u}_1$ and $\bar{\boldsymbol{u}}_2$. Denote by α the angle of the vectors AS_1 and $\boldsymbol{u}_2$. Then

$$\bar{\boldsymbol{u}}_1 = \frac{\sqrt{2}}{2}\,\boldsymbol{u}_1 + \frac{\sqrt{2}}{2}\,\boldsymbol{u}_2 \cos\alpha + \frac{\sqrt{2}}{2}\,\boldsymbol{u}_3 \sin\alpha\,.$$

Further, we have

$$\bar{\boldsymbol{u}}_2 = \boldsymbol{u}_1 \cos\mu + \boldsymbol{u}_2 \sin\mu\,.$$

It must be

$$(\bar{\boldsymbol{u}}_1, \bar{\boldsymbol{u}}_2) = 0\,, \quad \text{i.e.} \quad \frac{\sqrt{2}}{2}\cos\mu + \frac{\sqrt{2}}{2}\cos\alpha\sin\mu = 0\,,$$

i.e. $\cos\mu = -\cos\alpha\sin\mu$ or $\cos\alpha = -\cot\mu$. Then $\sin\alpha = (1 - \cot^2\mu)^{1/2}$. Thus, we obtain

$$\bar{\boldsymbol{u}}_1 = \frac{\sqrt{2}}{2}\left[\boldsymbol{u}_1 - \boldsymbol{u}_2\cot\mu + \boldsymbol{u}_3(1 - \cot^2\mu)^{1/2}\right],$$

$$\bar{\boldsymbol{u}}_2 = \boldsymbol{u}_1\cos\mu + \boldsymbol{u}_2\sin\mu\,,$$

$$\bar{\boldsymbol{u}}_3 = \bar{\boldsymbol{u}}_1 \times \bar{\boldsymbol{u}}_2\,,$$

i.e.

$$\bar{\boldsymbol{u}}_3 = \frac{\sqrt{2}}{2}\begin{vmatrix} \boldsymbol{u}_1, & \boldsymbol{u}_2, & \boldsymbol{u}_3 \\ 1, & -\cot\mu, & (1-\cot^2\mu)^{1/2} \\ \cos\mu, & \sin\mu, & 0 \end{vmatrix}$$

$$= \frac{\sqrt{2}}{2}\left[-\boldsymbol{u}_1\sin\mu(1-\cot^2\mu)^{1/2} + \boldsymbol{u}_2\cos\mu(1-\cot^2\mu)^{1/2}\right.$$

$$\left. + \boldsymbol{u}_3(\sin\mu + \cot\mu\cos\mu)\right]$$

$$= \frac{\sqrt{2}}{2}\{-\boldsymbol{u}_1[-\cos(2\mu)]^{1/2} + \boldsymbol{u}_2\cot\mu[-\cos(2\mu)]^{1/2} + \boldsymbol{u}_3\sin^{-1}\mu\}\,.$$

For the frame $\bar{\mathscr{R}} = \{\bar{\boldsymbol{u}}_1, \bar{\boldsymbol{u}}_2, \bar{\boldsymbol{u}}_3\}$ we thus have, on the whole,

$$\bar{\boldsymbol{u}}_1 = \frac{\sqrt{2}}{2}\left[\boldsymbol{u}_1 - \boldsymbol{u}_2\cot\mu + \boldsymbol{u}_3(1 - \cot^2\mu)^{1/2}\right],$$

$$\bar{\boldsymbol{u}}_2 = \frac{\sqrt{2}}{2}(\boldsymbol{u}_1\sqrt{2}\cos\mu + \boldsymbol{u}_2\sqrt{2}\sin\mu)\,,$$

$$\bar{\boldsymbol{u}}_3 = \frac{\sqrt{2}}{2}\{-\boldsymbol{u}_1[-\cos(2\mu)]^{1/2} + \boldsymbol{u}_2\cot\mu[-\cos(2\mu)]^{1/2} + \boldsymbol{u}_3\sin^{-1}\mu\}\,.$$

The matrix $g(\mu)$ of motion is then

$$g(\mu) = \frac{\sqrt{2}}{2}\begin{pmatrix} 1, & \sqrt{2}\cos\mu, & -[-\cos(2\mu)]^{1/2} \\ -\cot\mu, & \sqrt{2}\sin\mu, & \cot\mu[-\cos(2\mu)]^{1/2} \\ (1-\cot^2\mu)^{1/2}, & 0, & \sin^{-1}\mu \end{pmatrix}.$$

The directing function is $r(\mu) = \dot{g}(\mu)\, g^{\mathrm{T}}(\mu)$:

$$r(\mu) = \frac{1}{2}\begin{bmatrix} 0, & -\sqrt{2}\sin\mu, & \dfrac{-\sin(2\mu)}{-[\cos(2\mu)]^{1/2}} \\ \dfrac{1}{\sin^2\mu}, & \sqrt{2}\cos\mu, & \dfrac{\sin(2\mu)+2\cos(2\mu)}{2\sin^2\mu[-\cos(2\mu)]^{1/2}} \\ \dfrac{\cos\mu}{\sin^2\mu[-\cos(2\mu)]^{1/2}}, & 0, & \dfrac{-\cos\mu}{\sin^2\mu} \end{bmatrix} \cdot$$

$$\cdot \begin{pmatrix} 1, & -\cot\mu, & (1-\cot^2\mu)^{1/2} \\ \sqrt{2}\cos\mu, & \sqrt{2}\sin\mu, & 0 \\ -[-\cos(2\mu)]^{1/2}, & \cot\mu[-\cos(2\mu)]^{1/2}, & \sin^{-1}\mu \end{pmatrix}.$$

Denote $r(\mu) = (a^i_j)$. Then

$$a^1_2 = \tfrac{1}{2}[-2\sin^2\mu - \sin(2\mu)\cot\mu] = -1\,,$$

$$a^1_3 = -\frac{1}{2}\frac{\sin(2\mu)}{[-\cos(2\mu)]^{1/2}}\frac{1}{\sin\mu} = -\frac{\cos\mu}{[-\cos(2\mu)]^{1/2}},$$

$$a^3_2 = -\frac{1}{2}\left(\cot\mu\frac{\cos\mu}{\sin^2\mu[-\cos(2\mu)]^{1/2}} + \cot\mu\frac{\cos\mu[-\cos(2\mu)]^{1/2}}{\sin^2\mu}\right) = -\frac{\cos^2\mu}{\sin\mu[-\cos(2\mu)]^{1/2}}.$$

The directing function $r(\mu)$ in matrix form is

$$r(\mu) = \begin{bmatrix} 0, & -1, & \dfrac{-\cos\mu}{[-\cos(2\mu)]^{1/2}} \\ 1, & 0, & \dfrac{\cos^2\mu}{\sin\mu[-\cos(2\mu)]^{1/2}} \\ \dfrac{\cos\mu}{[-\cos(2\mu)]^{1/2}}, & \dfrac{-\cos^2\mu}{\sin\mu[-\cos(2\mu)]^{1/2}}, & 0 \end{bmatrix}.$$

In vector form we have

$$r(\mu) = -u_1\frac{\cos^2\mu}{\sin\mu[-\cos(2\mu)]^{1/2}} - u_2\frac{\cos\mu}{[-\cos(2\mu)]^{1/2}} + u_3\,.$$

To simplify further computations we denote

$$\cos\mu[-\cos(2\mu)]^{-1/2} = (\tan^2\mu - 1)^{-1/2} = A\,.$$

Then

$$\boldsymbol{r}(\mu) = \begin{pmatrix} -A\cot\mu \\ -A \\ 1 \end{pmatrix}, \quad \boldsymbol{r}^{\boldsymbol{\cdot}}(\mu) = \begin{pmatrix} -\dot{A}\cot\mu + A\sin^{-2}\mu \\ -\dot{A} \\ 0 \end{pmatrix},$$

$$\boldsymbol{r}^{\boldsymbol{\cdot\cdot}}(\mu) = \begin{pmatrix} -\ddot{A}\cot\mu + 2\dot{A}\sin^{-2}\mu - 2A\sin^{-3}\mu\cos\mu \\ -\ddot{A} \\ 0 \end{pmatrix}.$$

Further, we have

$$(\dot{\varphi})^2 = \boldsymbol{r}^2 = 1 + A^2(1 + \cot^2\mu) = 1 + A^2\sin^{-2}\mu\,,$$

$$\begin{aligned}(\boldsymbol{r}^{\boldsymbol{\cdot}})^2 &= (\dot{A})^2\cot^2\mu + A^2\sin^{-4}\mu - 2A\dot{A}\cos\mu\sin^{-3}\mu + (\dot{A})^2 \\ &= (\dot{A})^2\sin^{-2}\mu - 2A\dot{A}\cos\mu\sin^{-3}\mu + A^2\sin^{-4}\mu\,,\end{aligned}$$

$$\begin{aligned}(\boldsymbol{r}, \boldsymbol{r}^{\boldsymbol{\cdot}}) &= A\dot{A}\cot^2\mu - A^2\cos\mu\sin^{-3}\mu + A\dot{A} \\ &= A\dot{A}\sin^{-2}\mu + A^2\cos\mu\sin^{-3}\mu\,,\end{aligned}$$

$$\begin{aligned}|\boldsymbol{r}, \boldsymbol{r}^{\boldsymbol{\cdot}}, \boldsymbol{r}^{\boldsymbol{\cdot\cdot}}| &= -\ddot{A}(-\dot{A}\cot\mu + A\sin^{-2}\mu) \\ &\quad + \dot{A}(-\ddot{A}\cot\mu + 2\dot{A}\sin^{-2}\mu - 2A\sin^{-3}\mu\cos\mu) \\ &= -A\ddot{A}\sin^{-2}\mu + 2(\dot{A})^2\sin^{-2}\mu - 2A\dot{A}\sin^{-3}\mu\cos\mu\,.\end{aligned}$$

Denote

$$V = (\boldsymbol{r}^{\boldsymbol{\cdot}})^2(\dot{\varphi})^2 - (\boldsymbol{r}, \boldsymbol{r}^{\boldsymbol{\cdot}})^2\,.$$

Then

$$\kappa_1 = (\dot{\varphi})^{-3}V^{1/2}, \quad \kappa_2 = V^{-1}|\boldsymbol{r}, \boldsymbol{r}^{\boldsymbol{\cdot}}, \boldsymbol{r}^{\boldsymbol{\cdot\cdot}}|\,.$$

For V we obtain

$$\begin{aligned}V &= \sin^{-6}\mu\{[(\dot{A})^2\sin^{-2}\mu - 2A\dot{A}\cos\mu\sin\mu + A^2](\sin^2\mu + A^2) \\ &\quad - (A\dot{A}\sin\mu - A^2\cos\mu)^2\} \\ &= \sin^{-6}\mu[A^4\sin^2\mu + (\dot{A})^2\sin^4\mu - 2A\dot{A}\cos\mu\sin^3\mu \\ &\quad + A^2\sin^2\mu] = \sin^{-4}\mu[A^2 + A^4 + (\dot{A})^2\sin^2\mu - A\dot{A}\sin(2\mu)]\,.\end{aligned}$$

Further, we shall have

$$(\dot{\varphi})^{-3} = [(\dot{\varphi})^2]^{-3/2} = [\sin^{-2}\mu(\sin^2\mu + A^2)]^{-3/2}$$
$$= \sin^3\mu(\sin^2\mu + A^2)^{-3/2} .$$

Finally,

$$\kappa_1 = (\dot{\varphi})^{-3}\, V^{1/2}$$
$$= \sin\mu(\sin^2\mu + A^2)^{-3/2}\,[A^2 + A^4 + (\dot{A})^2\sin^2\mu$$
$$- A\dot{A}\sin(2\mu)]^{1/2} ,$$

$$\kappa_2 = V|\boldsymbol{r}, \boldsymbol{r}^{\cdot}, \boldsymbol{r}^{\cdot\cdot}|$$
$$= \sin\mu[A^2 + A^4 + (\dot{A})^2\sin^2\mu - A\dot{A}\sin(2\mu)]^{-1}\,[2(\dot{A})^2\sin\mu$$
$$-\ddot{A}A\sin\mu - 2A\dot{A}\cos\mu] .$$

For concrete computations these expressions are sufficient, it is not necessary to substitute for $A = (\tan^2\mu - 1)^{-1/2}$. This would make the expressions unnecessarily complicated. For illustration, let us compute κ_1 and κ_2 for $\mu = \frac{1}{3}\pi$. Then

$$\sin\mu = \frac{\sqrt{3}}{2}, \quad \cos\mu = \tfrac{1}{2}, \quad \tan\mu = \sqrt{3} .$$

$$A = \frac{\sqrt{2}}{2},$$

$$\dot{A} = [(\tan^2\mu - 1)^{-1/2}]^{\cdot}|_{\pi/3}$$
$$= \left[-\frac{1}{2}(\tan^2\mu - 1)^{-3/2}\, 2\tan\mu\,\frac{1}{\cos^2\mu}\right]\Bigg|_{\pi/3}$$
$$= -2^{-3/2}\sqrt{(3)}\,.\,\frac{1}{\frac{1}{4}} = -\sqrt{6},$$

$$\ddot{A} = \left[\frac{3}{2}(\tan^2\mu - 1)^{-5/2}\, 2\tan^2\mu\cos^{-4}\mu\right.$$
$$\left.- (\tan^2\mu - 1)^{-3/2}\left(\frac{1}{\cos^4\mu} + 2\tan\mu\cos^{-3}\mu\sin\mu\right)\right]\Bigg|_{\pi/3}$$
$$= \frac{3}{2}\,.\,2^{-5/2}\,.\,6\,.\,\frac{1}{\frac{1}{16}} - 2^{-3/2}\left(\frac{1}{\frac{1}{16}} + 2\sqrt{3}\,.\,\frac{1}{\frac{1}{8}}\,.\,\frac{\sqrt{3}}{2}\right) = 8\sqrt{2} .$$

Direct substitution yields

$$\kappa_1\left(\frac{\pi}{3}\right) = \frac{18\sqrt{5}}{25}, \quad \kappa_2\left(\frac{\pi}{3}\right) = \frac{4\sqrt{3}}{9}.$$

3.45 Exercise. Find the invariants of the fixed directing cone for the motion given by the matrix

$$g(\alpha) = \begin{pmatrix} \cos^2\alpha, & \sin\alpha, & \sin\alpha\cos\alpha \\ -\sin\alpha\cos\alpha, & \cos\alpha, & -\sin^2\alpha \\ -\sin\alpha, & 0, & \cos\alpha \end{pmatrix}.$$

3.46 Example. Let us find the formulae for the computation of κ_1 and κ_2 for a motion given by means of Euler angles. With regard to the fact that the resulting formulae are rather complicated we limit ourselves to the case $\psi_1 = 0$. Then

$$\boldsymbol{r} = \begin{pmatrix} \dot\psi_2 \\ -\dot\psi_3 \sin\psi_2 \\ \dot\psi_3\cos\psi_2 \end{pmatrix}, \quad \boldsymbol{r}^\bullet = \begin{pmatrix} \ddot\psi_2 \\ -\ddot\psi_3\sin\psi_2 - \dot\psi_2\dot\psi_3\cos\psi_2 \\ \ddot\psi_3\cos\psi_2 - \dot\psi_2\dot\psi_3\sin\psi_2 \end{pmatrix},$$

$$\boldsymbol{r}^{\bullet\bullet} = \begin{pmatrix} \dddot\psi_2 \\ -\dddot\psi_3\sin\psi_2 - 2\dot\psi_2\ddot\psi_3\cos\psi_2 - \dot\psi_3\ddot\psi_2\cos\psi_2 + \dot\psi_3(\dot\psi_2)^2\sin\psi_2 \\ \dddot\psi_3\cos\psi_2 - 2\dot\psi_2\ddot\psi_3\sin\psi_2 - \dot\psi_3\ddot\psi_2\sin\psi_2 - \dot\psi_3(\dot\psi_2)^2\cos\psi_2 \end{pmatrix}.$$

Further, we have

$$\begin{aligned}
(\dot\varphi)^2 &= (\dot\psi_2)^2 + (\dot\psi_3)^2, \\
(\boldsymbol{r}^\bullet)^2 &= (\ddot\psi_2)^2 + (\ddot\psi_3)^2 + (\dot\psi_2\dot\psi_3)^2, \\
(\boldsymbol{r}, \boldsymbol{r}^\bullet) &= \dot\psi_2\ddot\psi_2 + \dot\psi_3\ddot\psi_3, \\
|\boldsymbol{r}, \boldsymbol{r}^\bullet, \boldsymbol{r}^{\bullet\bullet}| &= 2(\dot\psi_2)^2(\ddot\psi_2)^2(\ddot\psi_3)^2 - (\ddot\psi_2)^2(\dot\psi_3)^2 \\
&\quad + \dot\psi_2\dot\psi_3[\dddot\psi_3\dot\psi_3 - \dddot\psi_3\dot\psi_2 + (\dot\psi_2)^3\dot\psi_3 - \ddot\psi_2\ddot\psi_3]. \\
\kappa_1 &= (\dot\varphi)^{-3}\{(\boldsymbol{r}^\bullet)^2(\dot\varphi)^2 - (\boldsymbol{r}, \boldsymbol{r}^\bullet)^2\}^{1/2}, \\
\kappa_1 &= \{(\boldsymbol{r}^\bullet)^2(\dot\varphi)^2 - (\boldsymbol{r}, \boldsymbol{r}^\bullet)^2\}^{-1}|\boldsymbol{r}, \boldsymbol{r}^\bullet, \boldsymbol{r}^{\bullet\bullet}|,
\end{aligned}$$

i.e.

$$\begin{aligned}
\kappa_1 &= [(\dot\psi_2)^2 + (\dot\psi_3)^2]^{-3/2}\{(\dot\psi_2\ddot\psi_2 - \ddot\psi_3\dot\psi_2)^2 \\
&\quad + (\dot\psi_2\dot\psi_3)^2[(\dot\psi_2)^2 + (\dot\psi_3)^2]\}^{1/2},
\end{aligned} \tag{3.5}$$

$$\kappa_2 = \{(\dot\psi_3\ddot\psi_2 - \dot\psi_2\ddot\psi_3)^2 + (\dot\psi_2\dot\psi_3)^2 [(\dot\psi_2)^2 + (\dot\psi_3)^2]\}^{-1} \cdot \{2(\ddot\psi_3)^2 (\dot\psi_2)^2 - (\dot\psi_3)^2 (\ddot\psi_2)^2 + \dot\psi_3\dot\psi_2[\dot\psi_3\dddot\psi_2 - \dddot\psi_3\dot\psi_2 + \dot\psi_3(\dot\psi_2)^3 - \ddot\psi_3\ddot\psi_2]\}\,. \tag{3.6}$$

3.47 Exercise. Determine the motion from Exercise 3.44 with the aid of Euler angles, and find the curvatures κ_1 and κ_2 from the formulae (3.5) and (3.6).

Hint. First of all, it is necessary to modify the matrix of motion to a form suitable for transformation to Euler angles (we need $\psi_1 = 0$). This can be achieved by suitably renumbering the base vectors, i.e. by their permutation, namely by the permutation $\{\boldsymbol{u}_1, \boldsymbol{u}_2, \boldsymbol{u}_3\} \to \{\boldsymbol{u}_3, \boldsymbol{u}_1, \boldsymbol{u}_2\}$.

3.48 Example. Let us find the curvatures of the motion from Example 2.5 (see also Example 3.25). There we had $\psi_1 = 0$, $\psi_2 = -1$, $\psi_3 = \arcsin(-\lambda \tan t)$, and by Example 3.25

$$\dot\psi_1 = 0\,, \quad \dot\psi_2 = -1\,, \quad \dot\psi_3 = -\lambda \cos^{-2} t(1 - \lambda^2 \tan^2 t)^{-1/2}\,.$$

The formula is simplified to

$$\kappa_1 = [(\dot\psi_3)^2 + 1]^{-3/2} \{(\ddot\psi_3)^2 + (\dot\psi_3)^2 [1 + (\dot\psi_3)^2]\}^{1/2}\,,$$
$$\kappa_2 = \{(\ddot\psi_3)^2 + (\dot\psi_3)^2 [1 + (\dot\psi_3)^2]\}^{-1} [2(\ddot\psi_3)^2 - \dot\psi_3(\dddot\psi_3 - \dot\psi_3)]\,,$$

where we substitute the derivatives of the function $\arcsin(-\lambda \tan t)$ for $\dot\psi_3$, $\ddot\psi_3$, and $\dddot\psi_3$.

3.49 Exercise. Find the values κ_1 and κ_2 for the motion from Example 3.47 at the instant $t = 0$.

Result. $\kappa_1 = (\lambda^2 + 1)^{-1} \lambda$, $\kappa_2 = (\lambda^2 + 1)^{-1} (3 - \lambda^2)$.

3.50 Exercise. Find the curvatures κ_1 and κ_2 for the motion from Example 3.26 with the aid of formulae (3.5) and (3.6).

3.51 Definition. Let $g(\varphi)$ be a spherical motion with fixed directing cone $\boldsymbol{R}(\varphi)$. Then $\boldsymbol{R}(\varphi)$ can be regarded as a curve on the sphere S^2. The parameter s of the spherical curve $\boldsymbol{R}(\varphi)$ such that $(d\boldsymbol{R}/ds, d\boldsymbol{R}/ds) = 1$ is called the *arc of the spherical curve.*

3.52 Theorem. *For the spherical curve* $\boldsymbol{R}(\varphi)$ *we have the Frenet formulae*

$$\frac{d\boldsymbol{R}}{ds} = \boldsymbol{T}, \quad \frac{d\boldsymbol{T}}{ds} = -\boldsymbol{R} + k\boldsymbol{N}, \quad \frac{d\boldsymbol{N}}{ds} = -k\boldsymbol{T};$$

k is called the *spherical curvature of the curve* $\boldsymbol{R}(\varphi)$.

Proof. We have

$$\frac{d\boldsymbol{R}}{d\varphi} = \frac{d\boldsymbol{R}}{ds}\frac{ds}{d\varphi} = \kappa_1\boldsymbol{T} = \frac{ds}{d\varphi}\boldsymbol{T}, \quad \text{i.e.} \quad \kappa_1 = \frac{ds}{d\varphi}.$$

Further,

$$\frac{d\boldsymbol{T}}{ds} = \frac{d\boldsymbol{T}}{d\varphi}\frac{d\varphi}{ds} = \kappa_1^{-1}(-\kappa_1\boldsymbol{R} + \kappa_2\boldsymbol{N}) = -\boldsymbol{R} + \kappa_1^{-1}\kappa_2\boldsymbol{N},$$

$$\frac{d\boldsymbol{N}}{ds} = -\kappa_1^{-1}\kappa_2\boldsymbol{T},$$

so that $k = \kappa_1^{-1}\kappa_2$. This proves simultaneously the following theorem:

3.53 Theorem. *Let* $g(\varphi)$ *be a motion with invariants* κ_1 *and* κ_2 *of the fixed directing cone,* $\bar{\kappa}_1$ *and* $\bar{\kappa}_2$ *of the moving directing cone. Let* s *and* $\bar{s}$ *be arcs and* k *and* $\bar{k}$ *spherical curvatures of the curves* $\boldsymbol{R}(\varphi)$ *and* $\bar{\boldsymbol{R}}(\varphi)$*, respectively. Then*

$$\frac{ds}{d\varphi} = \frac{d\bar{s}}{d\varphi} = \kappa_1, \quad k = \frac{\kappa_2}{\kappa_1}, \quad \bar{k} = \frac{\bar{\kappa}_2}{\kappa_1}.$$

3.54 Theorem. $k - \bar{k} = \kappa_1^{-1}$.

Proof. The theorem is a direct consequence of Theorems 3.52 and 3.36.

3.55 Theorem. *Spherical motion is realized by rolling the moving centrode along the fixed centrode,* i.e. $\bar{s} = s + c$, $c = const.$

Proof. The theorem follows immediately from Theorem 3.52.

3.56 Theorem. *Let* $\boldsymbol{x}(t_1)$ *and* $\bar{\boldsymbol{x}}(t_2)$ *be two arbitrary curves on a sphere which contact at the point* $t_1 = 0$ *and* $t_2 = 0$*. Then there exists a unique spherical motion such that* $\boldsymbol{x}(t_1)$ *is its fixed centrode while* $\bar{\boldsymbol{x}}(t_2)$ *is its moving centrode.*

Proof. We represent the curves $\boldsymbol{x}(t_1)$ and $\bar{\boldsymbol{x}}(t_2)$ as functions of the arc. Taking into account that the arc is determined up to an additive constant it can be assumed that for $t_1 = 0$ and $t_2 = 0$ we also have $s = 0$. We express the curvatures of the curves $\boldsymbol{x}(t_1)$ and $\bar{\boldsymbol{x}}(t_2)$ also as functions of the arc. By Theorem 3.52 we then have $k(s) - \bar{k}(s) = \kappa_1^{-1}(s)$. Here

$$\kappa_1^{-1}(s) = \frac{d\varphi}{ds}, \quad \text{so that} \quad \frac{d\varphi}{ds} = k(s) - \bar{k}(s).$$

Integration yields φ as a function of s, $\varphi = \varphi(s)$; we compute s as a function of φ, $s = s(\varphi)$, substitute into $\kappa_1(s)$ and $\kappa_2(s) = k(s)\,\kappa_1(s)$, and obtain $\kappa_1 = \kappa_1(s(\varphi))$, $\kappa_2 = k(s(\varphi))\,\kappa_1(s(\varphi))$, i.e. the curvatures of the fixed directing cone are found as functions of the canonical parameter. By Theorem 3.41 there exists a unique motion which satisfies the initial conditions on the contact of the directing cones and has the curvatures κ_1 and κ_2.

Briefly we say that spherical motion is realized as the rolling of the centrodes, and it is determined by these centrodes.

3.57 Example. Let us elucidate the proof of Theorem 3.55 using a simple example. Let two curves $\boldsymbol{x}(t_1)$ and $\bar{\boldsymbol{x}}(t_2)$ on the sphere S^2 be given by the parametric equations

$$x^1(t_1) = \frac{1}{2}(1 + \cos t_1), \quad \bar{x}^1(t_2) = \frac{1}{2}(1 - \cos t_2),$$

$$x^2(t_1) = \frac{\sqrt{2}}{2}\sin t_1, \qquad \bar{x}^2(t_2) = \frac{\sqrt{2}}{2}\sin t_2,$$

$$x^3(t_1) = \frac{1}{2}(1 - \cos t_1), \quad \bar{x}^3(t_2) = -\frac{1}{2}(1 - \cos t_2).$$

As we see from these equations, there are two circles with radius $\sqrt{2}/2$ lying symmetrically with respect to the plane $(\boldsymbol{u}_1\boldsymbol{u}_2)$. It is easily ascertained that $\boldsymbol{x}(0) = \bar{\boldsymbol{x}}(0) = \boldsymbol{u}_1$, $\boldsymbol{x}'(0) = \bar{\boldsymbol{x}}'(0) = \boldsymbol{u}_2$. Thus, $\dot{\boldsymbol{x}}$ and $\bar{\boldsymbol{x}}$ contact at the point $t_1 = 0$ and $t_2 = 0$. Consequently, the assumptions of the theorem are satisfied. Thus, let us look for the single motion $g(t)$ with centrodes $\boldsymbol{x}$ and $\bar{\boldsymbol{x}}$, i.e. with fixed directing cone $\boldsymbol{R} = \boldsymbol{x}$ and moving directing cone $\bar{\boldsymbol{R}} = \bar{\boldsymbol{x}}$. First, we find the arc s of the curves $\boldsymbol{x}$ and $\bar{\boldsymbol{x}}$:

$$(x^1)^\cdot = -\frac{\sin t_1}{2}; \quad (x^2)^\cdot = \frac{\sqrt{2}}{2}\cos t_1, \quad (x^3)^\cdot = \frac{\sin t_1}{2},$$

so that

$$\frac{\mathrm{ds}}{\mathrm{dt}_1} = \pm\left[\sum_{i=1}^{3}\left(\frac{\mathrm{d}x^i}{\mathrm{dt}_1}\right)^2\right]^{1/2} = \pm\frac{\sqrt{2}}{2}.$$

Similarly, we have $\mathrm{ds}/\mathrm{dt}_2 = \pm\sqrt{2}/2$. Choose the plus sign; from what follows it will be seen that this is the correct choice. Thus, $\mathrm{ds} = \frac{1}{2}\sqrt{(2)}\,\mathrm{dt}_1$, i.e. $\mathrm{s} = \frac{1}{2}\sqrt{(2)}\,\mathrm{t}_1$, i.e. $\mathrm{t}_1 = \mathrm{s}\sqrt{2}$. By a prime we denote the derivative with respect to the arc s. From the Frenet formulae in Theorem 3.51 we easily see that $k = |\boldsymbol{x}, \boldsymbol{x}', \boldsymbol{x}''|$, $\bar{k} = |\bar{\boldsymbol{x}}, \bar{\boldsymbol{x}}', \bar{\boldsymbol{x}}''|$. Since both these curves are circles, k and $\bar{k}$ are constant. Consequently, it suffices to compute $k(0)$ and $\bar{k}(0)$. We obtain $k = 1$, $\bar{k} = -1$. Now, we have $k - \bar{k} = \kappa_1^{-1}$, i.e. $\kappa_1 = \frac{1}{2} > 0$ which confirms our choice of sign. (Should $\kappa_1 < 0$ we would choose the arc $-$s.) Further, $\mathrm{d}\varphi/\mathrm{ds} = \kappa_1^{-1} = 2$, i.e. $\varphi = 2\mathrm{s}$. Thus, we have the fixed directing cone represented as a function of the canonical parameter φ, where $\mathrm{t}_1 = \frac{1}{2}\sqrt{(2)}\,\varphi$. Following the solution of the differential equation $\mathrm{d}g/\mathrm{d}\varphi = \boldsymbol{R}(\varphi)\,g$ with initial condition $\boldsymbol{g}(0) = e$, we obtain the uniquely determined motion $g(\varphi)$ which satisfies our conditions.

3.58 Remark. As a matter of fact, Theorem 3.56 is not suitable for practical purposes. If the centrodes are given, it is possible to find the motion directly without solving differential equations. Indeed, let the centrodes be given as functions of the arc s, i.e. let the directing cones $\boldsymbol{R}(\mathrm{s})$ and $\bar{\boldsymbol{R}}(\mathrm{s})$ be given. For the desired motion $g(\mathrm{s})$ we then have $g(\mathrm{s})\,\bar{\boldsymbol{R}}(\mathrm{s}) = \boldsymbol{R}(\mathrm{s})$, $g(\mathrm{s})\,\bar{\boldsymbol{T}}(\mathrm{s}) = \boldsymbol{T}(\mathrm{s})$, and thus also $g(\mathrm{s})\,\bar{\boldsymbol{N}}(\mathrm{s}) = \boldsymbol{N}(\mathrm{s})$. Denoting then by $\bar{\mathscr{F}}(\mathrm{s}) = \{\bar{\boldsymbol{R}}(\mathrm{s}), \bar{\boldsymbol{T}}(\mathrm{s}), \bar{\boldsymbol{N}}(\mathrm{s})\}$ and $\mathscr{F}(\mathrm{s}) = \{\boldsymbol{R}(\mathrm{s}), \boldsymbol{T}(\mathrm{s}), \boldsymbol{N}(\mathrm{s})\}$ the Frenet frames of the moving and fixed centrodes, and if $\mathscr{R} = \{\boldsymbol{u}_1, \boldsymbol{u}_2, \boldsymbol{u}_3\}$ is the basic frame in E_3, we have $\mathscr{F}(\mathrm{s}) = \mathscr{R}\,\gamma(\mathrm{s})$, $\bar{\mathscr{F}}(\mathrm{s}) = \mathscr{R}\,\bar{\gamma}(\mathrm{s})$ where the matrices γ and $\bar{\gamma}$ are constituted by columns of coordinates of the vectors from $\mathscr{F}$ and $\bar{\mathscr{F}}$, respectively. Further, we have

$$\mathscr{R} = \bar{\mathscr{F}}(\mathrm{s})\,\gamma(\mathrm{s})^{\mathrm{T}},$$

i.e.

$$\bar{\gamma}(\mathrm{s})^{\mathrm{T}}\,\bar{\boldsymbol{R}}(\mathrm{s}) = \boldsymbol{u}_1, \quad \bar{\gamma}(\mathrm{s})^{\mathrm{T}}\,\bar{\boldsymbol{T}}(\mathrm{s}) = \boldsymbol{u}_2, \quad \bar{\gamma}(\mathrm{s})^{\mathrm{T}}\,\bar{\boldsymbol{N}}(\mathrm{s}) = \boldsymbol{u}_3,$$

so that

$$\gamma(\mathrm{s})\,\bar{\gamma}(\mathrm{s})^{\mathrm{T}}\,\bar{\boldsymbol{R}}(\mathrm{s}) = \gamma(\mathrm{s})\,\boldsymbol{u}_1 = \boldsymbol{R}(\mathrm{s})$$

and similarly

$$\gamma(s)\,\bar{\gamma}(s)^T\,\bar{\boldsymbol{T}}(s) = \boldsymbol{T}(s)\,, \quad \gamma(s)\,\bar{\gamma}(s)^T\,\bar{\boldsymbol{N}}(s) = \boldsymbol{N}(s)\,;$$

$g(s) = \gamma(s)\,\bar{\gamma}(s)^T$ is thus the representation of the desired motion in the frame $\mathscr{R}$. Let us apply the above considerations to Example 3.57. We write the directing cones $\boldsymbol{R}$ and $\bar{\boldsymbol{R}}$ directly as functions of the arc s. (There was $t_1 = s\sqrt{2}$, $t_2 = s\sqrt{2}$.) We obtain

$$\boldsymbol{R}(s) = \frac{\boldsymbol{u}_1}{2}\,[1 + \cos(s\sqrt{2})] + \frac{\boldsymbol{u}_2\sqrt{2}}{2}\sin(s\sqrt{2}) + \frac{\boldsymbol{u}_3}{2}\,[1 - \cos(s\sqrt{2})]\,,$$

$$\bar{\boldsymbol{R}}(s) = \frac{\boldsymbol{u}_1}{2}\,[1 + \cos(s\sqrt{2})] + \frac{\boldsymbol{u}_2\sqrt{2}}{2}\sin(s\sqrt{2}) - \frac{\boldsymbol{u}_3}{2}\,[1 - \cos(s\sqrt{2})]\,.$$

Further, we have

$$\boldsymbol{T}(s) = -\frac{\boldsymbol{u}_1\sqrt{2}}{2}\sin(s\sqrt{2}) + \boldsymbol{u}_2\cos(s\sqrt{2}) + \frac{\boldsymbol{u}_3\sqrt{2}}{2}\sin(s\sqrt{2})\,,$$

$$\bar{\boldsymbol{T}}(s) = -\frac{\boldsymbol{u}_1\sqrt{2}}{2}\sin(s\sqrt{2}) + \boldsymbol{u}_2\cos(s\sqrt{2}) - \frac{\boldsymbol{u}_3\sqrt{2}}{2}\sin(s\sqrt{2})\,.$$

Finally, $\boldsymbol{N}(s) = \boldsymbol{R}(s) \times \boldsymbol{T}(s)$, $\bar{\boldsymbol{N}}(s) = \bar{\boldsymbol{R}}(s) \times \bar{\boldsymbol{T}}(s)$, i.e.

$$\boldsymbol{N} = \frac{\boldsymbol{u}_1}{2}\,[1 - \cos(s\sqrt{2})] - \frac{\boldsymbol{u}_2\sqrt{2}}{2}\sin(s\sqrt{2}) + \frac{\boldsymbol{u}_3}{2}\,[1 + \cos(s\sqrt{2})]\,,$$

$$\bar{\boldsymbol{N}} = -\frac{\boldsymbol{u}_1}{2}\,[1 - \cos(s\sqrt{2})] + \frac{\boldsymbol{u}_2\sqrt{2}}{2}\sin(s\sqrt{2}) + \frac{\boldsymbol{u}_3}{2}\,[1 + \cos(s\sqrt{2})]\,.$$

For the sake of simplicity we denote $s\sqrt{2} = t$. Then we obtain

$$\gamma(t) = \tfrac{1}{2}\begin{pmatrix} 1 + \cos t, & -\sqrt{2}\sin t, & 1 - \cos t \\ \sqrt{2}\sin t, & 2\cos t, & -\sqrt{2}\sin t \\ (1 - \cos t, & \sqrt{2}\sin t, & 1 + \cos t \end{pmatrix},$$

$$\bar{\gamma}(t) = \tfrac{1}{2}\begin{pmatrix} 1 + \cos t, & -\sqrt{2}\sin t, & \cos t - 1 \\ \sqrt{2}\sin t, & 2\cos t, & \sqrt{2}\sin t \\ \cos t - 1, & -\sqrt{2}\sin t, & 1 + \cos t \end{pmatrix},$$

so that

$$g(t) = \gamma(t)\,\bar{\gamma}^{T}(t)$$

$$= \frac{\sin t}{2}\begin{pmatrix} 2\cot t + \sin t, & \sqrt{2}(\cos t - 1), & -\sin t \\ \sqrt{2}(\cos t - 1), & 2\cot t\cos t, & -\sqrt{2}(\cos t + 1) \\ \sin t, & \sqrt{2}(\cos t + 1), & 2\cot t - \sin t \end{pmatrix}.$$

3.59 Theorem. *Let $\overline{\mathscr{F}}(s)$ and $\mathscr{F}(s)$ be Frenet frames of the moving and fixed centrodes of motion $g(s)$, respectively; let $\mathscr{R}$ be the basic frame in E_3. Denote $\mathscr{F}(s) = \mathscr{R}\,\gamma(s)$, $\overline{\mathscr{F}}(s) = \mathscr{R}\,\bar{\gamma}(s)$. Then*

$$g(s) = \gamma(s)\,\bar{\gamma}^{T}(s)$$

in the frame $\mathscr{R}$ and

$$\mathscr{F}(s) = \overline{\mathscr{F}}(s)\,h(s),$$

where

$$h(s) = \bar{\gamma}^{T}(s)\,\gamma(s).$$

3.60 Remark. Between the curvature K of a spherical curve $\boldsymbol{R}(s)$ and its spherical curvature k the relation $K^2 = 1 + k^2$ holds, i.e. the vector $k\boldsymbol{N}$ of spherical curvature is the projection of the curvature vector $K\boldsymbol{N}_0$, where

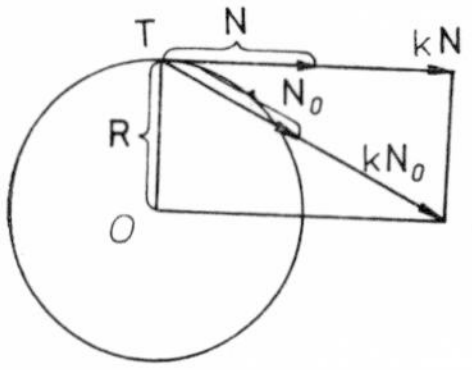

Fig. 23.

$\boldsymbol{N}_0$ is the unit vector of the principal normal of the curve $\boldsymbol{R}(s)$, into the tangent plane of the spherical surface S^2 at the point $\boldsymbol{R}(s)$ – see Fig. 23. As a matter of fact, we have

$$\frac{d\boldsymbol{T}}{ds} = K\boldsymbol{N}_0 = -\boldsymbol{R} + k\boldsymbol{N},$$

i.e. $(K\boldsymbol{N}_0, \boldsymbol{N}) = k$, i.e. the component of the vector $K\boldsymbol{N}_0$ to $\boldsymbol{N}$ has exactly the length of the spherical curvature k.

4. INVARIANTS OF THE TRAJECTORY OF A POINT, SPECIAL MOTIONS

Let $X(\mathrm{t})$ be a curve on a sphere. From Theorems 3.38 and 3.52 it is easily verified that for its spherical curvature k we have

$$k = (\dot{X}, \dot{X})^{-3/2} \left|X, \dot{X}, \ddot{X}\right|,$$

for arbitrary parameter t. Now, if $g(\varphi)$ is a spherical motion with directing cone $\boldsymbol{R}(\varphi)$, then for the tangent vector $X'(\varphi)$ of the trajectory $X(\varphi)$ of a point $\bar{X}$ it is, by formula (3.1b): $X' = \boldsymbol{R} \times X$. Differentiating this equation once more we obtain (the prime denotes the derivative with respect to φ)

$$\begin{aligned} X'' &= \boldsymbol{R}' \times X + \boldsymbol{R} \times X' = \boldsymbol{R}' \times X + \boldsymbol{R} \times (\boldsymbol{R} \times X) \\ &= \kappa_1 \boldsymbol{T} \times X + (\boldsymbol{R}, X)\boldsymbol{R} - X, \end{aligned}$$

exploiting the familiar identity $\boldsymbol{u} \times (\boldsymbol{v} \times \boldsymbol{w}) = (\boldsymbol{u}, \boldsymbol{w})\,\boldsymbol{v} - (\boldsymbol{u}, \boldsymbol{v})\,\boldsymbol{w}$ for arbitrary vectors $\boldsymbol{u}, \boldsymbol{v}, \boldsymbol{w}$. Further,

$$\begin{aligned} (X', X') &= (\boldsymbol{R} \times X, \boldsymbol{R} \times X) = -(X \times \boldsymbol{R}, \boldsymbol{R} \times X) \\ &= -(X, \boldsymbol{R} \times (\boldsymbol{R} \times X)) \\ &= -(X, (\boldsymbol{R}, X)\boldsymbol{R} - X) = 1 - (\boldsymbol{R}, X)^2, \end{aligned}$$

$$\begin{aligned} \left|X, X', X''\right| &= \left|X, \boldsymbol{R} \times X, \kappa_1 \boldsymbol{T} \times X + (\boldsymbol{R}, X)\boldsymbol{R} - X\right| \\ &= \left|X, \boldsymbol{R} \times X, \kappa_1 \boldsymbol{T} \times X + (\boldsymbol{R}, X)\boldsymbol{R}\right| = (X \times (\boldsymbol{R} \times X)), \end{aligned}$$

$$\begin{aligned} \kappa_1 \boldsymbol{T} \times X + (\boldsymbol{R}, X)\boldsymbol{R} &= (\boldsymbol{R} - (\boldsymbol{R}, X)X, \kappa_1 \boldsymbol{T} \times X + (\boldsymbol{R}, X)\boldsymbol{R}) \\ &= \kappa_1 \left|\boldsymbol{R}, \boldsymbol{T}, X\right| + (\boldsymbol{R}, X) - (\boldsymbol{R}, X)^3 \\ &= \kappa_1 (\boldsymbol{N}, X) + (\boldsymbol{R}, X) - (\boldsymbol{R}, X)^3. \end{aligned}$$

We note the result:

4.1 Theorem. *For the spherical curvature k_X of the trajectory of a point $\bar{X}$ at the point X we have*

$$k_X = \left[1 - (\boldsymbol{R}, X)^2\right]^{-3/2} \left[\kappa_1 (\boldsymbol{N}, X) + (\boldsymbol{R}, X) - (\boldsymbol{R}, X)^3\right].$$

Further, let us compute the spherical curvature k_X in coordinate form. We write

$$X = \alpha \boldsymbol{R} + \beta \boldsymbol{T} + \gamma \boldsymbol{N}, \quad \alpha, \beta, \gamma \in \mathbf{R}.$$

Then $\alpha^2 + \beta^2 + \gamma^2 = 1$. For k_X we obtain

$$k_X = (1 - \alpha^2)^{-3/2} (\kappa_1 \gamma + \alpha - \alpha^3)$$
$$= (\beta^2 + \gamma^2)^{-3/2} [\kappa_1 \gamma + \alpha(\beta^2 + \gamma^2)] .$$

Now, we define the centre S_X of the spherical curvature of the trajectory of point X (see Fig. 24): $S_X = X + k_X^{-1} N_X$, where N_X stands for the normal to the trajectory of the point X (i.e. the normal which lies in the tangent plane of the sphere). From Fig. 24 we see that S_X is the projection

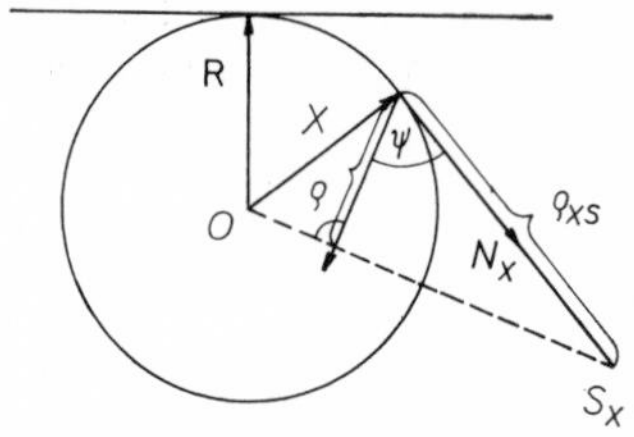

Fig. 24.

of the centre of the osculating circle of the curve $X(\varphi)$ into the tangent plane of the sphere S^2 at the point X. Indeed, if K_X is the curvature of the trajectory at the point X, N_{0X} the principal normal of the trajectory at the point X, N_X the normal of the trajectory at the point X which lies in the tangent plane of S^2, k_X the spherical curvature, then by Remark 3.60 we have $k_X = K_X \cos \psi$, where ψ is the angle from N_{0X} to N_X. Denote by ϱ_X the radius of curvature, $\varrho_X = K_X^{-1}$, and let ϱ_{XS} denote the spherical radius of curvature, i.e. $\varrho_{XS} = k_X^{-1}$. We have

$$\frac{1}{\varrho_{XS}} = \frac{1}{\varrho_X} \cos \psi , \quad \text{i.e.} \quad \varrho_X = \varrho_{XS} \cos \psi .$$

At this stage, it is possible to discuss the properties of the trajectory of a point. First, we find those points of the sphere S^2 whose osculating planes pass through the instantaneous pole, i.e. through the point $\boldsymbol{R}$ or $-\boldsymbol{R}$. The osculating plane of the trajectory at the point X has the equation $A = X + \lambda X' + \mu X''$ where $\lambda, \mu \in \mathbf{R}$. Let us substitute the expressions for X' and X''. Then

$$A = X + \lambda \boldsymbol{R} \times X + \mu[\kappa_1 \boldsymbol{T} \times X + (\boldsymbol{R}, X) \boldsymbol{R} - X]$$

and after computation

$$A = X + \lambda(-\gamma \boldsymbol{T} + \beta \boldsymbol{N}) + \mu(\kappa_1 \gamma \boldsymbol{R} - \kappa_1 \alpha \boldsymbol{N} + \alpha \boldsymbol{R} - X).$$

Choose $\varepsilon = \pm 1$. If the osculating circle of the trajectory is to pass through the point $\varepsilon \boldsymbol{R}$, then

$$\varepsilon \boldsymbol{R} = X + \lambda_0(-\gamma \boldsymbol{T} + \beta \boldsymbol{N}) + \mu_0(\kappa_1 \gamma \boldsymbol{R} - \kappa_1 \alpha \boldsymbol{N} + \alpha \boldsymbol{R} - X)$$

has to hold for certain $\lambda_0, \mu_0 \in \mathbf{R}$. Multiply this equation successively by the vectors $\boldsymbol{R}, \boldsymbol{T}, \boldsymbol{N}$. We obtain

$$\varepsilon = \alpha + \mu_0 \kappa_1 \gamma, \quad 0 = \beta - \lambda_0 \gamma - \mu_0 \beta, \quad 0 = \gamma + \lambda_0 \beta - \mu_0(\kappa_1 \alpha + \gamma).$$

Now, we multiply the second equation by β, the third equation by γ and add them. We obtain the equation

$$0 = \beta^2 + \gamma^2 - \mu_0(\beta^2 + \gamma^2 + \alpha \gamma \kappa_1);$$

multiply it by $\kappa_1 \gamma$ and substitute from the first equation.

We have

$$0 = (\beta^2 + \gamma^2)\kappa_1 \gamma + (\alpha - \varepsilon)(\beta^2 + \gamma^2 + \alpha \gamma \kappa_1),$$

i.e.

$$0 = (1 - \alpha^2)\kappa_1 \gamma + (\alpha - \varepsilon)(1 - \alpha^2 + \alpha \gamma \kappa_1).$$

Consider that $1 - \alpha^2 = \varepsilon^2 - \alpha^2 = (\varepsilon - \alpha)(\varepsilon + \alpha)$. Then

$$\begin{aligned} 0 &= (\varepsilon - \alpha)[(\varepsilon + \alpha)\kappa_1 \gamma - 1 + \alpha^2 - \alpha \gamma \kappa_1] \\ &= (\varepsilon - \alpha)(\alpha^2 - 1 + \varepsilon \kappa_1 \gamma). \end{aligned}$$

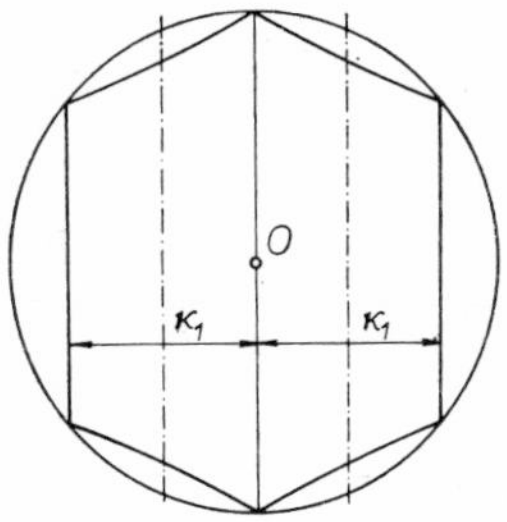

Fig. 25.

Two solutions are obtained: 1. $\alpha = \varepsilon$, $\beta = 0$, $\gamma = 0$; 2. $\beta^2 + \gamma^2 = \varepsilon\kappa_1\gamma$. It is sufficient to consider the second solution since it contains the first. We modify the equation $\beta^2 + \gamma^2 - \varepsilon\kappa_1\gamma = 0$ to the form $\beta^2 + (\gamma - \frac{1}{2}\varepsilon\kappa_1)^2 = \frac{1}{4}\kappa_1^2$. Consequently, the following holds:

The set of points of the sphere whose osculating planes pass through the point $\boldsymbol{R}$ lies on the cylindrical surface of revolution having the equation $\beta^2 + (\gamma - \frac{1}{2}\kappa_1)^2 = \frac{1}{4}\kappa_1^2$. The set of points whose osculating planes pass through the point $-\boldsymbol{R}$ lies on the cylindrical surface of revolution having the equation $\beta^2 + (\gamma + \frac{1}{2}\kappa_1)^2 = \frac{1}{4}\kappa_1^2$ (see Fig. 25).

Let us proceed further in the considerations started above. Denote by S_X the projection of the centre of the osculating circle from the centre O of the sphere S^2 into the tangent plane of the sphere S^2 at the point X, see Fig. 24. Obviously, we then have $S_X = X + \varrho_{XS}N_X$ where N_X is the normal of the trajectory at the point X. Let us find this normal. We have $X' = \boldsymbol{R} \times X$, i.e. the vector $\boldsymbol{R} \times X$ has the direction of the tangent. Consequently,

$$T_X = (\beta^2 + \gamma^2)^{-1/2}\, \boldsymbol{R} \times X$$

since the length of the vector $\boldsymbol{R} \times X$ is $(1 - \alpha^2)^{1/2} = (\beta^2 + \gamma^2)^{1/2}$. Thus, we have

$$\begin{aligned} N_X &= X \times T_X = (\beta^2 + \gamma^2)^{-1/2}\, X \times (\boldsymbol{R} \times X) \\ &= (\beta^2 + \gamma^2)^{-1/2}\, (\boldsymbol{R} - \alpha X)\,. \end{aligned}$$

4.2 Theorem. *The centre S_X of the spherical curvature of the trajectory of the point $\bar{X}$ at the point X is the point*

$$S_X = X + \frac{\beta^2 + \gamma^2}{\kappa_1\gamma + \alpha(\beta^2 + \gamma^2)}\,(\boldsymbol{R} - \alpha X)\,.$$

Proof. Substituting for ϱ_{XS} according to Theorem 4.1 we obtain the desired expression for S_X immediately.

4.3 Theorem. *The inflection points of the trajectory (i.e. the points for which $k_X = 0$, or rather the points whose osculation circles are great circles) lie on the spherical curve of third degree with the equation*

$$\kappa_1\gamma + \alpha(1 - \alpha^2) = 0\,.$$

Proof. The theorem again follows immediately from Theorem 4.1.

According to Theorem 4.2 a relation is obtained between the points on the sphere and their centres of spherical curvature, similarly as encountered in plane kinematics. However, the fact that the point S_X lies in different planes from case to case is inconvenient. So let us investigate next what would we obtain by projecting the points X as well as S_X from the

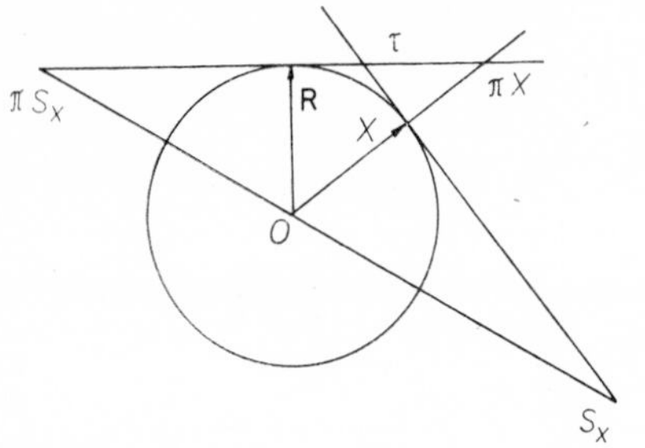

Fig. 26.

point O into the tangent plane τ of the sphere S^2 at the point $\boldsymbol{R}$. Denote this projection by π, see Fig. 26. If $X = \alpha\boldsymbol{R} + \beta\boldsymbol{T} + \gamma\boldsymbol{N}$, then $\pi X = \boldsymbol{R} + \beta\alpha^{-1}\boldsymbol{T} + \gamma\alpha^{-1}\boldsymbol{N}$. For the coordinate system in τ we take the origin $\boldsymbol{R}$ and the coordinate vectors $\boldsymbol{T}$ and $\boldsymbol{N}$. Then $\pi X = [\beta\alpha^{-1}, \gamma\alpha^{-1}]$. Further,

$$\begin{aligned} S_X &= \boldsymbol{R}\left[\alpha + \frac{(\beta^2+\gamma^2)(1-\alpha^2)}{\kappa_1\gamma + \alpha(\beta^2+\gamma^2)}\right] + \boldsymbol{T}\left[\beta - \frac{\alpha\beta(\beta^2+\gamma^2)}{\kappa_1\gamma + \alpha(\beta^2+\gamma^2)}\right] \\ &\quad + \boldsymbol{N}\left[\gamma - \frac{\alpha\gamma(\beta^2+\gamma^2)}{\kappa_1\gamma + \alpha(\beta^2+\gamma^2)}\right] \\ &= \frac{\boldsymbol{R}[\alpha\kappa_1\gamma + \alpha^2(1-\alpha^2) + (1-\alpha^2)^2] + \beta\kappa_1\gamma\boldsymbol{T} + \kappa_1\gamma^2\boldsymbol{N}}{\kappa_1\gamma + \alpha(\beta^2+\gamma^2)} \\ &= \frac{\boldsymbol{R}(\alpha\kappa_1\gamma + 1 - \alpha^2) + \kappa_1\beta\gamma\boldsymbol{T} + \kappa_1\gamma^2\boldsymbol{N}}{\kappa_1\gamma + \alpha(\beta^2+\gamma^2)}. \end{aligned}$$

Consequently, we have

$$\pi S_X = \left[\frac{\kappa_1\gamma}{\alpha\kappa_1\gamma + 1 - \alpha^2}\,\beta\,,\ \frac{\kappa_1\gamma}{\alpha\kappa_1\gamma + 1 - \alpha^2}\,\gamma\right];$$

if we denote $\beta/\alpha = x$, $\gamma/\alpha = y$, then to the point $\pi X = [x, y]$ corresponds the point

$$\pi S_{\mathbf{X}} = \frac{\kappa_1 y}{\alpha\kappa_1\gamma + \beta^2 + \gamma^2}[\beta, \gamma] = \frac{\kappa_1\left(\frac{\gamma}{\alpha}\right)}{\kappa_1\left(\frac{\gamma}{\alpha}\right) + \left(\frac{\beta}{\alpha}\right)^2 + \left(\frac{\gamma}{\alpha}\right)^2}\left[\frac{\beta}{\alpha}, \frac{\gamma}{\alpha}\right]$$

$$= \frac{\kappa_1\gamma}{\kappa_1 y + x^2 + y^2}[x, y] .$$

Writing $\pi S_{\mathbf{X}} = [x', y']$, the desired relationship assumes the form

$$x' = \frac{\kappa_1 y}{\kappa_1 y + x^2 + y^2}x , \quad y' = \frac{\kappa_1 y}{\kappa_1 y + x^2 + y^2}y .$$

Introducing polar coordinates by the relations

$$x = \varrho \sin\delta , \quad y = \varrho\cos\delta , \quad x' = \varrho'\sin\delta' , \quad y' = \varrho'\cos\delta' ,$$

we have

$$\delta = \delta' , \quad \varrho' = \frac{\varrho^2\kappa_1\cos\delta}{\varrho^2 + \varrho\kappa_1\cos\delta} = \frac{\varrho d}{\varrho + d} ,$$

where we write $d = \kappa_1\cos\delta$. Here it is necessary, however, to assume $0 \leqq \delta < \pi$, $-\infty < \varrho < \infty$. The relationship is then usually written in the form

$$\varrho\varrho' = d(\varrho - \varrho') , \quad d = \kappa_1\cos\delta \tag{4.1}$$

This is a relationship familiar from plane kinematic geometry: the relationship between points and the centres of the osculating circles of their trajectories. Sometimes we also encounter the form

$$\frac{1}{d} = \frac{1}{\varrho'} - \frac{1}{\varrho} , \quad d = \kappa_1\cos\delta . \tag{4.2}$$

Let us denote S_p or S_h the centre of the spherical curvature of the fixed or moving centrode, respectively. Their distance from $\boldsymbol{R}$ is equal to $1/k$ or $1/\bar{k}$, respectively. If we notice now that the tangent plane of the sphere at the point $\boldsymbol{R}$ is the plane τ and $\boldsymbol{R}$ is the origin of the coordinate system, we see that S_p corresponds to S_h in (4.1) because of $k - \bar{k} = 1/\kappa_1$ i.e. $1/r - 1/\bar{r} = 1/\kappa_1 = 1/d$, similarly as in plane kinematics. (See Fig. 27.)

From equation (4.2) we also see that $\varrho' = \pm\infty$ if and only if $\varrho = -d$, i.e. $\varrho = -\kappa_1 \cos \delta$ which is a circle with centre $[0, -\frac{1}{2}\kappa_1]$ and radius $\frac{1}{2}\kappa_1$. Conversely, if $\varrho = \pm\infty$, then $\varrho' = d$ which is a circle with centre $[0, \frac{1}{2}\kappa_1]$

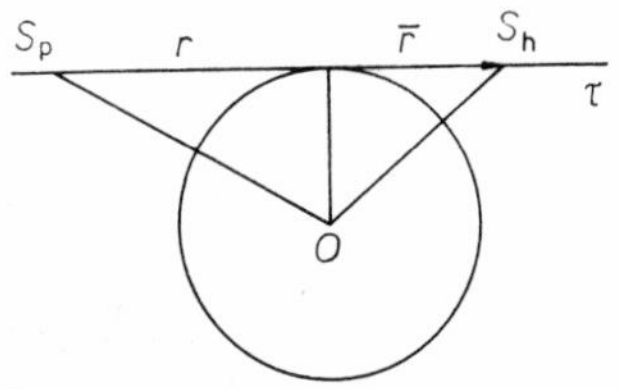

Fig. 27.

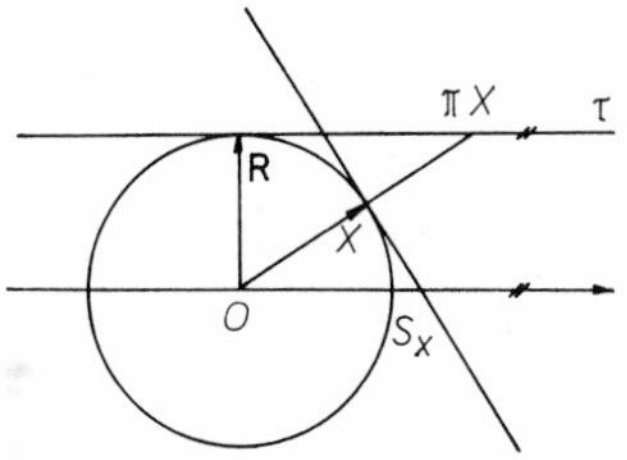

Fig. 28.

and radius $\frac{1}{2}\kappa_1$. In plane kinematic geometry we call these circles inflection circle and cuspidal circle. Their meaning is the following (see Fig. 28): the inflection circle ($\varrho = -\kappa_1 \cos \delta$) is the set of projections of all points of the sphere whose centres of spherical curvature have their projections at infinity, i.e. points whose centres of spherical curvature lie in the plane $\alpha = 0$. The cuspidal circle is the set of projections of the centres of spherical curvature of all the points of the great circle $\alpha = 0$ (see Fig. 29). What is more:

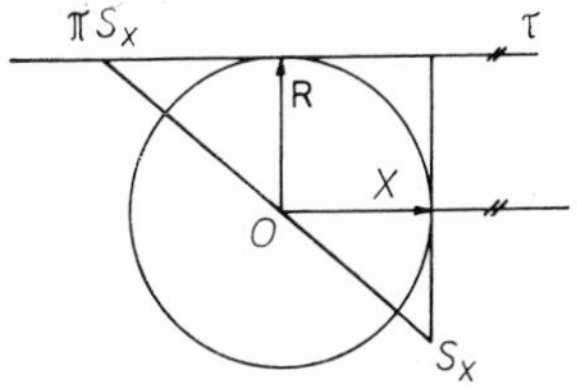

Fig. 29.

on the sphere there is a curve of inflection points, its equation is $\kappa_1 \gamma + \alpha(1 - \alpha^2) = 0$. Let us find its projection: for the coordinates of the projection points we have $x = \beta/\alpha$, $y = \gamma/\alpha$, i.e. $\beta = \alpha x$, $\gamma = \alpha y$. From the equation $\alpha^2 + \beta^2 + \gamma^2 = 1$ we now have $\alpha^2(1 + x^2 + y^2) = 1$, or rather $1 - \alpha^2 = (x^2 + y^2)(1 + x^2 + y^2)^{-1}$. Substitution yields

$$\kappa_1 y + \frac{x^2 + y^2}{1 + x^2 + y^2} = 0\,, \quad \text{i.e.} \quad \kappa_1 y(1 + x^2 + y^2) + x^2 + y^2 = 0\,,$$

which is a plane curve of third degree. We will not investigate its properties below; we note only that its image in our relationship in the plane τ is a curve with the equation

$$\kappa_1 y(1 + x^2 + y^2) - (x^2 + y^2) = 0 ,$$

this is easily verified by passing to the inverse motion.

Till now, the projection of the points X as well as S_X was into the plane τ, and a relationship in the plane was obtained. If we project S_X from the centre $\boldsymbol{O}$ on to the sphere S^2, we obtain a relationship on the sphere. Let us introduce polar coordinates on the sphere, with origin at $\boldsymbol{R}$, with the aid of the angle δ from the great circle passing through $\boldsymbol{R}$ and N to the great circle passing through $\boldsymbol{R}$ and X (as before) and of the angle ϑ

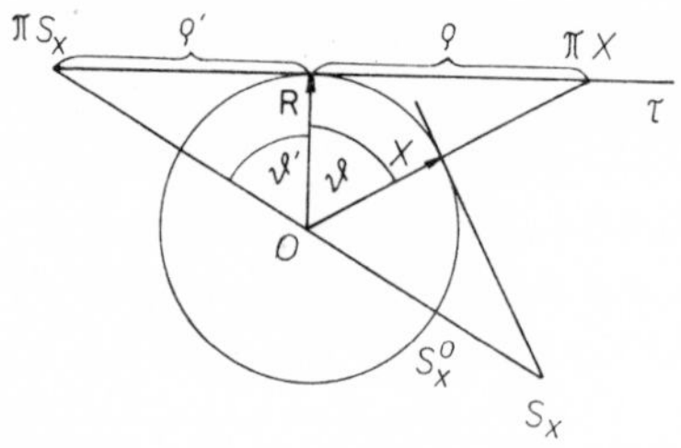

Fig. 30.

from $\boldsymbol{R}$ to X, $0 \leqq \delta < \pi$, $-\pi < \vartheta < \pi$. Then we have the relation $\varrho = \tan \vartheta$ between the polar coordinates (ϱ, δ) of the point πX in τ and the polar coordinates (ϑ, δ) of the point X on the sphere (see Fig. 30). Denote by S_X^0 the projection of the point S_X on to the sphere S^2, let its polar coordinates be ϑ', δ'. For the relation between the points X and S_X^0 we then have

$$\delta' = \delta , \quad \frac{1}{\kappa_1 \cos \delta} = \cot \vartheta' - \cot \vartheta ,$$

as can be verified by substitution into (4.2). The projections of the centres of spherical curvature of both centrodes are now in correspondence again. (The relationship is not unique, there are two points S_X^0; however, we shall not go into the details here.)

The exposition which follows is devoted to special motions. First, let us consider the motion composed of two rotations about fixed axes. Choose two unit vectors X, Y and denote $(X, Y) = \cos \vartheta$. Further, let $\varphi_1(t)$, $\varphi_2(t)$ be two arbitrary (differentiable) functions of the parameter t. Let $g_1(t)$ denote the rotation about the axis X through the angle $\varphi_1(t)$, let $g_2(t)$ be the rotation about the axis Y through the angle $\varphi_2(t)$. Obviously we have

$$\frac{dg_1}{d\varphi_1} g_1^{-1} = X, \quad \frac{dg_2}{d\varphi_2} g_2^{-1} = Y,$$

i.e.

$$\frac{dg_1(\varphi_1)}{dt} g_1^{-1} = \frac{dg_1(\varphi_1)}{d\varphi_1} \frac{d\varphi_1}{dt} g_1^{-1} = \dot{\varphi}_1 X;$$

similarly

$$\frac{dg_2}{dt} g_2^{-1} = \dot{\varphi}_2 Y.$$

Now, let us consider the motion $g(t) = g_1(t)\, g_2(t)$, i.e. $\bar{\mathscr{R}}(t) = [\mathscr{R}\, g_1(t)]\, g_2(t)$, i.e. first we perform the rotation about the axis X through the angle $\varphi_1(t)$ and then the rotation about the axis $g_1(t)\, Y$ through the angle $\varphi_2(t)$, or rather first about the axis Y through the angle $\varphi_2(t)$ and then about the axis X through the angle $\varphi_1(t)$. For the fixed directing function $\boldsymbol{r}(t)$ of the motion $g(t)$ we obtain

$$\begin{aligned} \boldsymbol{r}(t) &= (g_1 g_2)^{\boldsymbol{\cdot}} (g_1 g_2)^{-1} = \dot{g}_1 g_2 g_2^{-1} g_1^{-1} + g_1 \dot{g}_2 g_2^{-1} g_1^{-1} \\ &= \dot{g}_1 g_1^{-1} + \operatorname{ad} g_1(\dot{g}_2 g_2^{-1}) = \dot{\varphi}_1 X + \operatorname{ad} g_1(\dot{\varphi}_2 Y) \\ &= \operatorname{ad} g_1(\dot{\varphi}_1 X + \dot{\varphi}_2 Y), \end{aligned}$$

since $\operatorname{ad} g_1 X = X$ as g_1 is the rotation about the axis X. To be brief, let us denote $\dot{\varphi}_1 = \omega_1$, $\dot{\varphi}_2 = \omega_2$. Then $\boldsymbol{r}(t) = \operatorname{ad} g_1(\omega_1 X + \omega_2 Y)$. We derive the following simple lemma needed in the sequel.

4.4 Lemma. Let $A(t) = \operatorname{ad} g_1\, B(t)$, where $A(t)$, $B(t)$ are vector functions from $\mathfrak{O}(3)$. Then $\dot{A}(t) = [\dot{g}_1 g_1^{-1}, \operatorname{ad} g_1 B] + \operatorname{ad} g_1 \dot{B}$.

Proof.

$$A = g_1 B g_1^{-1},$$

$$\begin{aligned}\dot{A} &= \dot{g}_1 B g_1^{-1} + g_1 B (g_1^{-1})^{\bullet} + g_1 \dot{B} g_1^{-1} \\ &= \dot{g}_1 g_1^{-1} \operatorname{ad} g_1 B - \operatorname{ad} g_1 B \dot{g}_1 g_1^{-1} + \operatorname{ad} g_1 \dot{B} \\ &= [\dot{g}_1 g_1^{-1}, \operatorname{ad} g_1 B] + \operatorname{ad} g_1 \dot{B}\,.\end{aligned}$$

Let us compute the derivatives of the directing function $\boldsymbol{r}(t)$ according to Lemma 4.4:

$$\begin{aligned}\boldsymbol{r}^{\bullet}(t) &= [\omega_1 X, \operatorname{ad} g_1(\omega_1 X + \omega_2 Y)] + \operatorname{ad} g_1(\dot{\omega}_1 X + \dot{\omega}_2 Y) \\ &= \operatorname{ad} g_1[\omega_1 X, \omega_1 X + \omega_2 Y] + \operatorname{ad} g_1(\dot{\omega}_1 X + \dot{\omega}_2 Y) \\ &= \operatorname{ad} g_1(\dot{\omega}_1 X + \dot{\omega}_2 Y + \omega_1\omega_2[X, Y])\,,\end{aligned}$$

since

$$\omega_1 X = \operatorname{ad} g_1(\omega_1 X)\,, \quad [\operatorname{ad} g_1 A, \operatorname{ad} g_1 B] = \operatorname{ad} g_1[A, B]$$

for arbitrary A and B.

$$\begin{aligned}\boldsymbol{r}^{\bullet\bullet}(t) &= \operatorname{ad} g_1\{[\omega_1 X, \omega_1\omega_2[X, Y] + \dot{\omega}_1 X + \dot{\omega}_2 Y] \\ &\quad + (\omega_1\omega_2)^{\bullet}\,[X, Y] + \ddot{\omega}_1 X + \ddot{\omega}_2 Y\} \\ &= \operatorname{ad} g_1\{\omega_1^2\omega_2[X, [X, Y]] + \omega_1\dot{\omega}_2[X, Y] \\ &\quad + (\omega_1\omega_2)^{\bullet}\,[X, Y] + \ddot{\omega}_1 X + \ddot{\omega}_2 Y\} \\ &= \operatorname{ad} g_1\{\omega_1^2\omega_2(\cos\vartheta)\,X - \omega_1^2\omega_2 Y \\ &\quad + (\dot{\omega}_1\omega_2 + 2\omega_1\dot{\omega}_2)\,X \times Y + \ddot{\omega}_1 X + \ddot{\omega}_2 Y\} \\ &= \operatorname{ad} g_1\{(\ddot{\omega}_1 + \omega_1^2\omega_2\cos\vartheta)\,X + (\ddot{\omega}_2 - \omega_1^2\omega_2)\,Y \\ &\quad + (\dot{\omega}_1\omega_2 + 2\omega_1\dot{\omega}_2)\,X \times Y\}\,,\end{aligned}$$

where we use the relations

$$[X, Y] = X \times Y\,, \quad X \times (X \times Y) = X\cos\vartheta - Y\,.$$

Now it is easy to find the curvatures of the given motion $g(t)$:

$$\begin{aligned}(\dot{\varphi})^2 &= \omega_1^2 + \omega_2^2 + 2\omega_1\omega_2\cos\vartheta\,, \\ (\boldsymbol{r}^{\bullet})^2 &= \omega_1^2\omega_2^2\sin^2\vartheta + (\dot{\omega}_1)^2 + (\dot{\omega}_2)^2 + 2\dot{\omega}_1\dot{\omega}_2\cos\vartheta\,, \\ (\boldsymbol{r}, \boldsymbol{r}^{\bullet}) &= \dot{\omega}_1\omega_1 + \dot{\omega}_2\omega_2 + \omega_1\dot{\omega}_2\cos\vartheta + \dot{\omega}_1\omega_2\cos\vartheta\,.\end{aligned}$$

After some manipulation we obtain

$$(\boldsymbol{r}^{\bullet})^2 (\dot\varphi)^2 - (\boldsymbol{r}, \boldsymbol{r}^{\bullet})^2 = \{(\omega_1\dot\omega_2 - \dot\omega_1\omega_2)^2 + \omega_1^2\omega_2^2(\omega_1^2 + \omega_2^2 + 2\omega_1\omega_2 \cos \vartheta)\} \sin^2 \vartheta ,$$

$$|\boldsymbol{r}, \boldsymbol{r}^{\bullet}, \boldsymbol{r}^{\bullet\bullet}| = \{\omega_1\omega_2(\omega_1^3\omega_2 - \omega_1\ddot\omega_2 + \omega_2\ddot\omega_1 - \dot\omega_1\dot\omega_2 + \omega_1^2\omega_2^2 \cos \vartheta) + 2\omega_1^2\dot\omega_2^2 - \dot\omega_1^2\omega_2^2\} \sin^2 \vartheta ,$$

so that the following theorem holds:

4.5 Theorem. *Let $g_1(\varphi_1(t))$ and $g_2(\varphi_2(t))$ be rotations about the axes X and Y through the angles $\varphi_1(t)$ and $\varphi_2(t)$, respectively. Let $(X, Y) = \cos \vartheta = const.$ For the motion $g(t) = g_1(\varphi_1(t))\, g_2(\varphi t))$ we have*

$$(\dot\varphi)^2 = \omega_1^2 + \omega_2^2 + 2\omega_1\omega_2 \cos \vartheta ,$$

$$\kappa_1 = (\dot\varphi)^{-3} \sin \vartheta[(\omega_1\dot\omega_2 - \omega_2\dot\omega_1)^2 + (\dot\varphi)^2\, \omega_1^2\omega_2^2]^{1/2} ,$$

$$\kappa_2 = [(\omega_1\dot\omega_2 - \omega_2\dot\omega_1)^2 + (\dot\varphi)^2\, \omega_1^2\omega_2^2]^{-1} \cdot [\omega_1\omega_2(\omega_1^3\omega_2 - \omega_1\ddot\omega_2 + \omega_2\ddot\omega_1 - \dot\omega_1\dot\omega_2 + \omega_1^2\omega_2^2 \cos \vartheta) + 2\omega_1^2\dot\omega_2^2 - \dot\omega_1^2\omega_2^2] .$$

Proof. It suffices to substitute into the formulae from Theorem 3.39.

Let us note that $\bar\kappa_2$ is obtained easily from the equation $\kappa_2 - \bar\kappa_2 = 1$, or directly if we take into account that for the curvature κ_2^{in} of the inverse motion we have $\bar\kappa_2 = -\kappa_2^{in}$ and that the inverse motion to $g(t)$ is obtained by the exchange $\omega_1 \sim -\omega_2$, $\omega_2 \sim -\omega_1$. From Theorem 4.5 we derive easily the formula for the spherical curvature k and $\bar k$ of both centrodes $k = \kappa_2\kappa_1^{-1}$, $\bar k = \bar\kappa_2\bar\kappa_1^{-1}$. We write these formulae for the simpler case of rotation about two perpendicular axes X and Y. In this case $\cos \vartheta = 0$, $(\dot\varphi)^2 = \omega_1^2 + \omega_2^2$; if we assume that the motion is given as a function of the canonical parameter φ, i.e. that $(\dot\varphi)^2 = 1$, it is possible to write $\omega_1 = \cos \gamma(\varphi)$, $\omega_2 = \sin \gamma(\varphi)$ and we obtain

$$\omega_1' = -\gamma' \sin \gamma , \quad \omega_2' = \gamma' \cos \gamma ,$$

$$\omega_1'' = -(\gamma')^2 \cos \gamma - \gamma'' \sin \gamma , \quad \omega_2'' = -(\gamma')^2 \sin \gamma + \gamma'' \cos \gamma .$$

After substitution

$$\kappa_1 = [(\gamma')^2 + \cos^2\gamma \sin^2\gamma]^{1/2},$$

$$\kappa_2 = [(\gamma')^2 + \cos^2\gamma \sin^2\gamma]^{-1} \cdot [\cos\gamma \sin\gamma(\cos^3\gamma \sin\gamma - \gamma'') + (\gamma')^2 (3\cos^2\gamma - 1)] .$$

Further, we have

$$k = [(\gamma')^2 + \cos^2\gamma \sin^2\gamma]^{-3/2} \cdot [\cos\gamma \sin\gamma(\cos^3\gamma \sin\gamma - \gamma'') + (\gamma')^2 (3\cos^2\gamma - 1)],$$

$$\bar{k} = -[(\gamma')^2 + \cos^2\gamma \sin^2\gamma]^{-3/2} \cdot [\cos\gamma \sin\gamma(\cos\gamma \sin^3\gamma + \gamma'') + (\gamma')^2 (3\sin^2\gamma - 1)] .$$

However, the most important special case arises when the axes are in general position and $\varphi_1(\varphi)$ and $\varphi_2(\varphi)$ are linear functions. Such motions are called *cyclic* since their centrodes are circles (and, conversely, if the centrodes are circles then the respective motion is cyclic). For a cyclic motion we have $\varphi_1(\varphi) = a\varphi, \varphi_2(\varphi) = b\varphi$, where φ is the canonical parameter, i.e. a, b are such that $a^2 + b^2 + 2ab\cos\vartheta = 1$. This can be always achieved since only the ratio $a : b$ is characteristical for the motion. Into the formulae for invariants we then substitute

$$\omega_1 = a, \quad \omega_2 = b, \quad \dot{\omega}_1 = \dot{\omega}_2 = \ddot{\omega}_1 = \ddot{\omega}_2 = 0 .$$

After substitution we obtain

$$\kappa_1 = ab \sin\vartheta, \quad \kappa_2 = a(a + b\cos\vartheta), \quad \bar{\kappa}_2 = -b(b + a\cos\vartheta).$$

For the spherical curvatures of the centrodes we have

$$k = \frac{a + b\cos\vartheta}{b\sin\vartheta}, \quad \bar{k} = -\frac{b + a\cos\vartheta}{a\sin\vartheta},$$

for the radii of the spherical curvatures r_S, $\bar{r}_S$ we have

$$r_S = \frac{b\sin\vartheta}{a + b\cos\vartheta}, \quad \bar{r}_S = \frac{-a\sin\vartheta}{b + a\cos\vartheta}.$$

Here, if the signs of r_S and $\bar{r}_S$ are the same they are placed on the same side, if they differ they are placed on different sides. The centrodes are circles, the spherical radius is the projection of the radius into the tangent plane from the point O (see Fig. 31). From the figure we see how to compute r

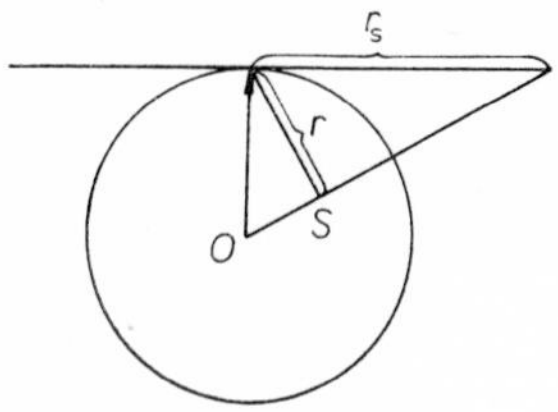

Fig. 31.

if r_S is known. In fact, $r : 1 = r_S : (r_S^2 + 1)^{1/2}$ follows from the homothety of triangles so that

$$r = r_S(r_S^2 + 1)^{-1/2} ,$$

i.e.

$$r = b \sin \vartheta [b^2 \sin^2 \vartheta + (a + b \cos \vartheta)^2]^{-1/2} = b \sin \vartheta ;$$

$$\bar{r} = -a \sin \vartheta$$

similarly.

4.6 Theorem. *For the motion* $g = g_1 g_2$, *where* $g_1(\varphi) = \exp(a\varphi X)$, $g_2(\varphi) = \exp(b\varphi Y)$, $(X, Y) = \cos\vartheta$, $(X, X) = (Y, Y) = 1$, $a^2 + b^2 + 2ab \cos\vartheta = 1$, *the fixed centrode is a circle with centre on* X *and radius* $r = b \sin\vartheta$, *the moving centrode is a circle with centre on* Y *and radius* $\bar{r} = -a \sin\vartheta$.

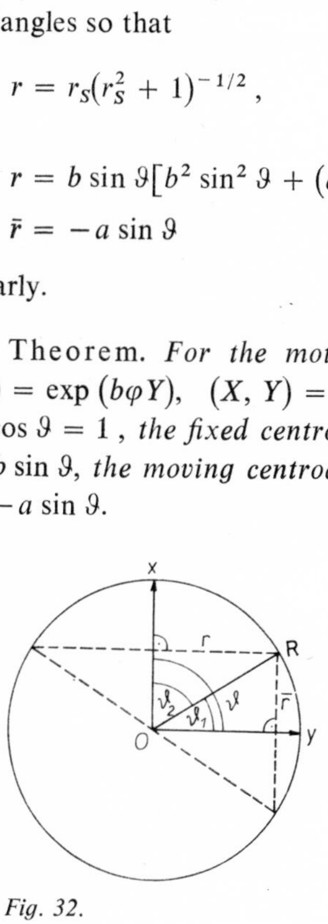

Fig. 32.

Conversely, if two circles are given with radii r and $\bar{r}$, we can find $\cos\vartheta$ by Fig. 32: $\bar{r} = \sin\vartheta_1$, $r = \sin\vartheta_2$, $\vartheta = \vartheta_1 - \vartheta_2$ (since if r and $\bar{r}$ have different signs we have $\vartheta = |\vartheta_1| + |\vartheta_2|$, if they have the same sign $\vartheta = |\vartheta_1| - |\vartheta_2|$). Hence, we obtain

$$\begin{aligned}\sin\vartheta &= \sin(\vartheta_1 - \vartheta_2) = \sin\vartheta_1\cos\vartheta_2 - \cos\vartheta_1\sin\vartheta_2\\ &= r(1-\bar{r}^2)^{1/2} - \bar{r}(1-r^2)^{1/2}\,.\end{aligned}$$

Further, from the equations

$$a = -\bar{r}\sin^{-1}\vartheta\,, \quad b = r\sin^{-1}\vartheta$$

we compute the constants a and b.

4.7 Exercise. Find the centrodes of the motion (and with their aid find also κ_1 and κ_2)

$$g(\mathrm{t}) = \tfrac{1}{5}\begin{pmatrix} -3 - 2\sin^2\mathrm{t}, & -2\cos\mathrm{t}\sin\mathrm{t}, & 4\cos\mathrm{t}\\ 2\cos\mathrm{t}\sin\mathrm{t}, & 3 + 2\cos^2\mathrm{t}, & 4\sin\mathrm{t}\\ -4\cos\mathrm{t}, & 4\sin\mathrm{t}, & -3\end{pmatrix}.$$

4.8 Exercise. Find the centrodes of the motion $g(\mathrm{t}) = g_1(\mathrm{t})\,g_2(\mathrm{t})$, where $g_1(\mathrm{t}) = \exp(\lambda\mathrm{t})\,\boldsymbol{u}_1$, $g_2(\mathrm{t}) = \exp(\mu\mathrm{t}^2)\,\boldsymbol{u}_2$, $\lambda, \mu \in \mathbf{R}$, for suitably chosen λ and μ find the invariants κ_1 and κ_2.

4.9 Example. If the radii r and $\bar{r}$ of the centrodes of cyclic motion are given, it is possible to find the invariants κ_1 and κ_2 directly. Indeed, we have

$$r^2(r_S^2 + 1) = r_S^2\,, \quad \text{i.e.} \quad r_S^2(1 - r^2) = r^2\,,$$

i.e.

$$r_S = r(1 - r^2)^{-1/2}\,, \quad \bar{r}_S = \bar{r}(1 - \bar{r}^2)^{-1/2}\,.$$

Further, we know that $k - \bar{k} = \kappa_1^{-1}$, i.e.

$$\frac{1}{r_S} - \frac{1}{\bar{r}_S} = \frac{1}{\kappa_1}\,, \quad \text{i.e.} \quad \frac{(1 - r^2)^{1/2}}{r} - \frac{(1 - \bar{r}^2)^{1/2}}{\bar{r}} = \frac{1}{\kappa_1}\,.$$

Finally,

$$\kappa_2 = \kappa_1 k = \frac{\kappa_1}{r_S}\,, \quad \bar{\kappa}_2 = \frac{\kappa_1}{\bar{r}_S}\,.$$

4.10 Exercise. Find κ_1 and κ_2 for the motion from Exercise 4.7 directly using the formulae from Example 4.9.

4.11 Example. A simple special case of cyclic motion is obtained if one of the centrodes is a great circle. Let us discuss this case in more detail. Thus, let e.g. $r = 1$. For the constants a, b, ϑ we then have the equations

$$r = b \sin \vartheta\,, \quad \bar{r} = -a \sin \vartheta\,, \quad a^2 + b^2 + 2ab \cos \vartheta = 1\,.$$

If $r = 1$, then

$$b = \frac{1}{\sin \vartheta}\,, \quad a = \frac{-\bar{r}}{\sin \vartheta}\,, \quad \left(\frac{1}{\sin \vartheta}\right)^2 + \left(\frac{\bar{r}}{\sin \vartheta}\right)^2 - \frac{2\bar{r} \cos \vartheta}{\sin^2 \vartheta} = 1\,,$$

i.e.

$$1 + (\bar{r})^2 - 2\bar{r} \cos \vartheta = \sin^2 \vartheta = 1 - \cos^2 \vartheta\,,$$

i.e.

$$\cos^2 \vartheta - 2\bar{r} \cos \vartheta + \bar{r}^2 = 0\,,$$

i.e.

$$(\bar{r} - \cos \vartheta)^2 = 0\,, \quad \text{i.e.} \quad \bar{r} = \cos \vartheta\,,$$

so that, finally,

$$b = (1 - \bar{r}_2)^{-1/2}\,, \quad a = -\bar{r}(1 - \bar{r}^2)^{-1/2}\,.$$

For the invariants we obtain

$$\kappa_1 = ab \sin \vartheta = a = -\bar{r}(1 - \bar{r}^2)^{-1/2}\,, \; \kappa_2 = 0\,, \quad \bar{\kappa}_2 = -1\,.$$

Analogous formulae are obtained for the second case when $\bar{r} = 1$.

4.12 Example. Let us find the matrix representation of all cyclic motions. This is useful in the case that we need the equation of the trajectory of some point. Thus, let the centrodes be given with radii r and $\bar{r}$. Assume that the constants a, b, $\cos \vartheta$, $\sin \vartheta$ are computed already. Choose

$$X = \boldsymbol{u}_3\,, \quad Y = \boldsymbol{u}_3 \cos \vartheta + \boldsymbol{u}_2 \sin \vartheta\,.$$

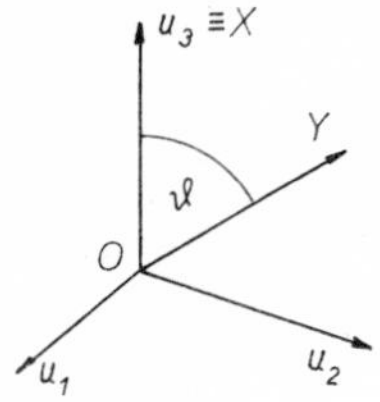

Fig. 33.

Then (see Fig. 33)

$$g_1(\varphi) = \begin{pmatrix} \cos(a\varphi), & -\sin(a\varphi), & 0 \\ \sin(a\varphi), & \cos(a\varphi), & 0 \\ 0, & 0, & 1 \end{pmatrix}.$$

Denote by γ the rotation about the axis $\boldsymbol{u}_1$ through the angle ϑ so that γ maps X into Y. Then rotation about Y can be written in the form $\gamma g_3 \gamma^{-1}$, where g_3 is the rotation about the axis $\boldsymbol{u}_3$ through the same angle. This is true since

$$\gamma g_3 \gamma^{-1}(Y) = \gamma g_3(\boldsymbol{u}_3) = \gamma(\boldsymbol{u}_3) = Y.$$

Thus $g_2(\varphi)$ is obtained in the form

$$g_2(\varphi) = \begin{pmatrix} 1, & 0, & 0 \\ 0, & \cos\vartheta, & -\sin\vartheta \\ 0, & \sin\vartheta, & \cos\vartheta \end{pmatrix} \begin{pmatrix} \cos(b\varphi), & -\sin(b\varphi), & 0 \\ \sin(b\varphi), & \cos(b\varphi), & 0 \\ 0, & 0, & 1 \end{pmatrix} \begin{pmatrix} 1, & 0, & 0 \\ 0, & \cos\vartheta, & \sin\vartheta \\ 0, & -\sin\vartheta, & \cos\vartheta \end{pmatrix}.$$

The computations yield

$$g_2(\varphi) = \begin{pmatrix} \cos(b\varphi), & -\sin(b\varphi)\cos\vartheta, & -\sin(b\varphi)\sin\vartheta \\ \sin(b\varphi)\cos\vartheta, & \cos(b\varphi)\cos^2\vartheta + \sin^2\vartheta, & [\cos(b\varphi) - 1]\cos\vartheta\sin\vartheta \\ \sin(b\varphi)\sin\vartheta, & [\cos(b\varphi) - 1]\cos\vartheta\sin\vartheta, & \cos(b\varphi)\sin^2\vartheta + \cos^2\vartheta \end{pmatrix}.$$

Naturally, $g(\varphi) = g_1(\varphi)\, g_2(\varphi)$. Obviously, only one of the possible representations of cyclic motions was thus obtained.

4.13 Example. In the end, let us discuss the case of the spherical quadrangle. Let four nonzero spherical segments $\alpha_1, \alpha_2, \alpha_3, \alpha_4$ be given on the unit sphere S^2 which form a quadrangle as shown in Fig. 34. Denote the angles of these segments $\varphi_1, \varphi_2, \varphi_3, \varphi_4$. The segment α_1 is considered fixed, the segments are hinge connected at the points X_1, X_2, X_3, X_4. We are interested in the motion of the sphere S^2 determined by the motion of the segment α_3. The motion of the segment α_3 can be composed of two rotations about fixed axes, first the rotation about X_3 through the angle φ_2 and then the rotation about X_2 through the angle φ_1. Moreover, φ_2 is a function of φ_1 which has to be found to determine the motion. The motion of the hinged quadrangle is represented as the composition of simple motions as follows:

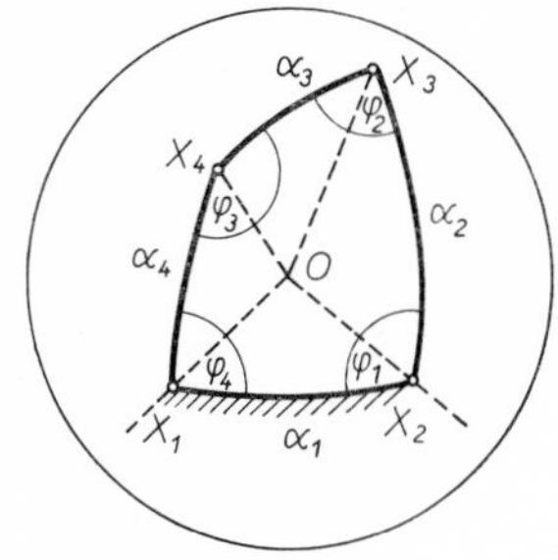

Fig. 34.

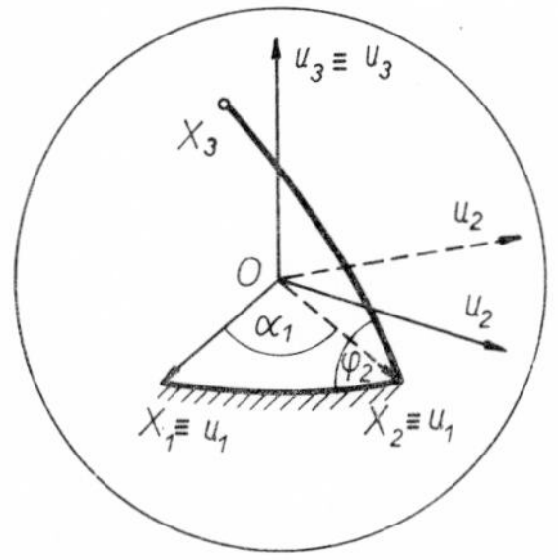

Fig. 35.

First, let us consider the segments α_1 and α_2 (see Fig. 35). Consider the frame $\mathscr{R}_1 = \{\boldsymbol{u}_1, \boldsymbol{u}_2, \boldsymbol{u}_3\}$ defined as follows: $\boldsymbol{u}_1 = X_1$, $\boldsymbol{u}_3$ is perpendicular to α_1, $\{X_1, X_2, \boldsymbol{u}_3\}$ constitutes a right-handed system. Define the frame $\mathscr{R}'_1 = \{\boldsymbol{u}'_1, \boldsymbol{u}'_2, \boldsymbol{u}'_3\}$ as follows: $\boldsymbol{u}'_3 = \boldsymbol{u}_3$, $\boldsymbol{u}'_1 = \boldsymbol{X}_2$, i.e. $\mathscr{R}'_1$ is obtained from $\mathscr{R}_1$ by rotation through the angle α_1 about the axis $\boldsymbol{u}_3$. Further, let

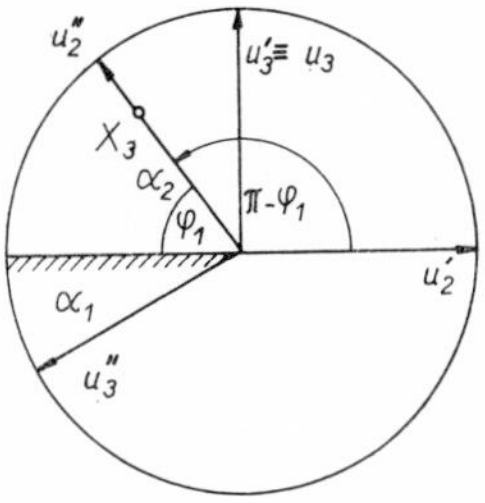

Fig. 36.

us define the frame $\mathscr{R}_2 = \{\boldsymbol{u}''_1, \boldsymbol{u}''_2, \boldsymbol{u}''_3\}$ (see Fig. 36): $\boldsymbol{u}''_1 = \boldsymbol{u}'_1 = X_2$, $\boldsymbol{u}_3$ is perpendicular to the segment α_2. The frame $\mathscr{R}_2$ is obtained from the frame $\mathscr{R}'_1$ by rotation about the axis $\boldsymbol{u}'_1$ through the angle $\pi - \varphi_1$. Thus, we have $\mathscr{R}'_1 = \mathscr{R}_1\gamma_1$ where γ_1 is the rotation about the axis $\boldsymbol{u}_3$ through the angle α_1, $\mathscr{R}_2 = \mathscr{R}'_1\gamma_2$ where γ_2 is the rotation about the axis $\boldsymbol{u}'_1$ through the angle $\pi - \varphi_1$. Finally, $\mathscr{R}_2 = \mathscr{R}_1\gamma_1\gamma_2$. The matrices of rotations γ_1 and γ_2 are found easily:

$$\gamma_1 = \begin{pmatrix} \cos\alpha_1, & -\sin\alpha_1, & 0 \\ \sin\alpha_1, & \cos\alpha_1, & 0 \\ 0, & 0, & 1 \end{pmatrix}, \quad \gamma_2 = \begin{pmatrix} 1, & 0, & 0 \\ 0, & -\cos\varphi_1, & -\sin\varphi_1 \\ 0, & \sin\varphi_1, & -\cos\varphi_1 \end{pmatrix}.$$

Denote $g_1 = \gamma_1\gamma_2$. Then $\mathscr{R}_2 = \mathscr{R}_1 g_1$ where

$$g_1 = \begin{pmatrix} \cos\alpha_1, & \sin\alpha_1\cos\varphi_1, & \sin\alpha_1\sin\varphi_1 \\ \sin\alpha_1, & -\cos\alpha_1\cos\varphi_1, & -\cos\alpha_1\sin\varphi_1 \\ 0, & \sin\varphi_1, & -\cos\varphi_1 \end{pmatrix}.$$

Further, we note that the position of the frame $\mathscr{R}_2$ with respect to the segment α_2 is the same as the position of $\mathscr{R}_1$ with respect to α_1. Consequently, it is possible to define the frames $\mathscr{R}_3, \mathscr{R}_4, \mathscr{R}_5 = \mathscr{R}_1$ in an entirely analogous manner. Denote

$$g_i = \begin{pmatrix} \cos\alpha_i, & \sin\alpha_i\cos\varphi_i, & \sin\alpha_i\sin\varphi_i \\ \sin\alpha_i, & -\cos\alpha_i\cos\varphi_i, & -\cos\alpha_i\sin\varphi_i \\ 0, & \sin\varphi_i, & -\cos\varphi_i \end{pmatrix}.$$

Obviously, we have $\mathscr{R}_{i+1} = \mathscr{R}_i g_i$, thus $\mathscr{R}_5 = \mathscr{R}_1 g_1 g_2 g_3 g_4 = \mathscr{R}_1$. The condition of closedness of the hinged quadrangle is thus represented by the equation $g_1 g_2 g_3 g_4 = e$. To avoid multiplying four matrices in a row, this condition is written in the form

$$g_1 g_2 = g_4^{\mathrm{T}} g_3^{\mathrm{T}}. \tag{4.3}$$

From equation (4.3) we have to determine the dependence between φ_1 and φ_2. To make the equation clearer let us denote

$$\sin\alpha_1 = A, \quad \sin\alpha_2 = B, \quad \sin\alpha_3 = C, \quad \sin\alpha_4 = D,$$
$$\cos\alpha_1 = a, \quad \cos\alpha_2 = b, \quad \cos\alpha_3 = c, \quad \cos\alpha_4 = d.$$

Now, we write equation (4.3) but in the place of the terms which are not needed a dot is placed. Thus, we obtain

$$\begin{pmatrix} a, & A\cos\varphi_1, & A\sin\varphi_1 \\ ., & ., & . \\ ., & ., & . \end{pmatrix} \begin{pmatrix} b, & B\cos\varphi_2, & . \\ B, & -b\cos\varphi_2, & . \\ 0, & \sin\varphi_2, & . \end{pmatrix} =$$

$$\begin{pmatrix} d, & D, & 0 \\ ., & ., & . \\ ., & ., & . \end{pmatrix} \begin{pmatrix} c, & C, & \\ C\cos\varphi_3, & -c\cos\varphi_3, & . \\ C\sin\varphi_3, & -c\sin\varphi_3, & . \end{pmatrix}.$$

Compute the product of the first row and the first and second columns, on both sides. We obtain the equations

$$ab + AB\cos\varphi_1 = dc + CD\cos\varphi_3\,,$$

$$aB\cos\varphi_2 - Ab\cos\varphi_1\cos\varphi_2 + A\sin\varphi_1\sin\varphi_2 = dC - cD\cos\varphi_3.$$

Multiply the first equation by c and the second by C. Then

$$abc + ABc\cos\varphi_1 - dc^2 = cCD\cos\varphi_3\,,$$

$$aBC\cos\varphi_2 - AbC\cos\varphi_1\cos\varphi_2 + AC\sin\varphi_1\sin\varphi_2 - dC^2 =$$
$$-cCd\cos\varphi_3\,;$$

the sum is

$$abc - d + ABc\cos\varphi_1 +$$
$$aBC\cos\varphi_2 - AC(b\cos\varphi_1\cos\varphi_2 - \sin\varphi_1\sin\varphi_2) = 0\,. \qquad (4.4)$$

This is the relation which φ_1 and φ_2 have to satisfy. However, to be able to apply the formulae from Theorem 4.5 we have to know φ_2 as an explicit function of φ_1. For the computation of the invariants for merely one position this is not necessary; the formulae for the derivative of an implicit function can be applied. Let us elucidate this last method using a simple example. Consider a hinged quadrangle for which $b = 0$, $B = 1$, $d = Ac$. Equation (4.4) assumes the form

$$-Ac + Ac\cos\varphi_1 + aC\cos\varphi_2 + AC\sin\varphi_1\sin\varphi_2 = 0\,. \qquad (4.5)$$

Let us compute the invariants of this motion at the position $\varphi_1 = 0$. We consider φ_2 as a function of φ_1, $\varphi_2 = \varphi_2(\varphi_1)$. Under this assumption we differentiate equation (4.5) and write $\dot\varphi_2 = \omega_2(\dot\varphi_1 = \omega_1 = 1)$. Then

$$Ac\sin\varphi_1 - AC\cos\varphi_1\sin\varphi_2$$
$$+ \omega_2 C(a\sin\varphi_2 - A\sin\varphi_1\cos\varphi_2) = 0\,. \qquad (4.6)$$

Further differentiation yields

$$Ac(\cos\varphi_1 + \sin\varphi_1\sin\varphi_2) - 2\omega_2 AC\cos\varphi_1\cos\varphi_2$$
$$+ \omega_2^2 C(a\cos\varphi_2 + A\sin\varphi_1\sin\varphi_2)$$
$$+ \dot\omega_2 C(a\sin\varphi_2 - A\sin\varphi_1\cos\varphi_2) = 0\,. \qquad (4.7)$$

Let us put $\varphi_1 = 0$ in (4.5). Then $aC \cos \varphi_2 = 0$, i.e. $\cos \varphi_2 = 0$, i.e. $\varphi_2 = \frac{1}{2}\pi$. We substitute into (4.6):

$$AC = aC\omega_2 = 0\,, \quad \text{i.e.} \quad \omega_2 = \frac{A}{a}\,.$$

Further, let us substitute into (4.7):

$$Ac + aC\dot{\omega}_2 = 0\,, \quad \text{i.e.} \quad \dot{\omega}_2 = -\frac{Ac}{aC}\,.$$

It remains to substitute from the formulae into Theorem 4.5. Taking into account that $\vartheta = \alpha_2 = \frac{1}{2}\pi$, we have

$$(\dot{\varphi})^2\,(0) = \omega_1^2 + \omega_2^2 = 1 + \frac{A^2}{a^2} = \frac{1}{a^2}\,, \quad \dot{\varphi}(0) = \frac{1}{a}\,.$$

$$\begin{aligned}\kappa_1(0) &= \frac{1}{(\dot{\varphi})^3}\,[(\omega_1\omega_2 - \omega_2\dot{\omega}_1)^2 + \omega_1^2\omega_2^2(\dot{\varphi})^2]^{1/2} \\ &= a^3\left[\left(-\frac{Ac}{aC}\right)^2 + \frac{A^2}{a^2}\frac{1}{a^2}\right]^{1/2} = \frac{Aa}{C}\,(C^2 + c^2a^2)^{1/2}\,.\end{aligned}$$

If we wish to find $\kappa_2(0)$ as well, we have to differentiate equation (4.7), after substitution for ω_2 and $\dot{\omega}_2$ from the preceding equations to compute $\ddot{\omega}_2(0)$ from there and to substitute into the formula for κ_2. We leave the computations to the kind attention of the reader.

As a matter of fact, from equation (4.4) it is possible to find an explicit expression for φ_2 as a function of φ_1. But the computations are rather lengthy, and since merely the solution of a quadratic equation is in question, we present the result directly:

$$\begin{aligned}\sin \varphi_2 = {} & C^{-1}[1 - (ab + AB\cos\varphi_1)^2]^{-1}\,. \\ & \{-A\sin\varphi_1(ABc\cos\varphi_1 + abc - d) \pm (aB - Ab\cos\varphi_1)\,. \\ & [C^2(1 - (ab + AB\cos\varphi_1)^2) - (ABc\cos\varphi_1 + abc - d)^2]^{1/2}\}\,.\end{aligned} \tag{4.8}$$

Two values for $\sin \varphi_2$ are obtained because in general two spherical quadrangles can be constructed (see Fig. 37). The resulting formula for $\sin \varphi_2$ is rather complicated for practical computations, taking into account especially that differentiation of the function arcsin of the expression on the right-hand side of equation (4.8) should be carried out. However, in practice one encounters quadrangles whose some sides have length $\frac{1}{2}\pi$. In such

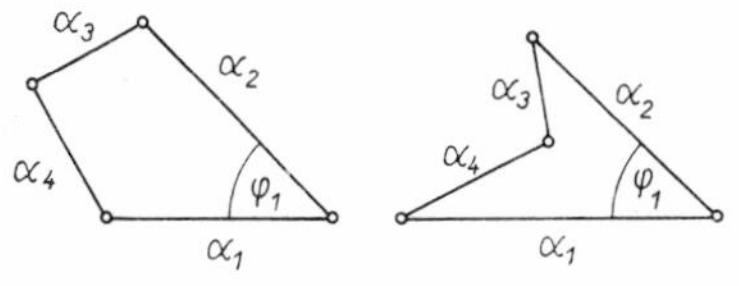

Fig. 37.

cases the formula for $\sin\varphi_2$ is simplified, sometimes even substantially. Thus, let us discuss the individual cases:

1. $\alpha_1 = \frac{1}{2}\pi$. Then $a = 0$, $A = 1$, and we have

$$\sin\varphi_2 = C^{-1}(1 - B^2\cos^2\varphi_1)^{-1} . \{-\sin\varphi_1(Bc\cos\varphi_1 - d) \pm b\cos\varphi_1[C^2(1 - B^2\cos^2\varphi_1) - (Bc\cos\varphi_1 - d)^2]^{1/2}\} .$$

2. $\alpha_1 = \alpha_2 = \frac{1}{2}\pi$. Then $a = b = 0$, $A = B = 1$, and we have

$$\sin\varphi_2 = \frac{d - c\cos\varphi_1}{C\sin\varphi_1} = \frac{\cos\alpha_4 - \cos\alpha_3\cos\varphi_1}{\sin\alpha_3\sin\varphi_1} .$$

3. $\alpha_1 = \alpha_2 = \alpha_4 = \frac{1}{2}\pi$. Then $\sin\varphi_2 = -\cot\alpha_3\cot\varphi_1$, which is the motion from Example 3.44, as a matter of fact.

4. $\alpha_1 = \alpha_2 = \alpha_3 = \frac{1}{2}\pi$. Then $\sin\varphi_2 = \cos\alpha_4\sin\varphi_1$.

5. $\alpha_1 = \alpha_3 = \frac{1}{2}\pi$. Then $a = c = 0$, $A = C = 1$, and we obtain

$$\sin\varphi_2 = \frac{d\sin\varphi_1 \pm b\cos\varphi_1(D^2 - B^2\cos^2\varphi_1)^{1/2}}{1 - B^2\cos^2\varphi_1} .$$

6. $\alpha_2 = \frac{1}{2}\pi$, $b = 0$, $B = 1$. Then

$$\sin\varphi_2 = C^{-1}(1 - A^2\cos^2\varphi_1)^{-1}\{-A\sin\varphi_1(Ac\cos\varphi_1 - d) \pm a[C^2(1 - A^2\cos^2\varphi_1) - (Ac\cos\varphi_1 - d)^2]^{1/2}\} .$$

7. $\alpha_2 = \alpha_3 = \frac{1}{2}\pi$, $b = c = 0$, $B = C = 1$. Then

$$\sin\varphi_2 = \frac{Ad\sin\varphi_1 \pm a(D^2 - A^2\cos^2\varphi_1)^{1/2}}{1 - A^2\cos^2\varphi_1} .$$

8. $\alpha_2 = \alpha_4 = \frac{1}{2}\pi$, $b = d = 0$, $B = D = 1$. Then

$$\sin \varphi_2 = \frac{-A^2 c \sin \varphi_1 \cos \varphi_1 \pm a(C^2 - A^2 \cos^2 \varphi_1)^{1/2}}{C(1 - A^2 \cos^2 \varphi_1)} .$$

9. $\alpha_2 = \alpha_3 = \alpha_4 = \frac{1}{2}\pi$, $b = c = d = 0$, $B = C = D = 1$. Then

$$\sin \varphi_2 = \pm a(1 - A^2 \cos^2 \varphi_1)^{-1/2} .$$

10. $\alpha_3 = \frac{1}{2}\pi$. Then $c = 0$, $C = 1$, so that

$$\sin \varphi_2 = [1 - (ab + AB \cos \varphi_1)^2]^{-1} .$$

$$\{Ad \sin \varphi_1 \pm (aB - Ab \cos \varphi_1)[1 - (ab + AB \cos \varphi_1)^2 - d^2]^{1/2}\} .$$

Let us return to the general equation (4.4) and try to find several cases of the spherical quadrangle which generate an elementary motion, i.e. rotation about a fixed axis. This case arises, e.g. if $\sin \varphi_2 = \text{const.}$ since then we have $g(\varphi_1) = g_1(\varphi_1)\, g_2$, where g_2 is a constant matrix and $g_1(\varphi_1)$ is rotation. Thus, let us consider equation (4.4), and let us investigate when this equation can be solved by constants $\cos \varphi_2 = \mu$, $\sin \varphi_2 = \lambda$. This is possible only if the coefficients of $\cos \varphi_1$ and $\sin \varphi_1$, and thus also the absolute term, are zero. We obtain the equations

$$abc - d + aBC\mu = 0 ,$$

$$A(Bc - bC\mu) = 0 ,$$

$$AC\lambda = 0 .$$

The following cases can arise:

1. $A = 0$, then $a = -1$ (we exclude quadrangles with zero sides). We have the equation

$$-bc - d - BC\mu = 0 , \quad \text{i.e.} \quad \mu = -\frac{bc + d}{BC} .$$

2. $A \neq 0$, $C = 0$. Then $A = 0$, $c = -1$, $b = -1$, $a = d$. We obtain the singular case of a quadrangle, $X_2 = X_4$.

3. $A \neq 0$, $C \neq 0$, $\lambda = 0$. Then $\mu = \varepsilon$ where $\varepsilon = \pm 1$. We obtain

$$Bc = Cb\varepsilon , \quad \frac{B}{b} = \frac{C}{c}\varepsilon ,$$

i.e.

$$b = \varepsilon c\,, \quad a\varepsilon b^2 - d + aB^2\varepsilon = 0\,,$$

i.e.

$$a\varepsilon - d = 0\,, \quad d = a\varepsilon\,,$$

which is again a singular case of a quadrangle: $X_2 = X_4$.

4.14 Exercise. Find the function $\varphi_2 = \varphi_2(\varphi_1)$ for quandrangles with the properties

a) $\alpha_1 = \alpha_3 = \frac{1}{2}\pi\,, \quad \alpha_2 = \alpha_4\,,$

b) $\alpha_1 = \alpha_3\,, \quad \alpha_2 = \alpha_4 = \frac{1}{2}\pi\,.$

Find the directing cones and invariants of these two motions.

4.15 Example. In the end we derive formulae for the approximate computation of the trajectory of a point. Assume that it is possible to compute easily the derivatives of the directing function for some value of the parameter t (e.g. for t = 0). Using these derivatives the trajectory of a given point is approximately found. In fact, we have

$$X(t) = X(0) + \dot{X}(0)\,t + \frac{\ddot{X}(0)\,t^2}{2!} + \ldots + \frac{X^{(n)}(0)\,t^n}{n!} + \varrho_{n+1}(t)\,,$$

$\varrho_{n+1}(t)$ is the remainder whose magnitude is estimated by the formula

$$\varrho_{n+1}(t) = \frac{X^{(n+1)}(\vartheta_i t)\,t^{n+1}}{(n+1)!}\,, \quad \text{where} \quad \vartheta_i \in (0, 1)\,.$$

Further, we have

$$\dot{X}(t) = \boldsymbol{r}(t)\,X(t)\,,$$
$$\ddot{X}(t) = \boldsymbol{r}^{\cdot}(t)\,X(t) + \boldsymbol{r}(t)\,\dot{X}(t) = \boldsymbol{r}^{\cdot}(t)\,X(t) + \boldsymbol{r}(t)\,.\,\boldsymbol{r}(t)\,X(t)$$

etc., for higher derivatives up to the order which is necessary. At the point $t = 0$ we then have $\dot{X}(0) = \boldsymbol{r}(0)\,X(0)$, $\ddot{X}(0) = [\boldsymbol{r}^{\cdot}(0) + \boldsymbol{r}^2(0)]\,X(0)$, etc. for higher orders. For the approximation of second order we thus have, in matrix representation,

$$X(t) \doteq \left\{e + t\,\boldsymbol{r}(0) + \frac{t^2}{2}\,[\boldsymbol{r}'(0) + \boldsymbol{r}^2(0)]\right\} X(0)\,.$$

It is possible to proceed with the approximations to arbitrary order. However, since it is difficult to find a general formula it is necessary to derive particular formulae for each of the considered orders.

Consider concretely the motion from Example 3.26. There we have

$$r(\alpha) = \begin{bmatrix} 0, & \dfrac{-\mu\cos^2\alpha}{\mu^2+\sin^2\alpha}, & \dfrac{\mu\sin\alpha\cos\alpha}{\mu^2+\sin^2\alpha} \\ \dfrac{\mu\cos^2\alpha}{\mu^2+\sin^2\alpha}, & 0, & 1 \\ \dfrac{-\mu\sin\alpha\cos\alpha}{\mu^2+\sin^2\alpha}, & -1, & 0 \end{bmatrix},$$

so that after differentiation and substitution the following holds at the point $\alpha = 0$:

$$r(0) = \begin{pmatrix} 0, & -\mu^{-1}, & 0 \\ \mu^{-1}, & 0, & 1 \\ 0, & -1, & 0 \end{pmatrix}, \quad r'(0) = \begin{pmatrix} 0, & 0, & \mu^{-1} \\ 0, & 0, & 0 \\ -\mu^{-1}, & 0, & 0 \end{pmatrix},$$

$$r^2(0) = \begin{pmatrix} -\mu^2, & 0, & -\mu^{-1} \\ 0, & -\mu^2-1, & 0 \\ -\mu^{-1}, & 0, & -1 \end{pmatrix}.$$

For the approximation of the trajectory to second order we have

$$X(\alpha) \doteq \begin{bmatrix} 1-\dfrac{\alpha^2}{2\mu^2}, & -\dfrac{\alpha}{\mu}, & 0 \\ \dfrac{\alpha}{\mu}, & 1-\dfrac{\alpha^2}{2}\left(1+\dfrac{1}{\mu^2}\right), & \alpha \\ -\dfrac{\alpha^2}{\mu}, & -\alpha, & 1-\dfrac{\alpha^2}{2} \end{bmatrix} X(0).$$

Thus, we consider the point $X(0) = u_1$ for the sake of definiteness, its approximate trajectory in parametric representation is

$$X(\alpha) = \begin{bmatrix} 1 - \dfrac{\alpha^2}{2\mu^2} \\ \dfrac{\alpha}{\mu} \\ -\dfrac{\alpha^2}{\mu} \end{bmatrix}.$$

However, we should not forget that the approximate trajectory need not lie on the unit sphere. This drawback can be removed by norming, which cannot make the approximation worse.

4.16 Example. A slightly different situation arises if we know at some instant of the motion the invariants κ_1 and κ_2 and their derivatives. Then it is possible to compute the approximate trajectory of a point in the canonical frame $\boldsymbol{R}$, $\boldsymbol{T}$, $\boldsymbol{N}$, which is simpler. Let us derive the corresponding formulae. The prime denotes derivatives with respect to the canonical parameter φ. Then

$$\begin{aligned} X' &= \boldsymbol{R}X\,, \\ X'' &= \boldsymbol{R}'X + \boldsymbol{R}^2X = (\kappa_1\boldsymbol{T} + \boldsymbol{R}^2)\,X\,, \\ X''' &= (\kappa_1'\boldsymbol{T} + \kappa_1\boldsymbol{T}' + \boldsymbol{R}'\boldsymbol{R} + \boldsymbol{R}\boldsymbol{R}')\,X + (\kappa_1\boldsymbol{T} + \boldsymbol{R}^2)\,\boldsymbol{R}X \\ &= (\kappa_1'\boldsymbol{T} - \kappa_1^2\boldsymbol{R} + \kappa_1\kappa_2\boldsymbol{N} + 2\kappa_1\boldsymbol{T}\boldsymbol{R} + \kappa_1\boldsymbol{R}\boldsymbol{T} + \boldsymbol{R}^3)\,X \\ &= (\kappa_1'\boldsymbol{T} - \kappa_1^2\boldsymbol{R} + \kappa_1(\kappa_2 + 1)\,\boldsymbol{N} + 3\kappa_1\boldsymbol{T}\boldsymbol{R} + \boldsymbol{R}^3)\,X \end{aligned}$$

in matrix notation. The relation $\boldsymbol{R}\boldsymbol{T} - \boldsymbol{T}\boldsymbol{R} = \boldsymbol{N}$ was exploited. For the approximation of the trajectory, e.g. to the fourth order, we have

$$\begin{aligned} X(\varphi) &\doteq X(0) + \varphi\,X'(0) + \varphi^2\,\frac{X''(0)}{2!} + \varphi^3\,\frac{X'''(0)}{3!} \\ &= \Big\{e + \varphi\boldsymbol{R} + \frac{\varphi^2}{2}\,(\kappa_1\boldsymbol{T} + \boldsymbol{R}^2) \\ &\quad + \frac{\varphi^3}{6}\,\big[\kappa_1'\boldsymbol{T} - \kappa_1^2\boldsymbol{R} + \kappa_1(\kappa_2 + 1)\,\boldsymbol{N} + 3\kappa_1\boldsymbol{T}\boldsymbol{R} + \boldsymbol{R}^3\big]\Big\}\,X(0)\,. \end{aligned}$$

In the Frenet frame $\mathscr{F} = \{\boldsymbol{R}, \boldsymbol{T}, \boldsymbol{N}\}$ we obtain

$$\boldsymbol{R} = \begin{pmatrix} 0, & 0, & 0 \\ 0, & 0, & -1 \\ 0, & 1, & 0 \end{pmatrix}, \quad \boldsymbol{T} = \begin{pmatrix} 0, & 0, & 1 \\ 1, & 0, & 0 \\ -1, & 0, & 0 \end{pmatrix}, \quad \boldsymbol{N} = \begin{pmatrix} 0, & -1, & 0 \\ 1, & 0, & 0 \\ 0, & 0, & 0 \end{pmatrix},$$

further, we have

$$\boldsymbol{R}^2 = \begin{pmatrix} 0, & 0, & 0 \\ 0, & -1, & 0 \\ 0, & 0, & -1 \end{pmatrix}, \quad \boldsymbol{R}^3 = -\boldsymbol{R}, \quad \boldsymbol{TR} = \begin{pmatrix} 0, & 1, & 0 \\ 0, & 0, & 0 \\ 0, & 0, & 0 \end{pmatrix}.$$

Substitution yields

$$X(\varphi) \doteq \left[\begin{pmatrix} 1, & 0, & 0 \\ 0, & 1, & 0 \\ 0, & 0, & 1 \end{pmatrix} + \varphi \begin{pmatrix} 0, & 0, & 0 \\ 0, & 0, & -1 \\ 0, & 1, & 0 \end{pmatrix} + \frac{\varphi^2}{2} \begin{pmatrix} 0, & 0, & \kappa_1 \\ 0, & -1, & 0 \\ -\kappa_1, & 0, & -1 \end{pmatrix}\right.$$
$$\left. + \frac{\varphi^3}{6} \begin{pmatrix} 0, & -\kappa_1(\kappa_2 + 1) + 3\kappa_1, & \kappa_1' \\ \kappa_1(\kappa_2 + 1), & 0, & \kappa_1^2 + 1 \\ -\kappa_1', & -\kappa_1^2 - 1, & 0 \end{pmatrix}\right] X(0),$$

i.e.

$$X(\varphi) \doteq \begin{bmatrix} 1, & \dfrac{\varphi^3}{6}\kappa_1(2 - \kappa_2), & \dfrac{\kappa_1\varphi^2}{2} + \dfrac{\kappa_1'\varphi^3}{6} \\ \dfrac{\varphi_3}{6}\kappa_1(\kappa_2 + 1), & 1 - \dfrac{\varphi^2}{2}, & -\varphi + \dfrac{\varphi^3}{6}(\kappa_1^2 + 1) \\ -\dfrac{\kappa_1\varphi^2}{2} - \dfrac{\kappa_1'\varphi^3}{6}, & \varphi - \dfrac{\varphi^3}{6}(\kappa_1^2 + 1), & 1 - \dfrac{\varphi^2}{2} \end{bmatrix} \cdot \begin{bmatrix} \alpha \\ \beta \\ \gamma \end{bmatrix},$$

where $X(0) = \alpha\boldsymbol{R} + \beta\boldsymbol{T} + \gamma\boldsymbol{N}$.

For the approximation of the trajectory to the fourth order the resulting formula is

$$X(\varphi) \doteq \boldsymbol{R}\left\{\alpha + \frac{\beta\kappa_1\varphi^3}{6}(2 - \kappa_2) + \frac{\gamma\varphi^2}{6}(3\kappa_1 + \kappa_1'\varphi)\right\}$$
$$+ \boldsymbol{T}\left\{\frac{\alpha\varphi^3\kappa_1}{6}(\kappa_2 + 1) + \frac{\beta}{2}(2 - \varphi^2) + \frac{\gamma\varphi}{6}\left[\varphi^2(\kappa_1^2 + 1) - 6\right]\right\}$$

$$+ \boldsymbol{N}\left\{-\frac{\alpha\varphi^2}{6}(3\kappa_1 + \kappa_1'\varphi) + \frac{\beta\varphi}{6}[6 - \varphi^2(\kappa_1^2 + 1)]\right.$$
$$\left. + \frac{\gamma}{2}(2 - \varphi^2)\right\}.$$

For the sake of definiteness, let us have a cyclic motion for which $\kappa_1 = \sqrt{2}$, $\kappa_2 = 2\,(\bar{\kappa}_2 = 1$, $k = \sqrt{2}$, $\bar{k} = \frac{1}{2}\sqrt{2})$. For the trajectory of the point $X = \alpha\boldsymbol{R} + \beta\boldsymbol{T} + \gamma\boldsymbol{N}$ we then have the approximate representation

$$X(\varphi) \doteq \tfrac{1}{2}\{\boldsymbol{R}(2\alpha + \gamma\varphi^2\sqrt{2}) + \boldsymbol{T}[\alpha\varphi^3\sqrt{2} + (\beta - \varphi\gamma)(2 - \varphi^2)]$$
$$+ \boldsymbol{N}[-\alpha\varphi^2\sqrt{2} + (\beta\varphi + \gamma)(2 - \varphi^2)]\}.$$

Consider, for instance, the trajectory of the point $X(0) = \boldsymbol{T}$, i.e. $\alpha = \gamma = 0$, $\beta = 1$. We have

$$X(\varphi) \doteq \frac{2 - \varphi^2}{2}(\boldsymbol{T} + \varphi\boldsymbol{N}).$$

Moreover, the curve $X(\varphi)$ can be normed to obtain a spherical curve:

$$\|X(\varphi)\| = \frac{2 - \varphi^2}{2}(1 + \varphi^2)^{1/2},$$

i.e., finally,

$$X(\varphi) \doteq (1 + \varphi^2)^{-1/2}(\boldsymbol{T} + \varphi\boldsymbol{N}).$$

Obviously, this is nothing else than a great circle. This indicates that the trajectory of the point $\boldsymbol{T}$ "approaches considerably" a great circle, i.e. that it is "very slightly curved". The trajectory of the point $\boldsymbol{N}$ is approximated by the curve

$$X(\varphi) = \tfrac{1}{2}[\boldsymbol{R}\varphi^2\sqrt{2} - \varphi\boldsymbol{T}(2 - \varphi^2) + \boldsymbol{N}(2 - \varphi^2)],$$

which will obviously not be a circle after norming.

5. KINEMATICS OF SPHERICAL MOTION

Let us assume that spherical motion is given as a function of time t. We suppose that motion $g(\varphi)$ is given as a function of the canonical parameter φ, and that in this representation φ is a certain function of time t.

Thus, the representation of motion is in the form $g(\varphi(\mathrm{t}))$. Let us denote the derivative with respect to time by a dot. In kinematics we are interested mainly in the velocities and accelerations of the individual points. In the computations below the topics discussed in the preceding exposition will be applied. For the velocity $\boldsymbol{v}_X$ of a point X we obtain first of all $\boldsymbol{v}_X = \dot{X}$, for the acceleration $\boldsymbol{a}_X$ we have $\boldsymbol{a}_X = \ddot{X}$. Further, we have

$$\dot{X}(\mathrm{t}) = \frac{\mathrm{d}X}{\mathrm{dt}} = \frac{\mathrm{d}X}{\mathrm{d}\varphi}\frac{\mathrm{d}\varphi}{\mathrm{dt}} = \frac{\mathrm{d}\varphi}{\mathrm{dt}} \boldsymbol{R}X\,,$$

where $\boldsymbol{R}$ is the fixed directing cone of the motion. The function $\mathrm{d}\varphi/\mathrm{dt}$ is called *angular velocity*, it is denoted by ω. Thus, we have

$$\boldsymbol{v}_X = \omega \boldsymbol{R}X \tag{5.1}$$

or, in equivalent notation,

$$\boldsymbol{v}_X = \omega \boldsymbol{R} \times X\,, \quad \text{eventually} \quad \boldsymbol{v}_X = \omega[\boldsymbol{R}, X]\,.$$

Acceleration is obtained by differentiation of equation (5.1):

$$\boldsymbol{a}_X = \dot{\boldsymbol{v}}_X = \dot{\omega}\boldsymbol{R}X + \omega(\boldsymbol{R}^{\boldsymbol{\cdot}}X + \boldsymbol{R}^2 X\omega) = \dot{\omega}\boldsymbol{R}X + \omega^2\kappa_1 \boldsymbol{T}X + \omega^2\boldsymbol{R}^2 X\,,$$

since

$$\boldsymbol{R}^{\boldsymbol{\cdot}} = \frac{\mathrm{d}\boldsymbol{R}}{\mathrm{dt}} = \frac{\mathrm{d}\boldsymbol{R}}{\mathrm{d}\varphi}\frac{\mathrm{d}\varphi}{\mathrm{dt}} = \omega\kappa_1 \boldsymbol{T}\,;$$

$\dot{\omega}$ is denoted ε and called *angular acceleration*. Altogether, we obtain

$$\boldsymbol{a}_X = (\varepsilon\boldsymbol{R} + \omega^2\kappa_1\boldsymbol{T} + \omega^2\boldsymbol{R}^2)\,X\,. \tag{5.2}$$

Acceleration is usually decomposed into two components: the tangent and the normal components. First, we find the component perpendicular to the tangent plane of the sphere, then we find the component lying in the plane of the sphere. For this purpose we write the formula (5.2) in vector form:

$$\begin{aligned}\boldsymbol{a}_X &= \varepsilon\boldsymbol{R} \times X + \omega^2\kappa_1\boldsymbol{T} \times X + \omega^2\boldsymbol{R} \times (\boldsymbol{R} \times X)\\ &= \varepsilon\boldsymbol{R} \times X + \omega^2\kappa_1\boldsymbol{T} \times X + \omega^2(\boldsymbol{R}, X)\,\boldsymbol{R} - \omega^2 X\,.\end{aligned}$$

Further, we have

$$\begin{aligned}(\boldsymbol{a}_X, X) &= \varepsilon(\boldsymbol{R} \times X, X) + \omega^2\kappa_1(\boldsymbol{T} \times X, X) + \omega^2(\boldsymbol{R}, X)^2 - \omega^2\\ &= \omega^2[(\boldsymbol{R}, X)^2 - 1]\,.\end{aligned}$$

Denote by ϑ the angle between $\boldsymbol{R}$ and X. Then $(\boldsymbol{R}, X) = \cos\vartheta$ so that $(\boldsymbol{a}_X, X) = \omega^2 \sin^2\vartheta$. For the component $\boldsymbol{n}_X$ of acceleration perpendicular to the sphere we thus obtain

$$\boldsymbol{n}_X = -\omega^2 X \sin^2\vartheta .$$

For the component $\boldsymbol{t}_X$ lying in the tangent plane of the sphere we have

$$\boldsymbol{t}_X = \varepsilon \boldsymbol{R} \times X + \omega^2 \kappa_1 \boldsymbol{T} \times X + \omega^2 \boldsymbol{R} \cos\vartheta - \omega^2 X \cos^2\vartheta .$$

The tangent component of acceleration $\boldsymbol{t}_X$ can be decomposed further into a component $\boldsymbol{t}'_X$ lying in the direction of the tangent of the trajectory and a component $\boldsymbol{t}''_X$ lying in the direction of the normal of the trajectory. The unit tangent vector of the trajectory is given by the formula

$$\tau_X = \frac{\boldsymbol{R} \times X}{\|\boldsymbol{R} \times X\|} .$$

Here

$$\|\boldsymbol{R} \times X\|^2 = (\boldsymbol{R} \times X, \boldsymbol{R} \times X) = 1 - (\boldsymbol{R}, X)^2 = \sin^2\vartheta ,$$

so that

$$\tau_X = (\boldsymbol{R} \times X) \sin^{-1}\vartheta .$$

Further, we have

$$\begin{aligned}(\tau_X, \boldsymbol{t}_X) &= (\varepsilon \boldsymbol{R} \times X + \omega^2 \kappa_1 \boldsymbol{T} \times X + \omega^2 \boldsymbol{R} \cos\vartheta \\ &\quad - \omega^2 X \cos^2\vartheta, \sin^{-1}\vartheta [\boldsymbol{R} \times X]) \\ &= \sin^{-1}\vartheta [\varepsilon \sin^2\vartheta - \omega^2 \kappa_1 (\boldsymbol{T}, X) \cos\vartheta] ,\end{aligned}$$

i.e.

$$\boldsymbol{t}'_X = \frac{\varepsilon \sin^2\vartheta - \omega^2 \kappa_1 (\boldsymbol{T}, X) \cos\vartheta}{\sin^2\vartheta} \boldsymbol{R} \times X .$$

The analogous holds for the component $\boldsymbol{t}''_X$: the normal of the trajectory of the point X has the direction of the vector $X \times (\boldsymbol{R} \times X) = \boldsymbol{R} - X \cos\vartheta$, its length is $\sin\vartheta$, so that the unit vector $\boldsymbol{v}_X$ of the normal of the trajectory at the point X is $\boldsymbol{v}_X = (\boldsymbol{R} - X \cos\vartheta) \sin^{-1}\vartheta$. Further,

$$\begin{aligned}(\boldsymbol{t}_X, \boldsymbol{v}_X) &= (\varepsilon \boldsymbol{R} \times X + \omega^2 \kappa_1 \boldsymbol{T} \times X + \boldsymbol{R} \omega^2 \cos\vartheta \\ &\quad - X \omega^2 \cos^2\vartheta, \sin^{-1}\vartheta [\boldsymbol{R} - X \cos\vartheta]) \\ &= \sin^{-1}\vartheta [\omega^2 \kappa_1 (\boldsymbol{N}, X) + \omega^2 \cos\vartheta \sin^2\vartheta] ,\end{aligned}$$

so that

$$t''_X = \frac{\omega^2\kappa_1(N, X) + \omega^2 \cos\vartheta \sin^2\vartheta}{\sin^2\vartheta}(R - X\cos\vartheta).$$

Now, we find those sets on the sphere in which one of the discussed components is zero.

1. $n_X = 0$.

Then $\sin\vartheta = 0$, i.e. either $\vartheta = 0$ or $\vartheta = \pi$, i.e. $X = \pm R$. Thus, acceleration lies in the tangent plane of the sphere at the instantaneous pole only, or $\omega = 0$ at the given instant and the acceleration of all points lies in the tangent plane of the sphere.

2. $a_X = 0$.

There has to be $n_X = 0$ as well, i.e. acceleration can be zero only at the point $\pm R$ or $\omega = 0$. Substituting $X = \pm R$ into the expression for acceleration we obtain the equation $\omega^2\kappa_1 = 0$. Thus, the point $\pm R$ has zero acceleration only if either $\omega = 0$ or $\kappa_1 = 0$.

3. $t'_X = 0$.

Let us write $X = \alpha R + \beta T + \gamma N$ again. Then $t'_X = 0$ has the form $\varepsilon \sin^2\vartheta - \omega^2\kappa_1(T, X)\cos\vartheta = \varepsilon(1 - \alpha^2) - \omega^2\kappa_1\alpha\beta = 0$. The following result is obtained: $t'_X = 0$ in the following cases:

a) $\varepsilon = 0,\ \kappa_1\omega = 0$ – at all points of the sphere
b) $\varepsilon = 0,\ \kappa_1 \neq 0,\ \omega \neq 0$ – on the circles $\alpha = 0$ and $\beta = 0$,
c) $\varepsilon \neq 0,\ \kappa_1\omega = 0$ – at the point $\pm R$,
d) $\varepsilon \neq 0,\ \kappa_1 \neq 0,\ \omega \neq 0$ – at the points of the curve

$$\varepsilon(\beta^2 + \gamma^2) - \omega^2\kappa_1\alpha\beta = 0. \tag{5.3}$$

Let us project the curve of case (d) from the point 0 into the tangent plane of the sphere at the point R. Then $\pi X = R + \beta\alpha^{-1}T + \gamma\alpha^{-1}N$. Denote $\beta\alpha^{-1} = x$, $\gamma\alpha^{-1} = y$. Divide equation (5.3) by the number α^2 (necessarily we have $\alpha \neq 0$). We have

$$\left(\frac{\beta}{\alpha}\right)^2 + \left(\frac{\gamma}{\alpha}\right)^2 - \frac{\omega^2\kappa_1}{\varepsilon}\frac{\beta}{\alpha} = 0, \quad \text{i.e.} \quad x^2 + y^2 - \frac{\omega^2\kappa_1}{\varepsilon}x = 0,$$

i.e. after manipulation

$$\left(x - \frac{\omega^2\kappa_1}{2\varepsilon}\right)^2 + y^2 = \frac{\omega^4\kappa_1^2}{4\varepsilon^2}.$$

Consequently, we obtain the following result: the projection of the set of points on the sphere for which the projection of acceleration into the direction of the tangent of the trajectory is zero is a circle with centre $\boldsymbol{S} = \boldsymbol{R} + \frac{1}{2}\omega^2\kappa_2\varepsilon^{-1}\boldsymbol{T}$ and radius $r = \frac{1}{2}\omega^2\kappa_1\varepsilon^{-1}$.

4. $\boldsymbol{t}''_X = \boldsymbol{0}$.

We obtain the equation $\omega^2\kappa_1(\boldsymbol{N}, X) + \omega^2 \cos\vartheta \sin^2\vartheta = 0$ so that: if (a) $\omega = 0$ we have $\boldsymbol{t}'_X = 0$ at all points of the sphere; if (b) $\omega \neq 0$ we have $\boldsymbol{t}'_X = 0$ at the points of the curve $\kappa_1\gamma + \alpha(1 - \alpha^2) = 0$ which is the already familiar curve of inflection points.

5. $\boldsymbol{t}_X = \boldsymbol{0}$.

The following cases arise:

a) $\omega = 0,\ \varepsilon = 0$ – at all points of the sphere;
b) $\omega = 0,\ \varepsilon \neq 0$ – at the points $\pm\boldsymbol{R}$;
c) $\omega \neq 0,\ \varepsilon = 0,\ \kappa_1 = 0$ – at the points $\pm\boldsymbol{R}$ and on the circle $\alpha = 0$;

d) $\omega \neq 0,\ \varepsilon = 0,\ \kappa_1 \neq 0$. The conditions are $\kappa_1\gamma + \alpha(1 - \alpha^2) = 0$ and $\alpha = 0$ or $\beta = 0$.

i) $\alpha = 0$ – at the points $\pm\boldsymbol{T}$.
ii) $\beta = 0$. Then we have the equation $\kappa_1 + \alpha\gamma = 0$ or $\gamma = 0$.

$\gamma = 0$ – at the points $\pm\boldsymbol{R}$,
$\kappa_1 + \alpha\gamma = 0$ yields $\kappa_1^2 = \alpha^2(1 - \alpha^2)$, i.e.
$\alpha_4 - \alpha^2 + \kappa_1^2 = 0$, i.e. $\alpha_{1,2}^2 = \frac{1}{2}[1 \pm (1 - 4\kappa_1^2)^{1/2}]$.

For $0 < |\kappa_1| < \frac{1}{2}$ we have the solution

$$\alpha_0 = \varepsilon_1\left[\frac{1 + \varepsilon_2(1 - 4\kappa_1^2)^{1/2}}{2}\right]^{1/2},$$

$$\gamma_0 = -\varepsilon_1 \left[\frac{1 - \varepsilon_2(1 - 4\kappa_1^2)^{1/2}}{2} \right]^{1/2},$$

$\varepsilon_1 = \pm 1$, $\varepsilon_2 = \pm 1$.

Consequently, in case (d) we have $\boldsymbol{t}_X = 0$ at the points $\pm \boldsymbol{R}$, $\pm \boldsymbol{T}$, $\alpha_0 \boldsymbol{R} + \gamma_0 \boldsymbol{N}$.

e) $\omega \neq 0$, $\varepsilon \neq 0$, $\kappa_1 = 0$ – only at the points $\pm \boldsymbol{R}$;

f) $\omega \neq 0$, $\varepsilon \neq 0$, $\kappa_1 \neq 0$. Then

$$1 - \alpha^2 - \frac{\omega^2 \kappa_1}{\varepsilon} \alpha\beta = 0\,, \quad \kappa_1 \gamma + \alpha(1 - \alpha^2) = 0\,,$$

must simultaneously hold; this leads to an equation of third degree. In all, we obtain the points $\pm \boldsymbol{R}$ and at most 6 further points for which it is difficult to find an explicit expression.

Finally, let us note that the acceleration of the point $\boldsymbol{R}$ is given by the expression $\boldsymbol{a}_R = -\omega^2 \kappa_1 \boldsymbol{N}$. The points at which acceleration is perpendicular to the axis of rotation lie on the circle $\gamma = 0$; the points at which acceleration is perpendicular to $\boldsymbol{T}$ lie on the great circle in the plane perpendicular to the vector $\omega^2 \kappa_1 \boldsymbol{R} - \varepsilon \boldsymbol{T} + \omega^2 \boldsymbol{N}$. The verification of these assertions is left to the reader as an exercise.

In the end, we investigate the case of two simultaneous general motions from the kinematic point of view. Thus, let $\mathscr{R}_1(t) = \mathscr{R}_0\, g_1(t)$, $\overline{\mathscr{R}}_1(t) = \mathscr{R}_0\, g_2(t)$ be two motions. Consider the motion $\mathscr{R}_2(t) = \mathscr{R}_0\, g_1(t)\,.\, g_2(t)$. The motion of the frame $\mathscr{R}_2$ is generated by composing two motions $\mathscr{R}_1(t) = \mathscr{R}_0\, g_1(t)$ and $\mathscr{R}_2(t) = \mathscr{R}_1(t)\, g_2(t)$. The motion $\mathscr{R}_1(t)$ is called *carrying motion*, the motion $\mathscr{R}_3(t) = \mathscr{R}_2(t_0)\, g_2(t)$ is called *relative motion* at the instant t_0 [essentially, it is the motion $\overline{\mathscr{R}}_1(t) = \mathscr{R}_0\, g_2(t)$ transferred to the position $g_1(t_0)$ by the carrying motion]. Note that there is an infinite number of relative motions. They are the motions $g_1(t_0)\, g_2(t)$, we have one for every instant t_0. Actually, it is the same motion all the time, considered at each moment with respect to some other position of the moving system of the carrying motion which is simultaneously the fixed system of our motion. For the motions $\mathscr{R}_1 = \mathscr{R}_0 g_1$, $\overline{\mathscr{R}}_1 = \mathscr{R}_0 g_2$, $\mathscr{R}_2 = \mathscr{R}_0 g_1 g_2$, we obtain the following directing functions:

$$\boldsymbol{r}_{21} = \omega_{21} \boldsymbol{R}_1 = \dot{g}_1 g_1^{-1}\,,$$

$$\boldsymbol{r}_{32} = \omega_{32}\boldsymbol{R}_2 = \dot{g}_2 g_2^{-1},$$
$$\boldsymbol{r}_{31} = \omega_{31}\boldsymbol{R}_3 = (g_2 g_2)^{\boldsymbol{\cdot}}\, g_2^{-1} g_1^{-1}.$$

Here ω_{21} is the angular velocity of the carrying motion, ω_{31} is the angular velocity of the resulting motion, ω_{32} is the angular velocity of the "specimen" of relative motion. Denote the angular velocity of relative motion by ω_{32}^r. The vectors $\boldsymbol{r}_{21}$, $\boldsymbol{r}_{31}$, $\boldsymbol{r}_{32}$ are also called the vectors of the angular velocities of the individual motions. Further, let us denote $\boldsymbol{r}_{32}^r = \omega_{32}^r \boldsymbol{R}_2^r = \mathrm{ad}g_1(\dot{g}_2 g_2^{-1})$. Then $\boldsymbol{r}_{32}^r$ is the vector of angular velocity of the relative motion. Indeed,

$$\boldsymbol{r}_{32}^r(\mathrm{t}_0) = \frac{\mathrm{d}}{\mathrm{dt}}\{g_1(\mathrm{t}_0)\, g_2(\mathrm{t})\}\,[g_1(\mathrm{t}_0)\, g_2(\mathrm{t})]^{-1}\big|_{t=t_0}$$
$$= g_1(\mathrm{t}_0)\, \dot{g}_2(\mathrm{t}_0)\, g_2^{-1}(\mathrm{t}_0)\, g_1^{-1}(\mathrm{t}_0) = \mathrm{ad}g_1(\mathrm{t}_0)\,(\dot{g}_2 g_2^{-1}).$$

Thus, we have $\boldsymbol{r}_{32}^r(\mathrm{t}) = \mathrm{ad}g_1(\mathrm{t})\, \boldsymbol{r}_{32}(\mathrm{t})$ so that

$$\boldsymbol{r}_{32}^r = \omega_{32}^r \boldsymbol{R}_2^r = \mathrm{ad}g_1(\omega_{32}\boldsymbol{R}_2), \quad \text{i.e.} \quad \omega_{32} = \omega_{32}^r, \quad \boldsymbol{R}_2^r = \mathrm{adg}_1 \boldsymbol{R}_2.$$

For $\boldsymbol{r}_{31} = \omega_{31}\boldsymbol{R}_3$ we now have

$$\boldsymbol{r}_{31} = [g_1(\mathrm{t})\, g_2(\mathrm{t})]^{\boldsymbol{\cdot}}\, (g_1 g_2)^{-1} = \mathrm{ad}g_1 \boldsymbol{r}_{32} + \boldsymbol{r}_{21},$$

so that $\boldsymbol{r}_{31} = \boldsymbol{r}_{32}^r + \boldsymbol{r}_{21}$. For the angular velocities we obtain

$$\omega_{31}\boldsymbol{R}_3 = \omega_{21}\boldsymbol{R}_1 + \omega_{32}\mathrm{ad}g_1\boldsymbol{R}_2.$$

Taking into account that $\mathrm{ad}g_1\boldsymbol{R}_1 = \boldsymbol{R}_1$ we have

$$\omega_{31}\boldsymbol{R}_3 = \mathrm{ad}g_1(\omega_2\boldsymbol{R}_1 + \omega_{32}\boldsymbol{R}_2).$$

For the magnitudes of the angular velocities we obtain

$$\omega_{31}^2 = \omega_{21}^2 + \omega_{32}^2 + 2\omega_{32}\omega_{21}(\boldsymbol{R}_1, \boldsymbol{R}_2^r).$$

Further, denote by $\boldsymbol{N}$ the unit vector in the direction of $\boldsymbol{R}_1 \times \boldsymbol{R}_2^r$, by ϑ_{31} the angle from $\boldsymbol{R}_3$ to $\boldsymbol{R}_1$, by ϑ_{23} the angle from $\boldsymbol{R}_2^r$ to $\boldsymbol{R}_{31}$, by ϑ_{21} the angle from $\boldsymbol{R}_2^r$ to $\boldsymbol{R}_1$. Then

$$\omega_{31}\boldsymbol{R}_3 \times \boldsymbol{R}_1 = \boldsymbol{N}\omega_{31}\sin\vartheta_{31} = (\omega_{21}\boldsymbol{R}_1 + \omega_{32}\mathrm{ad}g_1\boldsymbol{R}_2) \times \boldsymbol{R}_1$$
$$= \boldsymbol{N}\omega_{32}\sin\vartheta_{21},$$

$$\omega_{31}\boldsymbol{R}_3 \times \boldsymbol{R}_2^r = -N\omega_{31} \sin \vartheta_{23} = (\omega_1\boldsymbol{R}_1 + \omega_{32}\boldsymbol{R}_2^r) \times \boldsymbol{R}_2^r = -N\omega_{21} \sin \vartheta_{21} ,$$

so that

$$\omega_{31} \sin \vartheta_{31} = \omega_{32} \sin \vartheta_{21} , \quad \omega_{31} \sin \vartheta_{23} = \omega_{21} \sin \vartheta_{21} .$$

These equations yield

$$\omega_{32} : \omega_{31} : \omega_{21} = \sin \vartheta_{31} : \sin \vartheta_{21} : \sin \vartheta_{23} ;$$

here we have $\vartheta_{21} = \vartheta_{23} + \vartheta_{31}$, which is a relation between the configuration of the instantaneous poles and the angular velocities. The equation $\omega_{31}^2 = \omega_{32}^2 + \omega_{21}^2 + 2\omega_{32}\omega_{21} \cos \vartheta_{21}$ makes possible the direct computation of the angular velocity of the resulting motion.

It remains to find the relation between the velocities and accelerations of a point for the individual motions. For this we denote, as above, by $\boldsymbol{v}_{31}, \boldsymbol{v}_{32}^r, \boldsymbol{v}_{21}$ the velocities of the point X under the individual motions. Then

$$\boldsymbol{v}_{31} = \omega_{31}\boldsymbol{R}_3 \times X = (\omega_{21}\boldsymbol{R}_1 + \omega_{32}\boldsymbol{R}_2^r) \times X = \boldsymbol{v}_{21} + \boldsymbol{v}_{32}^r .$$

Thus, the resulting velocity is the simple sum of the carrying and relative velocities. Indeed, we have

$$\boldsymbol{v}_{31} = \boldsymbol{v}_{21} + \mathrm{ad} g_1 \boldsymbol{v}_{32} ;$$

differentiating this equation we obtain

$$\boldsymbol{a}_{31} = \boldsymbol{a}_{21} + \mathrm{ad} g_1 \boldsymbol{a}_{32} + (\mathrm{ad} g_1)^{\boldsymbol{\cdot}}\, \boldsymbol{v}_{32} = \boldsymbol{a}_{21} + \boldsymbol{a}_{32}^r + \boldsymbol{r}_{21} \times \boldsymbol{v}_{32}^r .$$

The additional term $\boldsymbol{r}_{21} \times \boldsymbol{v}_{32}^r$ is called the Coriolis acceleration. Thus: the resulting acceleration is the sum of the carrying, the relative, and the Coriolis accelerations.

Chapter III.

Kinematics of space motion

1. LIE GROUP OF CONGRUENCES OF E_3 AND ITS LIE ALGEBRA

Assume $\mathscr{R}_0 = \{O, \boldsymbol{u}_1, \boldsymbol{u}_2, \boldsymbol{u}_3\}$ and $\mathscr{R} = \{A, \boldsymbol{f}_1, \boldsymbol{f}_2, \boldsymbol{f}_3\}$ to be two orthonormal frames in the Euclidean space E_3. The congruence $\mathscr{S}$, which transforms frame $\mathscr{R}_0$ into frame $\mathscr{R}$, can be expressed with the aid of the 4×4 matrix $g = \begin{pmatrix} 1, & 0 \\ a, & \gamma \end{pmatrix}$, where a is the column of three coordinates of point A in frame $\mathscr{R}_0$ while $\gamma \in O(3)$ is formed by the columns of coordinates of vectors $\boldsymbol{f}_1, \boldsymbol{f}_2, \boldsymbol{f}_3$ in $\mathscr{R}_0$. Symbolically $\mathscr{R} = \mathscr{R}_0 g$, the right-hand side being the product of row $\mathscr{R}_0$ and of matrix g. If $M \in E_3$ is an arbitrary point, $M = O + x^i \boldsymbol{u}_i$, we assign to it the column $M_{\mathscr{R}_0} = \begin{pmatrix} 1 \\ x^i \end{pmatrix}$*, if $\boldsymbol{v} = v^i \boldsymbol{u}_i$ is a vector, we assign to it the column $\boldsymbol{v}_{\mathscr{R}_0} = \begin{pmatrix} 0 \\ v^i \end{pmatrix}$. We may then symbolically write $M = \mathscr{R}_0 M_{\mathscr{R}_0}$, $\boldsymbol{v} = \mathscr{R}_0 v_{\mathscr{R}_0}$, in every frame $\mathscr{R}_0$. As in the previous chapters, $(\mathscr{S}(M))_{\mathscr{R}_0} = g(M_{\mathscr{R}_0})$, $(\mathscr{S}(\boldsymbol{v}))_{\mathscr{R}_0} = g(v_{\mathscr{R}_0}))$ for the coordinates of point M transformed by congruence $\mathscr{S}$ to point $\mathscr{S}(M)$, and similarly for the vector $\boldsymbol{v}$.

We shall now consider only orthonormal frames, i.e. by the term frame we shall understand an orthonormal frame in E_3 and we shall assume that we have fixed one such frame in E_3. Each congruence in E_3 is then uniquely represented by a matrix of the form $\begin{pmatrix} 1, & 0 \\ a, & \gamma \end{pmatrix}$, $\gamma \in O(3)$, and in what follows we shall identify the congruences in E_3 with these matrices.

* Abbreviated notation for $\begin{pmatrix} 1 \\ x^i \end{pmatrix} = \begin{pmatrix} 1 \\ x^1 \\ x^2 \\ x^3 \end{pmatrix}$.

All congruences in E_3 form a group and, from what was said in Chapter I, it is clear that it is a Lie group. Its dimension is 6 and we shall denote it by $\mathscr{E}$. Its Lie algebra is a vector space of the same dimension. Let us denote it by $\mathfrak{E}$ and find out by which matrices it is constituted. Therefore, assume $g(t)$ to be a curve on the group $\mathscr{E}$ such that $g(0) = e$, i.e. let

$$g(t) = \begin{pmatrix} 1, & 0 \\ a(t), & \gamma(t) \end{pmatrix}, \quad a(t) \equiv \begin{pmatrix} a_1(t) \\ a_2(t) \\ a_3(t) \end{pmatrix},$$

$$\gamma(t) \equiv [a_i^j(t)], \quad a_i(0) = 0, \quad a_i^j(0) = \delta_i^j.$$

Moreover, $\gamma(t) \in O(3)$, so that $a_i^j(t)\, a_k^j(t) = \delta_{ik}$. If we differentiate this equation and substitute $t = 0$, we arrive at

$$(a_i^j)^{\bullet}(t)\, a_k^j(t) + a_i^j(t)\, (a_k^j)^{\bullet}(t) = 0,$$

i.e.

$$(a_i^j)^{\bullet}(0)\, \delta_k^j + \delta_i^j (a_k^j)^{\bullet}(0) = 0, \quad \text{and} \quad (a_i^k)^{\bullet}(0) + (a_k^i)^{\bullet}(0) = 0.$$

Therefore,

$$\dot{g}(0) = \begin{pmatrix} 0, & 0 \\ \dot{a}(0), & \dot{\gamma}(0) \end{pmatrix},$$

where $\dot{\gamma}(0)$ is a skew-symmetric matrix. Thus, Lie algebra $\mathfrak{E}$ is constituted by all matrices of the form

$$\boldsymbol{X} = \begin{pmatrix} 0, & 0 \\ \boldsymbol{t}, & \boldsymbol{x} \end{pmatrix},$$

where $\boldsymbol{t}$ is any column of three numbers and $\boldsymbol{x}$ is any skew-symmetric 3×3 matrix.

We shall now determine the adjoint mapping ad$\mathscr{E}$ of group $\mathscr{E}$ in algebra $\mathfrak{E}$. Let

$$g = \begin{pmatrix} 1, & 0 \\ a, & \gamma \end{pmatrix} \in \mathscr{E}, \quad \boldsymbol{X} = \begin{pmatrix} 0, & 0 \\ \boldsymbol{t}, & \boldsymbol{x} \end{pmatrix} \in \mathfrak{E}.$$

Let us calculate the matrix g^{-1} inverse to matrix g. Put

$$g^{-1} = \begin{pmatrix} 1, & 0 \\ b, & \delta \end{pmatrix}.$$

It must be

$$gg^{-1} = \begin{pmatrix} 1, & 0 \\ a, & \gamma \end{pmatrix} \begin{pmatrix} 1, & 0 \\ b, & \delta \end{pmatrix} = \begin{pmatrix} 1, & 0 \\ 0, & e \end{pmatrix},$$

where e is the unit 3×3 matrix. After multiplying we have

$$\begin{pmatrix} 1, & 0 \\ a, & \gamma \end{pmatrix} \begin{pmatrix} 1, & 0 \\ b, & \delta \end{pmatrix} = \begin{pmatrix} 1, & 0 \\ a + \gamma b, & \gamma\delta \end{pmatrix} = \begin{pmatrix} 1, & 0 \\ 0, & e \end{pmatrix},$$

i.e.

$$\gamma\delta = e, \quad a + \gamma b = 0;$$

thus $\delta = \gamma^{\mathrm{T}}$, since $\gamma \in O(3)$ and $\gamma b = -a$, i.e. $b = -\gamma^{\mathrm{T}} a$. Therefore,

$$g^{-1} = \begin{pmatrix} 1, & 0 \\ -\gamma^{\mathrm{T}} a, & \gamma^{\mathrm{T}} \end{pmatrix}. \tag{1.1}$$

Also,

$$\mathrm{ad} g \boldsymbol{X} = g\boldsymbol{X}g^{-1} = \begin{pmatrix} 0, & 0 \\ \gamma \boldsymbol{t} - \gamma \boldsymbol{x} \gamma^{\mathrm{T}} a, & \gamma \boldsymbol{x} \gamma^{\mathrm{T}} \end{pmatrix}.$$

Finally, we determine the bracket of two vectors in $\mathfrak{E}$. Let us put

$$\boldsymbol{Y} = \begin{pmatrix} 0, & 0 \\ \boldsymbol{s}, & \boldsymbol{y} \end{pmatrix} \in \mathfrak{E}.$$

Consequently,

$$[\boldsymbol{X}, \boldsymbol{Y}] = \boldsymbol{XY} - \boldsymbol{YX} = \begin{pmatrix} 0, & 0 \\ \boldsymbol{t}, & \boldsymbol{x} \end{pmatrix} \begin{pmatrix} 0, & 0 \\ \boldsymbol{s}, & \boldsymbol{y} \end{pmatrix} - \begin{pmatrix} 0, & 0 \\ \boldsymbol{s}, & \boldsymbol{y} \end{pmatrix} \begin{pmatrix} 0, & 0 \\ \boldsymbol{t}, & \boldsymbol{x} \end{pmatrix} =$$

$$\begin{pmatrix} 0, & 0 \\ \boldsymbol{xs} - \boldsymbol{yt}, & \boldsymbol{xy} - \boldsymbol{yx} \end{pmatrix}.$$

Realizing that $\boldsymbol{x}, \boldsymbol{y} \in \mathfrak{O}(3)$ and $\gamma \in O(3)$, we may write

$$\mathrm{ad} g \boldsymbol{X} = \begin{pmatrix} 0, & 0 \\ \gamma \boldsymbol{t} - (\mathrm{ad}\gamma \boldsymbol{x})\, a, & \mathrm{ad}\gamma \boldsymbol{x} \end{pmatrix}, \quad [\boldsymbol{X}, \boldsymbol{Y}] = \begin{pmatrix} 0, & 0 \\ \boldsymbol{xs} - \boldsymbol{yt}, & [\boldsymbol{x}, \boldsymbol{y}] \end{pmatrix}. \tag{1.2}$$

To simplify the notation of the vectors in $\mathfrak{E}$ we shall now use a procedure similar to that used at the beginning of Section 3 in Chapter II for

Lie algebra $\mathfrak{O}(3)$. In algebra $\mathfrak{E}$ we choose the base $\boldsymbol{X}_1, \boldsymbol{X}_2, \boldsymbol{X}_3, \boldsymbol{T}_1, \boldsymbol{T}_2, \boldsymbol{T}_3$ as follows:

$$\boldsymbol{X}_i = \begin{pmatrix} 0, & 0 \\ \boldsymbol{0}, & \boldsymbol{x}_i \end{pmatrix}, \quad \boldsymbol{T}_i = \begin{pmatrix} 0, & 0 \\ \boldsymbol{t}_i, & \boldsymbol{0} \end{pmatrix},$$

where

$$\boldsymbol{x}_1 = \begin{pmatrix} 0, & 0, & 0 \\ 0, & 0, & -1 \\ 0, & 1, & 0 \end{pmatrix}, \quad \boldsymbol{x}_2 = \begin{pmatrix} 0, & 0, & 1 \\ 0, & 0, & 0 \\ -1, & 0, & 0 \end{pmatrix}, \quad \boldsymbol{x}_3 = \begin{pmatrix} 0, & -1, & 0 \\ 1, & 0, & 0 \\ 0, & 0, & 0 \end{pmatrix};$$

$$\boldsymbol{t}_1 = \begin{pmatrix} 1 \\ 0 \\ 0 \end{pmatrix}, \quad \boldsymbol{t}_2 = \begin{pmatrix} 0 \\ 1 \\ 0 \end{pmatrix}, \quad \boldsymbol{t}_3 = \begin{pmatrix} 0 \\ 0 \\ 1 \end{pmatrix}.$$

If we now also bear in mind that each skew-symmetric 3×3 matrix can be expressed, as in Section 3 of Chapter II, in the form

$$\boldsymbol{x} = \begin{pmatrix} 0, & -x^3, & x^2 \\ x^3, & 0, & -x^1 \\ -x^2, & x^1, & 0 \end{pmatrix},$$

we see that for $\boldsymbol{X} \in \mathfrak{E}$ we may put $\boldsymbol{X} = x^i \boldsymbol{X}_i + t^i \boldsymbol{T}_i$. This we shall now write as $\boldsymbol{X} = [\boldsymbol{x}; \boldsymbol{t}]$ or, in coordinate notation, as

$$\boldsymbol{X} = \begin{bmatrix} x^1; & t^1 \\ x^2; & t^2 \\ x^3; & t^3 \end{bmatrix}.$$

The bracket of two vectors $\boldsymbol{X} = [\boldsymbol{x}; \boldsymbol{t}]$ and $\boldsymbol{Y} = [\boldsymbol{y}; \boldsymbol{s}]$ may now be written as

$$[\boldsymbol{X}, \boldsymbol{Y}] = [[\boldsymbol{x}; \boldsymbol{t}], [\boldsymbol{y}; \boldsymbol{s}]] = [\boldsymbol{x} \times \boldsymbol{y}; \ \boldsymbol{x} \times \boldsymbol{s} + \boldsymbol{t} \times \boldsymbol{y}], \tag{1.3}$$

where $\boldsymbol{x} \times \boldsymbol{y}$ is the ordinary cross product and $\boldsymbol{x} \times \boldsymbol{s}$ and $\boldsymbol{t} \times \boldsymbol{y}$ are analogously introduced cross products (true, between vectors of two different vector spaces but formally in the same way). For example, $\boldsymbol{X}_1 \times \boldsymbol{X}_2 = \boldsymbol{X}_3$ and $\boldsymbol{X}_1 \times \boldsymbol{T}_2 = \boldsymbol{T}_3$, $\boldsymbol{X}_1 \times \boldsymbol{X}_1 = 0$ and $\boldsymbol{X}_1 \times \boldsymbol{T}_1 = 0$. This is really so, because in the expression for $[\boldsymbol{X}, \boldsymbol{Y}]$ in (1.2) $[\boldsymbol{x}, \boldsymbol{y}] = \boldsymbol{xy} - \boldsymbol{yx}$ is the bracket of two skew-symmetric matrices which, in vector notation, becomes the cross product while the terms $\boldsymbol{xs}$ and $\boldsymbol{yt}$ are the products of the skew-symmetric matrix $\boldsymbol{x}$ and the column $\boldsymbol{s}$ and of the skew-symmetric matrix $\boldsymbol{y}$

and the column $\boldsymbol{t}$, respectively. In vector notation these are again cross products as we know from Section 3 of Chapter II. Moreover, if we note that in vector notation $\gamma x \gamma^{\mathrm{T}}$ becomes $\gamma \boldsymbol{x}^{\mathrm{V}}$, using the notation of Section 3 of Chapter II it is possible to express the adjoint mapping ad$g\boldsymbol{X}$ in (1.2) as

$$\mathrm{ad}g\boldsymbol{X} = [\gamma \boldsymbol{x}; \gamma \boldsymbol{t} + a \times \gamma \boldsymbol{x}]\,. \tag{1.4}$$

1.1 Lemma. Let

$$\boldsymbol{X} = [\boldsymbol{x}; \boldsymbol{t}]\,, \quad \boldsymbol{Y} = [\boldsymbol{y}; \boldsymbol{s}] \in \mathfrak{E}\,,$$

$$g = \begin{pmatrix} 1, & 0 \\ a, & \gamma \end{pmatrix} \in \mathscr{E}\,.$$

Then

$$[\boldsymbol{X}, \boldsymbol{Y}] = [\boldsymbol{x} \times \boldsymbol{y}; \boldsymbol{x} \times \boldsymbol{s} + \boldsymbol{y} \times \boldsymbol{t}]\,, \quad \mathrm{ad}g\boldsymbol{X} = [\gamma \boldsymbol{x}; \gamma \boldsymbol{t} + a \times \gamma \boldsymbol{x}]\,.$$

1.2 Definition. To simplify the notation we define an auxiliary inner product in $\mathfrak{E}$ as follows: $(\boldsymbol{X}_i, \boldsymbol{X}_j) = \delta_{ij}$, $(\boldsymbol{X}_i, \boldsymbol{T}_j) = \delta_{ij}$, $(\boldsymbol{T}_i, \boldsymbol{T}_j) = \delta_{ij}$, so that for $\boldsymbol{X} = [\boldsymbol{x}; \boldsymbol{t}]$ and $\boldsymbol{Y} = [\boldsymbol{y}; \boldsymbol{s}]$ we may write, e.g., $(\boldsymbol{x}, \boldsymbol{y}) = x^i y^i$, $(\boldsymbol{x}, \boldsymbol{s}) = x^i s^i$. In $\mathfrak{E}$ we define the following symmetric bilinear forms:

$\mathbf{K}(\boldsymbol{X}, \boldsymbol{Y}) = (\boldsymbol{x}, \boldsymbol{y})$		– *Killing bilinear form,*
$\mathscr{K}(\boldsymbol{X}, \boldsymbol{Y}) = (\boldsymbol{x}, \boldsymbol{s}) + (\boldsymbol{y}, \boldsymbol{t})$		– *Klein bilinear form,*
$\mathbf{T}(\boldsymbol{X}, \boldsymbol{Y}) = (\boldsymbol{t}, \boldsymbol{s})$	for $\boldsymbol{x} = \boldsymbol{y} = 0$	– ordinary inner product.

1.3 Theorem. *The bilinear forms* $\mathbf{K}$, $\mathscr{K}$ *and* $\mathbf{T}$ *are invariant with respect to* ad$\mathscr{E}$.

Proof. The bilinear form $\mathbf{B}(\boldsymbol{X}, \boldsymbol{Y})$ is called invariant with respect to ad$\mathscr{E}$ if $\mathbf{B}(\boldsymbol{X}, \boldsymbol{Y}) = \mathbf{B}(\mathrm{ad}g\boldsymbol{X}, \mathrm{ad}g\boldsymbol{Y})$ for all $g \in \mathscr{E}$. According to Lemma 1.1

$$\mathrm{ad}g\boldsymbol{X} = [\gamma \boldsymbol{x}; \gamma \boldsymbol{t} + a \times \gamma \boldsymbol{x}]\,,$$

where

$$\boldsymbol{X} = [\boldsymbol{x}; \boldsymbol{t}]\,, \quad \boldsymbol{Y} = [\boldsymbol{y}; \boldsymbol{s}]\,, \quad g = \begin{pmatrix} 1, & 0 \\ a, & \gamma \end{pmatrix}.$$

Therefore,

$$\mathbf{K}(\boldsymbol{X}, \boldsymbol{Y}) = (\boldsymbol{x}, \boldsymbol{y})\,, \quad \mathbf{K}(\mathrm{ad}g\boldsymbol{X}, \mathrm{ad}g\boldsymbol{Y}) = (\gamma \boldsymbol{x}, \gamma \boldsymbol{y}) = (\boldsymbol{x}, \boldsymbol{y})\,,$$

because $\gamma \in O(3)$ and the ordinary inner product is invariant with respect to orthogonal transformations. [By direct computation we can easily find that $\mathbf{K}(X, Y)$, up to a proportionality factor, is actually the Killing invariant symmetric bilinear form of algebra $\mathfrak{E}$; we shall not need this fact below, therefore its computation is not given here but left to the reader as an exercise.] Moreover, we have

$$\begin{aligned}\mathcal{K}(X, Y) &= (\boldsymbol{x}, \boldsymbol{s}) + (\boldsymbol{y}, \boldsymbol{t}),\\ \mathcal{K}(\mathrm{ad}gX, \mathrm{ad}gY) &= (\gamma\boldsymbol{x}, \gamma\boldsymbol{s} + a \times \gamma\boldsymbol{y}) + (\gamma\boldsymbol{y}, \gamma\boldsymbol{t} + a \times \gamma\boldsymbol{x})\\ &= (\gamma\boldsymbol{x}, \gamma\boldsymbol{s}) + (\gamma\boldsymbol{t}, \gamma\boldsymbol{y}) + |\gamma\boldsymbol{x}, a, \gamma\boldsymbol{y}| + |\gamma\boldsymbol{y}, a, \gamma\boldsymbol{x}|\\ &= (\boldsymbol{x}, \boldsymbol{s}) + (\boldsymbol{t}, \boldsymbol{y}),\end{aligned}$$

again as a result of the invariance of the inner product to orthogonal transformations. Finally, if $\boldsymbol{x} = \boldsymbol{y} = 0$, $\mathrm{ad}gX = [0; \gamma\boldsymbol{t})$, $\mathrm{ad}gY = [0; \gamma\boldsymbol{s}]$, so that the form $\mathbf{T}$ is invariant for the same reasons.

1.4 Example. We shall explain the formal apparatus just introduced with a simple example. Choose $X, Y \in \mathfrak{E}$, $g \in \mathscr{E}$ as follows:

$$X = \begin{pmatrix} 0, & 0, & 0, & 0\\ 2, & 0, & 2, & 0\\ 1, & -2, & 0, & -1\\ 0, & 0, & 1, & 0\end{pmatrix}, \quad Y = \begin{pmatrix} 0, & 0, & 0, & 0\\ 0, & 0, & 3, & 1\\ -1, & -3, & 0, & 0\\ 1, & -1, & 0, & 0\end{pmatrix},$$

$$g = \begin{pmatrix} 1, & 0, & 0, & 0\\ 0, & 1, & 0, & 0\\ -2, & 0, & \frac{1}{2}, & \frac{\sqrt{3}}{2}\\ 0, & 0, & -\frac{\sqrt{3}}{2}, & \frac{1}{2}\end{pmatrix}.$$

Then $X = X_1 - 2X_3 + 2T_1 + T_2$, $Y = X_2 - 3X_3 - T_2 + T_3$; formally

$$X = \begin{bmatrix} 1; & 2\\ 0; & 1\\ -2; & 0\end{bmatrix}, \quad Y = \begin{bmatrix} 0; & 0\\ 1; & -1\\ -3; & 1\end{bmatrix},$$

$$\boldsymbol{x} = \begin{pmatrix} 1\\ 0\\ -2\end{pmatrix}, \quad \boldsymbol{t} = \begin{pmatrix} 2\\ 1\\ 0\end{pmatrix}, \quad \boldsymbol{y} = \begin{pmatrix} 0\\ 1\\ -3\end{pmatrix}, \quad \boldsymbol{s} = \begin{pmatrix} 0\\ -1\\ 1\end{pmatrix}.$$

Further, we have

$$x \times y = \begin{pmatrix} 2 \\ 3 \\ 1 \end{pmatrix}, \quad x \times s = \begin{pmatrix} -2 \\ -1 \\ -1 \end{pmatrix},$$

$$t \times y = \begin{pmatrix} -3 \\ 6 \\ 2 \end{pmatrix}, \quad [X, Y] = \begin{pmatrix} 2; & -5 \\ 3; & 5 \\ 1; & 1 \end{pmatrix}.$$

Direct computation yields

$$[X, Y] = XY - YX =$$

$$= \begin{pmatrix} 0, & 0, & 0, & 0 \\ 2, & 0, & 2, & 0 \\ 1, & -2, & 0, & -1 \\ 0, & 0, & 1, & 0 \end{pmatrix} \begin{pmatrix} 0, & 0, & 0, & 0 \\ 0, & 0, & 3, & 1 \\ -1, & -3, & 0, & 0 \\ 1, & -1, & 0, & 0 \end{pmatrix} -$$

$$\begin{pmatrix} 0, & 0, & 0, & 0 \\ 0, & 0, & 3, & 1 \\ -1, & -3, & 0, & 0 \\ 1, & -1, & 0, & 0 \end{pmatrix} \begin{pmatrix} 0, & 0, & 0, & 0 \\ 2, & 0, & 2, & 0 \\ 1, & -2, & 0, & -1 \\ 0, & 0, & 1, & 0 \end{pmatrix}$$

$$= \begin{pmatrix} 0, & 0, & 0, & 0 \\ -5, & 0, & -1, & 3 \\ 5, & 1, & 0, & -2 \\ 1, & -3, & 2, & 0 \end{pmatrix} = \begin{bmatrix} 2; & -5 \\ 3; & 5 \\ 1; & 1 \end{bmatrix},$$

which agrees.

Further,

$$\mathbf{K}(X, Y) = (x, y) = 6,$$

$$\mathcal{K}(X, Y) = (x, s) + (t, y) = -2 + 1 = -1.$$

If we put

$$g = \begin{pmatrix} 1, & 0 \\ a, & \gamma \end{pmatrix},$$

we have

$$a = \begin{bmatrix} 0 \\ -2 \\ 0 \end{bmatrix}, \quad \gamma = \begin{bmatrix} 1, & 0, & 0 \\ 0, & \frac{1}{2}, & \frac{\sqrt{3}}{2} \\ 0, & -\frac{\sqrt{3}}{2}, & \frac{1}{2} \end{bmatrix}, \quad g^{-1} = \begin{pmatrix} 1, & 0 \\ -\gamma^{\mathrm{T}}\boldsymbol{a}, & \gamma^{\mathrm{T}} \end{pmatrix},$$

so that

$$-\gamma^{\mathrm{T}}\boldsymbol{a} = \begin{pmatrix} 0 \\ 1 \\ \sqrt{3} \end{pmatrix}, \quad \text{i.e.} \quad g^{-1} = \begin{bmatrix} 1, & 0, & 0, & 0 \\ 0, & 1, & 0, & 0 \\ 1, & 0, & \frac{1}{2}, & -\frac{\sqrt{3}}{2} \\ \sqrt{3}, & 0, & \frac{\sqrt{3}}{2}, & \frac{1}{2} \end{bmatrix}.$$

Now, let us compute $\mathrm{ad}g\boldsymbol{X} = [\gamma\boldsymbol{x}; \gamma\boldsymbol{t} + a \times \gamma\boldsymbol{x}]$:

$$\gamma\boldsymbol{x} = \begin{bmatrix} 1 \\ -\sqrt{3} \\ -1 \end{bmatrix}, \quad \gamma\boldsymbol{t} = \begin{bmatrix} 2 \\ \frac{1}{2} \\ -\frac{\sqrt{3}}{2} \end{bmatrix}, \quad a \times \gamma\boldsymbol{x} = \begin{bmatrix} 2 \\ 0 \\ 2 \end{bmatrix},$$

i.e.

$$\mathrm{ad}g\boldsymbol{X} = \begin{bmatrix} 1; & 4 \\ -\sqrt{3}; & \frac{1}{2} \\ -1; & 2 - \frac{\sqrt{3}}{2} \end{bmatrix}.$$

Direct computation yields

$$\mathrm{ad}\,gX = \begin{bmatrix} 1, 0, & 0, & 0 \\ 0, 1, & 0, & 0 \\ -2, 0, & \frac{1}{2}, & \frac{\sqrt{3}}{2} \\ 0, 0, & -\frac{\sqrt{3}}{2}, & \frac{1}{2} \end{bmatrix} \begin{bmatrix} 0, & 0, 0, & 0 \\ 2, & 0, 2, & 0 \\ 1, & -2, 0, & -1 \\ 0, & 0, 1, & 0 \end{bmatrix} \begin{bmatrix} 1, & 0, 0, & 0 \\ 0, & 1, 0, & 0 \\ 1, & 0, \frac{1}{2}, & -\frac{\sqrt{3}}{2} \\ \sqrt{3}, & 0, \frac{\sqrt{3}}{2}, & \frac{1}{2} \end{bmatrix}$$

$$= \begin{bmatrix} 0, & 0, 0, & 0 \\ 4, & 0, 1, & -\sqrt{3} \\ \frac{1}{2}, & -1, 0, & -1 \\ 2-\frac{\sqrt{3}}{2}, & \sqrt{3}, 1, & 0 \end{bmatrix} = \begin{bmatrix} 1; 4 \\ -\sqrt{3}; \frac{1}{2} \\ -1; 2-\frac{\sqrt{3}}{2} \end{bmatrix},$$

which again agrees.

1.5 Exercise. Using the vectors X, Y, and the matrix g from Example 1.4, show that the forms $\mathbf{K}(X, Y)$, $\mathscr{K}(X, Y)$ and $\mathbf{T}(X, Y)$ are invariant.

Finally, we introduce the group of all congruences of E_3 which preserve a given direction.

1.6 Definition. The Lie algebra $\mathfrak{A}$ is a *direct sum* of its subalgebras $\mathfrak{B}$ and $\mathfrak{C}$ if every $z \in \mathfrak{A}$ can be written uniquely as $z = x + y$, where $x \in \mathfrak{B}$, $y \in \mathfrak{C}$ and $[x, y] = 0$.

1.7 Definition. Assume $X \in \mathfrak{E}$ to be a non-zero vector such that $\mathbf{K}(X, X) = 0$. Denote $\mathfrak{Z}(X)$ the set of all $Y \in \mathfrak{E}$ such that $[X, Y] = 0$. $\mathfrak{Z}(X)$ is called the *centralizer* of vector X in $\mathfrak{E}$.

1.8 Theorem. *Let* $\mathbf{K}(X, X) = 0$, $X \neq 0$. *Then* $\mathfrak{Z}(X)$ *is a subalgebra in* $\mathfrak{E}$ *which has dimension* 4. *It is the Lie algebra of the group* $Z(X)$ *of all congruences of* E_3 *which preserve the direction determined by the vector* X. *Further*: $\mathfrak{Z}(X) = \{\lambda X\} \dotplus \mathfrak{P}(X)$, *where* $\{\lambda X\}$ *represents the set of all* λ*-multiples of vector* X, *and* $\mathfrak{P}(X)$ *is a three-dimensional subalgebra in* $\mathfrak{Z}(X)$ *in which* $\mathscr{K}$ *is identically equal to zero. Algebra* $\mathfrak{P}(X)$ *is defined* *uniquely and it is the Lie algebra of the group $P(X)$ of all congruences which preserve each plane perpendicular to the direction determined by vector X. [$P(X)$ is isomorphic with the group of all congruences in the plane.]*

Proof. $\mathfrak{Z}(X)$ is a Lie algebra because $[X,[Y_1,Y_2]]+[Y_1,[Y_2,X]]+[Y_2,[X,Y_1]]=[X,[Y_1,Y_2]]=0$, i.e. $[Y_1,Y_2]\in\mathfrak{Z}(X)$, if $[X,Y_1]=[X,Y_2]=0$. Since the algebra $\mathfrak{Z}(X)$ is defined invariantly with respect to the adjoint mapping in $\mathfrak{E}$, i.e. $\mathfrak{Z}(\mathrm{ad}gX)=\mathrm{ad}g(\mathfrak{Z}(X))$, we may assume that $X=[0;\boldsymbol{u}_1]$. For $Y=[\boldsymbol{y};\boldsymbol{s}]\in\mathfrak{Z}(X)$ we then have $[X,Y]=[[0;\boldsymbol{u}_1],[\boldsymbol{y};\boldsymbol{s}]]=[0;\boldsymbol{u}_1\times\boldsymbol{y}]$, so that $\boldsymbol{y}=\lambda\boldsymbol{u}_1$ and $\boldsymbol{s}$ is arbitary. Therefore, $\mathfrak{Z}(X)$ has dimension 4. If we put $\mathfrak{P}(X)=\{[\alpha\boldsymbol{u}_1;\beta\boldsymbol{u}_2+\gamma\boldsymbol{u}_3]\}$, $\alpha,\beta,\gamma\in\mathbf{R}$, we see that $\mathfrak{P}(X)$ satisfies the statement of the theorem. Conversely, let $\mathfrak{Z}(X)=\{\lambda X\}\dotplus\mathfrak{P}_1(X)$. There must exist a vector $Y=(\alpha\boldsymbol{u}_1;\ a^i\boldsymbol{u}_i)$ in $\mathfrak{P}_1(X)$ such that $\alpha\neq 0$ because $X\in\mathfrak{P}_1(X)$ otherwise, in virtue of the dimension of $\mathfrak{P}_1(X)$, and that is impossible. If $Y_1=[\beta\boldsymbol{u}_1;\ b^i\boldsymbol{u}_i]$ from $\mathfrak{P}_1(X)$, we arrive at $\mathscr{K}(Y,Y_1)=\alpha b^1+\beta a^1=0$, $\mathscr{K}(Y,Y)=\alpha a^1=0$. Since $\alpha\neq 0$, we have $a^1=0$, and, therefore, $b^1=0$. Consequently, $\mathfrak{P}_1(X)=\mathfrak{P}(X)$. The rest is simple.

2. KLEIN QUADRICS

Consider the Euclidean space E_3 and an arbitrary straight line p in it. The line p can be defined parametrically, e.g., as follows: $p\equiv A(\lambda)=\boldsymbol{m}+\lambda\boldsymbol{x}$, where $\boldsymbol{x}$ is the unit vector of the line p and $\boldsymbol{m}$ is the radius-vector of the foot P of the perpendicular dropped from point O onto line p, i.e. $P=O+\boldsymbol{m}$, $(\boldsymbol{m},\boldsymbol{x})=0$, $(\boldsymbol{x},\boldsymbol{x})=1$. The pair of vectors $[\boldsymbol{x};\boldsymbol{t}]=[\boldsymbol{x};\boldsymbol{m}\times\boldsymbol{x}]$ is then called the Plücker coordinates of the line p, or, to be precise, two triads of coordinates of vectors $\boldsymbol{x}$ and $\boldsymbol{t}$ yield the six Plücker coordinates of the line p, as we know from Chapter I. Conversely, if the Plücker coordinates of the line p are given as $[\boldsymbol{x};\boldsymbol{t}]$, its equation reads $p\equiv A(\lambda)=\boldsymbol{x}\times\boldsymbol{t}+\lambda\boldsymbol{x}$.

Denote by H_p the group of all congruences of the space E_3 which preserve line p as a whole, by H'_p the group of all congruences which preserve all the points of line p. i.e., H_p are all helical motions with the axis p, H'_p are all rotations about the axis p. Denote by $\mathfrak{H}_p$, $\mathfrak{H}'_p$ the Lie algebras of groups

H_p and H'_p, respectively. Let us express them in coordinate form. Assume, therefore, that

$$g = \begin{pmatrix} 1, & 0 \\ a, & \gamma \end{pmatrix} \in \mathscr{E} \quad \text{and} \quad g(p) = p\,.$$

For all $\lambda \in \mathbf{R}$ it must be

$$\begin{pmatrix} 1, & 0 \\ a, & \gamma \end{pmatrix} \begin{pmatrix} 1 \\ \boldsymbol{m} + \lambda \boldsymbol{x} \end{pmatrix} = \begin{pmatrix} 1 \\ \boldsymbol{m} + \mu \boldsymbol{x} \end{pmatrix},$$

where $\mu \in \mathbf{R}$ and depends on λ, i.e. $a + \gamma \boldsymbol{m} + \lambda \gamma \boldsymbol{x} = \boldsymbol{m} + \mu \boldsymbol{x}$. Moreover, for $g \in H'_p$, $\lambda = \mu$. If $\lambda = 0$, we arrive at the equation $a + \gamma \boldsymbol{m} = \boldsymbol{m} + \mu_0 \boldsymbol{x}$, where $\mu_0 = \mu_0(a, \gamma)$. If we substitute back into the original equation, we obtain $\lambda \gamma \boldsymbol{x} = (\mu - \mu_0)\, \boldsymbol{x}$, i.e. $\gamma \boldsymbol{x} = [(\mu - \mu_0)/\lambda]\, \boldsymbol{x}$. Since γ must preserve the length of vectors, $(\mu - \mu_0)/\lambda$ must be equal to ± 1, i.e. $\mu - \mu_0 = \pm\lambda$. It is sufficient to consider $\mu - \mu_0 = \lambda$ because H_p is not a connected group and consists of two components; in one the orientation of the space E_3 does not change, in the other it does; $\mu - \mu_0 = \lambda$ yields the first, $\mu - \mu_0 = -\lambda$ yields the second.

This means that the component of group H_p containing the unit element is described by the equations $\gamma \boldsymbol{x} = \boldsymbol{x}$, $a + \gamma \boldsymbol{m} = \boldsymbol{m} + \mu_0(a, \gamma)\, \boldsymbol{x}$. Let $g(t)$ be a curve in the group H_p such that $g(0) = e$. For the tangent vector of $g(t)$ at $t = 0$ we have $\gamma' \boldsymbol{x} = 0$, $a' + \gamma' \boldsymbol{m} = \mu'_0 \boldsymbol{x}$. γ' is a skew-symmetric matrix and so $\gamma' \boldsymbol{x} = 0$ can be written in vector notation as $\gamma' \times \boldsymbol{x} = 0$, thus $\gamma' = \alpha \boldsymbol{x}$, $\alpha \in \mathbf{R}$.

Then also $a'(0) + \alpha \boldsymbol{x} \times \boldsymbol{m} = \mu'_0(0)\, \boldsymbol{x}$, so that $a'(0) = -\alpha \boldsymbol{x} \times \boldsymbol{m} + \beta \boldsymbol{x} = \alpha \boldsymbol{m} \times \boldsymbol{x} + \beta \boldsymbol{x}$, where we have put $\mu'_0(0) = \beta$.

Now

$$g'(0) = \begin{pmatrix} 0, & 0 \\ a'(0), & \gamma'(0) \end{pmatrix}$$

is any vector from $\mathfrak{H}_p$, so that $\mathfrak{H}_p = \{[\alpha \boldsymbol{x};\, \alpha \boldsymbol{m} \times \boldsymbol{x} + \beta \boldsymbol{x}]\}$, where $\alpha, \beta \in \mathbf{R}$. Moreover, for H'_p we have $\lambda = \mu$, i.e. $\mu_0 = 0$, i.e. $\mu'_0(0) = 0$, i.e. $\beta = 0$, so that $\mathfrak{H}'_p = \{[\alpha \boldsymbol{x};\, \alpha \boldsymbol{m} \times \boldsymbol{x}]\}$. We record the result as a lemma:

2.1 Lemma. Let $p \equiv A(\lambda) = \boldsymbol{m} + \lambda \boldsymbol{x}$ be a straight line, where $(\boldsymbol{x}, \boldsymbol{x}) = 1$, $(\boldsymbol{m}, \boldsymbol{x}) = 0$. Then the group H_p is of dimension 2 and its Lie algebra $\mathfrak{H}_p$

is constituted by all vectors $X = [\alpha x; \alpha m \times x + \beta x]$, where $\alpha, \beta \in \mathbf{R}$ are arbitrary numbers. The group H'_p has dimension 1 and its Lie algebra is constituted by all vectors $X = [\alpha x; \alpha m \times x]$, where $\alpha \in \mathbf{R}$ is an arbitrary number.

Lemma 2.1 also presents the geometrical interpretation of the Klein form $\mathscr{K}(X, Y)$: $\mathscr{K}(X, X) = 0$ for a vector $X \in \mathfrak{E}$, $\mathbf{K}(X, X) \neq 0$, if and only if the one-parametric subgroup $g(t)$ with the tangent vector X is the group of all rotations about a fixed straight line. This is evident, since $(x, t) = 0$ if $X = [x; t] \in \mathfrak{E}$, $x \neq 0$, $\mathscr{K}(X, X) = 0$. We may obviously also assume that $\mathbf{K}(X, X) = (x, x) = 1$. If we put $t = m \times x$, the set of all vectors αX, $\alpha \in \mathbf{R}$, is clearly the Lie algebra of the group of all rotations about line $p \equiv A(\lambda) = x \times t + \lambda x$, because $(x \times t) \times x = t$. Conversely, $\mathscr{K}([\alpha x; \alpha m \times x], [\alpha x; \alpha m \times x]) = 2\alpha^2 |x, m, x| = 0$. We again note the result of this computation:

2.2 Lemma. Straight lines in E_3 are in one-to-one correspondence with one-dimensional subspaces $\{\alpha X\}$ in $\mathfrak{E}$ such that $\mathbf{K}(X, X) \neq 0$, $\mathscr{K}(X, X) = 0$. Moreover, if $[x; t]$ are the Plücker coordinates of a straight line, $\{[\alpha x; \alpha t]\}$ is the corresponding one-dimensional subspace in $\mathfrak{E}$.

Let us now characterize the algebras $\mathfrak{H}_p$:

2.3 Lemma. $\mathfrak{H}_p$ are the maximal commutative subalgebras of dimension 2 in $\mathfrak{E}$. There exists a one-to-one correspondence between the straight lines in E_3 and the maximal commutative subalgebras of dimension 2 in $\mathfrak{E}$.

Let us recall that Lie algebra $\mathfrak{A}$ is called commutative if $X, Y \in \mathfrak{A}$ implies that $[X, Y] = 0$. A commutative subalgebra is said to be maximal if it is not a part of a larger commutative subalgebra.

Proof. We shall first prove that $\mathfrak{H}_p$ are the maximal commutative subalgebras of dimension 2 in $\mathfrak{E}$. Therefore, assume that a subalgebra $\mathfrak{H}_p$ is given which is the set of all vectors from $\mathfrak{E}$ of the form $X = [\alpha x; \alpha t + \beta x]$, where $(x, x) = 1$, $(x, t) = 0$ $\alpha, \beta \in \mathbf{R}$. $\mathfrak{H}_p$ is commutative since if $X_1 = [\alpha_1 x; \alpha_1 t + \beta_1 x]$, $X_2 = [\alpha_2 x; \alpha_2 t + \beta_2 x]$ are two vectors from $\mathfrak{H}_p$,

$$[X_1, X_2] =$$
$$= [\alpha_1\alpha_2 x \times x; \alpha_1 x \times (\alpha_2 t + \beta_2 x) + (\alpha_1 t + \beta_1 x) \times \alpha_2 x]$$
$$= [0; \alpha_1\alpha_2 x \times t + \alpha_1\alpha_2 t \times x] = [0; 0] .$$

Therefore, $\mathfrak{H}_p$ is a commutative subalgebra of dimension 2. We shall prove that it is maximal. Choose a $\boldsymbol{Y} = (\boldsymbol{y}; \boldsymbol{s})$ in $\mathfrak{E}$ and assume that $[\boldsymbol{X}, \boldsymbol{Y}] = 0$ for all $\boldsymbol{X} \in \mathfrak{H}_p$. Then

$$\boldsymbol{0} = [[\alpha\boldsymbol{x}; \alpha\boldsymbol{t} + \beta\boldsymbol{x}], [\boldsymbol{y}; \boldsymbol{s}]] = [\alpha\boldsymbol{x} \times \boldsymbol{y}; \alpha\boldsymbol{x} \times \boldsymbol{s} + \alpha\boldsymbol{t} \times \boldsymbol{y} + \beta\boldsymbol{x} \times \boldsymbol{y}],$$

thus $\boldsymbol{x} \times \boldsymbol{y} = \boldsymbol{0}$ and $\boldsymbol{x} \times \boldsymbol{s} + \boldsymbol{t} \times \boldsymbol{y} = \boldsymbol{0}$. The former equation yields $\boldsymbol{y} = \mu\boldsymbol{x}$, $\mu \in \mathbf{R}$. If we substitute into the latter equation, we obtain $\boldsymbol{x} \times (\boldsymbol{s} - \mu\boldsymbol{t}) = \boldsymbol{0}$, so that $\boldsymbol{s} - \mu\boldsymbol{t} = \nu\boldsymbol{x}$, $\nu \in \mathbf{R}$, i.e. $\boldsymbol{s} = \nu\boldsymbol{x} + \mu\boldsymbol{t}$ so that $\boldsymbol{Y} = [\mu\boldsymbol{x}; \mu\boldsymbol{t} + \nu\boldsymbol{x}] \in \mathfrak{H}_p$.

Conversely, let $\mathfrak{H}$ be a maximal commutative subalgebra of dimension 2 in $\mathfrak{E}$. Choose a base $\boldsymbol{X} = [\boldsymbol{x}; \boldsymbol{t}]$, $\boldsymbol{Y} = [\boldsymbol{y}; \boldsymbol{s}]$ in $\mathfrak{H}$. It must then be true that

$$[\boldsymbol{X}, \boldsymbol{Y}] = [[\boldsymbol{x}; \boldsymbol{t}], [\boldsymbol{y}; \boldsymbol{s}]] = [\boldsymbol{x} \times \boldsymbol{y}; \boldsymbol{x} \times \boldsymbol{s} + \boldsymbol{t} \times \boldsymbol{y}] = \boldsymbol{0}.$$

Two alternatives may now occur: 1. $\boldsymbol{x} = \boldsymbol{y} = 0$, 2. at least one of the two vectors $\boldsymbol{x}$ and $\boldsymbol{y}$ is non-zero. In the 1st case, $\boldsymbol{X} = [\boldsymbol{0}; \boldsymbol{t}]$, $\boldsymbol{Y} = [\boldsymbol{0}; \boldsymbol{s}]$. Moreover, the vectors $\boldsymbol{t}$ and $\boldsymbol{s}$ are linearly independent because $\mathfrak{H}$ is of dimension 2. The vector $\boldsymbol{Z} = [\boldsymbol{0}; \boldsymbol{t} \times \boldsymbol{s}]$, together with $\boldsymbol{X}$ and $\boldsymbol{Y}$, then defines a commutative subalgebra of dimension 3 and $\mathfrak{H}$ is not maximal. Therefore, case 1 cannot occur. In the 2nd case we may assume that $\boldsymbol{x} \neq \boldsymbol{0}$. Then $\boldsymbol{y} = \lambda\boldsymbol{x}$ and we obtain the equation $\boldsymbol{x} \times (\boldsymbol{s} - \lambda\boldsymbol{t}) = \boldsymbol{0}$, i.e. $\boldsymbol{s} = \lambda\boldsymbol{t} + \mu\boldsymbol{x}$. Therefore, $\boldsymbol{X} = [\boldsymbol{x}; \boldsymbol{t}]$, $\boldsymbol{x} \neq 0$ and $\boldsymbol{Y} = [\lambda\boldsymbol{x}; \lambda\boldsymbol{t} + \mu\boldsymbol{x}]$, where $\mu \neq 0$, because $\boldsymbol{X}$ and $\boldsymbol{Y}$ constitute a base. The vector $\boldsymbol{Y} - \lambda\boldsymbol{X} = [\boldsymbol{0}; \mu\boldsymbol{x}]$, i.e. the vector $[\boldsymbol{0}; \boldsymbol{x}]$, also lies in $\mathfrak{H}$, the vectors $\boldsymbol{X}$ and $[\boldsymbol{0}; \boldsymbol{x}]$ constitute a base of $\mathfrak{H}$; moreover, we may assume that $(\boldsymbol{x}, \boldsymbol{x}) = 1$. Now, consider the vector $\boldsymbol{Z} = [\boldsymbol{x}; \boldsymbol{t} + \alpha\boldsymbol{x}]$, $\alpha \in \mathbf{R}$, and try to solve the equation $\mathscr{K}(\boldsymbol{Z}, \boldsymbol{Z}) = 2[(\boldsymbol{x}, \boldsymbol{t}) + \alpha] = 0$. We arrive at $\alpha = -(\boldsymbol{x}, \boldsymbol{t})$, and $\mathfrak{H}$ is the set of all vectors of the form $\boldsymbol{X} = [\alpha\boldsymbol{x}; \alpha[\boldsymbol{t} - (\boldsymbol{t}, \boldsymbol{x})\,\boldsymbol{x}] + \beta\boldsymbol{x}] = [\alpha\boldsymbol{x}; \alpha\boldsymbol{t}' + \beta\boldsymbol{x}]$, where $(\boldsymbol{x}, \boldsymbol{t}') = 0$, so that $\mathfrak{H} = \mathfrak{H}_p$ for the line $\boldsymbol{x} \times \boldsymbol{t} + \lambda\boldsymbol{x}$.

2.4 Definition. Denote by S_5 the set of all vectors $\boldsymbol{X} \in \mathfrak{E}$ for which $\mathbf{K}(\boldsymbol{X}, \boldsymbol{X}) = 1$. S_5 is called the *pseudospherical space of dimension* 5. Further, denote by $\mathscr{K}$ the set of all vectors $\boldsymbol{X} \in S_5$ for which $\mathscr{K}(\boldsymbol{X}, \boldsymbol{X}) = 0$. $\mathscr{K}$ is called the *Klein quadric* in S_5.

Remark. If $\boldsymbol{X} = [\boldsymbol{x}; \boldsymbol{t}]$, the equation of S_5 reads $(x^1)^2 + (x^2)^2 + (x^3)^2 = 1$ and the equation of $\mathscr{K}$ in S_5 reads $x^1t^1 + x^2t^2 + x^3t^3 = 0$.

Lemmas 2.1 and 2.2 now imply the following:

2.5 Lemma. To every straight line in E_3 there correspond two opposite straight lines in S_5. Each of these lines intersects $\mathscr{K}$ at a single point. These two points are called the *Klein image of line p*. Conversely, each pair of opposite points in $\mathscr{K}$ corresponds to just one straight line in E_3.

2.6 Lemma. Let X and Y be Klein images of straight lines p and q, φ the angle and d the distance between lines p and q. Then we have $\mathbf{K}(X, Y) = \pm \cos \varphi$, $\mathscr{K}(X, Y) = \pm d \sin \varphi$. Therefore, $\mathbf{K}(X, Y) = 0$ if and only if p and q are perpendicular, $\mathscr{K}(X, Y) = 0$ if and only if they lie in the same plane.

Proof. Denote by $\boldsymbol{X} = [\boldsymbol{x}; A \times \boldsymbol{x}]$, $\boldsymbol{Y} = [\boldsymbol{y}; B \times \boldsymbol{y}]$, where $(\boldsymbol{x}, \boldsymbol{x}) = 1$, $(\boldsymbol{y}, \boldsymbol{y}) = 1$, the Klein images of lines p and q. Here $A \in p$, $B \in q$. Then $\mathbf{K}(X, Y) = (\boldsymbol{x}, \boldsymbol{y}) = \cos \varphi$, $\mathscr{K}(X, Y) = (\boldsymbol{x}, B \times \boldsymbol{y}) + (\boldsymbol{y}, A \times \boldsymbol{x}) = -|\boldsymbol{x}, \boldsymbol{y}, B| + |\boldsymbol{x}, \boldsymbol{y}, A| = |\boldsymbol{x}, \boldsymbol{y}, A - B| = (\boldsymbol{x} \times \boldsymbol{y}, A - B)$. Now, assume that A and B are the feet of the perpendicular common to both lines. (This is possible, because the value of the expression does not depend on the choice of points A and B.) $\boldsymbol{x} \times \boldsymbol{y}$ then has the direction of the vector $A - B$ and length $\sin \varphi$, i.e. if we denote the distance of the two lines $\|A - B\| = d$, then

$$\pm \boldsymbol{x} \times \boldsymbol{y} = \frac{A - B}{d} \sin \varphi ,$$

so that

$$(\boldsymbol{x} \times \boldsymbol{y}, A - B) = \pm \left(\frac{A - B}{d} \sin \varphi, A - B \right) = \pm d \sin \varphi ,$$

which is what we had to prove.

2.7 Lemma. Let p and q be two straight lines which are not parallel, and r their shortest transversal (i.e. a line concurrent and perpendicular to both). Denote by $\mathfrak{H}_p, \mathfrak{H}_q, \mathfrak{H}_r$ the corresponding commutative algebras. Then $\mathbf{K}(\mathfrak{H}_p, \mathfrak{H}_r) = 0$, $\mathscr{K}(\mathfrak{H}_p, \mathfrak{H}_r) = 0$, and $[\mathfrak{H}_p, \mathfrak{H}_q] = \mathfrak{H}_r$. If $[\boldsymbol{x}; \boldsymbol{t}]$, $[\boldsymbol{y}; \boldsymbol{s}]$, and $[\boldsymbol{z}; \boldsymbol{u}]$ are Klein images of lines p, q, and r, then

$$[\boldsymbol{z}; \boldsymbol{u}] = \pm[\boldsymbol{x} \times \boldsymbol{y}; \boldsymbol{x} \times \boldsymbol{s} + \boldsymbol{t} \times \boldsymbol{y} + \cot \varphi[(\boldsymbol{x}, \boldsymbol{s}) + (\boldsymbol{y}, \boldsymbol{t})]\, \boldsymbol{x} \times \boldsymbol{y}] \,.\, \sin^{-1} \varphi ,$$

where φ is the angle between the vectors $\boldsymbol{x}$ and $\boldsymbol{y}$.

Proof. We first prove that $[\mathfrak{H}_p, \mathfrak{H}_q]$ is a maximal commutative subalgebra of dimension 2. Put

$$\mathfrak{H}_p = \{[\alpha_1 \boldsymbol{x}; \alpha_1 \boldsymbol{t} + \beta_1 \boldsymbol{x}]\}, \quad \mathfrak{H}_q = \{[\alpha_2 \boldsymbol{y}; \alpha_2 \boldsymbol{s} + \beta_2 \boldsymbol{y}]\}.$$

Then

$$\begin{aligned}
&[\mathfrak{H}_p, \mathfrak{H}_q] \\
&= \{[[\alpha_1 \boldsymbol{x}; \alpha_1 \boldsymbol{t} + \beta_1 \boldsymbol{x}], [\alpha_2 \boldsymbol{y}; \alpha_2 \boldsymbol{s} + \beta_2 \boldsymbol{y}]]\} \\
&= \{[\alpha_1 \alpha_2 \boldsymbol{x} \times \boldsymbol{y}; \alpha_1 \boldsymbol{x} \times (\alpha_2 \boldsymbol{s} + \beta_2 \boldsymbol{y}) + (\alpha_1 \boldsymbol{t} + \beta_1 \boldsymbol{x}) \times \alpha_2 \boldsymbol{y}]\} \\
&= \{[\alpha_1 \alpha_2 \boldsymbol{x} \times \boldsymbol{y}; \alpha_1 \alpha_2 (\boldsymbol{x} \times \boldsymbol{s} + \boldsymbol{t} \times \boldsymbol{y}) + (\alpha_1 \beta_2 + \beta_1 \alpha_2) \boldsymbol{x} \times \boldsymbol{y}]\} \\
&= \{[\alpha_3 \boldsymbol{x} \times \boldsymbol{y}; \alpha_3 (\boldsymbol{x} \times \boldsymbol{s} + \boldsymbol{t} \times \boldsymbol{y}) + \beta_3 \boldsymbol{x} \times \boldsymbol{y}]\},
\end{aligned}$$

which defines a two-dimensional subspace in $\mathfrak{E}$. Also

$$[[\boldsymbol{x} \times \boldsymbol{y}; \boldsymbol{x} \times \boldsymbol{s} + \boldsymbol{t} \times \boldsymbol{y}], [\boldsymbol{0}; \boldsymbol{x} \times \boldsymbol{y}]] = \boldsymbol{0},$$

so that $[\mathfrak{H}_p, \mathfrak{H}_q]$ is a commutative subalgebra of dimension 2; from the proof of Lemma 2.3 we see that it is also maximal. Therefore, according to Lemma 2.3, $[\mathfrak{H}_p, \mathfrak{H}_q] = \mathfrak{H}_r$, where r is a straight line. The statement $\mathbf{K}(\mathfrak{H}_p, \mathfrak{H}_r) = 0$ is evidently true. Further,

$$\begin{aligned}
&\mathscr{K}(\mathfrak{H}_p, \mathfrak{H}_r) \\
&= \mathscr{K}([\alpha_1 \boldsymbol{x}; \alpha_1 \boldsymbol{t} + \beta_1 \boldsymbol{x}], [\alpha_3 \boldsymbol{x} \times \boldsymbol{y}; \alpha_3 (\boldsymbol{x} \times \boldsymbol{s} + \boldsymbol{t} \times \boldsymbol{y}) + \beta_3 \boldsymbol{x} \times \boldsymbol{y}]) \\
&= (\alpha_1 \boldsymbol{x}, \alpha_3 [\boldsymbol{x} \times \boldsymbol{s} + \boldsymbol{t} \times \boldsymbol{y}] + \beta_3 \boldsymbol{x} \times \boldsymbol{y}) + (\alpha_1 \boldsymbol{t} + \beta_1 \boldsymbol{x}, \alpha_3 \boldsymbol{x} \times \boldsymbol{y}) \\
&= \alpha_1 \alpha_3 |\boldsymbol{x}, \boldsymbol{t}, \boldsymbol{y}| + \alpha_1 \alpha_3 |\boldsymbol{t}, \boldsymbol{x}, \boldsymbol{y}| = 0.
\end{aligned}$$

Since Klein images lie in the appropriate subalgebras, e.g., $\mathfrak{H}'_p \subset \mathfrak{H}_p$, we also have $\mathbf{K}(\mathfrak{H}'_p, \mathfrak{H}'_r) = 0$, $\mathscr{K}(\mathfrak{H}'_p, \mathfrak{H}'_r) = 0$, and similarly for line q; this means that the line r is concurrent to both lines p and q and also perpendicular to them. Therefore, r is really the shortest transversal of lines p and q. It now remains to determine the Klein image of line r. The vector $[[\boldsymbol{x}; \boldsymbol{t}], [\boldsymbol{y}; \boldsymbol{s}]] = [\boldsymbol{x} \times \boldsymbol{y}; \boldsymbol{x} \times \boldsymbol{s} + \boldsymbol{t} \times \boldsymbol{y}]$ clearly belongs to $\mathfrak{H}_r$. If we now decompose the vector $\boldsymbol{x} \times \boldsymbol{s} + \boldsymbol{t} \times \boldsymbol{y}$ into the component in the direction of $\boldsymbol{x} \times \boldsymbol{y}$ and into the component perpendicular to $\boldsymbol{x} \times \boldsymbol{y}$, we arrive at the required result easily.

2.8 Consequence. Let $\boldsymbol{X}, \boldsymbol{Y}, \boldsymbol{Z}$ be Klein images of lines p, q, r; let p be perpendicular to q, let r be the shortest transversal of p and q. Then $\boldsymbol{Z} = \pm[\boldsymbol{X}, \boldsymbol{Y}]$.

2.9 Exercise. Assume $X, Y \in \mathscr{K}$. Prove that $\mathscr{K}([X,Y],[X,Y]) = -2\mathbf{K}(X,Y)\,\mathscr{K}(X,Y)$. Find a similar formula for any two vectors $X, Y \in \mathfrak{E}$.

3. REPRESENTATION OF SPACE MOTION, ASSOCIATED SPHERICAL MOTION

Assume that two collocated Euclidean spaces are given: moving space $\bar{E}_3$ and fixed space E_3. Select frames $\bar{\mathscr{R}} = \{\bar{O}, \bar{u}_1, \bar{u}_2, \bar{u}_3\}$ in $\bar{E}_3$ and $\mathscr{R} = \{O, u_1, u_2, u_3\}$ in E_3. The motion of $\bar{E}_3$ in E_3 can be determined by frame $\bar{\mathscr{R}}$ being given as a function of some parameter t, i.e. $\bar{\mathscr{R}}(t) = \{\bar{O}(t), \bar{u}_i(t)\}$. If we express frame $\bar{\mathscr{R}}(t)$ in terms of frame $\mathscr{R}$ we arrive at

$$\bar{\mathscr{R}}(t) = \mathscr{R}\, g(t)\,, \quad \text{where} \quad g(t) = \begin{pmatrix} 1, & 0 \\ a(t), & \gamma(t) \end{pmatrix} \in \mathscr{E}\,,$$

so that the space motion is given as a curve $g(t)$ on the group $\mathscr{E}$. As in Chapter II we can show that if frame $\bar{\mathscr{R}}$ or $\mathscr{R}$ changes, the curve $g(t)$ changes by left or right translations. Before going on to examples, let us clarify the association between space and spherical motion. Thus, let

$$g(t) = \begin{pmatrix} 1, & 0 \\ a(t), & \gamma(t) \end{pmatrix}$$

be space motion, and denote by V_3 the set of all vectors from E_3, by S^2 the set of all unit vectors from V_3; $\bar{V}_3$ and $\bar{S}^2$ are defined analogously. The frames $\bar{\mathscr{R}}(t) = \{\bar{O}(t), \bar{u}_i(t)\}$ and $\mathscr{R} = \{O, u_i\}$ immediately determine frames $\bar{\mathscr{R}}_0(t) = \{\bar{u}_i(t)\}$ and $\mathscr{R}_0 = \{u_i\}$ in $\bar{V}_3$ and V_3. By developing equation $\bar{\mathscr{R}}(t) = \mathscr{R}\, g(t)$ we obtain

$$\begin{aligned} \bar{\mathscr{R}}(t) &= \{\bar{O}(t), \bar{\mathscr{R}}_0(t)\} \\ &= \{O, \mathscr{R}_0\} \begin{pmatrix} 1, & 0 \\ a(t), & \gamma(t) \end{pmatrix} = \{O + \mathscr{R}_0\, a(t), \mathscr{R}_0\, \gamma(t)\}\,, \end{aligned}$$

i.e.

$$\bar{\mathscr{R}}_0(t) = \mathscr{R}_0\, \gamma(t)\,,$$

so that the vector spaces $\bar{V}_3$ and V_3 (the collocation of which is determined by the collocation of $\bar{E}_3$ and E_3) are subject to spherical motion, which can be considered on the spheres $\bar{S}^2$ and S^2 only. We call the motion $\gamma(t)$ the *spherical motion associated with space motion* $g(t)$.

3.1 Example. Let us determine the matrix of rotation of the space about line $p = A + \lambda \boldsymbol{u}_3$, where A is an arbitrary point. The associated spherical motion is evidently rotation about the axis $\boldsymbol{u}_3$, i.e.

$$\gamma(\varphi) = \begin{pmatrix} \cos\varphi, & -\sin\varphi, & 0 \\ \sin\varphi, & \cos\varphi, & 0 \\ 0, & 0, & 1 \end{pmatrix},$$

where we have adopted the angle of rotation as the parameter. Put

$$A = O + m^i \boldsymbol{u}_i, \quad g(\varphi) = \begin{pmatrix} 1, & 0 \\ a(\varphi), & \gamma(\varphi) \end{pmatrix};$$

it must be $g(\varphi)\,A = A$, i.e.

$$\begin{pmatrix} 1, & 0 \\ a, & \gamma \end{pmatrix} \begin{pmatrix} 1 \\ m^i \end{pmatrix} = \begin{pmatrix} 1 \\ a + \gamma m^i \end{pmatrix} = \begin{pmatrix} 1 \\ m^i \end{pmatrix},$$

i.e.

$$a = \begin{pmatrix} m^1 \\ m^2 \\ m^3 \end{pmatrix} - \begin{pmatrix} \cos\varphi, & -\sin\varphi, & 0 \\ \sin\varphi, & \cos\varphi, & 0 \\ 0, & 0, & 1 \end{pmatrix} \begin{pmatrix} m^1 \\ m^2 \\ m^3 \end{pmatrix}$$

$$= \begin{pmatrix} m^1(1 - \cos\varphi) + m^2 \sin\varphi \\ -m^1 \sin\varphi + m^2(1 - \cos\varphi) \\ 0 \end{pmatrix},$$

therefore,

$$g(\varphi) = \begin{pmatrix} 1, & 0, & 0, & 0 \\ m^1(1 - \cos\varphi) + m^2 \sin\varphi, & \cos\varphi, & -\sin\varphi, & 0 \\ -m^1 \sin\varphi + m^2(1 - \cos\varphi), & \sin\varphi, & \cos\varphi, & 0 \\ 0, & 0, & 0 & 1 \end{pmatrix}.$$

3.2 Exercise. Determine the matrix of helical motion about the axis $p = A + \lambda \boldsymbol{u}_2$, where $A = O + 2\boldsymbol{u}_1 - \boldsymbol{u}_3$, which has the reduced thread pitch (the parameter) $v_0 = 3$.

3.3 Exercise. Use the result from Example II, 1.12 to determine the matrix of helical motion about the axis determined by the vector $\boldsymbol{v} = 2\boldsymbol{u}_1 +$

$3\boldsymbol{u}_2 + 6\boldsymbol{u}_3$ and passing through the origin, which has the parameter $v_0 = \frac{1}{2}$.

3.4 Example. Let us consider the following mechanism (see Fig. 38). A line segment $m = AB$, of length d, is moving so that its endpoint A keeps to axis $\boldsymbol{u}_2$ while its endpoint B keeps to axis $\boldsymbol{u}_3$. At a certain point, a line segment n, which passes constantly through fixed point M, is connected

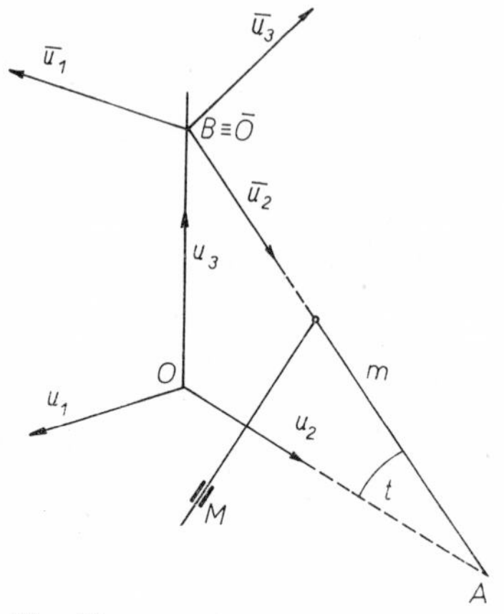

Fig. 38.

movably to line segment m. Let us choose the moving frame $\bar{\mathscr{R}}(t) = \{\bar{O}(t), \bar{\boldsymbol{u}}_i(t)\}$ as follows: $\bar{O}(t) = B$, $\bar{\boldsymbol{u}}_2$ lies on the line segment m, $\bar{\boldsymbol{u}}_1$ is assumed to lie in the plane of line segments m and n. As the parameter of motion we choose the angle t between the line segment m and the axis $\boldsymbol{u}_2$. Then $\bar{O}(t) = d \sin t$. Line segment m also intersects the axis $\boldsymbol{u}_2$, i.e. there exists μ such that $\bar{O} + d\bar{\boldsymbol{u}}_2 = \mu\boldsymbol{u}_2$. Put $\bar{\boldsymbol{u}}_2 = \alpha_2\boldsymbol{u}_2 + \alpha_3\boldsymbol{u}_3$. Then

$$\boldsymbol{u}_3 d \sin t + d(\alpha_2\boldsymbol{u}_2 + \alpha_3\boldsymbol{u}_3) = \mu\boldsymbol{u}_2 .$$

The inner product of this equation with the vector $\boldsymbol{u}_3$ yields

$$d \sin t + d\alpha_3 = 0 , \quad \text{i.e.} \quad \alpha_3 = -\sin t , \quad \alpha_2 = \cos t .$$

We must also use the fact that the point M is located in the plane $(\bar{\boldsymbol{u}}_1\bar{\boldsymbol{u}}_2)$, which means that the vectors $\bar{\boldsymbol{u}}_1$, $\bar{\boldsymbol{u}}_2$, and $M - \bar{O}$ are coplanar, i.e. it must be

$$|\bar{\boldsymbol{u}}_1, \bar{\boldsymbol{u}}_2, M - \bar{O}| = 0 .$$

If we put $\bar{\boldsymbol{u}}_1 = \beta_1 \boldsymbol{u}_1 + \beta_2 \boldsymbol{u}_2 + \beta_3 \boldsymbol{u}_3$, $M = m_i \boldsymbol{u}_i$, and if we notice that

$$\bar{e}_2 = e_2 \cos t - e_3 \sin t \,,$$

we arrive at the equation

$$\begin{vmatrix} \beta_1, & 0, & m_1 \\ \beta_2, & \cos t, & m_2 \\ \beta_3, & -\sin t, & m_3 - d \sin t \end{vmatrix}$$
$$= \beta_1[\cos t(m_3 - d \sin t) + m_2 \sin t] + m_1[-\beta_2 \sin t - \beta_3 \cos t] = 0 .$$

Also $(\bar{\boldsymbol{u}}_1, \bar{\boldsymbol{u}}_2) = 0$, i.e. $\beta_2 \cos t - \beta_3 \sin t = 0$, i.e. $\beta_2 = \lambda \sin t$, $\beta_3 = \lambda \cos t$. Substitution into the previous equation yields

$$\beta_1[\cos t(m_3 - d \sin t) + m_2 \sin t] - m_1 \lambda = 0 \,,$$

i.e.

$$\beta_1 = \lambda \frac{m_1}{\cos t(m_3 - d \sin t) + m_2 \sin t} \,.$$

After substitution the equation $\beta_1^2 + \beta_2^2 + \beta_3^2 = 1$ yields

$$1 = \lambda^2 \left(1 + \frac{m_1^2}{\cos t(m_3 - d \sin t) + m_2 \sin t}\right)^2 .$$

For the sake of brevity, let us put $M = (m_1^2 + N^2)^{-1/2}$, $N = m_3 \cos t - d \cos t \sin t + m_2 \sin t$. Then

$$1 = \frac{\lambda^2}{M^2 N^2} \,, \quad \text{i.e.} \quad \lambda^2 = M^2 N^2 \,, \quad \lambda = MN \,.$$

It follows that

$$\beta_1 = \frac{MNm_1}{N} = Mm_1 \,, \quad \beta_2 = MN \sin t \,, \quad \beta_3 = MN \cos t \,.$$

For $\bar{\boldsymbol{u}}_3$ we obtain

$$\bar{\boldsymbol{u}}_3 = \begin{vmatrix} \boldsymbol{u}_1, & \boldsymbol{u}_2, & \boldsymbol{u}_3 \\ Mm_1, & MN \sin t, & MN \cos t \\ 0, & \cos t, & -\sin t \end{vmatrix}$$
$$= -\boldsymbol{u}_1 MN + \boldsymbol{u}_2 Mm_1 \sin t + \boldsymbol{u}_3 Mm_1 \cos t \,.$$

Thus the resulting matrix of motion is

$$g(t) = \begin{pmatrix} 1, & 0, & 0, & 0 \\ 0, & Mm_1, & 0, & -MN \\ 0, & MN \sin t, & \cos t, & Mm_1 \sin t \\ d \sin t, & MN \cos t, & -\sin t, & Mm_1 \cos t \end{pmatrix}.$$

3.5 Example. As an example of determining the matrix of general space motion we consider *Bennet's mechanism*. Before studying this mechanism we show how to bind a frame with two skew lines, and how to proceed from one such frame to another. Consider three skew lines p_1, p_2, p_3

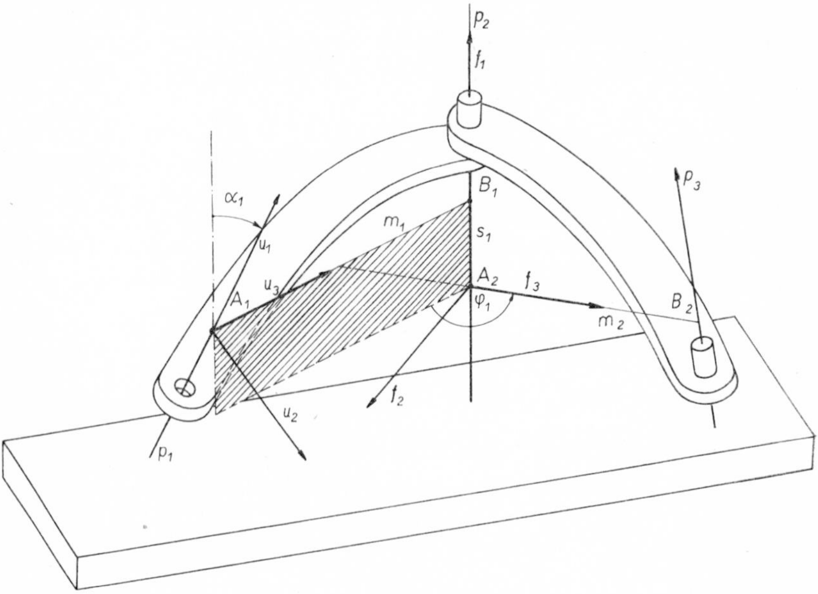

Fig. 39.

(Fig. 39). Denote their shortest transversals m_1 and m_2. Further, denote the points of intersection of the straight lines $m_1 \times p_1 = A_1$, $m_1 \times p_2 = B_1$, $m_2 \times p_2 = A_2$, $m_2 \times p_3 = B_2$. Now, choose the frame $\mathscr{R}_1 = \{A_1, \boldsymbol{u}_1, \boldsymbol{u}_2, \boldsymbol{u}_3\}$ as follows: $\boldsymbol{u}_1$ lies on the line p_1, $\boldsymbol{u}_3$ on the line m_1 in the direction from A_1 to B_1. Analogously we choose the frame $\mathscr{R}_2 = \{A_2, \boldsymbol{f}_1, \boldsymbol{f}_2, \boldsymbol{f}_3\}$, with f_1 lying on p_2, f_3 on m_2 in the direction from A_2 to B_2. Thus it is possible to assign a frame to each pair of skew lines uniquely (but for the orientation of the first vector). The construction remains valid even for concurrent lines; if parallels are involved, the origin of the frame may be arbitrary. Now, we have to describe how to proceed from frame $\mathscr{R}_1$ to frame $\mathscr{R}_2$. Denote by α_1 the angle between the vectors $\boldsymbol{u}_1$ and $\boldsymbol{f}_1$, by $\pi - \varphi_1$ the angle between the vectors $\boldsymbol{u}_3$ and $\boldsymbol{f}_3$, by m_1 the length of the transversal m_1, by s_1 the distance between transversals m_1 and m_2. From frame $\mathscr{R}_1$ to frame $\mathscr{R}_2$ we proceed exploiting the following two simple transformations:

1. Rotation about the third axis A_1B_1 through the angle α_1 and translation along the third axis by m_1.

2. Rotation about the first axis p_2 through the angle $\pi - \varphi_1$ and translation along the first axis by s_1. For the purposes of computation put $\mathscr{R}'_1 = \{B_1, \boldsymbol{u}'_1, \boldsymbol{u}'_2, \boldsymbol{u}'_3\}$, where $\boldsymbol{u}'_1 = \boldsymbol{f}_1$, $\boldsymbol{u}'_2 = \boldsymbol{u}_2$, i.e. $\mathscr{R}'_1$ is obtained from frame $\mathscr{R}_1$ by transformation 1. Also, denote by g'_1 the matrix of transformation 1, by g''_1 the matrix of transformation 2. Then $\mathscr{R}'_1 = \mathscr{R}_1 g'_1$, $\mathscr{R}_2 = \mathscr{R}'_1 g''_1$, i.e. $\mathscr{R}_2 = \mathscr{R}_1 g'_1 g''_1$. Moreover,

$$g'_1 = \begin{pmatrix} 1, & 0, & 0, & 0 \\ 0, & \cos\alpha_1, & -\sin\alpha_1, & 0 \\ 0, & \sin\alpha_1, & \cos\alpha_1, & 0 \\ \mathrm{m}_1, & 0, & 0, & 1 \end{pmatrix}, \quad g''_1 = \begin{pmatrix} 1, & 0, & 0, & 0 \\ \mathrm{s}_1, & 1, & 0, & 0 \\ 0, & 0, & -\cos\varphi_1, & -\sin\varphi_1 \\ 0, & 0, & \sin\varphi_1, & -\cos\varphi_1 \end{pmatrix}.$$

Thus we have $\mathscr{R}_2 = \mathscr{R}_1 g_1$ and $g_1 = g'_1 g''_1$, i.e.

$$g_1 = \begin{pmatrix} 1, & 0, & 0, & 0 \\ \mathrm{s}_1\cos\alpha_1, & \cos\alpha_1, & \sin\alpha_1\cos\varphi_1, & \sin\alpha_1\sin\varphi_1 \\ \mathrm{s}_1\sin\alpha_1, & \sin\alpha_1, & -\cos\alpha_1\cos\varphi_1, & -\cos\alpha_1\sin\varphi_1 \\ \mathrm{m}_1, & 0, & \sin\varphi_1, & -\cos\varphi_1 \end{pmatrix}.$$

The reader's attention should be drawn to the fact that Example II, 4.13 is a precise spherical analogy of the considerations presented here, and also the formulae agree. We may now proceed as in the said example, i.e. if the mechanism is composed of four straight lines (not necessarily skew) p_1, p_2, p_3, p_4, we define the matrices g_2, g_3, g_4 similarly to matrix g_1. The condition for the mechanism to be closed reads $\boldsymbol{e} = g_1 g_2 g_3 g_4$. Consider now particularly the mechanism of Bennet. Bennet's mechanism is a mechanism of the type just described, with four rotational pairs, i.e. for which $\mathrm{s}_i = 0$,

$i = 1, 2, 3, 4$, in which $\alpha_1 = \alpha_3$, $\alpha_2 = \alpha_4$. See Fig. 40. During the computation we shall see that under these assumptions Bennet's mechanism must

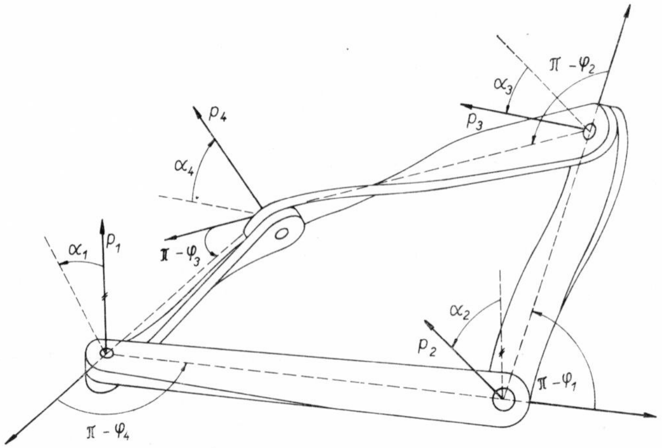

Fig. 40.

satisfy yet another condition. We are thus considering a mechanism for which we have

$$g_i(\varphi_i) = \begin{pmatrix} 1, & 0, & 0, & 0 \\ 0, & \cos\alpha_i, & \sin\alpha_i \cos\varphi_i, & \sin\alpha_i \sin\varphi_i \\ 0, & \sin\alpha_i, & -\cos\alpha_i \cos\varphi_i, & -\cos\alpha_i \sin\varphi_i \\ m_i, & 0, & \sin\varphi_i, & -\cos\varphi_i \end{pmatrix},$$

$$\alpha_1 = \alpha_3\,, \quad \alpha_2 = \alpha_4\,, \quad g_1 g_2 g_3 g_4 = e\,.$$

Put

$$g_i = \begin{pmatrix} 1, & 0 \\ t_i, & \gamma_i \end{pmatrix}.$$

Equation $g_1 g_2 g_3 g_4 = e$ yields

$$\begin{pmatrix} 1, & 0 \\ 0, & e \end{pmatrix} = \begin{pmatrix} 1, & 0 \\ t_1, & \gamma_1 \end{pmatrix} \begin{pmatrix} 1, & 0 \\ t_2, & \gamma_2 \end{pmatrix} \begin{pmatrix} 1, & 0 \\ t_3, & \gamma_3 \end{pmatrix} \begin{pmatrix} 1, & 0 \\ t_4, & \gamma_4 \end{pmatrix}$$

$$= \begin{pmatrix} 1, & 0 \\ t_1 + \gamma_1 t_2 + \gamma_1 \gamma_2 t_3 + \gamma_1 \gamma_2 \gamma_3 t_4, & \gamma_1 \gamma_2 \gamma_3 \gamma_4 \end{pmatrix}.$$

For the appropriate spherical mechanism we have the condition $\gamma_1\gamma_2\gamma_3\gamma_4 = e$. Take another look at Example II, 4.13. We see that the matrix g_i in II, 4.13 corresponds to our matrices γ_i, which means that the spherical motion assigned to Bennet's mechanism is a spherical quadrangle with identical opposite sides. If we use the notation of II, 4.13, i.e. $\cos\alpha_1 = a$, $\cos\alpha_2 = b$, $\cos\alpha_3 = c$, $\cos\alpha_4 = d$; $\sin\alpha_1 = A$, $\sin\alpha_2 = B$, $\sin\alpha_3 = C$, $\sin\alpha_4 = D$, we have the conditions $a = c$, $A = C$, $b = d$ and $B = D$. If we substitute these conditions into the expression for $\sin\varphi_2$ in formula (II, 4.8), we arrive at

$$\sin\varphi_2 = A^{-1}[1 - (ab + AB\cos\varphi_1)^2]^{-1} .$$
$$\{-A\sin\varphi_1(ABa\cos\varphi_1 + a^2b - b) \pm (aB - Ab\cos\varphi_1) .$$
$$[A^2(1 - (ab + AB\cos\varphi_1)^2) - (Aba\cos\varphi_1 + a^2b - b^2)^2]^{1/2}\} .$$

The expression under the square root can be modified to read $A^2B^2\sin^2\varphi_1$. If we put $\varepsilon = \pm 1$, the expression in the numerator becomes

$$-A\sin\varphi_1(ABa\cos\varphi_1 - bA^2) + \varepsilon(aB - Ab\cos\varphi_1)\,AB\sin\varphi_1$$
$$= A\sin\varphi_1(\varepsilon aB^2 - \varepsilon AbB\cos\varphi_1 - ABa\cos\varphi_1 + bA^2)$$
$$= A\sin\varphi_1(a + b\varepsilon)(\varepsilon - ab - AB\cos\varphi_1) .$$

Decomposing the denominator into

$$A(\varepsilon + ab + AB\cos\varphi_1)(\varepsilon - ab - AB\cos\varphi_1) ,$$

we find that

$$\sin\varphi_2 = \frac{a + b\varepsilon}{\varepsilon + ab + AB\cos\varphi_1}\sin\varphi_1 .$$

The condition for the translations reads

$$t_1 + \gamma_1 t_2 + \gamma_1\gamma_2 t_3 + \gamma_1\gamma_2\gamma_3 t_4 = \boldsymbol{0} .$$

This condition contains the angle φ_3 so that we have to find a condition for φ_3 as well. For the case considered, we write the equation $\gamma_1\gamma_2 = \gamma_3^{\mathrm{T}}\gamma_4^{\mathrm{T}}$ as in II, 4.13. We arrive at

$$\begin{pmatrix} a, & A\cos\varphi_1, & A\sin\varphi_1 \\ A, & -a\cos\varphi_1, & -a\sin\varphi_1 \\ 0, & \sin\varphi_1, & -\cos\varphi_1 \end{pmatrix} \begin{pmatrix} b, & B\cos\varphi_2, & B\sin\varphi_2 \\ B, & -b\cos\varphi_2, & -b\sin\varphi_2 \\ 0, & \sin\varphi_2, & -\cos\varphi_2 \end{pmatrix} =$$

$$\begin{pmatrix} b, & B, & 0 \\ B\cos\varphi_4, & -b\cos\varphi_4, & \sin\varphi_4 \\ B\sin\varphi_4, & -b\sin\varphi_4, & -\cos\varphi_4 \end{pmatrix} \begin{pmatrix} a, & A, & 0 \\ A\cos\varphi_3, & -a\cos\varphi_3, & \sin\varphi_3 \\ A\sin\varphi_3, & -a\sin\varphi_3, & -\cos\varphi_3 \end{pmatrix}.$$

The product of the 1st row and 1st column yields

$$ab + AB\cos\varphi_1 = ab + AB\cos\varphi_3\,, \quad \text{i.e.} \quad \cos\varphi_1 = \cos\varphi_3\,,$$

the product of the 1st row and 2nd column yields

$$\begin{aligned} &aB\cos\varphi_2 - Ab\cos\varphi_1\cos\varphi_2 + A\sin\varphi_1\sin\varphi_2 \\ &= AB - aB\cos\varphi_1\,. \end{aligned}$$

Upon substitution for $\sin\varphi_2$ and modification, this equation becomes

$$\cos\varphi_2 = -\frac{AB + (ab + \varepsilon)\cos\varphi_1}{\varepsilon + ab + AB\cos\varphi_1}\,.$$

The product of the 1st row and 3rd column is

$$aB\sin\varphi_2 - Ab\cos\varphi_1\sin\varphi_2 - A\sin\varphi_1\cos\varphi_2 = B\sin\varphi_3\,.$$

If we put $\sin\varphi_3 = \eta\sin\varphi_1$, and if we substitute for $\sin\varphi_2$ and $\cos\varphi_2$, we find that $\eta = \varepsilon$, i.e.

$$\sin\varphi_3 = \varepsilon\sin\varphi_1\,.$$

If we calculate the products of the 3rd row and the 1st and 2nd columns, we arrive at

$$\begin{aligned} &B\sin\varphi_1 \\ &= aB\sin\varphi_4 - Ab\cos\varphi_1\sin\varphi_4 - A\varepsilon\sin\varphi_1\cos\varphi_4 \\ &\quad - b\sin\varphi_1\cos\varphi_2 - \cos\varphi_1\sin\varphi_2 \\ &= AB\sin\varphi_4 + ab\cos\varphi_1\sin\varphi_4 + a\varepsilon\sin\varphi_1\cos\varphi_4\,. \end{aligned}$$

If we multiply the first of these equations by a and the second by A and add the resulting equations, we obtain

$$aB\sin\varphi_1 - Ab\sin\varphi_1\cos\varphi_2 - A\cos\varphi_1\sin\varphi_2 = B\sin\varphi_4\,.$$

Upon substitution for $\cos\varphi_2$ and $\sin\varphi_2$ and some manipulation we arrive at

$$\sin\varphi_4 = \sin\varphi_1 \frac{a\varepsilon + b}{\varepsilon + ab + AB\cos\varphi_1} = \varepsilon\sin\varphi_2\,.$$

The product of the 3rd row and 1st column yields $\cos\varphi_4 = \cos\varphi_2$. (The relations between φ_4 and φ_2 follow immediately from the relations for φ_3 and φ_1 if we consider that the roles of φ_1 and φ_3 will be exchanged for those of φ_2 and φ_4, respectively, in passing to inverse motion; however, just to be quite sure, we gave preference to direct computation.)

Let us write the translation equation in a slightly more advantageous form:

$$t_1 + \gamma_1 t_2 + \gamma_1\gamma_2 t_3 + \gamma_4^{\mathrm{T}} t_4 = 0\,.$$

Upon substitution we obtain the following equations

$$\mathrm{m}_1 \begin{pmatrix} 0 \\ 0 \\ 1 \end{pmatrix} + \mathrm{m}_2 \begin{pmatrix} A\sin\varphi_1 \\ -a\sin\varphi_1 \\ -\cos\varphi_1 \end{pmatrix}$$
$$+ \mathrm{m}_3 \begin{pmatrix} aB\sin\varphi_2 - Ab\cos\varphi_1\sin\varphi_2 - A\sin\varphi_1\cos\varphi_2 \\ AB\sin\varphi_2 + ab\cos\varphi_1\sin\varphi_2 + a\sin\varphi_1\cos\varphi_2 \\ -b\sin\varphi_1\sin\varphi_2 + \cos\varphi_1\cos\varphi_2 \end{pmatrix}$$
$$+ \mathrm{m}_4 \begin{pmatrix} 0 \\ \varepsilon\sin\varphi_2 \\ -\cos\varphi_2 \end{pmatrix} = \begin{pmatrix} 0 \\ 0 \\ 0 \end{pmatrix}.$$

The first equation reads

$$\mathrm{m}_2 A\sin\varphi_1$$
$$+ \mathrm{m}_3(aB\sin\varphi_2 - Ab\cos\varphi_1\sin\varphi_2 - A\sin\varphi_1\cos\varphi_2) = 0\,.$$

If we substitute for $\sin\varphi_2$ and $\cos\varphi_2$ once more, modification yields

$$\mathrm{m}_2 A\sin\varphi_1 + \mathrm{m}_3\,\varepsilon B\sin\varphi_1 = 0\,, \quad \text{i.e.} \quad \mathrm{m}_2 A + \mathrm{m}_3\,\varepsilon B = 0\,.$$

The second equation reads

$$-\mathrm{m}_2 a\sin\varphi_1$$
$$+ \mathrm{m}_3[(AB + ab\cos\varphi_1)\sin\varphi_2 + a\sin\varphi_1\cos\varphi_2]$$
$$+ \mathrm{m}_4\varepsilon\sin\varphi_2 = 0\,.$$

If we adjust the expression in the brackets, we arrive at

$$-\mathrm{m}_2 a(\varepsilon + ab + AB\cos\varphi_1) + \mathrm{m}_3 B\varepsilon(Ab - aB\cos\varphi_1) + \mathrm{m}_4\varepsilon(a + \varepsilon b) = 0\,.$$

If we substitute from the previous equation for $\mathrm{m}_3\varepsilon B$, we find that

$$\mathrm{m}_2 = \mathrm{m}_4\,.$$

The last equation is

$$\mathrm{m}_1 - \mathrm{m}_2\cos\varphi_1 + \mathrm{m}_3(-b\sin\varphi_1\sin\varphi_2 + \cos\varphi_1\cos\varphi_2) - \mathrm{m}_4\cos\varphi_2 = 0\,.$$

Upon modifications similar to those applied to the second equation, we find that

$$\mathrm{m}_1 = \mathrm{m}_3\,.$$

Finally, we summarize the results:

Bennet's mechanism is thus defined by the conditions $g_1 g_2 g_3 g_4 = e$, $\mathrm{s}_i = 0$, $\alpha_1 = \alpha_3$, $\alpha_2 = \alpha_4$, $\mathrm{m}_1 = \mathrm{m}_3$, $\mathrm{m}_2 = \mathrm{m}_4$, $\mathrm{m}_2\sin\alpha_1 + \mathrm{m}_1\varepsilon\sin\alpha_2 = 0$. It is $\varphi_3 = \varepsilon\varphi_1$, $\varphi_4 = \varepsilon\varphi_2$. The matrix of motion of the frame $\mathcal{R}_3(\varphi_1)$ is given by the relation $\mathcal{R}_3(\varphi_1) = \mathcal{R}_1\, g(\varphi_1)$, where $g(\varphi_1) = g_1(\varphi_1)\,.\,g_2(\varphi_2)$ and

$$\sin\varphi_2 = \sin\varphi_1 \frac{a + b\varepsilon}{\varepsilon + ab + AB\cos\varphi_1}\,,$$

$$\cos\varphi_2 = -\frac{AB + (ab + \varepsilon)\cos\varphi_1}{\varepsilon + ab + AB\cos\varphi_1}\,.$$

It is now no longer difficult to express the matrix $g(\varphi_1)$:

$$g(\varphi_1) = \begin{pmatrix} 1, & 0, & 0, & 0 \\ 0, & a, & A\cos\varphi_1, & A\sin\varphi_1 \\ 0, & A, & -a\cos\varphi_1, & -a\sin\varphi_1 \\ \mathrm{m}_1, & 0, & \sin\varphi_1, & -\cos\varphi_1 \end{pmatrix} \begin{pmatrix} 1, & 0, & 0, & 0 \\ 0, & b, & B\cos\varphi_2, & B\sin\varphi_2 \\ 0, & B, & -b\cos\varphi_2, & -b\sin\varphi_2 \\ \mathrm{m}_2, & 0, & \sin\varphi_2, & -\cos\varphi_2 \end{pmatrix}.$$

Here we have to substitute

$$\mathrm{m}_2 = -\frac{\varepsilon B \mathrm{m}_1}{A}, \quad \sin \varphi_2 = \sin \varphi_1 \frac{a + b\varepsilon}{\varepsilon + ab + AB \cos \varphi_1},$$

$$\cos \varphi_2 = -\frac{AB + (ab + \varepsilon) \cos \varphi_1}{\varepsilon + ab + AB \cos \varphi_1}.$$

3.6 Exercise. Determine the explicit expression of the matrix $g(\varphi_1)$ from Example 3.5.

Below, we shall not deal in greater detail with the various ways of finding the matrix of motion. There are many of them and they have been described at sufficient length in the literature. We shall rather tackle the methods of computing the invariants of motion, or of determining some of their other properties. We shall assume, therefore, that the matrix of motion has already been determined, or we shall devote ourselves to motions which do not require the matrix of motion to be determined (e.g. double-rotational and double-helical motions). Notice that Bennet's mechanism which we are investigating belongs to this group of motions. Indeed, a combination of two rotations around skew axes is involved, so that Bennet's mechanism is in fact fully described by the constants α_1, m_1, and by the function

$$\varphi_2 = \arcsin\left(\frac{a + b\varepsilon}{\varepsilon + ab + AB \cos \varphi_1} \sin \varphi_1\right),$$

as we shall see further on.

4. DIRECTING CONES OF SPACE MOTION, AXOIDS, ELEMENTARY MOTIONS

Since the considerations of this section are similar to those of Section 3 of Chapter II, we shall proceed more rapidly.

Let

$$g(\mathrm{t}) = \begin{pmatrix} 1, & 0 \\ a(\mathrm{t}), & \gamma(\mathrm{t}) \end{pmatrix}$$

be a space motion; $\gamma(\mathrm{t})$ is then the associated spherical motion.

4.1 Definition. The conical surface in $\mathfrak{E}$ determined by the vector function $\boldsymbol{r}(\mathrm{t}) = g'(\mathrm{t})\, g^{-1}(\mathrm{t})$ is called the *fixed directing conical surface of motion* $g(\mathrm{t})$; the conical surface in $\mathfrak{E}$ determined by the vector function $\bar{\boldsymbol{r}}(\mathrm{t}) = g^{-1}(\mathrm{t})\, g'(\mathrm{t})$ is called the *moving directing conical surface of motion* $g(\mathrm{t})$. The vector functions $\boldsymbol{r}(\mathrm{t})$ and $\bar{\boldsymbol{r}}(\mathrm{t})$ are respectively called the *fixed* and *moving directing functions of motion* $g(\mathrm{t})$. (The directing function is a kinematic concept, the directing conical surface is a kinematico-geometrical concept and, therefore, we have two definitions.)

We shall first determine the relation between the directing function of motion $g(\mathrm{t})$ and the directing function of the associated spherical motion. For this purpose we need the expressions for the directing functions of space motion.

4.2 Theorem. *Let* $g(\mathrm{t}) = \begin{pmatrix} 1, & 0 \\ a(\mathrm{t}), & \gamma(\mathrm{t}) \end{pmatrix}$ *be a space motion. Then*

$$\boldsymbol{r}(\mathrm{t}) = \begin{pmatrix} 0, & 0 \\ a' - \gamma'\gamma^{\mathrm{T}}a, & \gamma'\gamma^{\mathrm{T}} \end{pmatrix}, \quad \bar{\boldsymbol{r}}(\mathrm{t}) = \begin{pmatrix} 0, & 0 \\ \gamma^{\mathrm{T}}a', & \gamma^{\mathrm{T}}\gamma' \end{pmatrix}.$$

Proof. $g^{-1}(\mathrm{t}) = \begin{pmatrix} 1, & 0 \\ -\gamma^{\mathrm{T}}a, & \gamma^{\mathrm{T}} \end{pmatrix}$. Then

$$\boldsymbol{r}(\mathrm{t}) = g'g^{-1} = \begin{pmatrix} 0, & 0 \\ a', & \gamma' \end{pmatrix} \begin{pmatrix} 1, & 0 \\ -\gamma^{\mathrm{T}}a, & \gamma^{\mathrm{T}} \end{pmatrix} = \begin{pmatrix} 0, & 0 \\ a' - \gamma'\gamma^{\mathrm{T}}a, & \gamma'\gamma^{\mathrm{T}} \end{pmatrix},$$

$$\bar{\boldsymbol{r}}(\mathrm{t}) = g^{-1}g' = \begin{pmatrix} 1, & 0 \\ -\gamma^{\mathrm{T}}a, & \gamma^{\mathrm{T}} \end{pmatrix} \begin{pmatrix} 0, & 0 \\ a', & \gamma' \end{pmatrix} = \begin{pmatrix} 0, & 0 \\ \gamma^{\mathrm{T}}a', & \gamma^{\mathrm{T}}\gamma' \end{pmatrix}.$$

The desired relation can now be easily found. If we use vector notation for the directing functions of motion, as introduced in Section 2, we have $\boldsymbol{r}(\mathrm{t}) = [\boldsymbol{x}(\mathrm{t}); \boldsymbol{t}(\mathrm{t})]$, $\bar{\boldsymbol{r}}(\mathrm{t}) = [\bar{\boldsymbol{x}}(\mathrm{t}); \bar{\boldsymbol{t}}(\mathrm{t})]$, where $\boldsymbol{x}(\mathrm{t})$ is the matrix $\gamma'(\mathrm{t})\, \gamma^{\mathrm{T}}(\mathrm{t})$ in vector notation, $\bar{\boldsymbol{x}}(\mathrm{t})$ is the matrix $\gamma^{\mathrm{T}}(\mathrm{t})\, \gamma'(\mathrm{t})$ in vector notation, $\boldsymbol{t}(\mathrm{t}) = a'(\mathrm{t}) - \boldsymbol{x} \times a$, $\bar{\boldsymbol{t}}(\mathrm{t}) = \gamma^{\mathrm{T}}a'$. Moreover, $\boldsymbol{x}(\mathrm{t})$ and $\bar{\boldsymbol{x}}(\mathrm{t})$ are the fixed and moving directing functions of the associated spherical motion, respectively.

The canonical parameter of associated motion can thus be adopted as the canonical parameter φ (we assume all the time that it exists, which means that we have tacitly excluded translations from the investigation).

In this case the directing function is called the directing cone. This brings us to the following definition:

4.3 Definition. Let $g(\varphi)$ be space motion given as a function of the canonical parameter φ. We call the vector functions $\boldsymbol{R}(\varphi) = g'(\varphi)\, g^{-1}(\varphi)$ and $\bar{\boldsymbol{R}}(\varphi) = g^{-1}(\varphi)\, g'(\varphi)$ the *fixed* and *moving directing cones of motion* $g(\varphi)$, respectively.

4.4 Lemma. $\mathbf{K}(\boldsymbol{R}, \boldsymbol{R}) = \mathbf{K}(\bar{\boldsymbol{R}}, \bar{\boldsymbol{R}}) = 1$.

Proof. Let $\boldsymbol{R}(\varphi) = [\boldsymbol{x}(\varphi); \boldsymbol{t}(\varphi)]$, $\bar{\boldsymbol{R}}(\varphi) = [\bar{\boldsymbol{x}}(\varphi); \bar{\boldsymbol{t}}(\varphi)]$. Then $\mathbf{K}(\boldsymbol{R}, \boldsymbol{R}) = (\boldsymbol{x}, \boldsymbol{x})$, $\mathbf{K}(\bar{\boldsymbol{R}}, \bar{\boldsymbol{R}}) = (\bar{\boldsymbol{x}}, \bar{\boldsymbol{x}})$. $\boldsymbol{x}$ and $\bar{\boldsymbol{x}}$ are the fixed and moving directing cones of the associated spherical motion and, therefore, they are unit by definition.

4.5 Lemma. $\boldsymbol{R} = \{\mathbf{K}(\boldsymbol{r}, \boldsymbol{r})\}^{-1/2}\, \boldsymbol{r}$, $\bar{\boldsymbol{R}} = \{\mathbf{K}(\bar{\boldsymbol{r}}, \bar{\boldsymbol{r}})\}^{-1/2}\, \bar{\boldsymbol{r}}$, where $\boldsymbol{r}$ and $\bar{\boldsymbol{r}}$ are directing functions. Further, we have

$$\frac{\mathrm{d}\varphi}{\mathrm{d}t} = \{\mathbf{K}(\boldsymbol{r}, \boldsymbol{r})\}^{1/2} = \{\mathbf{K}(\bar{\boldsymbol{r}}, \bar{\boldsymbol{r}})\}^{1/2}.$$

Proof.

$$\boldsymbol{r} = \frac{\mathrm{d}g(\varphi)}{\mathrm{d}\varphi}\frac{\mathrm{d}\varphi}{\mathrm{d}t}\, g^{-1} = \frac{\mathrm{d}\varphi}{\mathrm{d}t}\frac{\mathrm{d}g}{\mathrm{d}\varphi}\, g^{-1} = \frac{\mathrm{d}\varphi}{\mathrm{d}t}\boldsymbol{R};$$

also

$$\mathbf{K}(\boldsymbol{r}, \boldsymbol{r}) = \left(\frac{\mathrm{d}\varphi}{\mathrm{d}t}\right)^2 \mathbf{K}(\boldsymbol{R}, \boldsymbol{R}) = \left(\frac{\mathrm{d}\varphi}{\mathrm{d}t}\right)^2,$$

and similarly for the moving directing cone; moreover, the directing cone is determined up to its sign so that we may always assume that $\mathrm{d}\varphi/\mathrm{d}t > 0$.

4.6 Example. Let us demonstrate the above considerations on a simple example.

Consider the motion described by the matrix

$$g(\alpha) = \begin{pmatrix} 1, & 0, & 0, & 0 \\ m_1(\alpha), & \cos^2\alpha, & \sin\alpha, & \sin\alpha\cos\alpha \\ m_2(\alpha), & -\sin\alpha\cos\alpha, & \cos\alpha, & -\sin^2\alpha \\ m_3(\alpha), & -\sin\alpha, & 0, & \cos\alpha \end{pmatrix},$$

where $m_1(\alpha), m_2(\alpha), m_3(\alpha)$ are arbitrary functions of the parameter α. Notice that the associated spherical motion is the motion from Example II, 3.44. Let us first compute $g^{-1}(\alpha)$ and $g'(\alpha)$:

$$g^{-1}(\alpha) = \begin{pmatrix} 1, & 0 \\ -\gamma^{\mathrm{T}}a, & \gamma^{\mathrm{T}} \end{pmatrix},$$

where

$$\begin{aligned} -\gamma^{\mathrm{T}}a &= -\begin{pmatrix} \cos^2\alpha, & -\sin\alpha\cos\alpha, & -\sin\alpha \\ \sin\alpha, & \cos\alpha, & 0 \\ \sin\alpha\cos\alpha, & -\sin^2\alpha, & \cos\alpha \end{pmatrix} \begin{pmatrix} m_1 \\ m_2 \\ m_3 \end{pmatrix} \\ &= -\begin{pmatrix} m_1\cos^2\alpha - m_2\sin\alpha\cos\alpha - m_3\sin\alpha \\ m_1\sin\alpha + m_2\cos\alpha \\ m_1\sin\alpha\cos\alpha - m_2\sin^2\alpha + m_3\cos\alpha \end{pmatrix}. \end{aligned}$$

$$g'(\alpha) = \begin{pmatrix} 0, & 0, & 0, & 0 \\ m_1', & -2\sin\alpha\cos\alpha, & \cos\alpha, & \cos^2\alpha - \sin^2\alpha \\ m_2', & -\cos^2\alpha + \sin^2\alpha, & -\sin\alpha, & -2\sin\alpha\cos\alpha \\ m_3', & -\cos\alpha, & 0, & -\sin\alpha \end{pmatrix},$$

$$\begin{aligned} &g^{-1}(\alpha) \\ &= \begin{pmatrix} 1, & 0, & 0 & 0 \\ -m_1\cos^2\alpha + m_2\sin\alpha\cos\alpha + m_3\sin\alpha, & \cos^2\alpha, & -\sin\alpha\cos\alpha, & -\sin\alpha \\ -m_1\sin\alpha - m_2\cos\alpha, & \sin\alpha, & \cos\alpha, & 0 \\ -m_1\sin\alpha\cos\alpha + m_2\sin^2\alpha - m_3\cos\alpha, & \sin\alpha\cos\alpha, & -\sin^2\alpha, & \cos\alpha \end{pmatrix}, \end{aligned}$$

$$\boldsymbol{r}(\alpha) = g'g^{-1} = \begin{pmatrix} 0 & 0, & 0, & 0 \\ m_1' - m_2 - m_3\cos\alpha, & 0, & 1, & \cos\alpha \\ m_2' + m_1 + m_3\sin\alpha, & -1, & 0, & -\sin\alpha \\ m_3' + m_1\cos\alpha - m_2\sin\alpha, & -\cos\alpha, & \sin\alpha, & 0 \end{pmatrix},$$

$$\begin{aligned} &\bar{\boldsymbol{r}}(\alpha) = g^{-1}g' \\ &= \begin{pmatrix} 0 & 0, & 0, & 0 \\ m_1'\cos^2\alpha - m_2'\sin\alpha\cos\alpha - m_3'\sin\alpha, & 0, & \cos\alpha, & 1 \\ m_1'\sin\alpha + m_2'\cos\alpha, & -\cos\alpha, & 0, & -\sin\alpha \\ m_1'\sin\alpha\cos\alpha - m_2'\sin^2\alpha + m_3'\cos\alpha, & -1, & \sin\alpha, & 0 \end{pmatrix}. \end{aligned}$$

Remark. We undertook the direct computation for the sake of demonstration only, otherwise the application of formula (4.1) is more advantageous.

In vector notation we obtain

$$\bar{\boldsymbol{r}}(\alpha) = [\bar{\boldsymbol{x}}(\alpha); \bar{\boldsymbol{t}}(\alpha)] = \begin{bmatrix} \sin\alpha; & m_1' \cos^2\alpha - m_2' \sin\alpha\cos\alpha - m_3' \sin\alpha \\ 1; & m_1' \sin\alpha + m_2' \cos\alpha \\ -\cos\alpha; & m_1' \sin\alpha\cos\alpha - m_2' \sin^2\alpha + m_3' \cos\alpha \end{bmatrix},$$

$$\boldsymbol{r}(\alpha) = [\boldsymbol{x}(\alpha); \boldsymbol{t}(\alpha)] = \begin{bmatrix} \sin\alpha; & m_1' - m_2 - m_3 \cos\alpha \\ \cos\alpha; & m_2' + m_1 + m_3 \sin\alpha \\ -1; & m_3' + m_1 \cos\alpha - m_2 \sin\alpha \end{bmatrix}.$$

Further, $\mathbf{K}(\boldsymbol{r}, \boldsymbol{r}) = \mathbf{K}(\bar{\boldsymbol{r}}, \bar{\boldsymbol{r}}) = \sin^2\alpha + \cos^2\alpha + 1 = 2$, so that

$$\frac{d\varphi}{d\alpha} = \sqrt{2}\,, \quad \text{i.e.} \quad \varphi = \alpha\sqrt{2}\,, \quad \alpha = \varphi\sqrt{(2)}/2\,.$$

Our motion, as a function of the canonical parameter, is thus described by the matrix

$$g(\varphi) = \begin{bmatrix} 1, & 0, & 0, & 0 \\ m_1\left(\frac{\varphi\sqrt{2}}{2}\right), & \cos^2\left(\frac{\varphi\sqrt{2}}{2}\right), & \sin\left(\frac{\varphi\sqrt{2}}{2}\right), & \sin\left(\frac{\varphi\sqrt{2}}{2}\right)\cos\left(\frac{\varphi\sqrt{2}}{2}\right) \\ m_2\left(\frac{\varphi\sqrt{2}}{2}\right), & -\sin\left(\frac{\varphi\sqrt{2}}{2}\right)\cos\left(\frac{\varphi\sqrt{2}}{2}\right), & \cos\left(\frac{\varphi\sqrt{2}}{2}\right), & -\sin^2\left(\frac{\varphi\sqrt{2}}{2}\right) \\ m_3\left(\frac{\varphi\sqrt{2}}{2}\right), & -\sin\left(\frac{\varphi\sqrt{2}}{2}\right), & 0, & \cos\left(\frac{\varphi\sqrt{2}}{2}\right) \end{bmatrix}.$$

For the directing cone we have

$$R(\alpha) = \frac{\sqrt{2}}{2}\, r(\alpha)\,, \quad \bar{R}(\alpha) = \frac{\sqrt{2}}{2}\, \bar{r}(\alpha)\,.$$

As mentioned already, we shall consider only motions for which $x(\mathrm{t}) \neq 0$ for all t, where $r(\mathrm{t}) = [x(\mathrm{t}); t(\mathrm{t})]$. Then, also, $\bar{x}(\mathrm{t}) \neq 0$ for $\bar{r}(\mathrm{t}) = [\bar{x}(\mathrm{t}); \bar{t}(\mathrm{t})]$. Thus both directing cones exist for such motions. Consider such a motion, therefore, and put $R(\varphi) = [x(\varphi); t(\varphi)]$, $\bar{R}(\varphi) = [\bar{x}(\varphi); \bar{t}(\varphi)]$. $R(\varphi)$ and $\bar{R}(\varphi)$ then belong, for every φ, to some maximal commutative subalgebras of dimension 2 in $\mathfrak{E}$. Denote these subalgebras by $\mathfrak{H}_R(\varphi)$ and $\bar{\mathfrak{H}}_R(\varphi)$, respectively. Every subalgebra $\bar{\mathfrak{H}}_R(\varphi)$ and $\mathfrak{H}_R(\varphi)$ contains a vector $R_1(\varphi)$ and $\bar{R}_1(\varphi)$, respectively, for which $\mathrm{K}(R_1, R_1) = 1$, $\mathscr{K}(R_1, R_1) = 0$, and $\mathrm{K}(\bar{R}_1, \bar{R}_1) = 1$, $\mathscr{K}(\bar{R}_1, \bar{R}_1) = 0$ as we know from Section 2. Just one straight line in E_3 corresponds to each subalgebra $\mathfrak{H}_R(\varphi)$ and $\bar{\mathfrak{H}}_R(\varphi)$, and its Plücker coordinates are determined by the vectors $R_1(\varphi)$ and $\bar{R}_1(\varphi)$, respectively.

The ruled surface $\mathfrak{A}$, which is constituted by all straight lines which correspond to the subalgebras $\mathfrak{H}_R(\varphi)$, is called the *fixed axoid*. Analogously, $\bar{\mathfrak{A}}$ is the set of all straight lines corresponding to subalgebras $\bar{\mathfrak{H}}_R(\varphi)$ for all values of φ; it is called the *moving axoid*. The straight line corresponding to $\mathfrak{H}_R(\varphi_0)$ or $\bar{\mathfrak{H}}_R(\varphi_0)$ for a certain $\varphi = \varphi_0$ is called the *generating line* of the fixed or moving axoid, respectively. The relation between maximal commutative subalgebras from $\mathfrak{E}$ and straight lines from E_3 was discussed in Section 2. Let us look only at how the axoids will be expressed in coordinate notation. Once again, let $R(\varphi) = [x(\varphi); t(\varphi)]$. Algebra $\mathfrak{H}_R(\varphi)$ is then constituted by the vectors $X(\varphi, \alpha, \beta) = [\alpha\, x(\varphi); \alpha\, t(\varphi) + \beta\, x(\varphi)]$. As in the proof of Lemma 2.3 we find that $R_1(\varphi) = [x(\varphi), t(\varphi) - (x(\varphi), t(\varphi))\, x(\varphi)]$, so that for $\lambda \in \mathbf{R}$ we have

$$\begin{aligned}\mathfrak{A}(\varphi, \lambda) &= x(\varphi) \times [t(\varphi) - (x(\varphi), t(\varphi))\, x(\varphi)] + \lambda\, x(\varphi)\\ &= x(\varphi) \times t(\varphi) + \lambda\, x(\varphi)\,.\end{aligned}$$

The equations of the fixed and moving axoids thus read

$$\begin{aligned}\mathfrak{A}(\varphi, \lambda) &= x(\varphi) \times t(\varphi) + \lambda\, x(\varphi)\,,\\ \bar{\mathfrak{A}}(\varphi, \lambda) &= \bar{x}(\varphi) \times \bar{t}(\varphi) + \lambda\, \bar{x}(\varphi)\,, \quad \lambda \in \mathbf{R}\,.\end{aligned} \tag{4.2}$$

We have thus constructed two ruled surfaces, namely the fixed and the moving axoids, to each space motion. We shall deal with their geometrical meaning later, after introducing instantaneous motion. Let us give two examples for demonstration first:

4.7 Example. Let us determine the axoids of the motion from Example 4.6. For simplicity and to be quite explicit, let us put $m_1(\alpha) = 0$, $m_2(\alpha) = q \cos \alpha$, $m_3(\alpha) = q \sin \alpha$, where $q \in \mathbf{R}$, for the motion $g(\alpha)$ of Example 4.6. Upon substituting into the expressions for $\boldsymbol{r}(\alpha)$ and $\bar{\boldsymbol{r}}(\alpha)$ we obtain

$$\boldsymbol{r}(\alpha) = \begin{pmatrix} \sin \alpha; & -q \cos \alpha - q \sin \alpha \cos \alpha \\ \cos \alpha; & -q \sin \alpha + q \sin^2 \alpha \\ -1; & q \cos \alpha - q \sin \alpha \cos \alpha \end{pmatrix},$$

$$\bar{\boldsymbol{r}}(\alpha) = \begin{pmatrix} \sin \alpha; & q \cos \alpha \sin^2 \alpha - q \cos \alpha \sin \alpha \\ 1; & -q \sin \alpha \cos \alpha \\ -\cos \alpha; & q \sin^3 \alpha + q \cos^2 \alpha \end{pmatrix},$$

so that

$$\boldsymbol{R}(\alpha) = \frac{\sqrt{2}}{2} \begin{pmatrix} \sin \alpha; & -q \cos \alpha(1 + \sin \alpha) \\ \cos \alpha; & q \sin \alpha(\sin \alpha - 1) \\ -1; & q \cos \alpha(1 - \sin \alpha) \end{pmatrix},$$

$$\bar{\boldsymbol{R}}(\alpha) = \frac{\sqrt{2}}{2} \begin{pmatrix} \sin \alpha; & q \sin \alpha \cos \alpha(\sin \alpha - 1) \\ 1; & -q \sin \alpha \cos \alpha \\ -\cos \alpha; & q(\sin^3 \alpha + \cos^2 \alpha) \end{pmatrix}.$$

Also,

$$\boldsymbol{x} \times \boldsymbol{t} = \frac{q}{2}.$$

$$\cdot \begin{vmatrix} \boldsymbol{u}_1, & \sin \alpha, & -\cos \alpha(1 + \sin \alpha) \\ \boldsymbol{u}_2, & \cos \alpha, & \sin \alpha(\sin \alpha - 1) \\ \boldsymbol{u}_3, & -1, & \cos \alpha(1 - \sin \alpha) \end{vmatrix} = \frac{q}{2} \begin{pmatrix} 1 - \sin \alpha(1 + \cos^2 \alpha) \\ \cos \alpha(1 + \sin^2 \alpha) \\ \sin \alpha + \cos (2\alpha) \end{pmatrix}.$$

The equation of the fixed axoid then reads

$$\mathfrak{A}(\alpha, \lambda) = \frac{q}{2} \begin{pmatrix} 1 - \sin \alpha(1 + \cos^2 \alpha) \\ \cos \alpha(1 + \sin^2 \alpha) \\ \sin \alpha + \cos (2\alpha) \end{pmatrix} + \lambda \begin{pmatrix} \sin \alpha \\ \cos \alpha \\ -1 \end{pmatrix}.$$

Fig. 41 shows a part of the fixed axoid in oblique axonometry for $q = 60$, $\lambda \in \langle -20, 20 \rangle$, $\alpha \in \langle 80°, 260° \rangle$. The computation was carried out by computer, and the figure was plotted by an automatic plotter.

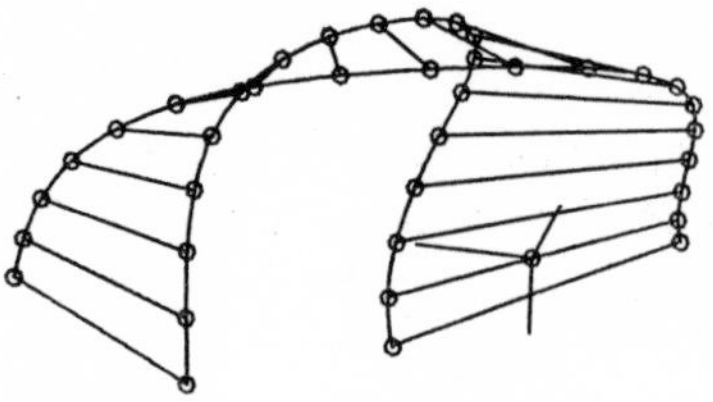

Fig. 41.

Similarly we can compute

$$\bar{\boldsymbol{x}} \times \bar{\boldsymbol{t}} = \frac{q}{2} \begin{pmatrix} \cos^2 \alpha - \sin \alpha \cos (2\alpha) \\ -\sin^2 \alpha \\ \sin \alpha \cos \alpha (1 - 2 \sin \alpha) \end{pmatrix},$$

$$\overline{\mathfrak{A}}(\alpha, \lambda) = \frac{q}{2} \begin{pmatrix} \cos^2 \alpha - \sin \alpha \cos (2\alpha) \\ -\sin^2 \alpha \\ \sin \alpha \cos \alpha (1 - 2 \sin \alpha) \end{pmatrix} + \lambda \begin{pmatrix} \sin \alpha \\ 1 \\ -\cos \alpha \end{pmatrix}.$$

4.8 Example. Let us determine the axoids of the motion from Example 3.4. We first repeat the problem: the space motion is given by the matrix $g(\mathrm{t})$,

$$g(\mathrm{t}) = \begin{pmatrix} 1, & 0 \\ a(\mathrm{t}), & \gamma(\mathrm{t}) \end{pmatrix}, \quad a(\mathrm{t}) = \begin{pmatrix} 0 \\ 0 \\ d \sin \mathrm{t} \end{pmatrix},$$

$$\gamma(\mathrm{t}) = \begin{pmatrix} Mm_1, & 0, & -MN \\ MN \sin t, & \cos \mathrm{t}, & Mm_1 \sin \mathrm{t} \\ MN \cos \mathrm{t}, & -\sin \mathrm{t}, & Mm_1 \cos \mathrm{t} \end{pmatrix},$$

where $N = m_3 \cos \mathrm{t} - d \cos \mathrm{t} \sin \mathrm{t}$, $M = (m_1^2 + N^2)^{-1/2}$. According to (4.1) we have

$$\boldsymbol{r}(\mathrm{t}) = \begin{pmatrix} 0, & 0 \\ a' - \gamma'\gamma^{\mathrm{T}} a, & \gamma'\gamma^{\mathrm{T}} \end{pmatrix}, \quad \bar{\boldsymbol{r}}(\mathrm{t}) = \begin{pmatrix} 0, & 0 \\ \gamma^{\mathrm{T}} a', & \gamma^{\mathrm{T}}\gamma' \end{pmatrix}.$$

Therefore, let us first determine the directing functions $\boldsymbol{r}(\mathrm{t})$ and $\bar{\boldsymbol{r}}(\mathrm{t})$. For this purpose we have to compute the matrix $\gamma'(\mathrm{t})$. In this connection, note that

$$M' = \frac{\mathrm{d}}{\mathrm{dt}}\{(m_1^2 + N^2)^{-1/2}\} = -\frac{1}{2}(m_1^2 + N^2)^{-3/2}\,2NN'$$
$$= -M^3NN'\,.$$

By differentiating matrix $\gamma(\mathrm{t})$ we arrive at

$$\gamma'(\mathrm{t}) = \begin{pmatrix} -M^3N'Nm_1, & 0 & 0, & M^3N^2N' - MN' \\ MN'\sin \mathrm{t}(1 - M^2N^2) + MN\cos \mathrm{t}, & -\sin \mathrm{t}, & & Mm_1(\cos \mathrm{t} - M^2NN'\sin \mathrm{t}) \\ MN'\cos \mathrm{t}(1 - M^2N^2) - MN\sin \mathrm{t}, & -\cos \mathrm{t}, & & Mm_1(-\sin \mathrm{t} - M^2NN'\cos \mathrm{t}) \end{pmatrix}$$

$$a'(\mathrm{t}) = \begin{pmatrix} 0 \\ 0 \\ d\cos \mathrm{t} \end{pmatrix}.$$

We now have to compute $\gamma'\gamma^{\mathrm{T}}$.

It would be tedious to compute the entire matrix. We know, of course, that the result is a skew-symmetric matrix and, therefore, we shall compute only some of its elements. Put $\gamma'\gamma^{\mathrm{T}} = (\mu_j^i)$. Then

$$\mu_2^1 = -M^4N^2N'm_1\sin \mathrm{t} + M^4N^2N'm_1\sin \mathrm{t} - M^2N'm_1\sin \mathrm{t}$$
$$= -M^2N'm_1\sin \mathrm{t}\,,$$

$$\mu_3^1 = -M^4N^2N'm_1\cos \mathrm{t} + M^4N^2N'm_1\cos \mathrm{t} - M^2N'm_1\cos \mathrm{t}$$
$$= -M^2N'm_1\cos \mathrm{t}\,,$$

$$\mu_3^2 = M^2N^2\cos^2 \mathrm{t} - M^4N^3N'\sin \mathrm{t}\cos \mathrm{t} + M^2NN'\sin \mathrm{t}\cos \mathrm{t}$$
$$+ \sin^2 \mathrm{t} - M^4NN'm_1^2\sin \mathrm{t}\cos \mathrm{t} + M^2m_1^2\cos^2 \mathrm{t}$$
$$= M^2(m_1^2 + N^2)\cos^2 \mathrm{t} + \sin^2 \mathrm{t}$$
$$+ M^2NN'(-M^2N^2 + 1 - M^2m_1^2)\sin \mathrm{t}\cos \mathrm{t}$$
$$= \cos^2 \mathrm{t} + \sin^2 \mathrm{t} + M^2NN'[1 - M^2(m_1^2 + N^2)]\sin \mathrm{t}\cos \mathrm{t} = 1\,.$$

Analogously, put $\gamma^{\mathrm{T}}\gamma' = (v_j^i)$. Then

$$v_2^1 = -MN \sin^2 \mathrm{t} - MN \cos^2 \mathrm{t} = -MN\,,$$

$$v_2^3 = -Mm_1 \sin^2 \mathrm{t} - Mm_1 \cos^2 \mathrm{t} = -Mm_1\,,$$

$$\begin{aligned} v_3^1 = {} & M^4N^2N'm_1 - M^2N'm_1 - M^4N^2N'm_1 \sin^2 \mathrm{t} \\ & + M^2Nm_1 \sin \mathrm{t} \cos \mathrm{t} \\ & - M^4N^2N'm_1 \cos^2 \mathrm{t} - M^2Nm_1 \sin \mathrm{t} \cos \mathrm{t} = -M^2N'm_1\,. \end{aligned}$$

For the directing functions of the associated motion we have thus determined that

$$\gamma'\gamma^{\mathrm{T}} = \begin{pmatrix} 0, & -M^2N'm_1 \sin \mathrm{t}, & -M^2N'm_1 \cos \mathrm{t} \\ M^2N'm_1 \sin \mathrm{t}, & 0, & 1 \\ M^2N'm_1 \cos \mathrm{t}, & -1, & 0 \end{pmatrix},$$

$$\gamma^{\mathrm{T}}\gamma' = \begin{pmatrix} 0, & -MN, & -M^2N'm_1 \\ MN, & 0, & Mm_1 \\ M^2N'm_1, & -Mm_1, & 0 \end{pmatrix}.$$

If we continue the computations we obtain

$$\gamma'\gamma^{\mathrm{T}}a = \begin{pmatrix} -M^2N'm_1 d \sin \mathrm{t} \cos \mathrm{t} \\ d \sin \mathrm{t} \\ 0 \end{pmatrix},$$

$$\gamma^{\mathrm{T}}a' = \begin{pmatrix} MNd \cos^2 \mathrm{t} \\ -d \sin \mathrm{t} \cos \mathrm{t} \\ Mm_1 d \cos^2 \mathrm{t} \end{pmatrix}.$$

The directing functions of our motion are thus

$$\boldsymbol{r}(\mathrm{t}) = \begin{pmatrix} 0, & 0, & 0, & 0 \\ M^2N'm_1 d \sin \mathrm{t} \cos \mathrm{t}, & 0, & -M^2N'm_1 \sin \mathrm{t}, & -M^2N'm_1 \cos \mathrm{t} \\ -d \sin \mathrm{t}, & M^2N'm_1 \sin \mathrm{t}, & 0, & 1 \\ d \cos \mathrm{t}, & M^2N'm_1 \cos \mathrm{t}, & -1, & 0 \end{pmatrix},$$

$$\bar{r}(t) = \begin{pmatrix} 0, & 0, & 0, & 0 \\ MNd\cos^2 t, & 0, & -MN, & -M^2N'm_1 \\ -d\sin t\cos t, & MN & 0, & Mm_1 \\ Mm_1 d\cos^2 t, & M^2N'm_1, & -Mm_1, & 0 \end{pmatrix}.$$

If we rewrite $\boldsymbol{r}$ and $\bar{\boldsymbol{r}}$ in vector notation, we have

$$\boldsymbol{r}(t) = [\boldsymbol{x}_0(t); \boldsymbol{t}_0(t)] = \begin{bmatrix} -1; & M^2N'm_1 d\sin t\cos t \\ -M^2N'm_1\cos t; & -d\sin t \\ M^2N'm_1\sin t; & d\cos t \end{bmatrix},$$

$$\bar{\boldsymbol{r}}(t) = [\bar{\boldsymbol{x}}_0(t); \bar{\boldsymbol{t}}_0(t)] = \begin{bmatrix} -Mm_1; & MNd\cos^2 t \\ -M^2N'm_1; & -d\sin t\cos t \\ MN; & Mm_1 d\cos^2 t \end{bmatrix}.$$

Also, $(\boldsymbol{x}_0(t), \boldsymbol{x}_0(t)) = 1 + M^4(N')^2 m_1^2$. Therefore, put $L = [1 + M^4(N')^2 m_1^2]^{-1/2}$. Then

$$\boldsymbol{R}(t) = (\boldsymbol{x}(t); \boldsymbol{t}(t)) = \begin{bmatrix} -L; & M^2N'Lm_1 d\sin t \\ -M^2LN'm_1\cos t; & -Ld\sin t \\ M^2LN'm_1\sin t; & Ld\cos t \end{bmatrix},$$

$$\bar{\boldsymbol{R}}(t) = (\bar{\boldsymbol{x}}(t); \bar{\boldsymbol{t}}(t)) = \begin{bmatrix} -LMm_1; & LMNd\cos^2 t \\ -LM^2N'm_1; & -Ld\sin t\cos t \\ LMN; & LMm_1 d\cos^2 t \end{bmatrix}.$$

The equations of the axoids read

$$\mathfrak{A}(t,\lambda) = \boldsymbol{x}(t) \times \boldsymbol{t}(t) + \lambda\,\boldsymbol{x}(t), \quad \bar{\mathfrak{A}}(t,\lambda) = \bar{\boldsymbol{x}}(t) \times \bar{\boldsymbol{t}}(t) + \lambda\,\bar{\boldsymbol{x}}(t).$$

$$\boldsymbol{x} \times \boldsymbol{t} = L^2 \begin{pmatrix} -M^2N'm_1 d\cos(2t) \\ d(M^4N'm_1^2\sin^2 t + 1)\cos t \\ d(M^4N'm_1^2\cos^2 t + 1)\sin t \end{pmatrix},$$

$$\bar{\boldsymbol{x}} \times \bar{\boldsymbol{t}} = L^2 \begin{pmatrix} Md(N\sin t - M^2N'm_1^2\cos t)\cos t \\ d\cos^2 t \\ Mm_1 d(\sin t + M^2NN'\cos t)\cos t \end{pmatrix},$$

so that the parametric equations of the two axoids are

$$\mathfrak{A}(t,\lambda) = \begin{pmatrix} -\lambda - M^2N'm_1L^2 d\cos(2t) \\ -\lambda M^2N'm_1\cos t + dL^2(M^4N'm_1^2\sin^2 t + 1)\cos t \\ \lambda M^2N'm_1\sin t + dL^2(M^4N'm_1^2\cos^2 t + 1)\sin t \end{pmatrix},$$

$$\mathfrak{A}(\mathrm{t}, \lambda) = \begin{pmatrix} -\lambda m_1 + MdL^2(N \sin \mathrm{t} - M^2N'm_1^2 \cos \mathrm{t}) \cos \mathrm{t} \\ -\lambda MN'm_1 + dL^2 \cos^2 \mathrm{t} \\ \lambda N + Mm_1dL^2(\sin \mathrm{t} + M^2NN' \cos \mathrm{t}) \cos \mathrm{t} \end{pmatrix}.$$

(For the sake of simplification we have multiplied the parameter λ by a suitable factor.)

We define elementary motions in a way similar to that used in Chapter II, Section 3 for spherical kinematics:

4.9 Definition. Space motion is called elementary if its fixed directing cone is a constant vector from $\mathfrak{E}$.

4.10 Lemma. Let $g(\mathrm{t})$ be a motion and $\boldsymbol{R}(\mathrm{t})$ and $\bar{\boldsymbol{R}}(\mathrm{t})$ its directing cones. Then $\boldsymbol{R}(\mathrm{t}) = \mathrm{ad}g(\mathrm{t})\,\bar{\boldsymbol{R}}(\mathrm{t})$, $\boldsymbol{R}'(\mathrm{t}) = \mathrm{ad}g(\mathrm{t})\,\bar{\boldsymbol{R}}'(\mathrm{t})$.

Proof. The proof is identical with the proofs of II, 3.14 and II, 3.12.

Lemma 4.10 implies: If $\boldsymbol{R}(\mathrm{t}) = const.$, we also have $\bar{\boldsymbol{R}}(\mathrm{t}) = const.$ We shall now determine all the elementary motions.

4.11 Theorem. *Elementary motions are all motions of the form* $g_1\, k(\varphi)\, g_2$, *where* $g_1, g_2 \in \mathscr{E}$, φ *is the canonical parameter and* $k(\varphi_1 + \varphi_2) = k(\varphi_1)\, k(\varphi_2)$.

The proof is the same as that of Theorem II, 3.15.

Now we shall show that the elementary motions are helical motions about a fixed axis, rotation being considered a special case of helical motion with the parameter equal to zero. Translations are not considered because they have no directing cone (see I, 5.27).

Therefore, let $\boldsymbol{R} \in \mathfrak{E}$, $\mathbf{K}(\boldsymbol{R}, \boldsymbol{R}) = 1$. In a suitable base we may write

$$\boldsymbol{R} = \begin{pmatrix} 0, & 0 \\ \boldsymbol{t}, & \boldsymbol{x} \end{pmatrix}, \quad \text{where} \quad \boldsymbol{t} = \begin{pmatrix} v_0 \\ 0 \\ 0 \end{pmatrix}, \quad \boldsymbol{x} = \begin{pmatrix} 0, & 0, & 0 \\ 0, & 0, & -1 \\ 0, & 1, & 0 \end{pmatrix}.$$

As in Chapter II, put $\exp X = \sum_{n=0}^{\infty} \frac{X^n}{n!}$ for any square matrix. Direct computation yields $\boldsymbol{R}^n = \begin{pmatrix} 0, & 0 \\ 0, & \boldsymbol{x}^n \end{pmatrix}$ for $n \geqq 2$. Therefore, as in II, 3.15

$$k(\varphi) = \exp(\varphi \boldsymbol{R}) = \begin{pmatrix} 1, & 0, \ 0, & 0 \\ v_0\varphi, & 1, \ 0, & 0 \\ 0, & 0, \ \cos\varphi, & -\sin\varphi \\ 0, & 0, \ \sin\varphi, & \cos\varphi \end{pmatrix} \tag{4.3}$$

Note also that $\exp(\varphi \boldsymbol{R})$ is the unique motion $g(\varphi)$ for which $g(0) = e$ and the fixed directing cone of which is $\boldsymbol{R}$. Then the moving directing cone is also $\boldsymbol{R}$.

4.12 Lemma. Let $\boldsymbol{R} \in \mathfrak{E}$, $\mathbf{K}(\boldsymbol{R}, \boldsymbol{R}) = 1$. Then $\exp(\varphi \boldsymbol{R})$ is a helical motion about the axis determined by the vector $\boldsymbol{R}$ whose parameter is $v_0 = \mathscr{K}(\boldsymbol{R}, \boldsymbol{R})$.

Proof. First, let $\boldsymbol{R}$ have the form $\boldsymbol{S} = (\boldsymbol{u}_1; v_0\boldsymbol{u}_1)$. Then $\exp(\varphi \boldsymbol{S})$ is a helical motion about the axis $o = \mu \boldsymbol{u}_1$ with the parameter $v_0 = \mathscr{K}(\boldsymbol{S}, \boldsymbol{S})$ according to (4.3). Further, assume $\boldsymbol{R} = \mathrm{ad}\, g_1 \boldsymbol{S}$. Then $\exp(\varphi \boldsymbol{R}) = g_1 \exp(\varphi \boldsymbol{S}) g_1^{-1}$ is a helical motion about the axis $g_1 o$ with the parameter $v_0 = \mathscr{K}(\boldsymbol{R}, \boldsymbol{R})$, because $\mathscr{K}$ is invariant and $\exp(\varphi \boldsymbol{R})(g_1 o) = g_1 o$. Also, we have $\boldsymbol{R} \in \mathfrak{H}_{g_1 o}$, so that the line $g_1 o$ is determined by the vector $\boldsymbol{R}$.

4.13 Definition. Let $g(\mathrm{t})$ be a motion with the fixed directing cone $\boldsymbol{R}(\mathrm{t})$. The motion $g(\varphi) = \{\exp[\varphi \boldsymbol{R}(\mathrm{t}_0)]\}\, g(\mathrm{t}_0)$ is called the *instantaneous motion* of motion $g(\mathrm{t})$ at time t_0. The straight line determined by the vector $\boldsymbol{R}(\mathrm{t}_0)$ is called the *axis of instantaneous motion* and it is the generating line of the fixed axoid for $\mathrm{t} = \mathrm{t}_0$. Up to a constant matrix, the instantaneous motion is a helical motion about this axis.

We are now able to prove properties analogous to the properties II, 3.18 to 3.22. We restrict ourselves to the formulations; the proofs are for the reader to carry out as an exercise.

4.14 Theorem. *Let $g(\mathrm{t})$ be a motion with directing cone $\boldsymbol{R}(\mathrm{t})$, and $\bar{X}$ a point from E_3. For the tangent vector $X'(\mathrm{t})$ of the trajectory $X(\mathrm{t})$ of point $\bar{X}$ at point $X(\mathrm{t})$ we have*

$$X'(\mathrm{t}) = \boldsymbol{R}(\mathrm{t})\, X(\mathrm{t}) \tag{4.4}$$

The field of tangent vectors of the trajectories is a Killing vector field in E_3, induced by the vector $\boldsymbol{R}$ of the directing cone in the sense of Chapter I (this concerns the tangent vectors as derivatives of the trajectories with respect to the canonical parameter φ; if the parameter is general, all tangent vectors must be multiplied by the factor $\mathrm{dt}/\mathrm{d}\varphi$). *The integral curves of* *the vector field of the tangent vectors of the trajectories are helices if* $\mathcal{K}(\boldsymbol{R}, \boldsymbol{R}) \neq 0$, *and circles if* $\mathcal{K}(\boldsymbol{R}, \boldsymbol{R}) = 0$. *If* $g_1(\varphi) = \{\exp[\varphi \boldsymbol{R}(\mathrm{t}_0)]\}\, g(\mathrm{t}_0)$ *is the instantaneous motion of motion* $g(\mathrm{t})$ *at time* $\mathrm{t} = \mathrm{t}_0$ *and if* $\boldsymbol{S}(\varphi)$ *and* $\bar{\boldsymbol{S}}(\varphi)$ *are its directing cones, then* $\boldsymbol{S}(\varphi) = \boldsymbol{R}(\mathrm{t}_0)$, $\bar{\boldsymbol{S}}(\varphi) = \bar{\boldsymbol{R}}(\mathrm{t}_0)$. *Moreover, the given motion at time* t_0 *and the instantaneous motion for all* φ *have the same tangents to the trajectories.*

4.15 Remark. As in Chapter II, we can compute the tangent vector of the trajectory in two ways: using either the matrix or the vector notations. Let us find the formula for the vector notation. Let $M \in E_3$. Then $X'_M = \boldsymbol{R}M$ according to (4.4) in matrix notation. If we put

$$M = \begin{pmatrix} 1 \\ m \end{pmatrix}, \quad X'_M = \begin{pmatrix} 0 \\ \boldsymbol{v}_M \end{pmatrix},$$

we arrive at

$$\begin{pmatrix} 0 \\ \boldsymbol{v}_M \end{pmatrix} = \begin{pmatrix} 0, & 0 \\ \boldsymbol{t}, & \boldsymbol{x} \end{pmatrix} \begin{pmatrix} 1 \\ m \end{pmatrix} = \begin{pmatrix} 0 \\ \boldsymbol{t} + \boldsymbol{x}m \end{pmatrix}.$$

Therefore, in vector notation we have:
If

$$\boldsymbol{R} = [\boldsymbol{x}; \boldsymbol{t}], \quad \text{then} \quad v_M = \boldsymbol{x} \times m + \boldsymbol{t}. \tag{4.5}$$

4.16 Example. Let us determine the instantaneous motion $h(\varphi)$ of the motion $g(\alpha)$ from Example 4.7 for $\alpha = \pi$ and $q = 2r$. Thus $h(\varphi) = \exp[\varphi\, \boldsymbol{R}(\pi)]\, g(\pi)$, where

$$g(\pi) = \begin{pmatrix} 1, & 0, & 0, & 0 \\ 0, & 0, & 0, & 0 \\ -2r, & 0, & -1, & 0 \\ 0, & 0, & 0, & -1 \end{pmatrix}, \quad R(\pi) = 1/\sqrt{2} \begin{bmatrix} 0; & 2r \\ -1; & 0 \\ -1; & -2r \end{bmatrix} = [x; t].$$

We have $\mathcal{K}(\boldsymbol{R}(\pi), \boldsymbol{R}(\pi)) = r$, so that $\exp[\varphi\, \boldsymbol{R}(\pi)]$ is a helical motion with the parameter r. Put $\boldsymbol{R} = \mathrm{ad} g_1\, \boldsymbol{S}$, where $\boldsymbol{S} = [\boldsymbol{u}_1; r\boldsymbol{u}_1]$ and $g_1 = \begin{pmatrix} 1, & 0 \\ a, & \gamma \end{pmatrix}$ is a suitably selected matrix. Then $\exp[\varphi\, \boldsymbol{R}(\pi)] = \exp[\mathrm{ad} g_1(\varphi \boldsymbol{S})] = g_1 \exp(\varphi \boldsymbol{S}) g_1^{-1}$. $\exp(\varphi \boldsymbol{S})$ is given by relation (4.3).

Let us solve equation $\boldsymbol{R} = \mathrm{ad}g_1\boldsymbol{S}$. By (1.4) we have $[\boldsymbol{x}; \boldsymbol{t}] = [\gamma\boldsymbol{u}_1; r\gamma\boldsymbol{u}_1 + a \times \gamma\boldsymbol{u}_1]$, i.e.

$$\boldsymbol{x} = \gamma\boldsymbol{u}_1\,, \quad \boldsymbol{t} = r\boldsymbol{x} + a \times \boldsymbol{x}\,. \tag{4.6}$$

The first equation in (4.6) is

$$\begin{pmatrix} 0 \\ -1/\sqrt{2} \\ -1/\sqrt{2} \end{pmatrix} = \begin{pmatrix} \gamma_1^1, & \gamma_2^1, & \gamma_3^1 \\ \gamma_1^2, & \gamma_2^2, & \gamma_3^2 \\ \gamma_1^3, & \gamma_2^3, & \gamma_3^3 \end{pmatrix} \begin{pmatrix} 1 \\ 0 \\ 0 \end{pmatrix} = \begin{pmatrix} \gamma_1^1 \\ \gamma_1^2 \\ \gamma_1^3 \end{pmatrix},$$

so that $\gamma_1^1 = 0$, $\gamma_1^2 = \gamma_1^3 = -1/\sqrt{2}$. The remaining coefficients of matrix γ must be chosen so that $\gamma \in O(3)$. Choose $\gamma_2^1 = 1$, $\gamma_2^2 = \gamma_2^3 = 0$. Then

$$\gamma = 1/\sqrt{2} \begin{pmatrix} 0, & \sqrt{2}, & 0 \\ -1, & 0, & -1 \\ -1, & 0, & 1 \end{pmatrix}$$

The second equation of (4.6) reads

$$r\sqrt{2}\begin{pmatrix} 1 \\ 0 \\ 1 \end{pmatrix} = -r/\sqrt{2}\begin{pmatrix} 0 \\ 1 \\ 1 \end{pmatrix} - 1/\sqrt{2}\begin{pmatrix} a^1 \\ a^2 \\ a^3 \end{pmatrix} \times \begin{pmatrix} 0 \\ 1 \\ 1 \end{pmatrix},$$

After some manipulation we obtain $a^1 = r$, $2r = a^3 - a^2$. Put $a^3 = 2r$, $a^2 = 0$. Then

$$g_1 = \begin{pmatrix} 1, & 0, & 0, & 0 \\ r, & 0, & 1, & 0 \\ 0, & -1/\sqrt{2}, & 0, & -1/\sqrt{2} \\ 2r, & -1/\sqrt{2}, & 0, & 1/\sqrt{2} \end{pmatrix}$$

After multiplying by $g_1 \exp(\varphi\boldsymbol{S})\, g_1^{-1}$ we arrive at

$$\exp[\varphi\, \boldsymbol{R}(\pi)] = \tfrac{1}{2}\,\cdot$$

$$\cdot \begin{pmatrix} 2, & 0, & 0, & 0 \\ 2r(1 - \cos\varphi + \sqrt{2}\sin\varphi), & 2\cos\varphi, & 2\sin\varphi, & -\sqrt{2}\sin\varphi \\ r(2\cos\varphi + \sqrt{2}\sin\varphi - 2 - \varphi\sqrt{2}), & -\sqrt{2}\sin\varphi, & 1 + \cos\varphi, & 1 - \cos\varphi \\ r(2 - \varphi\sqrt{2} - \sqrt{2}\sin\varphi - 2\cos\varphi), & \sqrt{2}\sin\varphi, & 1 - \cos\varphi, & 1 + \cos\varphi \end{pmatrix}.$$

$\exp[\varphi\, \boldsymbol{R}(\pi)]$ is a helical motion the axis of which is the generating line $\mathfrak{A}(\pi, \lambda)$ of the fixed axoid, and according to the result of Example 4.7 we have

$$\mathfrak{A}(\pi, \lambda) = \begin{pmatrix} r \\ -r - \lambda \\ r - \lambda \end{pmatrix}.$$

The reader is left to prove that this statement is true as an exercise. Finally, for the instantaneous motion $h(\varphi) = \exp[\varphi\, \boldsymbol{R}(\pi)]\, g(\pi)$ we get

$$h(\varphi) = \tfrac{1}{2} \cdot$$

$$\cdot \begin{pmatrix} 2, & 0, & 0, & 0 \\ 2r(1 - \cos\varphi), & 2\cos\varphi, & -\sqrt{2}\sin\varphi, & \sqrt{2}\sin\varphi \\ \sqrt{2}\, r(\sin\varphi - \varphi - 2\sqrt{2}), & -2\sin\varphi, & -1 - \cos\varphi, & -1 + \cos\varphi \\ -\sqrt{2}\, r(\sin\varphi + \varphi), & 2\sin\varphi, & -1 + \cos\varphi, & -1 - \cos\varphi \end{pmatrix}$$

The instantaneous motion transforms the line $\overline{\mathfrak{A}}(\pi, \lambda)$ of the moving axoid to the line $\mathfrak{A}(\pi, \lambda)$ of the fixed axoid for all φ. With a view to the result of Example 4.7, $\overline{\mathfrak{A}}(\pi, \lambda) = \begin{pmatrix} r \\ \lambda \\ \lambda \end{pmatrix}$. The reader is again left to prove this property.

Finally, we shall determine the vector field of the tangent vectors of the trajectories for $\alpha = \pi$. If $M \in E_3$ is a point with the coordinates m^1, m^2, m^3, then by (4.5) we have $v_M = \boldsymbol{x} \times m + \boldsymbol{t}$, i.e.

$$\begin{pmatrix} v_M^1 \\ v_M^2 \\ v_M^3 \end{pmatrix} = 1/\sqrt{2} \begin{pmatrix} 0 \\ -1 \\ -1 \end{pmatrix} \times \begin{pmatrix} m^1 \\ m^2 \\ m^3 \end{pmatrix} + 1/\sqrt{2} \begin{pmatrix} 2r \\ 0 \\ -2r \end{pmatrix} = 1/\sqrt{2} \begin{pmatrix} m^2 - m^3 + 2r \\ -m^1 \\ m^1 - 2r \end{pmatrix}.$$

Further, we verify that the tangent vectors of the trajectories of instantaneous motion $h(\varphi)$ belong to the vector field (4.7) for every φ. Therefore, let $\bar{A} \in \bar{E}_3$ be any point, a^i its coordinates in $\bar{E}_3$. The trajectory of the point $\bar{A}$ under motion $h(\varphi)$ has the parametric equation $X(\varphi) = h(\varphi)\, \bar{A}$. Let $X(\varphi)$ have the coordinates $x^i(\varphi)$ in E_3. Computation yields

$$x^1(\varphi) = \tfrac{1}{2}[2r(1 - \cos\varphi) + 2a^1 \cos\varphi - \sqrt{2}\, a^2 \sin\varphi + \sqrt{2}\, a^3 \sin\varphi\,, \qquad (4.8)$$

$$x^2(\varphi) = \tfrac{1}{2}[\sqrt{2}\, r(\sin\varphi - \varphi - 2\sqrt{2}) - \sqrt{2}\, a^1 \sin\varphi - (1 + \cos\varphi)\, a^2 + (-1 + \cos\varphi)\, a^3\,,$$

$$x^3(\varphi) = \tfrac{1}{2}[-\sqrt{2}\, r(\sin\varphi + \varphi) + \sqrt{2}\, a^1 \sin\varphi + (-1 + \cos\varphi)\, a^2 - (1 + \cos\varphi)\, a^3\,.$$

Now, if we differentiate the functions $x^i(\varphi)$ in (4.8) with respect to φ, we arrive at the same result we would obtain by substituting these functions into (4.7). The details are left once again to the reader to sort out.

5. INVARIANTS OF MOTION

The invariants of motion can be determined with the aid of directing cones as in the case of spherical motion. First, assume that the motion $g(\varphi)$ is defined as a function of the canonical parameter φ, $\boldsymbol{R}(\varphi) = g'(\varphi)\, g^{-1}(\varphi)$, and let $\mathrm{K}(\boldsymbol{R}, \boldsymbol{R}) \neq 0$. We know that the vector $\boldsymbol{R}$ determines the two-dimensional commutative subalgebra in $\mathfrak{E}$ described in Section 2; we have denoted it $\mathfrak{H}_{\boldsymbol{R}}$. If $\boldsymbol{R} = [\boldsymbol{x}; \boldsymbol{t}]$, then $\mathfrak{H}_{\boldsymbol{R}} = \{[\lambda \boldsymbol{x}; \lambda \boldsymbol{t} + \mu \boldsymbol{x}]\}$, $\lambda, \mu \in \mathbf{R}$. We also know that the algebra $\mathfrak{H}_{\boldsymbol{R}}$ determines a unique line in E_3, the generating line of the fixed axoid. This line is $\mathfrak{A}(\varphi, \lambda) = \boldsymbol{x}(\varphi) \times \boldsymbol{t}(\varphi) + \lambda\, \boldsymbol{x}(\varphi)$. Further, we know that $\exp(\varphi\, \boldsymbol{R}(\varphi_0))$ is a helical motion about the line $\mathfrak{A}(\varphi_0, \lambda)$ of the fixed axoid and that its parameter is $v_0 = (\boldsymbol{x}, \boldsymbol{t})$. From Section 4 it is also clear that the group with Lie algebra $\mathfrak{H}_{\boldsymbol{R}}$ is the group of all congruences of E_3 which preserve the line $\mathfrak{A}(\varphi, \lambda)$. This follows immediately from Lemma 4.13. According to 4.13 $\exp\{\varphi[\boldsymbol{x}; \boldsymbol{t} + \mu \boldsymbol{x}]\}$ is the helical motion about the line $Y = \boldsymbol{x} \times (\boldsymbol{t} + \mu \boldsymbol{x}) + \lambda \boldsymbol{x} = \boldsymbol{x} \times \boldsymbol{t} + \lambda \boldsymbol{x} = \mathfrak{A}(\varphi, \lambda)$. This means that all congruences from $\mathfrak{H}_{\boldsymbol{R}}$ always preserve the same line.

In algebra $\mathfrak{H}_{\boldsymbol{R}}$ there are two significant vectors; denote them $\boldsymbol{R}_1$ and $\boldsymbol{R}_2$. The former satisfies the condition $\mathscr{K}(\boldsymbol{R}_1, \boldsymbol{R}_1) = 0$, the latter the condition $\mathrm{K}(\boldsymbol{R}_2, \boldsymbol{R}_2) = 0$; the former corresponds to rotations about the line $\mathfrak{A}(\varphi, \lambda)$, the latter to translations along this line. Let us prove that such vectors exist in $\mathfrak{H}_{\boldsymbol{R}}$. Therefore, select $\boldsymbol{S} \in \mathfrak{H}_{\boldsymbol{R}}$. Then $\boldsymbol{S} = [\alpha \boldsymbol{x}; \alpha \boldsymbol{t} + \beta \boldsymbol{x}]$. If $\mathscr{K}(\boldsymbol{S}, \boldsymbol{S})$ is to equal zero, $(\alpha \boldsymbol{x}, \alpha \boldsymbol{t} + \beta \boldsymbol{x})$ must equal zero, i.e. $\alpha^2(\boldsymbol{x}, \boldsymbol{t}) + \alpha\beta = 0$, i.e. $\beta = -\alpha(\boldsymbol{x}, \boldsymbol{t})$, and this equation always has a solution. Similarly, if $\mathrm{K}(\boldsymbol{S}, \boldsymbol{S})$ is to equal zero, then $(\alpha \boldsymbol{x}, \alpha \boldsymbol{x}) = \alpha^2 = 0$, i.e. α must equal zero, so that the second equation always has a solution as well. We also see that the vectors $\boldsymbol{R}_1$ and $\boldsymbol{R}_2$ have been determined up to a factor $\lambda \in \boldsymbol{R}$. Choose $\boldsymbol{R}_1 = [\boldsymbol{x}; \boldsymbol{t} - (\boldsymbol{x}, \boldsymbol{t})\, \boldsymbol{x}]$, $\boldsymbol{R}_2 = [0; \boldsymbol{x}]$, i.e. so that

$$\mathrm{K}(\boldsymbol{R}_1, \boldsymbol{R}_1) = 1\,, \quad \mathscr{K}(\boldsymbol{R}_1, \boldsymbol{R}_1) = 0\,, \quad \mathscr{K}(\boldsymbol{R}_1, \boldsymbol{R}_2) = 1\,,$$

$$\mathrm{K}(\boldsymbol{R}_2, \boldsymbol{R}_2) = 0\,, \quad \mathrm{T}(\boldsymbol{R}_2, \boldsymbol{R}_2) = 1\,, \quad \boldsymbol{R} = \boldsymbol{R}_1 + v_0 \boldsymbol{R}_2\,,$$

where $\mathbf{T}$ is the invariant form introduced in Section 1, and $v_0 = (\boldsymbol{x}, \boldsymbol{t})$ is the parameter of instantaneous helical motion. Moreover, $[\boldsymbol{R}_1, \boldsymbol{R}_2] = \boldsymbol{0}$.

Until the end of this chapter the prime will denote the derivative with respect to the canonical parameter φ, while the derivative with respect to a general parameter will again be denoted by a dot.

The vector $\boldsymbol{R}_1'$ determines a two-dimensional commutative subalgebra in $\mathfrak{E}$ under the assumption that $\mathbf{K}(\boldsymbol{R}_1', \boldsymbol{R}_1') \neq 0$. Assume, therefore, that for all φ we have $\mathbf{K}(\boldsymbol{R}_1', \boldsymbol{R}_1') \neq 0$; we shall omit the case of singular instants, i.e. instants at which $\mathbf{K}(\boldsymbol{R}_1', \boldsymbol{R}_1') = 0$; the case $\mathbf{K}(\boldsymbol{R}_1', \boldsymbol{R}_1') = 0$ for all φ will be treated later. Further, notice that $[\boldsymbol{R}_1, \boldsymbol{R}_2] = \boldsymbol{0}$ implies also $[\boldsymbol{R}_1', \boldsymbol{R}_2'] = \boldsymbol{0}$; if we pass to coordinate notation, this statement is evident. This means that $\boldsymbol{R}_2'$ belongs to the maximal commutative subalgebra determined by the vector $\boldsymbol{R}_1'$, and thus we may put

$$\boldsymbol{R}_1' = \kappa_1 \boldsymbol{T}_1 + \mu_1 \boldsymbol{T}_2\,, \quad \boldsymbol{R}_2' = \kappa_1 \boldsymbol{T}_2\,,$$

where $\boldsymbol{T}_1$ and $\boldsymbol{T}_2$ are defined in $\mathfrak{H}_{\boldsymbol{R}'}$ analogously as $\boldsymbol{R}_1$ and $\boldsymbol{R}_2$ in $\mathfrak{H}_{\boldsymbol{R}}$. Finally we define $\boldsymbol{N}_1 = [\boldsymbol{R}_1, \boldsymbol{T}_1]$, $\boldsymbol{N}_2 = [\boldsymbol{R}_1, \boldsymbol{T}_2] = [\boldsymbol{R}_2, \boldsymbol{T}_1]$. With a view to Exercise 2.9 we then have

$$\mathscr{K}(\boldsymbol{N}_1, \boldsymbol{N}_1) = 0\,,$$

because

$$\mathscr{K}(\boldsymbol{N}_1, \boldsymbol{N}_1) = \mathscr{K}([\boldsymbol{R}_1, \boldsymbol{T}_1], [\boldsymbol{R}_1, \boldsymbol{T}_1]) = -2\mathbf{K}(\boldsymbol{R}_1, \boldsymbol{T}_1)\,\mathscr{K}(\boldsymbol{R}_1, \boldsymbol{T}_1) = 0.$$

Also, from the proof of Lemma 2.7 we see that $[\boldsymbol{N}_1, \boldsymbol{N}_2] = 0$, so that $\boldsymbol{N}_1$, $\boldsymbol{N}_2$ constitute a maximal commutative subalgebra of dimension 2.

5.1 Theorem. *Let $g(\varphi)$ be a motion with the canonical parameter φ and fixed directing cone $\boldsymbol{R}(\varphi)$; let $\mathbf{K}(\boldsymbol{R}', \boldsymbol{R}') \neq 0$. Then we can write the Frenet formulae*

$$\begin{aligned}
\boldsymbol{R} &= \boldsymbol{R}_1 + v_0 \boldsymbol{R}_2\,, \\
\boldsymbol{R}_1' &= \kappa_1 \boldsymbol{T}_1 + \mu_1 \boldsymbol{T}_2\,, \quad \kappa_1 > 0\,, & \boldsymbol{R}_2' &= \kappa_1 \boldsymbol{T}_2\,, \\
\boldsymbol{T}_1' &= -\kappa_1 \boldsymbol{R}_1 + \kappa_2 \boldsymbol{N}_1 - \mu_1 \boldsymbol{R}_2 + \mu_2 \boldsymbol{N}_2\,, & \boldsymbol{T}_2' &= -\kappa_1 \boldsymbol{R}_2 + \kappa_2 \boldsymbol{N}_2\,, \\
\boldsymbol{N}_1' &= -\kappa_2 \boldsymbol{T}_1 - \mu_2 \boldsymbol{T}_2\,, & \boldsymbol{N}_2' &= -\kappa_2 \boldsymbol{T}_2\,,
\end{aligned}$$

where $\kappa_1, \kappa_2, \mu_1, \mu_2, v_0$ are the invariants of motion, v_0 being the parameter of instantaneous helical motion; the formulae in the right-hand column are the Frenet formulae for the fixed directing cone of the associated spherical motion, and κ_1, κ_2 are the invariants of the associated spherical motion.

Proof. Put $\boldsymbol{R}_1 = [\boldsymbol{x}; \boldsymbol{t}]$, $\boldsymbol{R}_2 = [\boldsymbol{0}; \boldsymbol{x}]$, so that $(\boldsymbol{x}, \boldsymbol{t}) = 0$, $(\boldsymbol{x}, \boldsymbol{x}) = 1$. The formulae from the second column are the Frenet formulae for the associated spherical motion, to which the same formulae for the first component correspond in the first column (i.e. the part dealing with only κ_1 and κ_2). The remainder will be proved by direct computation, and at the same time we shall derive the formulae for computing the invariants. To simplify the computation, let us introduce a new parameter s; it is the spherical arc of the centrode of the associated spherical motion. As we know from Theorem 3.51, $\kappa_1 = \mathrm{d}s/\mathrm{d}\varphi$; if $X = X(\varphi(s))$ is an arbitrary vector function, then

$$\frac{\mathrm{d}X}{\mathrm{d}s} = \frac{\mathrm{d}X}{\mathrm{d}\varphi}\frac{\mathrm{d}\varphi}{\mathrm{d}s} = \frac{1}{\kappa_1}\frac{\mathrm{d}X}{\mathrm{d}\varphi}.$$

If we differentiate with respect to s, we obtain the invariants κ_2, μ_1, μ_2 divided by the invariant κ_1. To distinguish the derivative with respect to s we denote it by a Roman numerical superscript.

By assumption on the parameter s, $(\boldsymbol{x}^{\mathrm{I}}, \boldsymbol{x}^{\mathrm{I}}) = 1$. Consequently, $(\boldsymbol{x}^{\mathrm{I}}, \boldsymbol{x}^{\mathrm{II}}) = 0$. Equation $(\boldsymbol{x}, \boldsymbol{x}) = 1$ implies $(\boldsymbol{x}, \boldsymbol{x}^{\mathrm{I}}) = 0$; further differentiation yields $(\boldsymbol{x}^{\mathrm{I}}, \boldsymbol{x}^{\mathrm{I}}) + (\boldsymbol{x}, \boldsymbol{x}^{\mathrm{II}}) = 0$, i.e. $(\boldsymbol{x}, \boldsymbol{x}^{\mathrm{II}}) = -1$. The vectors $\boldsymbol{x}, \boldsymbol{x}^{\mathrm{I}}, \boldsymbol{x} \times \boldsymbol{x}^{\mathrm{I}}$ form an orthonormal frame [the vectors $[\boldsymbol{0}; \boldsymbol{x}]$, $[\boldsymbol{0}; \boldsymbol{x}^{\mathrm{I}}]$, $[\boldsymbol{0}; \boldsymbol{x} \times \boldsymbol{x}^{\mathrm{I}}]$ are in fact the vectors $\boldsymbol{R}_2, \boldsymbol{T}_2, \boldsymbol{N}_2$]. Assume that the vector $\boldsymbol{x}^{\mathrm{II}}$ has in this frame components α, β, γ, i.e. let $\boldsymbol{x}^{\mathrm{II}} = \alpha\boldsymbol{x} + \beta\boldsymbol{x}^{\mathrm{I}} + \gamma(\boldsymbol{x} \times \boldsymbol{x}^{\mathrm{I}})$. If we take the inner product of this equation in turn with the vectors $\boldsymbol{x}, \boldsymbol{x}^{\mathrm{I}}, \boldsymbol{x}^{\mathrm{II}}$, we obtain $\alpha = -1$, $\beta = 0$, $\gamma = |\boldsymbol{x}, \boldsymbol{x}^{\mathrm{I}}, \boldsymbol{x}^{\mathrm{II}}|$. Of course, $\boldsymbol{x}^{\mathrm{II}} = (\boldsymbol{T}_2)^{\mathrm{I}}$, so that $\kappa_2/\kappa_1 = k = |\boldsymbol{x}, \boldsymbol{x}^{\mathrm{I}}, \boldsymbol{x}^{\mathrm{II}}|$ where k is the spherical curvature of the fixed centrode of the associated spherical motion, as we already know from Chapter II. Thus, let us proceed with the computations:

$$\boldsymbol{R}_1^{\mathrm{I}} = [\boldsymbol{x}^{\mathrm{I}}; \boldsymbol{t}^{\mathrm{I}}],$$

so that

$$\boldsymbol{T}_1 = [\boldsymbol{x}^{\mathrm{I}}; \boldsymbol{t}^{\mathrm{I}} - (\boldsymbol{x}^{\mathrm{I}}, \boldsymbol{t}^{\mathrm{I}})\,\boldsymbol{x}^{\mathrm{I}}],$$

$$R_1^{\rm I} = T_1 + (x^{\rm I}, t^{\rm I})\, T_2\,.$$

Then,

$$\mu_1/\kappa_1 = (x^{\rm I}, t^{\rm I})\,.$$

We now compute the second component of the vector $T_1^{\rm I} - R_1 - kN_1$; first of all, we have

$$\begin{aligned} N_1 &= [R_1, T_1] = [[x; t], [x^{\rm I}; t^{\rm I} - (x^{\rm I}, t^{\rm I})\, x^{\rm I}]] \\ &= [x \times x^{\rm I}; x \times t^{\rm I} - (x^{\rm I}, t^{\rm I})\, x \times x^{\rm I} + t \times x^{\rm I}]\,. \end{aligned}$$

Then the desired second component is

$$\begin{aligned} &t^{\rm II} - (x^{\rm II}, t^{\rm I})\, x^{\rm I} - (x^{\rm I}, t^{\rm II})\, x^{\rm I} \\ &- (x^{\rm I}, t^{\rm I})\, x^{\rm II} + t - k[x \times t^{\rm I} - (x^{\rm I}, t^{\rm I})\, x \times x^{\rm I} + t \times x^{\rm I}]\,. \end{aligned}$$

Taking its inner product in turn with the vectors x, $x^{\rm I}$, and $x \times x^{\rm I}$, we obtain:

a) $(t^{\rm II}, x) + (x^{\rm I}, t^{\rm I}) - k|x, t, x^{\rm I}| = \mu_1/\kappa_1$, where we have to use the equation $x^{\rm II} = -x + kx \times x^{\rm I}$ and the equation $(t^{\rm II}, x) + 2(t^{\rm I}, x^{\rm I}) + (t, x^{\rm II}) = 0$, which is obtained by differentiating $(x, t) = 0$ twice.

b) $(x^{\rm I}, t^{\rm II}) - (x^{\rm II}, t^{\rm I}) - (x^{\rm I}, t^{\rm II}) + (t, x^{\rm I}) - k|x, t^{\rm I}, x^{\rm I}| = 0$, where the equations $x^{\rm II} + x = kx \times x^{\rm I}$ and $(x^{\rm I}, t) + (x, t^{\rm I}) = 0$ were used.

c) $|t^{\rm II}, x, x^{\rm I}| - (x^{\rm I}, t^{\rm I})\, k + |t, x, x^{\rm I}| - |x, x^{\rm I}, x^{\rm II}|\, [(x \times t^{\rm I},\ x \times x^{\rm I}) - (x^{\rm I}, t^{\rm I}) + (t \times x^{\rm I}, x \times x^{\rm I})] = |t^{\rm II} + t, x, x^{\rm I}| - k(x^{\rm I}, t^{\rm I}) = \mu_2/\kappa_1$.

It remains to determine the second component of vector $N_1^{\rm I} + kT_1$. This component is equal to

$$\begin{aligned} &x^{\rm I} \times t^{\rm I} + x \times t^{\rm II} - (x^{\rm II}, t^{\rm I})\, x \times x^{\rm I} - (x^{\rm I}, t^{\rm II})\, x \times x^{\rm I} \\ &- (x^{\rm I}, t^{\rm I})\, x \times x^{\rm II} + t^{\rm I} \times x^{\rm I} + t \times x^{\rm II} + kt^{\rm I} - k(x^{\rm I}, t^{\rm I})\, x^{\rm I}\,. \end{aligned}$$

Take again its inner product in turn with x, $x^{\rm I}$, and $x \times x^{\rm I}$:

a) $|x, x^{\rm I}, t^{\rm I}| + |x, t^{\rm I}, x^{\rm I}| + |x, t, x^{\rm II}| + k(t^{\rm I}, x) = k(x \times t, x \times x^{\rm I}) + k(t^{\rm I}, x) = k[(t^{\rm I}, x) + (x, t^{\rm I})] = 0\,,$

b) $|x^{\rm I}, x, t^{\rm II}| + (x^{\rm I}, t^{\rm I})\, k + |t, x^{\rm II}, x^{\rm I}| + k(x^{\rm I}, t^{\rm I}) - k(x^{\rm I}, t^{\rm I}) = -\mu_2/\kappa_1\,,$

c) $(x^{\mathrm{I}} \times t^{\mathrm{I}}, x \times x^{\mathrm{I}}) + (x \times t^{\mathrm{II}}, x \times x^{\mathrm{I}}) - (x^{\mathrm{II}}, t^{\mathrm{I}}) - (x^{\mathrm{I}}, t^{\mathrm{II}})$

$$
\begin{aligned}
&- (x^{\mathrm{I}}, t^{\mathrm{I}})(x \times x^{\mathrm{II}}, x \times x^{\mathrm{I}}) + (t^{\mathrm{I}} \times x^{\mathrm{I}}, x \times x^{\mathrm{I}}) \\
&+ (t \times x^{\mathrm{II}}, x \times x^{\mathrm{I}}) + k|t^{\mathrm{I}}, x, x^{\mathrm{I}}| \\
&= -(t^{\mathrm{I}}, x) + (t^{\mathrm{II}}, x^{\mathrm{I}}) - (x^{\mathrm{II}}, t^{\mathrm{I}}) - (x^{\mathrm{I}}, t^{\mathrm{II}}) + (t^{\mathrm{I}}, x) \\
&- (t, x^{\mathrm{I}})(x, x^{\mathrm{II}}) + k|t^{\mathrm{I}}, x, x^{\mathrm{I}}| = (x - kx \times x^{\mathrm{I}}, t^{\mathrm{I}}) \\
&- (t, x^{\mathrm{I}})(x, -x + kx \times x^{\mathrm{I}}) + k|t^{\mathrm{I}}, x, x^{\mathrm{I}}| = 0\,.
\end{aligned}
$$

The theorem is thus proved. Let us write down the formulae obtained:

$$
\kappa_1 = \frac{\mathrm{d}s}{\mathrm{d}\varphi}, \quad \kappa_2 = \kappa_1 |x, x^{\mathrm{I}}, x^{\mathrm{II}}|, \quad \mu_1 = \kappa_1 (x^{\mathrm{I}}, t^{\mathrm{I}}),
$$

$$
\mu_2 = \kappa_1 \left[|t^{\mathrm{II}} + t, x, x^{\mathrm{I}}| - |x, x^{\mathrm{I}}, x^{\mathrm{II}}| (x^{\mathrm{I}}, t^{\mathrm{I}}) \right].
$$

5.2 Theorem. *Let $g(\varphi)$ be a motion with the canonical parameter φ and the fixed directing cone $R(\varphi)$. Put $R(\varphi) = R_1(\varphi) + v_0(\varphi) R_2(\varphi)$; denote $R_1(\varphi) = [x; t]$. Then the following formulae hold for the invariants of motion:*

$$
\kappa_1 = (x', x')^{1/2}, \quad \kappa_2 = \kappa_1^{-2} |x, x', x''|, \quad \mu_1 = \kappa_1^{-1} (x', t'),
$$

$$
\mu_2 = |\kappa_1^{-2} t'' - \kappa_1^{-3} \kappa_1' t' + t, x, x'| - \kappa_1^{-1} \kappa_2 \mu_1\,.
$$

Proof. We have

$$
\frac{\mathrm{d}x}{\mathrm{d}s} = \frac{\mathrm{d}x}{\mathrm{d}\varphi} \frac{\mathrm{d}\varphi}{\mathrm{d}s}.
$$

This implies that $x^{\mathrm{I}} = x' \kappa_1^{-1}$, i.e. $\kappa_1 = (x', x')^{1/2}$.

Further differentiation yields

$$
\frac{\mathrm{d}^2 x}{\mathrm{d}s^2} = \frac{1}{\kappa_1} \frac{\mathrm{d}}{\mathrm{d}\varphi} \left(\frac{x'}{\kappa_1} \right) = \frac{1}{\kappa_1} \left(\frac{\mathrm{d}^2 x}{\mathrm{d}\varphi^2} \frac{1}{\kappa_1} - \frac{\mathrm{d}x}{\mathrm{d}\varphi} \frac{\kappa_1'}{\kappa_1^2} \right),
$$

so that

$$
x^{\mathrm{II}} = \kappa_1^{-2} x'' - x' \kappa_1^{-3} \kappa_1'\,.
$$

Upon substitution into the formulae from the proof of Theorem 5.1 we have

$$
\kappa_2 = \kappa_1 |x, x^{\mathrm{I}}, x^{\mathrm{II}}| = \kappa_1 |x, \kappa_1^{-1} x', \kappa_1^{-2} x'' - x' \kappa_1^{-3} \kappa_1'| = \kappa_1^{-2} |x, x', x''|,
$$

as we already know from Chapter II. Further,

$$\begin{aligned}\mu_1 &= \kappa_1(x^{\mathrm{I}}, t^{\mathrm{I}}) = \kappa_1(\kappa_1^{-1}x', \kappa_1^{-1}t') = \kappa_1^{-1}(x', t'),\\ \mu_2 &= \kappa_1\left|t^{\mathrm{II}} + t, x, x^{\mathrm{I}}\right| - \kappa_1^{-1}\kappa_2\mu_1\\ &= \kappa_1\left|\kappa_1^{-2}t'' - \kappa_1^{-3}\kappa_1' t' + t, x, \kappa_1^{-1}x'\right| - \kappa_1^{-1}\kappa_2\mu_1\\ &= \left|\kappa_1^{-2}t'' - \kappa_1^{-3}\kappa_1' t' + t, x, x'\right| - \kappa_1^{-1}\kappa_2\mu_1\\ &= \left|\kappa_1^{-2}t'' - \kappa_1^{-4}(x', x'')\, t' + t, x, x'\right| - \kappa_1^{-1}\kappa_2\mu_1 .\end{aligned}$$

5.3 Exercise. Prove that

$$\mu_2 = |x, x', t| - \left(\kappa_1^{-2}|x, t', x'|\right)' .$$

Hint. Express the vector x'' as a linear combination of the vectors $x, x', x \times x'$.

5.4 Example. Let us again consider the motion from Example 4.6. Find out how should the functions $m_1(\alpha)$, $m_2(\alpha)$, $m_3(\alpha)$ be chosen so that all the instantaneous motions be rotations. For this it is necessary and sufficient that $\mathscr{K}(R, R) = 0$. Since the directing function is proportional to the directing cone, this condition is equivalent to the condition $\mathscr{K}(r, r) = 0$. If we write $r(\alpha) = [x(\alpha); t(\alpha)]$, then this means that $(x(\alpha), t(\alpha)) = 0$, where

$$r(\alpha) = \begin{bmatrix} \sin\alpha; & m_1' - m_2 - m_3\cos\alpha \\ \cos\alpha; & m_2' + m_1 + m_3\sin\alpha \\ -1; & m_3' + m_1\cos\alpha - m_2\sin\alpha \end{bmatrix}.$$

Therefore, it must be

$$\begin{aligned}&\sin\alpha(m_1' - m_2 - m_3\cos\alpha) + \cos\alpha(m_2' + m_1 + m_3\sin\alpha)\\ &- (m_3' + m_1\cos\alpha - m_2\sin\alpha) = m_1'\sin\alpha + m_2'\cos\alpha - m_3' = 0 ,\end{aligned}$$

so that

$$m_3' = m_1'\sin\alpha + m_2'\cos\alpha , \quad \text{i.e.} \quad m_3 = \int (m_1'\sin\alpha + m_2'\cos\alpha)\, d\alpha .$$

Thus, the functions m_1 and m_2 may be chosen arbitrarily, and then

$$m_3 = \int (m_1'\sin\alpha + m_2'\cos\alpha)\, d\alpha .$$

5.5 Example. Let us find the invariants of the motion from Example 4.7. In this case, the directing cone as a function of the canonical parameter is given as follows:

$$\boldsymbol{R}(\varphi) \equiv (\boldsymbol{x}_0; \boldsymbol{t}_0) = \frac{\sqrt{2}}{2}\begin{bmatrix} \sin\dfrac{\varphi}{\sqrt{2}}; & -q\cos\dfrac{\varphi}{\sqrt{2}}\left(1+\sin\dfrac{\varphi}{\sqrt{2}}\right) \\ \cos\dfrac{\varphi}{\sqrt{2}}; & q\sin\dfrac{\varphi}{\sqrt{2}}\left(\sin\dfrac{\varphi}{\sqrt{2}}-1\right) \\ -1; & q\cos\dfrac{\varphi}{\sqrt{2}}\left(1-\sin\dfrac{\varphi}{\sqrt{2}}\right)\end{bmatrix}.$$

To be able to use the formulae from Theorem 5.2, we must first determine the vector $\boldsymbol{R}_1(\varphi)$. We obtain

$$\begin{aligned} v_0 &= (\boldsymbol{x}_0, \boldsymbol{t}_0) \\ &= \frac{1}{2}\left[-q\cos\frac{\varphi}{\sqrt{2}}\left(1+\sin\frac{\varphi}{\sqrt{2}}\right)\sin\frac{\varphi}{\sqrt{2}}\right. \\ &\quad \left. + q\cos\frac{\varphi}{\sqrt{2}}\sin\frac{\varphi}{\sqrt{2}}\left(\sin\frac{\varphi}{\sqrt{2}}-1\right) - q\cos\frac{\varphi}{\sqrt{2}}\left(1-\sin\frac{\varphi}{\sqrt{2}}\right)\right] \\ &= \frac{q}{2}\cos\frac{\varphi}{\sqrt{2}}\left(1+\sin\frac{\varphi}{\sqrt{2}}\right). \end{aligned}$$

Also, $\boldsymbol{R}_1 = \boldsymbol{R} - v_0\boldsymbol{R}_2$, i.e.

$$\boldsymbol{R}_1 = \frac{\sqrt{2}}{2}\begin{bmatrix} \sin\dfrac{\varphi}{\sqrt{2}}; & \dfrac{q}{2}\cos\dfrac{\varphi}{\sqrt{2}}\left(\sin^2\dfrac{\varphi}{\sqrt{2}} - \sin\dfrac{\varphi}{\sqrt{2}} - 2\right) \\ \cos\dfrac{\varphi}{\sqrt{2}}; & \dfrac{q}{2}\left(1+\sin^2\dfrac{\varphi}{\sqrt{2}} - 2\sin\dfrac{\varphi}{\sqrt{2}} + \cos^2\dfrac{\varphi}{\sqrt{2}}\sin\dfrac{\varphi}{\sqrt{2}}\right) \\ -1; & \dfrac{q}{2}\left(\cos\dfrac{\varphi}{\sqrt{2}} - 3\cos\dfrac{\varphi}{\sqrt{2}}\sin\dfrac{\varphi}{\sqrt{2}}\right)\end{bmatrix}.$$

Put $\boldsymbol{R}_1 = [\boldsymbol{x}; \boldsymbol{t}]$. Differentiating we arrive at

$$\boldsymbol{x} = \frac{\sqrt{2}}{2}\begin{bmatrix} \sin\dfrac{\varphi}{\sqrt{2}} \\ \cos\dfrac{\varphi}{\sqrt{2}} \\ -1 \end{bmatrix}, \quad \boldsymbol{x}' = \frac{1}{2}\begin{bmatrix} \cos\dfrac{\varphi}{\sqrt{2}} \\ -\sin\dfrac{\varphi}{\sqrt{2}} \\ 0 \end{bmatrix}, \quad \boldsymbol{x}'' = \frac{1}{2\sqrt{2}}\begin{bmatrix} -\sin\dfrac{\varphi}{\sqrt{2}} \\ -\cos\dfrac{\varphi}{\sqrt{2}} \\ 0 \end{bmatrix}.$$

$$\boldsymbol{t} = \frac{q}{2\sqrt{2}}\begin{bmatrix} \cos\dfrac{\varphi}{\sqrt{2}}\left(\sin^2\dfrac{\varphi}{\sqrt{2}} - \sin\dfrac{\varphi}{\sqrt{2}} - 2\right) \\ 1 + \sin^2\dfrac{\varphi}{\sqrt{2}} - 2\sin\dfrac{\varphi}{\sqrt{2}} + \cos^2\dfrac{\varphi}{\sqrt{2}}\sin\dfrac{\varphi}{\sqrt{2}} \\ \cos\dfrac{\varphi}{\sqrt{2}} - 3\cos\dfrac{\varphi}{\sqrt{2}}\sin\dfrac{\varphi}{\sqrt{2}} \end{bmatrix},$$

$$\boldsymbol{t}' = \frac{q}{4}\begin{bmatrix} -1 + 2\sin^2\dfrac{\varphi}{\sqrt{2}} + 4\sin\dfrac{\varphi}{\sqrt{2}} - 3\sin^3\dfrac{\varphi}{\sqrt{2}} \\ 2\sin\dfrac{\varphi}{\sqrt{2}}\cos\dfrac{\varphi}{\sqrt{2}} - \cos\dfrac{\varphi}{\sqrt{2}} - 3\cos\dfrac{\varphi}{\sqrt{2}}\sin^2\dfrac{\varphi}{\sqrt{2}} \\ 6\sin^2\dfrac{\varphi}{\sqrt{2}} - \sin\dfrac{\varphi}{\sqrt{2}} - 3 \end{bmatrix},$$

$$\boldsymbol{t}'' = \frac{q}{4\sqrt{2}}\begin{bmatrix} 4\sin\dfrac{\varphi}{\sqrt{2}}\cos\dfrac{\varphi}{\sqrt{2}} + 4\cos\dfrac{\varphi}{\sqrt{2}} - 9\sin^2\dfrac{\varphi}{\sqrt{2}}\cos\dfrac{\varphi}{\sqrt{2}} \\ -2\sin^2\dfrac{\varphi}{\sqrt{2}} + 2\cos^2\dfrac{\varphi}{\sqrt{2}} + \sin\dfrac{\varphi}{\sqrt{2}} + 3\sin^3\dfrac{\varphi}{\sqrt{2}} - 6\sin\dfrac{\varphi}{\sqrt{2}}\cos^2\dfrac{\varphi}{\sqrt{2}} \\ 12\sin\dfrac{\varphi}{\sqrt{2}}\cos\dfrac{\varphi}{\sqrt{2}} - \cos\dfrac{\varphi}{\sqrt{2}} \end{bmatrix}.$$

Since we shall differentiate no more in the course of further computation, we write again $\varphi/\sqrt{2} = \alpha$. According to Theorem 5.2 we obtain

$$\kappa_1 = \frac{1}{2}(\cos^2\alpha + \sin^2\alpha)^{1/2} = \frac{1}{2},$$

$$\kappa_2 = 4 \begin{vmatrix} \sin\alpha, & \cos\alpha, & -\sin\alpha \\ \cos\alpha, & -\sin\alpha, & -\cos\alpha \\ -1, & 0, & 0 \end{vmatrix} \cdot \frac{\sqrt{2}}{2} \cdot \frac{1}{2} \cdot \frac{1}{2\sqrt{2}} = \frac{1}{2}$$

(see II, 3.45). Further,

$$\mu_1 = 2 \cdot \frac{1}{2} \cdot \frac{q}{4} \left[\cos\alpha(-1 + 2\sin^2\alpha + 4\sin\alpha - 3\sin^3\alpha)\right.$$
$$\left. - \sin\alpha(2\sin\alpha\cos\alpha - \cos\alpha - 3\cos\alpha\sin^2\alpha)\right]$$
$$= \frac{q}{4}(-\cos\alpha + 5\sin\alpha\cos\alpha),$$

$$\mu_2 = -\mu_1 + \frac{q}{2\sqrt{2}} \cdot \frac{\sqrt{2}}{2} \cdot \frac{1}{2} \cdot$$
$$\begin{vmatrix} 8\sin\alpha\cos\alpha + 8\cos\alpha - 18\sin^2\alpha\cos\alpha + \\ \quad + \cos\alpha\sin^2\alpha - \cos\alpha\sin\alpha - 2\cos\alpha, & \sin\alpha, & \cos\alpha \\ 4\cos^2\alpha - 4\sin^2\alpha + 2\sin\alpha + 6\sin^3\alpha - 12\cos^2\alpha\sin\alpha + \\ \quad + 1 + \sin^2\alpha - 2\sin\alpha + \cos^2\alpha\sin\alpha, & \cos\alpha, & -\sin\alpha \\ -2\cos\alpha + 24\sin\alpha\cos\alpha + \cos\alpha - 3\cos\alpha\sin\alpha, & -1, & 0 \end{vmatrix}$$

$$= -\mu_1 + \frac{q}{8} \cdot$$
$$\begin{vmatrix} 6\cos\alpha - 17\sin^2\alpha\cos\alpha + 7\sin\alpha\cos\alpha, & \sin\alpha, & \cos\alpha \\ 1 + 4\cos^2\alpha - 3\sin^2\alpha - 11\cos^2\alpha\sin\alpha + 6\sin^3\alpha, & \cos\alpha, & -\sin\alpha \\ -\cos\alpha + 21\sin\alpha\cos\alpha, & -1, & 0 \end{vmatrix}$$

$$= \frac{q}{8}\{(-\cos\alpha + 21\sin\alpha\cos\alpha)(-\sin^2\alpha - \cos^2\alpha)$$
$$- (-1)[(-\sin\alpha)(6\cos\alpha - 17\sin^2\alpha\cos\alpha + 7\sin\alpha\cos\alpha)$$
$$- \cos\alpha(1 + 4\cos^2\alpha - 3\sin^2\alpha - 11\cos^2\alpha\sin\alpha + 6\sin^3\alpha)]\} - \mu_1$$
$$= -\frac{q}{4}(2\cos\alpha + 8\sin\alpha\cos\alpha) - \frac{q}{4}(-\cos\alpha + 5\sin\alpha\cos\alpha)$$
$$= -\frac{q}{4}(\cos\alpha + 13\sin\alpha\cos\alpha).$$

5.6 Theorem. *Let $\boldsymbol{R}(\varphi)$ be the fixed directing cone of motion with the canonical parameter φ. Put $\boldsymbol{R}(\varphi) = [\boldsymbol{x}_0; \boldsymbol{t}_0]$. Then*

$$\begin{aligned} v_0 &= (\boldsymbol{x}_0, \boldsymbol{t}_0)\,, \\ \kappa_1 &= (\boldsymbol{x}_0', \boldsymbol{x}_0')^{1/2}\,, \quad \kappa_2 = \kappa_1^{-2}|\boldsymbol{x}_0, \boldsymbol{x}_0', \boldsymbol{x}_0''|\,, \\ \mu_1 &= \kappa_1^{-1}(\boldsymbol{x}_0', \boldsymbol{t}_0') - v_0\kappa_1\,, \\ \mu_2 &= |\kappa_1^{-2}\boldsymbol{t}_0'' - \kappa_1^{-3}\kappa_1'\boldsymbol{t}_0' + \boldsymbol{t}_0, \boldsymbol{x}_0, \boldsymbol{x}_0'| - \kappa_1^{-2}(\boldsymbol{x}_0', \boldsymbol{t}_0')\,\kappa_2\,. \end{aligned}$$

Proof. If $\boldsymbol{R}(\varphi) = [\boldsymbol{x}_0; \boldsymbol{t}_0]$, then $\boldsymbol{R}_1 = (\boldsymbol{x}_0; \boldsymbol{t}_0 - (\boldsymbol{x}_0, \boldsymbol{t}_0)\,\boldsymbol{x}_0)$, so that for $\boldsymbol{R}_1 = [\boldsymbol{x}; \boldsymbol{t}]$ we obtain

$$\boldsymbol{x} = \boldsymbol{x}_0\,, \quad \boldsymbol{t} = \boldsymbol{t}_0 - (\boldsymbol{x}_0, \boldsymbol{t}_0)\,\boldsymbol{x}_0\,.$$

Differentiating we arrive at

$$\begin{aligned} \boldsymbol{t}' &= \boldsymbol{t}_0' - (\boldsymbol{x}_0', \boldsymbol{t}_0)\,\boldsymbol{x}_0 - (\boldsymbol{x}_0, \boldsymbol{t}_0')\,\boldsymbol{x}_0 - (\boldsymbol{x}_0, \boldsymbol{t}_0)\,\boldsymbol{x}_0'\,, \\ \boldsymbol{t}'' &= \boldsymbol{t}_0'' - 2(\boldsymbol{x}_0', \boldsymbol{t}_0')\,\boldsymbol{x}_0 - (\boldsymbol{x}_0, \boldsymbol{t}_0'')\,\boldsymbol{x}_0 - 2(\boldsymbol{x}_0, \boldsymbol{t}_0')\,\boldsymbol{x}_0' - (\boldsymbol{x}_0'', \boldsymbol{t}_0)\,\boldsymbol{x}_0 \\ &\quad - 2(\boldsymbol{x}_0', \boldsymbol{t}_0)\,\boldsymbol{x}_0' - (\boldsymbol{x}_0, \boldsymbol{t}_0)\,\boldsymbol{x}_0''\,, \end{aligned}$$

so that upon substitution into the formulae of Theorem 5.2 we have

$$\begin{aligned} \kappa_1 &= (\boldsymbol{x}_0', \boldsymbol{x}_0')^{1/2}\,, \quad \kappa_4 = \kappa_1^{-2}|\boldsymbol{x}_0, \boldsymbol{x}_0', \boldsymbol{x}_0''|\,, \\ \mu_1 &= \kappa_1^{-1}(\boldsymbol{x}_0', \boldsymbol{t}_0' - (\boldsymbol{x}_0', \boldsymbol{t}_0)\,\boldsymbol{x}_0 - (\boldsymbol{x}_0, \boldsymbol{t}_0')\,\boldsymbol{x}_0 - (\boldsymbol{x}_0, \boldsymbol{t}_0)\,\boldsymbol{x}_0') \\ &= \kappa_1^{-1}[(\boldsymbol{x}_0', \boldsymbol{t}_0') - (\boldsymbol{x}_0, \boldsymbol{t}_0)\,(\boldsymbol{x}_0', \boldsymbol{x}_0')] = \kappa_1^{-1}[(\boldsymbol{x}_0', \boldsymbol{t}_0') - v_0\kappa_1^2] \\ &= \kappa_1^{-1}(\boldsymbol{x}_0', \boldsymbol{t}_0') - v_0\kappa_1\,. \end{aligned}$$

Further,

$$\begin{aligned} \mu_2 &= |\kappa_1^{-2}[\boldsymbol{t}_0'' - 2(\boldsymbol{x}_0', \boldsymbol{t}_0')\,\boldsymbol{x}_0 - (\boldsymbol{x}_0, \boldsymbol{t}_0'')\,\boldsymbol{x}_0 - 2(\boldsymbol{x}_0, \boldsymbol{t}_0')\,\boldsymbol{x}_0' \\ &\quad - (\boldsymbol{x}_0'', \boldsymbol{t}_0)\,\boldsymbol{x}_0 - 2(\boldsymbol{x}_0', \boldsymbol{t}_0)\,\boldsymbol{x}_0' - (\boldsymbol{x}_0, \boldsymbol{t}_0)\,\boldsymbol{x}_0''] \\ &\quad - \kappa_1^{-3}\kappa_1'[\boldsymbol{t}_0' - (\boldsymbol{x}_0', \boldsymbol{t}_0)\,\boldsymbol{x}_0 - (\boldsymbol{x}_0, \boldsymbol{t}_0')\,\boldsymbol{x}_0 - (\boldsymbol{x}_0, \boldsymbol{t}_0)\,\boldsymbol{x}_0'] \\ &\quad + |\boldsymbol{t}_0, \boldsymbol{x}_0, \boldsymbol{x}_0'| - \kappa_1^{-1}\kappa_1^{-2}|\boldsymbol{x}_0, \boldsymbol{x}_0', \boldsymbol{x}_0''|\,[\kappa_1^{-1}(\boldsymbol{x}_0', \boldsymbol{t}_0') - v_0\kappa_1] \\ &= |\kappa_1^{-2}\boldsymbol{t}_0'' - \kappa_1^{-3}\kappa_1'\boldsymbol{t}_0' + \boldsymbol{t}_0, \boldsymbol{x}_0, \boldsymbol{x}_0'| - \kappa_1^{-2}v_0|\boldsymbol{x}_0'', \boldsymbol{x}_0, \boldsymbol{x}_0'| \\ &\quad + \kappa_1^{-2}v_0|\boldsymbol{x}_0, \boldsymbol{x}_0', \boldsymbol{x}_0''| - \kappa_1^{-2}\kappa_2(\boldsymbol{x}_0', \boldsymbol{t}_0') \\ &= |\kappa_1^{-2}\boldsymbol{t}_0'' - \kappa_1^{-3}\kappa_1'\boldsymbol{t}_0' + \boldsymbol{t}_0, \boldsymbol{x}_0, \boldsymbol{x}_0'| - \kappa_1^{-2}\kappa_2(\boldsymbol{x}_0', \boldsymbol{t}_0')\,. \end{aligned}$$

5.7 Exercise. Determine the invariants from Example 5.5 using the formula of Theorem 5.6.

5.8 Theorem. *Let $\boldsymbol{R}(\mathrm{t}) = [\boldsymbol{x}_0; \boldsymbol{t}_0]$ be the fixed directing cone of given motion $g(\mathrm{t})$, where* t *is an arbitrary parameter. Then the following formulae hold for the invariants of motion:*

$$\frac{\mathrm{d}\varphi}{\mathrm{dt}} = [\mathrm{K}(\boldsymbol{r}, \boldsymbol{r})]^{1/2},$$

where $\boldsymbol{r}$ *is the fixed directing function, i.e.*

$$\boldsymbol{r} = \frac{\mathrm{d}g}{\mathrm{dt}} g^{-1}, \quad v_0 = (\boldsymbol{x}_0, \boldsymbol{t}_0),$$

$$\kappa_1 = (\dot{\varphi})^{-1} (\dot{\boldsymbol{x}}_0, \dot{\boldsymbol{x}}_0)^{1/2}, \quad \kappa_2 = \kappa_1^{-2}(\dot{\varphi})^{-3} |\boldsymbol{x}_0, \dot{\boldsymbol{x}}_0, \ddot{\boldsymbol{x}}_0|,$$

$$\mu_1 = \kappa_1^{-1}(\dot{\varphi})^{-2} (\dot{\boldsymbol{x}}_0, \dot{\boldsymbol{t}}_0) - v_0\kappa_1,$$

$$\mu_2 = (\dot{\varphi})^{-1} |\kappa_1^{-2}(\dot{\varphi})^{-2} \ddot{\boldsymbol{t}}_0 - \kappa_1^{-2}\dot{\boldsymbol{t}}_0 \ddot{\varphi}(\dot{\varphi})^{-3} - \kappa_1^{-3} \dot{\kappa}_1(\dot{\varphi})^{-2} \dot{\boldsymbol{t}}_0 + \boldsymbol{t}_0, \boldsymbol{x}_0, \dot{\boldsymbol{x}}_0| - \kappa_1^{-2}(\dot{\varphi})^{-2} (\dot{\boldsymbol{x}}_0, \dot{\boldsymbol{t}}_0) \kappa_2,$$

where φ is the canonical parameter and the dot indicates differentiation with respect to t.

Proof.

$$\frac{\mathrm{d}g}{\mathrm{dt}} g^{-1} = \boldsymbol{r} = \frac{\mathrm{d}g}{\mathrm{d}\varphi}\frac{\mathrm{d}\varphi}{\mathrm{dt}} g^{-1} = \dot{\varphi}\frac{\mathrm{d}g}{\mathrm{d}\varphi} g^{-1} = \dot{\varphi}\boldsymbol{R},$$

so that

$$K(\boldsymbol{r}, \boldsymbol{r}) = (\dot{\varphi})^2 K(\boldsymbol{R}, \boldsymbol{R}) = (\dot{\varphi})^2, \quad \text{i.e.} \quad \dot{\varphi} = [K(\boldsymbol{r}, \boldsymbol{r})]^{1/2}.$$

Further,

$$\dot{\boldsymbol{x}}_0 = \boldsymbol{x}_0'\dot{\varphi}, \qquad \text{i.e.} \quad \boldsymbol{x}_0' = \dot{\boldsymbol{x}}_0(\dot{\varphi})^{-1};$$

$$\ddot{\boldsymbol{x}}_0 = \boldsymbol{x}_0''(\dot{\varphi})^2 + \boldsymbol{x}_0'\ddot{\varphi}, \quad \text{i.e.} \quad \boldsymbol{x}_0'' = \ddot{\boldsymbol{x}}_0(\dot{\varphi})^{-2} - \dot{\boldsymbol{x}}_0 \ddot{\varphi}(\dot{\varphi})^{-3};$$

analogous relations hold also for $\boldsymbol{t}_0$. Upon substitution we obtain

$$\kappa_1 = (\boldsymbol{x}_0', \boldsymbol{x}_0')^{1/2} = [(\dot{\varphi})^{-2} (\dot{\boldsymbol{x}}_0, \dot{\boldsymbol{x}}_0)]^{1/2} = (\dot{\varphi})^{-1} (\dot{\boldsymbol{x}}_0, \dot{\boldsymbol{x}}_0)^{1/2},$$

$$\begin{aligned}
\kappa_2 &= \kappa_1^{-2}\left|x_0, x_0^{\bullet}(\dot{\varphi})^{-1}, x_0^{\bullet\bullet}(\dot{\varphi})^{-2} - x_0^{\bullet}\,\ddot{\varphi}(\dot{\varphi})^{-3}\right| \\
&= \kappa_1^{-2}(\dot{\varphi})^{-3}\left|x_0, x_0^{\bullet}, x_0^{\bullet\bullet}\right|, \\
\mu_1 &= \kappa_1^{-2}(x_0^{\bullet}(\dot{\varphi})^{-1}, t_0^{\bullet}(\dot{\varphi})^{-1}) - v_0\kappa_1 = \kappa_1^{-2}(\dot{\varphi})^{-2}(x_0^{\bullet}, t_0^{\bullet}) - v_0\kappa_1, \\
\mu_2 &= \left|\kappa_1^{-2}\, t_0^{\bullet\bullet}(\dot{\varphi})^{-2} - \kappa_1^{-2} t_0^{\bullet}\,\ddot{\varphi}(\dot{\varphi})^{-3} - \kappa_1^{-3}\,\dot{\kappa}_1(\dot{\varphi})^{-2}\, t_0^{\bullet}\right. \\
&\quad \left. + t_0, x_0, x_0^{\bullet}(\dot{\varphi})^{-1}\right| - \kappa_1^{-2}(\dot{\varphi})^{-2}(x_0^{\bullet}, t_0^{\bullet})\,\kappa_2 .
\end{aligned}$$

5.9 Theorem. *Let* $r(\mathrm{t}) = [x; t]$ *be the fixed directing function of motion* $g(\mathrm{t})$; *the following formulae hold for the invariants of motion* $g(\mathrm{t})$:

$$\begin{aligned}
\dot{\varphi} &= \frac{\mathrm{d}\varphi}{\mathrm{dt}} = (x, x)^{1/2}, \quad v_0 = (x, t)(\dot{\varphi})^{-2}, \\
\kappa_1 &= (\dot{\varphi})^{-3}\, V^{1/2}, \quad \kappa_2 = V^{-1}\left|x, x^{\bullet}, x^{\bullet\bullet}\right|, \\
\mu_1 &= (\dot{\varphi})^{-5}\, V^{-1/2}[Z - (x, t)\, V], \\
\mu_2 &= (\dot{\varphi})^{-2}\, V^{-2}\Big[V\left|x, x^{\bullet}, t^{\bullet\bullet}(\dot{\varphi})^2 - (x, x^{\bullet\bullet})\, t\right| - \tfrac{1}{2}\dot{V}\left|x, x^{\bullet}, t^{\bullet}(\dot{\varphi})^2\right. \\
&\quad \left. - (x, x^{\bullet})\, t\right| - Z\left|x, x^{\bullet}, x^{\bullet\bullet}\right|\Big],
\end{aligned}$$

where

$$\begin{aligned}
V &= [(\dot{\varphi})^2 (x^{\bullet}, x^{\bullet}) - (x, x^{\bullet})^2], \\
Z &= [(x^{\bullet}, t^{\bullet})(\dot{\varphi})^4 - (x, t)^{\bullet}(\dot{\varphi})^2 (x, x^{\bullet}) + (x, t)(x, x^{\bullet})^2].
\end{aligned}$$

Proof. We shall use Theorem 5.8. If $R(\mathrm{t}) = [x_0; t_0]$ is the directing cone of our motion, we have

$$[x_0; t_0] = \frac{\mathrm{d}g}{\mathrm{d}\varphi} g^{-1}, \quad \varphi = \varphi(\mathrm{t}),$$

$$[x; t] = \frac{\mathrm{d}g}{\mathrm{dt}} g^{-1},$$

so that

$$[x_0; t_0]\frac{\mathrm{d}\varphi}{\mathrm{dt}} = [x; t], \quad \text{i.e.} \quad x_0\dot{\varphi} = x, \quad t_0\dot{\varphi} = t .$$

We thus have

$$x_0 = (\dot{\varphi})^{-1}\, x, \quad t_0 = (\dot{\varphi})^{-1}\, t;$$

further differentiation yields

$$x_0^{\bullet} = (\dot{\varphi})^{-1} x^{\bullet} - (\dot{\varphi})^{-2} \ddot{\varphi} x\,, \quad t_0^{\bullet} = (\dot{\varphi})^{-1} t^{\bullet} - (\dot{\varphi})^{-2\prime} \ddot{\varphi} t\,;$$

also $(\dot{\varphi})^2 = (x, x)$, so that $\dot{\varphi}\ddot{\varphi} = (x, x^{\bullet})$. Therefore,

$$x_0^{\bullet} = (\dot{\varphi})^{-1} x^{\bullet} - (\dot{\varphi})^{-3} (x, x^{\bullet})\, x\,, \quad t_0^{\bullet} = (\dot{\varphi})^{-1} t^{\bullet} - (\dot{\varphi})^{-3} (x, x^{\bullet})\, t\,;$$

the second derivatives are

$$\begin{aligned} x_0^{\bullet\bullet} &= (\dot{\varphi})^{-1} x^{\bullet\bullet} - 2(\dot{\varphi})^{-3} (x, x^{\bullet})\, x^{\bullet} + 3(\dot{\varphi})^{-5} (x, x^{\bullet})^2\, x \\ &\quad - (\dot{\varphi})^{-3} (x, x^{\bullet})^{\bullet}\, x\,, \\ t_0^{\bullet\bullet} &= (\dot{\varphi})^{-1} t^{\bullet\bullet} - 2(\dot{\varphi})^{-3} (x, x^{\bullet})\, t^{\bullet} + 3(\dot{\varphi})^{-5} (x, x^{\bullet})^2\, t \\ &\quad - (\dot{\varphi})^{-3} (x, x^{\bullet})^{\bullet}\, t\,. \end{aligned}$$

For the sake of simplicity put $(x, x^{\bullet}) = \alpha$. Consequently,

$$\begin{aligned} x_0 &= (\dot{\varphi})^{-1} x\,, \quad t_0 = (\dot{\varphi})^{-1} t\,, \\ x_0^{\bullet} &= (\dot{\varphi})^{-1} x^{\bullet} - (\dot{\varphi})^{-3} \alpha x\,, \quad t_0^{\bullet} = (\dot{\varphi})^{-1} t^{\bullet} - (\dot{\varphi})^{-3} \alpha t\,, \\ x_0^{\bullet\bullet} &= (\dot{\varphi})^{-1} x^{\bullet\bullet} - 2(\dot{\varphi})^{-3} \alpha x^{\bullet} + 3(\dot{\varphi})^{-5} \alpha^2 x - (\dot{\varphi})^{-3} \dot{\alpha} x\,, \\ t_0^{\bullet\bullet} &= (\dot{\varphi})^{-1} t^{\bullet\bullet} - 2(\dot{\varphi})^{-3} \alpha t^{\bullet} + 3(\dot{\varphi})^{-5} \alpha^2 t - (\dot{\varphi})^{-3} \dot{\alpha} t\,. \end{aligned}$$

After substitution we obtain

$$v_0 = (x_0, t_0) = (\dot{\varphi})^{-2} (x, t)\,;$$

we know the formulae for κ_1 and κ_2 from Chapter 2,

$$\begin{aligned} \mu_1 &= \kappa_1^{-1} (\dot{\varphi})^{-2} (x_0^{\bullet}, t_0^{\bullet}) - v_0 \kappa_1 \\ &= \kappa_1^{-1} [(\dot{\varphi})^{-2} ((\dot{\varphi})^{-1} x^{\bullet} - (\dot{\varphi})^{-3} \alpha x, (\dot{\varphi})^{-1} t^{\bullet} - (\dot{\varphi})^{-3} \alpha t)] \\ &\quad - v_0 \kappa_1 = \kappa_1^{-1} \{(\dot{\varphi})^{-8} [(\dot{\varphi})^4 (x^{\bullet}, t^{\bullet}) - (\dot{\varphi})^2 \alpha (t, x)^{\bullet} + \alpha^2 (x, t)] \\ &\quad - (x, t) (\dot{\varphi})^{-2} (\dot{\varphi})^{-6} [(\dot{\varphi})^2 (x^{\bullet}, x^{\bullet}) - \alpha^2]\} \\ &= \kappa_1^{-1} (\dot{\varphi})^{-8} [(\dot{\varphi})^4 (x^{\bullet}, t^{\bullet}) - (\dot{\varphi})^2 \alpha (t, x)^{\bullet} - (\dot{\varphi})^2 (x^{\bullet}, x^{\bullet}) (x, t) \\ &\quad + 2\alpha^2 (x, t)] = (\dot{\varphi})^{-5} [(\dot{\varphi})^2 (x^{\bullet}, x^{\bullet}) - \alpha^2]^{-1/2} [(x^{\bullet}, t^{\bullet}) (\dot{\varphi})^4 \\ &\quad - (x, t)^{\bullet} (x, x^{\bullet}) (\dot{\varphi})^2 - (x, t) (\dot{\varphi})^2 (x^{\bullet}, x^{\bullet}) + 2(x, t) (x, x^{\bullet})^2]\,, \end{aligned}$$

which agrees. Differentiating the formula for κ_1 we obtain

$$\kappa_1^{\bullet} = -3(\dot{\varphi})^{-5}\,\alpha V^{1/2} + \tfrac{1}{2}(\dot{\varphi})^{-3}\,V^{-1/2}\dot{V}.$$

Further,

$$\begin{aligned}
\mu_2 &= (\dot{\varphi})^{-1}\,\big|\kappa_1^{-2}(\dot{\varphi})^{-2}\,t_0^{\bullet\bullet} - \kappa_1^{-2} t_0\,\ddot{\varphi}(\dot{\varphi})^{-3} - \kappa_1^{-3}\,\dot{\kappa}_1(\dot{\varphi})^{-2}\,t_0^{\bullet} \\
&\quad + t_0, x_0, x_0^{\bullet}\big| - \kappa_1^{-2}(\dot{\varphi})^{-2}\,(x_0^{\bullet}, t_0^{\bullet})\,\kappa_2 \\
&= (\dot{\varphi})^{-3}\,\kappa_1^{-2}\big[\big|(\dot{\varphi})^{-2}\,t_0^{\bullet\bullet} - t_0^{\bullet}\,\alpha(\dot{\varphi})^{-4} - \kappa_1^{-1}\,\dot{\kappa}_1(\dot{\varphi})^{-2}\,t_0^{\bullet} \\
&\quad + t_0\kappa_1^2, x, x^{\bullet}\big| - \dot{\varphi}(x_0^{\bullet}, t_0^{\bullet})\,\kappa_2\big] \\
&= (\dot{\varphi})^{-7}\,\kappa_1^{-2}\big[\big|x, x^{\bullet}, (\dot{\varphi})^2\,t_0^{\bullet\bullet} - t_0^{\bullet}\alpha + t_0(\dot{\varphi})^4\,\kappa_1^2 - \kappa_1^{-1}\,\dot{\kappa}_1(\dot{\varphi})^2\,t_0^{\bullet}\big| \\
&\quad - (\dot{\varphi})^5\,(x_0^{\bullet}, t_0^{\bullet})\,\kappa_2\big] \\
&= (\dot{\varphi})^{-7}\,\kappa_1^{-2}\{\big|x, x^{\bullet}, \dot{\varphi}t^{\bullet\bullet} - 2(\dot{\varphi})^{-1}\,\alpha t^{\bullet} + 3(\dot{\varphi})^{-3}\,\alpha^2 t - (\dot{\varphi})^{-1}\,. \\
&\quad t[(x^{\bullet}, x^{\bullet}) + (x, x^{\bullet\bullet})] - (\dot{\varphi})^{-1}\,t^{\bullet}\alpha + (\dot{\varphi})^{-3}\,\alpha^2 t \\
&\quad + t(\dot{\varphi})^{-3}\,[(\dot{\varphi})^2\,(x^{\bullet}, x^{\bullet}) - \alpha^2] \\
&\quad - (\dot{\varphi})^5\,V^{-1/2}[(\dot{\varphi})^{-1}\,t^{\bullet} - (\dot{\varphi})^{-3}\,\alpha t]\,. \\
&\quad [-3(\dot{\varphi})^{-5}\,\alpha V^{1/2} + \tfrac{1}{2}(\dot{\varphi})^{-3}\,V^{-1/2}\dot{V}]\big| - (\dot{\varphi})^5\,(x_0^{\bullet}, t_0^{\bullet})\,\kappa_2\} \\
&= (\dot{\varphi})^{-7}\,\kappa_1^{-2}\{\big|x, x^{\bullet}, \dot{\varphi}t^{\bullet\bullet} - (\dot{\varphi})^{-1}\,(x, x^{\bullet\bullet})\,t \\
&\quad - \tfrac{1}{2}(\dot{\varphi})^2\,V^{-1}\dot{V}[(\dot{\varphi})^{-1}\,t^{\bullet} - (\dot{\varphi})^{-3}\,\alpha t]\big| - (\dot{\varphi})^5\,(x_0^{\bullet}, t_0^{\bullet})\,\kappa_2\} \\
&= (\dot{\varphi})^{-8}\,\kappa_1^{-2}\big|x, x^{\bullet}, (\dot{\varphi})^2\,t^{\bullet\bullet} - (x, x^{\bullet\bullet})\,t\,\tfrac{1}{2}V^{-1}\dot{V}[t^{\bullet}(\dot{\varphi})^2 - \alpha t]\big| \\
&\quad - (\dot{\varphi})^6\,V^{-1}\big|x, x^{\bullet}, x^{\bullet\bullet}\big|\,(x^{\bullet}(\dot{\varphi})^{-1} \\
&\quad - (\dot{\varphi})^{-3}\,\alpha x, t^{\bullet}(\dot{\varphi})^{-1} - (\dot{\varphi})^{-3}\,\alpha t) \\
&= (\dot{\varphi})^{-8}\,(\dot{\varphi})^6\,V^{-1}\big|x, x^{\bullet}, (\dot{\varphi})^2\,t^{\bullet\bullet} - (x, x^{\bullet\bullet})\,t - \tfrac{1}{2}V^{-1}\dot{V}[t(\dot{\varphi})^2 - \alpha t]\big| \\
&\quad - (\dot{\varphi})^6\,V^{-1}\big|x, x^{\bullet}, x^{\bullet\bullet}\big|\,[(x^{\bullet}, x^{\bullet})\,(\dot{\varphi})^{-2} - \alpha(x, t)^{\bullet}\,(\dot{\varphi})^{-4} \\
&\quad + (\dot{\varphi})^{-6}\,\alpha^2(x, t)] = (\dot{\varphi})^{-2}\,V^{-2}[V\big|x, x^{\bullet}, (\dot{\varphi})^2\,t^{\bullet\bullet} - (x, x^{\bullet\bullet})\,t\big| \\
&\quad - \tfrac{1}{2}\dot{V}\big|x, x^{\bullet}_{.}, (\dot{\varphi})^2\,t^{\bullet} - \alpha t\big| - Z\big|x, x^{\bullet}, x^{\bullet\bullet}\big|]\,,
\end{aligned}$$

which is what we had to prove.

5.10 Exercise. Determine the invariants of motion from Example 5.5 using the formulae of Theorem 5.9.

5.11 Example. We shall determine the invariants of motion from Example 4.8 using the formulae of Theorem 5.9. First, we recall the formulae of Example 4.8:

$$\boldsymbol{r}(\mathrm{t}) = \begin{bmatrix} -1; & M^2N^{\boldsymbol{\cdot}}m_1 d \sin \mathrm{t} \cos \mathrm{t} \\ -M^2N^{\boldsymbol{\cdot}}m_1 \cos \mathrm{t}; & -d \sin \mathrm{t} \\ M^2N^{\boldsymbol{\cdot}}m_1 \sin \mathrm{t}; & d \cos \mathrm{t} \end{bmatrix},$$

where

$$N = m_3 \cos \mathrm{t} - d \cos \mathrm{t} \sin \mathrm{t} + m_2 \sin \mathrm{t}\,, \quad M = (m_1^2 + N^2)^{-1/2}\,.$$

Put $A = M^2Nm_1$. Then

$$\boldsymbol{r}(\mathrm{t}) = [\boldsymbol{x}(\mathrm{t}); \boldsymbol{t}(\mathrm{t})] = \begin{bmatrix} -1; & Ad \sin \mathrm{t} \cos \mathrm{t} \\ -A \cos \mathrm{t}; & -d \sin \mathrm{t} \\ A \sin \mathrm{t}; & d \cos \mathrm{t} \end{bmatrix},$$

$$\boldsymbol{r}^{\boldsymbol{\cdot}}(\mathrm{t}) = \begin{bmatrix} 0; & \dot{A}d \sin \mathrm{t} \cos \mathrm{t} + Ad \cos 2\mathrm{t} \\ -\dot{A} \cos \mathrm{t} + A \sin \mathrm{t}; & -d \cos \mathrm{t} \\ \dot{A} \sin \mathrm{t} + A \cos \mathrm{t}; & -d \sin \mathrm{t} \end{bmatrix},$$

$$\boldsymbol{r}^{\boldsymbol{\cdot\cdot}}(\mathrm{t}) = \begin{bmatrix} 0; & 2\dot{A}d \cos 2\mathrm{t} + (\ddot{A}d/2 - 2Ad) \sin 2\mathrm{t} \\ (A - \ddot{A}) \cos \mathrm{t} + 2\dot{A} \sin \mathrm{t}; & d \sin \mathrm{t} \\ (\ddot{A} - A) \sin \mathrm{t} + 2\dot{A} \cos \mathrm{t}; & -d \cos \mathrm{t} \end{bmatrix}$$

Further,

$$\begin{aligned}
(\boldsymbol{x}, \boldsymbol{x}) &= (\dot{\varphi})^2 = 1 + A^2\,, \quad (\boldsymbol{x}, \boldsymbol{t}) = Ad \sin \mathrm{t} \cos \mathrm{t}\,, \\
(\boldsymbol{x}, \boldsymbol{t})^{\boldsymbol{\cdot}} &= \dot{A}d \sin \mathrm{t} \cos \mathrm{t} + Ad \cos 2\mathrm{t}\,, \\
(\boldsymbol{x}, \boldsymbol{x}^{\boldsymbol{\cdot}}) &= A\dot{A}\,, \quad (\boldsymbol{x}^{\boldsymbol{\cdot}}, \boldsymbol{x}^{\boldsymbol{\cdot}}) = (\dot{A})^2 + A^2\,, \\
(\boldsymbol{x}^{\boldsymbol{\cdot}}, \boldsymbol{t}^{\boldsymbol{\cdot}}) &= \dot{A}d \cos 2\mathrm{t} - Ad \sin 2\mathrm{t}\,, \\
(\boldsymbol{x}, \boldsymbol{x}^{\boldsymbol{\cdot\cdot}}) &= A\ddot{A} - A^2\,, \quad V = (\boldsymbol{x}^{\boldsymbol{\cdot}}, \boldsymbol{x}^{\boldsymbol{\cdot}})\,(\boldsymbol{x}, \boldsymbol{x}) - (\boldsymbol{x}, \boldsymbol{x}^{\boldsymbol{\cdot}})^2 \\
&= A^2 + A^4 + (\dot{A})^2\,.
\end{aligned}$$

$$|\boldsymbol{x}, \boldsymbol{x}^{\boldsymbol{\cdot}}, \boldsymbol{x}^{\boldsymbol{\cdot\cdot}}|$$

$$= \begin{vmatrix} -1, & 0, & 0 \\ -A \cos \mathrm{t}, & -\dot{A} \cos \mathrm{t} + A \sin \mathrm{t}, & -\ddot{A} \cos \mathrm{t} + 2\dot{A} \sin \mathrm{t} + A \cos \mathrm{t} \\ A \sin \mathrm{t}, & \dot{A} \sin \mathrm{t} + A \cos \mathrm{t}, & \ddot{A} \sin \mathrm{t} + 2\dot{A} \cos \mathrm{t} - A \sin \mathrm{t} \end{vmatrix}$$

$$= 2(\dot{A})^2 - A\ddot{A} + A^2\,.$$

$$\tfrac{1}{2}\dot{V} = A\dot{A} + 2A^3\dot{A} + \dot{A}\ddot{A},$$

$$\begin{aligned} Z &= (\boldsymbol{x}^\cdot, \boldsymbol{t}^\cdot)(\dot{\varphi})^4 - (\boldsymbol{x}, \boldsymbol{t})^\cdot(\dot{\varphi})^2(\boldsymbol{x}, \boldsymbol{x}^\cdot) + (\boldsymbol{x}, \boldsymbol{t})(\boldsymbol{x}, \boldsymbol{x}^\cdot)^2 \\ &= \dot{A}d(1 + A^2)\cos 2\mathrm{t} - Ad[2(1 + A^2)^2 + (\dot{A})^2]\sin \mathrm{t}\cos \mathrm{t}. \end{aligned}$$

For the invariants we thus obtain

$$\begin{aligned} v_0 &= (1 + A^2)^{-1} Ad \sin \mathrm{t} \cos \mathrm{t}, \\ \kappa_1 &= (1 + A^2)^{-3/2}[A^2 + A^4 + (\dot{A})^2]^{1/2}, \\ \kappa_2 &= [A^2 + A^4 + (\dot{A})^2]^{-1}[2(\dot{A})^2 - A\ddot{A} + A^2], \\ \mu_1 &= (\dot{\varphi})^{-5} V^{-1/2}[Z - (\boldsymbol{x}, \boldsymbol{t}) V] \\ &= (1 + A^2)^{-5/2}[A^2 + A^4 + (\dot{A})^2]^{-1/2} d\{\dot{A}(1 + A^2)^2 \cos 2\mathrm{t} \\ &\qquad - A[2(1 + A^2)^2 + 2(\dot{A})^2 + A^2 + A^4]\sin \mathrm{t}\cos \mathrm{t}\}. \end{aligned}$$

We still have to express μ_2. Consequently, we proceed with the computations:

$$\boldsymbol{t}^{\cdot\cdot}(\dot{\varphi})^2 - \boldsymbol{t}(\boldsymbol{x}, \boldsymbol{x}^{\cdot\cdot}) = d \cdot \begin{bmatrix} 2\dot{A}(1 + A^2)\cos 2\mathrm{t} + (\ddot{A} - 4A - 3A^3)\sin \mathrm{t}\cos \mathrm{t} \\ (1 + A\ddot{A})\sin \mathrm{t} \\ -(1 + A\ddot{A})\cos \mathrm{t} \end{bmatrix},$$

$$\boldsymbol{t}(\dot{\varphi})^2 - \boldsymbol{t}(\boldsymbol{x}, \boldsymbol{x}^{\cdot\cdot}) = d \begin{bmatrix} \dot{A}\sin \mathrm{t}\cos \mathrm{t} + A(1 + A^2)\cos 2\mathrm{t} \\ -(1 + A^2)\cos \mathrm{t} + A\dot{A}\sin \mathrm{t} \\ -(1 + A^2)\sin \mathrm{t} - A\dot{A}\cos \mathrm{t} \end{bmatrix}.$$

$$\begin{aligned} D_1 &= |\boldsymbol{x}, \boldsymbol{x}^\cdot, \boldsymbol{t}^{\cdot\cdot}(\dot{\varphi})^2 - \boldsymbol{t}(\boldsymbol{x}, \boldsymbol{x}^{\cdot\cdot})| \\ &= -d\dot{A}[(1 + A^2)^2 + A^4 + A\ddot{A}]\cos 2\mathrm{t} \\ &\quad + d[A^2\ddot{A} + 2A(1 + A^2)^2 + A^5]\sin \mathrm{t}\cos \mathrm{t}. \end{aligned}$$

$$\begin{aligned} D_2 &= |\boldsymbol{x}, \boldsymbol{x}^\cdot, \boldsymbol{t}^\cdot(\dot{\varphi})^2 - \boldsymbol{t}(\boldsymbol{x}, \boldsymbol{x}^\cdot)| \\ &= -dA[(\dot{A})^2 + (1 + A^2)^2]\cos 2\mathrm{t} - d\dot{A}(2 + A^2)\sin \mathrm{t}\cos \mathrm{t}. \end{aligned}$$

After lengthy manipulation we arrive at

$$VD_1 - \tfrac{1}{2}\dot{V}D_2 = d\dot{A}(1 + A^2)[A\ddot{A} + A^4 - (\dot{A})^2]\cos 2\mathrm{t}$$

$$+ d(1 + A^2)\left[A^4\ddot{A} + 2(\dot{A})^2\ddot{A} + 2A^3(1 + A^2)^2 + A^7 + 4A(\dot{A})^2 + 5A^3(\dot{A})^2\right]\sin t\cos t\,,$$

so that, finally,

$$\mu_2 = (1 + A^2)^{-1}[A^2 + A^4 + (\dot{A})^2]^{-2}\, d\{\dot{A}(1 + A^2)\,.$$
$$[A\ddot{A} + A^4 - (\dot{A})^2]\cos 2t$$
$$+ (1 + A^2)\left[A^4\ddot{A} + 2(\dot{A})^2\ddot{A} + 2A^3(1 + A^2)^2 + A^7\right.$$
$$\left.+ 4A(\dot{A})^2 + 5A^3(\dot{A})^2\right]\sin t\cos t - \left[2(\dot{A})^2 - A\ddot{A} + A^2\right].$$
$$\left[\dot{A}(1 + A^2)\cos 2t - A(2(1 + A^2)^2 + (\dot{A})^2)\sin t\cos t\right]\}\,.$$

For the sake of clarity, let us also determine the values of the invariants at time $t = 0$. First, compute $\dot{N}(0)$, $\ddot{N}(0)$ and $\dddot{N}(0)$. To simplify the notation put $b = m_2 - d$. Then

$$N = m_3\cos t - (m_2 - b)\cos t\sin t + m_2\sin t\,,\quad \text{i.e. } N(0) = m_3\,,$$
$$\dot{N} = -m_3\sin t - (m_2 - b)\cos 2t + m_2\cos t\,,\quad \text{i.e. } \dot{N}(0) = b\,,$$
$$\ddot{N} = -m_3\cos t + 2(m_2 - b)\sin 2t - m_2\sin t\,,\quad \text{i.e. } \ddot{N}(0) = -m_3\,,$$
$$\dddot{N} = m_3\sin t + 4(m_2 - b)\cos 2t - m_2\cos t\,,\quad \text{i.e. } \dddot{N}(0) = 3m_2 - 4b.$$

Let us express $A(0)$, $\dot{A}(0)$ and $\ddot{A}(0)$ in terms of $N(0)$, $\dot{N}(0)$, $\ddot{N}(0)$ and $\dddot{N}(0)$:

$$A = m_1(m_1^2 + N^2)^{-1}\dot{N}\,.$$

Upon differentiation we have

$$\dot{A} = m_1(m_1^2 + N^2)^{-2}\left[\ddot{N}(m_1^2 + N^2) - 2N(\dot{N})^2\right],$$
$$\ddot{A} = m_1(m_1^2 + N^2)^{-3}\{\dddot{N}(m_1^2 + N^2)^2$$
$$- 2(m_1^2 + N^2)\left[3N\dot{N}\ddot{N} + (\dot{N})^3\right] + 8N^2(\dot{N})^3\}\,.$$

To be brief, put also $m_1^2 + m_3^2 = a$. Consequently,

$$A(0) = \frac{m_1 b}{a}\,,$$

$$\dot{A}(0) = -\frac{m_1 m_3}{a^2}(a + 2b^2)\,,$$

$$\ddot{A}(0) = \frac{m_1}{a^3}\left[(3m_2 - 4b)\,a^2 + 2ab(3m_3^2 - b^2) + 8m_3^2 b^3\right].$$

If we wish to arrive at a numerical result, the constants which determine the size of the mechanism must of course be given. Assume, e.g., that $m_1 = m_2 = m_3 = d = 1$. Then $b = 0$, $a = 2$, so that

$$A(0) = 0\,, \quad \dot{A}(0) = -\tfrac{1}{2}\,, \quad \ddot{A}(0) = \tfrac{1}{8}\,.\,(3\,.\,4) = \tfrac{3}{2}\,.$$

We then easily obtain

$$\dot{\varphi}(0) = 1\,, \quad v_0(0) = 0\,, \quad \kappa_1(0) = \tfrac{1}{2}\,, \quad \kappa_2(0) = 2\,, \quad \mu_1(0) = -1\,,$$

$$\mu_2(0) = \tfrac{3}{4}\,.$$

In defining the invariants of motion in Theorem 5.1, the fixed directing cone was considered. However, we could have as well considered the moving directing cone. Let us demonstrate that this second approach would yield practically the same result.

5.12 Theorem. *Let $g(\varphi)$ be a motion with the moving directing cone $\bar{\boldsymbol{R}}(\varphi)$, and let $\mathbf{K}(\bar{\boldsymbol{R}}'(\varphi), \bar{\boldsymbol{R}}'(\varphi)) \neq 0$. Then we can write the Frenet formulae as follows:*

$$\begin{aligned}
\bar{\boldsymbol{R}} &= \bar{\boldsymbol{R}}_1 + \bar{v}_0 \bar{\boldsymbol{R}}_2\,, \\
\bar{\boldsymbol{R}}_1' &= \bar{\kappa}_1 \bar{\boldsymbol{T}}_1 + \bar{\mu}_1 \bar{\boldsymbol{T}}_2, \quad \bar{\kappa}_1 > 0\,, & \bar{\boldsymbol{R}}_2' &= \bar{\kappa}_1 \bar{\boldsymbol{T}}_2\,, \\
\bar{\boldsymbol{T}}_1' &= -\bar{\kappa}_1 \bar{\boldsymbol{R}}_1 + \bar{\kappa}_2 \bar{\boldsymbol{N}}_1 - \bar{\mu}_1 \bar{\boldsymbol{R}}_2 + \bar{\mu}_2 \bar{\boldsymbol{N}}_2\,, & \bar{\boldsymbol{T}}_2' &= -\bar{\kappa}_1 \bar{\boldsymbol{R}}_2 + \bar{\kappa}_2 \bar{\boldsymbol{N}}_2\,, \\
\bar{\boldsymbol{N}}_1' &= -\bar{\kappa}_2 \bar{\boldsymbol{T}}_1 - \bar{\mu}_2 \bar{\boldsymbol{T}}_2\,, & \bar{\boldsymbol{N}}_2' &= -\bar{\kappa}_2 \bar{\boldsymbol{T}}_2\,,
\end{aligned}$$

where $\bar{v}_0, \bar{\kappa}_1, \bar{\kappa}_2, \bar{\mu}_1, \bar{\mu}_2$ are the invariants of motion.

5.13 Theorem. *Let $g(\varphi)$ be a motion with the invariants $v_0, \kappa_1, \kappa_2, \mu_1, \mu_2$ of the fixed directing cone, and with the invariants $\bar{v}_0, \bar{\kappa}_1, \bar{\kappa}_2, \bar{\mu}_1, \bar{\mu}_2$ of the moving directing cone. Then we have*

$$v_0 = \bar{v}_0\,, \quad \kappa_1 = \bar{\kappa}_1\,, \quad \kappa_2 - \bar{\kappa}_2 = 1\,, \quad \mu_1 = \bar{\mu}_1\,, \quad \mu_2 - \bar{\mu}_2 = v_0\,.$$

Proof. We have

$$\boldsymbol{R}(\varphi) = \operatorname{ad} g(\varphi)\, \bar{\boldsymbol{R}}(\varphi)\,,$$

$$\boldsymbol{R}(\varphi) = \boldsymbol{R}_1(\varphi) + v_0(\varphi)\, \boldsymbol{R}_2(\varphi)\,,$$

$$\bar{\boldsymbol{R}}(\varphi) = \bar{\boldsymbol{R}}_1(\varphi) + \bar{v}_0(\varphi)\, \bar{\boldsymbol{R}}_2(\varphi)\,.$$

Further, $\operatorname{ad} g \mathfrak{h}_{\bar{R}}$ is the maximal commutative subalgebra of dimension 2 in $\mathfrak{E}$ containing $\boldsymbol{R}$, so that $\mathfrak{h}_{\boldsymbol{R}} = \operatorname{ad} g \mathfrak{h}_{\bar{R}}$. Consequently,

$$\operatorname{ad} g \bar{\boldsymbol{R}}_1 = \boldsymbol{R}_1 \quad \text{and} \quad \operatorname{ad} g \bar{\boldsymbol{R}}_2 = \boldsymbol{R}_2 .$$

These relations yield

$$\operatorname{ad} g \bar{\boldsymbol{R}} = \operatorname{ad} g(\bar{\boldsymbol{R}}_1 + \bar{v}_0 \bar{\boldsymbol{R}}_2) = \operatorname{ad} g_1 \bar{\boldsymbol{R}}_1 + \bar{v}_0 \operatorname{ad} g \bar{\boldsymbol{R}}_2 = \boldsymbol{R}_1 + v_0 \boldsymbol{R}_2 ,$$

i.e.

$$\bar{v}_0 = v_0 .$$

If we differentiate the equation $\operatorname{ad} g_1 \bar{\boldsymbol{R}}_1 = \boldsymbol{R}_1$ with respect to φ, we obtain

$$\boldsymbol{R}_1' = \operatorname{ad} g \bar{\boldsymbol{R}}_1' + \operatorname{ad} g[\bar{\boldsymbol{R}}, \bar{\boldsymbol{R}}_1] .$$

Indeed, since

$$\boldsymbol{R}_1 = g \bar{\boldsymbol{R}}_1 g^{-1} ,$$

i.e.

$$\begin{aligned} \boldsymbol{R}_1' &= g \bar{\boldsymbol{R}}_1' g^{-1} + g' \bar{\boldsymbol{R}}_1 g^{-1} + g \bar{\boldsymbol{R}}_1 (g^{-1})' \\ &= \operatorname{ad} g \bar{\boldsymbol{R}}_1' + g g^{-1} g' \bar{\boldsymbol{R}}_1 g^{-1} - g \bar{\boldsymbol{R}}_1 g^{-1} g' g^{-1} \\ &= \operatorname{ad} g \bar{\boldsymbol{R}}_1 + g[g^{-1} g' \bar{\boldsymbol{R}}_1 - \bar{\boldsymbol{R}}_1 g^{-1} g'] g^{-1} = \operatorname{ad} g \bar{\boldsymbol{R}}_1' + \operatorname{ad} g[\bar{\boldsymbol{R}}, \bar{\boldsymbol{R}}_1] . \end{aligned}$$

Considering that $[\bar{\boldsymbol{R}}, \bar{\boldsymbol{R}}_1] = 0$, we have $\boldsymbol{R}_1' = \operatorname{ad} g_1 \bar{\boldsymbol{R}}_1'$. Substituting from the Frenet formulae, we obtain

$$\kappa_1 \boldsymbol{T}_1 + \mu_1 \boldsymbol{T}_2 = \operatorname{ad} g_1(\bar{\kappa}_1 \bar{\boldsymbol{T}}_1 + \bar{\mu}_1 \bar{\boldsymbol{T}}_2) = \bar{\kappa}_1 \boldsymbol{T}_1 + \bar{\mu}_1 \boldsymbol{T}_2 ,$$

i.e.

$$\kappa_1 = \bar{\kappa}_1 , \quad \mu_1 = \bar{\mu}_1 .$$

The last equation yields $\boldsymbol{T}_1 = \operatorname{ad} g \bar{\boldsymbol{T}}_1$. Further differentiation results in

$$\boldsymbol{T}_1' = \operatorname{ad} g \bar{\boldsymbol{T}}_1' + \operatorname{ad} g[\bar{\boldsymbol{R}}, \bar{\boldsymbol{T}}_1] .$$

Moreover,

$$[\bar{\boldsymbol{R}}, \bar{\boldsymbol{T}}_1] = [\bar{\boldsymbol{R}}_1 + v_0 \bar{\boldsymbol{R}}_2, \bar{\boldsymbol{T}}_1] = \bar{\boldsymbol{N}}_1 + v_0 \bar{\boldsymbol{N}}_2 ,$$

so that

$$-\kappa_1 \boldsymbol{R}_1 + \kappa_2 \boldsymbol{N}_1 - \mu_1 \boldsymbol{R}_2 + \mu_2 \boldsymbol{N}_2$$
$$= \mathrm{ad}g(-\bar{\kappa}_1 \bar{\boldsymbol{R}}_1 + \bar{\kappa}_2 \bar{\boldsymbol{N}}_1 - \bar{\mu}_1 \bar{\boldsymbol{R}}_2 + \bar{\mu}_2 \bar{\boldsymbol{N}}_2) + \mathrm{ad}g(\bar{\boldsymbol{N}}_1 + v_0 \bar{\boldsymbol{N}}_2)$$
$$-\kappa_1 \boldsymbol{R}_1 + \bar{\kappa}_2 \boldsymbol{N}_1 - \mu_1 \boldsymbol{R}_2 + \bar{\mu}_2 \boldsymbol{N}_2 + \boldsymbol{N}_1 + v_0 \boldsymbol{N}_2 ,$$

i.e.

$$\kappa_2 \boldsymbol{N}_1 + \mu_2 \boldsymbol{N}_2 = \bar{\kappa}_2 \boldsymbol{N}_1 + \bar{\mu}_2 \boldsymbol{N}_2 + \boldsymbol{N}_1 - v_0 \boldsymbol{N}_2 ,$$

i.e.

$$\kappa_2 - \bar{\kappa}_2 = 1 , \quad \mu_2 - \bar{\mu}_2 = v_0 .$$

The theorem is thus proved. For future application we should keep in mind the following:

$$\boldsymbol{R}_1 = \mathrm{ad}g\bar{\boldsymbol{R}}_1 , \quad \boldsymbol{R}_2 = \mathrm{ad}g\bar{\boldsymbol{R}}_2 , \quad \boldsymbol{T}_1 = \mathrm{ad}g\bar{\boldsymbol{T}}_1 , \quad \boldsymbol{T}_2 = \mathrm{ad}g\bar{\boldsymbol{T}}_2 , \tag{5.1}$$

5.14 Example. Let us determine the invariants of the moving directing cone from Example 4.8 and verify Theorem 5.13. We have

$$\bar{\boldsymbol{r}}(\mathrm{t}) = [\bar{\boldsymbol{x}}(\mathrm{t}); \bar{\boldsymbol{t}}(\mathrm{t})] ,$$

where

$$\bar{\boldsymbol{x}} = \begin{pmatrix} -Mm_1 \\ -A \\ MN \end{pmatrix}, \quad \bar{\boldsymbol{t}} = d \begin{pmatrix} MN\cos^2 \mathrm{t} \\ -\sin \mathrm{t} \cos \mathrm{t} \\ Mm_1 \cos^2 \mathrm{t} \end{pmatrix} .$$

In the computations recall that

$$A = M^2 \dot{N} m_1 , \quad m_1^2 + N^2 = M^{-2} ,$$
$$\dot{M} = -m_1^{-1} MNA , \quad \dot{N} = m_1^{-1} AM^{-2} , \quad (MN)^{\boldsymbol{\cdot}} = m_1 AM .$$

$$\bar{\boldsymbol{x}}^{\boldsymbol{\cdot}} = \begin{pmatrix} MNA \\ -\dot{A} \\ m_1 MA \end{pmatrix}, \quad \bar{\boldsymbol{x}}^{\boldsymbol{\cdot\cdot}} = \begin{pmatrix} MN\dot{A} + m_1 MA^2 \\ -\ddot{A} \\ m_1 M\dot{A} - MNA^2 \end{pmatrix},$$

$$\bar{\boldsymbol{t}}^{\boldsymbol{\cdot}} = d \begin{pmatrix} m_1 AM \cos^2 \mathrm{t} - MN \sin 2\mathrm{t} \\ -\cos 2t \\ -MNA \cos^2 \mathrm{t} - Mm_1 \sin 2\mathrm{t} \end{pmatrix},$$

$$\bar{\boldsymbol{t}}^{\boldsymbol{\cdot\cdot}} = d \,.$$

$$\cdot \begin{pmatrix} m_1 \dot{A} M \cos^2 \mathrm{t} - MNA^2 \cos^2 \mathrm{t} - 2m_1 AM \sin 2\mathrm{t} - 2MN \cos 2t \\ 2 \sin 2t \\ -N\dot{A}M \cos^2 \mathrm{t} - m_1 MA^2 \cos^2 \mathrm{t} + 2NAM \sin 2\mathrm{t} - 2Mm_1 \cos 2\mathrm{t} \end{pmatrix} .$$

Further computation yields

$$(\bar{x}, \bar{x}) = (\dot{\varphi})^2 = 1 + A^2\,, \quad (\bar{x}, \bar{t}) = Ad \sin \mathrm{t} \cos \mathrm{t}\,,$$
$$(\bar{x}, \bar{x}^{\bullet}) = A\dot{A}\,, \quad (\bar{x}^{\bullet}, \bar{x}^{\bullet}) = (\dot{A})^2 + A^2\,,$$
$$(\bar{x}^{\bullet}, \bar{t}^{\bullet}) = \dot{A}d \cos 2\mathrm{t} - Ad \sin 2t\,, \quad (\bar{x}, \bar{x}^{\bullet\bullet}) = A\ddot{A} - A^2\,,$$

which are the same values as obtained with the fixed directing function, so that we also have $\bar{V} = V$, $\bar{V}^{\bullet} = \dot{V}$, $\bar{Z} = Z$. Further, we have

$$|\bar{x}, \bar{x}^{\bullet}, \bar{x}^{\bullet\bullet}| = M^2 \begin{vmatrix} A, & \dot{A}, & \ddot{A} \\ -m_1, & NA, & N\dot{A} + m_1 A^2 \\ N, & m_1 A, & m_1 \dot{A} - NA^2 \end{vmatrix}$$
$$= M^2[AA^3(-N^2 - m_1^2) - (\dot{A})^2(-m_1^2 - N^2) + \ddot{A}A(-m_1^2 - N^2)$$
$$= -A^4 + (\dot{A})^2 - \ddot{A}A\,.$$

Similarly, we compute

$$|\bar{x}, \bar{x}^{\bullet}, \bar{t}| = d(\dot{A} \cos^2 \mathrm{t} - A \cos \mathrm{t} \sin \mathrm{t})\,,$$
$$|\bar{x}, \bar{x}^{\bullet}, \bar{t}^{\bullet}| = -d(A^3 \cos^2 \mathrm{t} + \dot{A} \sin 2t + A \cos 2\mathrm{t})\,,$$
$$|\bar{x}, \bar{x}^{\bullet}, \bar{t}^{\bullet\bullet}| = d[-2A^2\dot{A} \cos^2 \mathrm{t} + 2A(1 + A^2) \sin 2\mathrm{t} - 2\dot{A} \cos 2\mathrm{t}]\,.$$

Further, we have

$$\bar{D}_1 = |\bar{x}, \bar{x}^{\bullet}, \bar{t}|\,(\dot{\varphi})^2 - |\bar{x}, \bar{x}^{\bullet}, \bar{t}|\,(x, x^{\bullet\bullet})$$
$$= -d\{(A^2\ddot{A} + 2A^4\dot{A} + A\dot{A}\ddot{A}) \cos^2 \mathrm{t} + 2\dot{A}(1 + A^2) \cos 2\mathrm{t}$$
$$+ [-4A(1 + A^2)^2 + A(A^2 - A\ddot{A})] \sin \mathrm{t} \cos \mathrm{t}\}\,.$$

$$\bar{D}_2 = |\bar{x}, \bar{x}^{\bullet}, \bar{t}^{\bullet}|\,(\dot{\varphi})^2 - |\bar{x}, \bar{x}^{\bullet}, \bar{t}|\,(x, x^{\bullet})$$
$$= -d\{A[A^2 + A^4 + (\dot{A})^2] \cos^2 \mathrm{t} + A(1 + A^2) \cos 2\mathrm{t}$$
$$+ \dot{A}(2 + A^2) \sin t \cos \mathrm{t}\}\,.$$

$$V\bar{D}_1 - \tfrac{1}{2}\dot{V}\bar{D}_2 = d\{-\dot{A}(1 + A^2)[A^2 + 2(\dot{A})^2 - A\ddot{A}] \cos 2\mathrm{t}$$
$$+ (1 + A^2)[A^4\ddot{A} + 6A(\dot{A})^2(1 + A^2) + 2(\dot{A})^2\ddot{A} + 4A^3$$
$$+ 7A^5 + 4A^7] \cos \mathrm{t} \sin \mathrm{t}\}\,.$$

For the invariants of the moving directing cone we obtain

$$\bar{v}_0 = (1 + A^2)^{-1} Ad \sin \mathrm{t} \cos \mathrm{t}\,,$$

$$\bar{\kappa}_1 = (1 + A^2)^{-3/2}\,[A^2 + A^4 + (\dot{A})^2]^{1/2}\,,$$
$$\bar{\kappa}_2 = [A^2 + A^4 + (\dot{A})^2]^{-1}\,[(\dot{A})^2 - A^4 - A\ddot{A}]\,,$$
$$\bar{\mu}_1 = (1 + A^2)^{-5/2}\,[A^2 + A^4 + (\dot{A})^2]^{-1/2}\, d\{\dot{A}(1 + A^2)^2 \cos 2t - A[2(1 + A^2)^2 + 2(\dot{A})^2 + A^2 + A^4] \sin t \cos t\}\,,$$
$$\bar{\mu}_2 = (1 + A^2)^{-1}\,[A^2 + A^4 + (\dot{A})^2]^{-2}\, d\{-\dot{A}(1 + A^2)\,. [A^2 + 2(\dot{A})^2 - A\ddot{A}] \cos 2t + (1 + A^2)\,[A^4\ddot{A} + 6A(\dot{A})^2\,(1 + A^2) + 2(\dot{A})^2\,\ddot{A} + 4A^3 + 7A^5 + 4A^7] \cos t \sin t + [A^4 - (\dot{A})^2 + A\ddot{A}]\,. [\dot{A}(1 + A^2) \cos 2t - A(2(1 + A^2)^2 + (\dot{A})^2) \sin t \cos t]\}$$

We see at first sight that indeed

$$\bar{v}_0 = v_0\,, \quad \bar{\kappa}_1 = \varkappa_1\,, \quad \bar{\mu}_1 = \mu_1\,.$$

Further,

$$\kappa_2 - \bar{\kappa}_2 = [A^2 + A^4 + (\dot{A})^2]^{-1}\,\{[2(\dot{A})^2 - A\ddot{A} + A^2] - [(\dot{A})^2 - A^4 - A\ddot{A}^2]\} = 1\,.$$

For the sake of brevity put

$$d(1 + A^2)^{-1}\,[A^2 + A^4 + (\dot{A})^2]^{-2} = K\,.$$

Then

$$\mu_2 - \bar{\mu}_2 = K\{\dot{A}(1 + A^2)\,[A\ddot{A} + A^4 - (\dot{A})^2] \cos 2t + (1 + A^2)\,[A^4\ddot{A} + 2(\dot{A})^2\,\ddot{A} + 2A^3(1 + A^2)^2 + 4A(\dot{A})^2 + 5A^3(\dot{A})^2 + A^7] \sin t \cos t - [2(\dot{A})^2 - A\ddot{A} + A^2]\,. [\dot{A}(1 + A^2) \cos 2t - A(2(1 + A^2)^2 + (\dot{A})^2) \sin t \cos t]\} - K\{-\dot{A}(1 + A^2)\,[A^2 + 2(\dot{A})^2 - A\ddot{A}] \cos 2t + (1 + A^2)\,[A^4\ddot{A} + 6A(\dot{A})^2\,(1 + A^2) + 2(\dot{A})^2\,\ddot{A} + 4A^3 + 7A^5 + 4A^7] \cos t \sin t + [A^4 - (\dot{A})^2 + A\ddot{A}]\,. [\dot{A}(1 + A^2) \cos 2t - A(2(1 + A^2)^2 + (\dot{A})^2) \sin t \cos t]\} = Ad(1 + A^2)^{-1} \sin t \cos t = v_0\,.$$

5.15 Exercise. Determine the invariants of the moving directing cone and verify Theorem 5.13 for the motion from Example 4.7.

Finally, we shall treat the excluded case of $\mathbf{K}(\boldsymbol{R}', \boldsymbol{R}') = 0$.

5.16 Definition. Space motion $g(t)$ is called *cylindrical*, if $g(t) \subset Z(\boldsymbol{X})$ for some $\boldsymbol{X} \neq 0$, $\mathbf{K}(\boldsymbol{X}, \boldsymbol{X}) = 0$ (see 1.8). This means that cylindrical motion preserves a certain direction.

5.17 Theorem. *Space motion* $g(t)$, *where* $g(0) = e$ *and* $\mathbf{K}(\boldsymbol{R}, \boldsymbol{R}) \neq 0$, *is cylindrical if and only if* $\mathbf{K}(\boldsymbol{R}', \boldsymbol{R}') = 0$.

Proof. If $g(t) \subset Z(\boldsymbol{X})$, then $\boldsymbol{R}(t) \in \mathfrak{Z}(\boldsymbol{X})$, i.e. $\boldsymbol{R} = [\boldsymbol{x}; \boldsymbol{t}]$, where $\boldsymbol{x}$ is constant. Therefore, $\boldsymbol{R}' = [\boldsymbol{0}; \boldsymbol{t}']$ and, thus, $\mathbf{K}(\boldsymbol{R}', \boldsymbol{R}') = 0$. Conversely, if $\mathbf{K}(\boldsymbol{R}', \boldsymbol{R}') = 0$, the associated spherical motion preserves the vector $\boldsymbol{x}$. The lemma now follows from Theorem 1.8.

If $g(t)$ is a cylindrical motion with the directing cone $\boldsymbol{R}$, $\mathbf{K}(\boldsymbol{R}, \boldsymbol{R}) \neq 0$, $\boldsymbol{R}' \neq 0$, we define $\boldsymbol{R}_1, \boldsymbol{R}_2, \boldsymbol{T}_2 = \lambda \boldsymbol{R}_1'$, $\lambda > 0$, $\mathbf{T}(\boldsymbol{T}_2, \boldsymbol{T}_2) = 1$, $\boldsymbol{N}_2 = [\boldsymbol{R}_1, \boldsymbol{T}_2]$ as in Theorem 5.1. We may then express the Frenet formulae in a form analogous to the general case.

5.18 Theorem. *Let* $g(\varphi)$ *be a cylindrical motion with the canonical parameter* φ *and the fixed directing cone* $\boldsymbol{R}$. *Then*

$$\boldsymbol{R} = \boldsymbol{R}_1 + v_0 \boldsymbol{R}_2\,, \quad \boldsymbol{R}_1' = \mu_1 \boldsymbol{T}_2\,, \quad \mu_1 > 0\,, \quad \boldsymbol{T}_2' = \kappa_2 \boldsymbol{N}_2\,, \quad \boldsymbol{N}_2' = -\kappa_2 \boldsymbol{T}_2\,,$$

$$\boldsymbol{R}_2' = 0\,,$$

where v_0, μ_1, κ_2 *are the invariants of motion.*

The proof is similar to that of Theorem 5.1 and is left to the reader to perform.

5.19 Remark. Cylindrical motion is decomposed into translation in the direction of $\boldsymbol{R}_2$ and into plane motion in any plane perpendicular to the direction of $\boldsymbol{R}_2$, as can be seen from Theorem 1.8. The first row of the Frenet formulae in Theorem 5.18 constitutes the Frenet formulae for this plane motion, the second row contains the Frenet formulae for the translation (even if trivial). The first formula then gives the relation between the two motions. Consequently, we shall only deal with the properties of cylindrical motions which cannot be derived from the properties of plane motion. As regards plane motion, we refer the reader to the appropriate literature, e.g.

[41]. Let us only mention that the invariants μ_1 and κ_2 have the usual meaning as in plane kinematics, i.e. μ_1 is the diameter of the circle of inflection while $\kappa_2\mu_1^{-1}$ is the curvature of the fixed centrode. Also, $\kappa_2 - \bar{\kappa}_2 = 1$.

5.20 Example. Let us represent cylindrical motions in matrix notation. Without loss of generality, we may assume that $\boldsymbol{R}_2 = [\boldsymbol{0}; \boldsymbol{u}_1]$ because we are able to transform any direction into the direction determined by the vector $\boldsymbol{u}_1$ with the aid of group $O(3)$. If $g(\varphi)$ is the matrix of cylindrical motion, then

$$g(\varphi) = \begin{bmatrix} 1, & 0, & 0, & 0 \\ t^1(\varphi), & 1, & 0, & 0 \\ t^2(\varphi), & 0, & \cos\varphi, & -\sin\varphi \\ t^3(\varphi), & 0, & \sin\varphi, & \cos\varphi \end{bmatrix}$$

where $t^1(\varphi)$ characterizes the translation and the matrix

$$g_1(\varphi) = \begin{pmatrix} 1, & 0, & 0 \\ t^2(\varphi), & \cos\varphi, & -\sin\varphi \\ t^3(\varphi), & \sin\varphi, & \cos\varphi \end{pmatrix}$$

determines the appropriate plane motion. From this expression it is easy to find the formulae for the invariants v_0, μ_1, and κ_2.

6. INVARIANTS OF AXOIDS AND RELATIONS BETWEEN THE INVARIANTS OF MOTION AND THE INVARIANTS OF AXOIDS

Consider the motion $g(t)$ with the directing cone $\boldsymbol{R}(t)$. Take the arc of the spherical image of the directing cone of the associated spherical motion as the parameter of motion, which is what we did in the proof of Theorem 5.1. Also, put $\boldsymbol{R}_1(s) = [\boldsymbol{x}(s); \boldsymbol{t}(s)]$. The fixed axoid is then described by the equation

$$\mathfrak{A}(s, \lambda) = \boldsymbol{x}(s) \times \boldsymbol{t}(s) + \lambda\, \boldsymbol{t}(s)\,, \quad \lambda \in \mathbf{R}\,;$$

here $(\boldsymbol{x}(s), \boldsymbol{x}(s)) = 1$, $(\boldsymbol{x}(s), \boldsymbol{t}(s)) = 0$.

By Chapter I, the Frenet formulae for a ruled surface $\mathscr{A}(\mathrm{s}, \lambda)$ may be represented in the following form: let $S(\mathrm{s})$ be a point of the striction curve, denote $\boldsymbol{x}(\mathrm{s})$ as $\boldsymbol{u}_1$, $\boldsymbol{x}^{\mathrm{I}}(\mathrm{s})$ as $\boldsymbol{u}_2$, and $\boldsymbol{u}_3 = \boldsymbol{x}(\mathrm{s}) \times \boldsymbol{x}^{\mathrm{I}}(\mathrm{s})$. Then

$$\frac{\mathrm{d}S}{\mathrm{ds}} = q\boldsymbol{u}_1 + d\boldsymbol{u}_3 ,$$

$$\frac{\mathrm{d}\boldsymbol{u}_1}{\mathrm{ds}} = \boldsymbol{u}_2 , \quad \frac{\mathrm{d}\boldsymbol{u}_2}{\mathrm{ds}} = -\boldsymbol{u}_1 + k\boldsymbol{u}_3 , \quad \frac{\mathrm{d}\boldsymbol{u}_3}{\mathrm{ds}} = -k\boldsymbol{u}_2 . \tag{6.1}$$

Here, $k = |\boldsymbol{x}, \boldsymbol{x}^{\mathrm{I}}, \boldsymbol{x}^{\mathrm{II}}|$, $k = \kappa_2/\kappa_1$, according to the proof of Theorem 5.1, further $d = (\boldsymbol{x}^{\mathrm{I}}, \boldsymbol{t}^{\mathrm{I}})$, i.e. $d = \mu_1/\kappa_1$, and, finally, if we consider the expression of the invariant q, we have $q = \mu_2/\kappa_1$.

6.1 Theorem. *Let $g(\varphi)$ be a motion with the invariants $v_0, \kappa_1, \kappa_2, \mu_1, \mu_2$ of the fixed directing cone and with the invariants $v_0, \kappa_1, \bar{\kappa}_2, \mu_1, \bar{\mu}_2$ of the moving directing cone. Further, let $\mathfrak{A}(\varphi, \lambda)$ be the fixed axoid of motion with the invariants* s, *q, d, k, and let $\bar{\mathfrak{A}}(\varphi, \lambda)$ be the moving axoid of motion with the invariants* $\bar{\mathrm{s}}$, *$\bar{q}$, $\bar{d}$, $\bar{k}$. Then we have*

$$\frac{\mathrm{ds}}{\mathrm{d}\varphi} = \kappa_1 , \quad \frac{\kappa_2}{\kappa_1} = k , \quad \frac{\mu_1}{\kappa_1} = d , \quad \frac{\mu_2}{\kappa_1} = q ;$$

$$\frac{\mathrm{d}\bar{\mathrm{s}}}{\mathrm{d}\varphi} = \kappa_1 , \quad \frac{\bar{\kappa}_2}{\kappa_1} = \bar{k} , \quad \frac{\mu_1}{\kappa_1} = \bar{d} , \quad \frac{\bar{\mu}_2}{\kappa_1} = \bar{q} .$$

Also, $\bar{\mathrm{s}} = \mathrm{s}$ (more accurately $\bar{\mathrm{s}} = \mathrm{s} + $ *const.*, we take the constant to be zero),

$$\bar{d} = d , \quad k - \bar{k} = \frac{1}{\kappa_1} , \quad q - \bar{q} = \frac{v_0}{\kappa_1} .$$

6.2 Theorem. *For the fixed axoid $\mathfrak{A}(\varphi, \lambda)$ let us use the notation from the beginning of this section. The following is then true: the vectors $\boldsymbol{R}_1, \boldsymbol{T}_1, \boldsymbol{N}_1$ are Klein images of the straight lines $S + \lambda\boldsymbol{u}_1$, $S + \lambda\boldsymbol{u}_2$, $S + \lambda\boldsymbol{u}_3$, $\lambda \in \mathbf{R}$. The line $S + \lambda\boldsymbol{u}_2$ is the normal of the fixed axoid at point S.*

Proof. For $\boldsymbol{R}_1$ the statement is clear. Put $\boldsymbol{R}_1 = [\boldsymbol{x}; \boldsymbol{t}]$. Then

$$\boldsymbol{T}_1 = [\boldsymbol{x}^{\mathrm{I}}; \boldsymbol{t}^{\mathrm{I}} - (\boldsymbol{x}^{\mathrm{I}}, \boldsymbol{t}^{\mathrm{I}})\,\boldsymbol{x}^{\mathrm{I}}] = [\boldsymbol{u}_2; \boldsymbol{t}^{\mathrm{I}} - (\boldsymbol{u}_2, \boldsymbol{t}^{\mathrm{I}})\,\boldsymbol{u}_2] .$$

Let us determine the striction curve of the fixed axoid:

$$S(s) = \boldsymbol{x} \times \boldsymbol{t} + \lambda(s)\,\boldsymbol{x}\,; \quad S^{I}(s) = \boldsymbol{u}_2 \times \boldsymbol{t} + \boldsymbol{u}_1 \times \boldsymbol{t}^{I} + \lambda^{I}\boldsymbol{u}_1 + \lambda \boldsymbol{u}_2\,;$$

it should be

$$(S^{I}, \boldsymbol{u}_2) = 0\,, \quad \text{i.e.} \quad (S^{I}, \boldsymbol{u}_2) = |\boldsymbol{u}_2, \boldsymbol{u}_1, \boldsymbol{t}^{I}| + \lambda = 0\,,$$

$$\text{i.e.} \quad \lambda = (\boldsymbol{u}_3, \boldsymbol{t}^{I})\,,$$

so that

$$S(s) = \boldsymbol{x} \times \boldsymbol{t} + (\boldsymbol{u}_3, \boldsymbol{t}^{I})\,\boldsymbol{x} = \boldsymbol{u}_1 \times \boldsymbol{t} + (\boldsymbol{u}_3, \boldsymbol{t}^{I})\,\boldsymbol{u}_1\,.$$

Now we shall show that rotations which preserve the point S correspond to the vector $\boldsymbol{T}_1$. Therefore, we have

$$\begin{aligned} \boldsymbol{T}_1(S) &= \boldsymbol{t}^{I} - (\boldsymbol{u}_2, \boldsymbol{t}^{I})\,\boldsymbol{u}_2 + \boldsymbol{u}_2 \times [\boldsymbol{u}_1 \times \boldsymbol{t} + (\boldsymbol{u}_3, \boldsymbol{t}^{I})\,\boldsymbol{u}_1] \\ &= \boldsymbol{t}^{I} - (\boldsymbol{u}_2, \boldsymbol{t}^{I})\,\boldsymbol{u}_2 + \boldsymbol{u}_2 \times (\boldsymbol{u}_1 \times \boldsymbol{t}) - (\boldsymbol{u}_3, \boldsymbol{t}^{I})\,\boldsymbol{u}_3 \\ &= (\boldsymbol{u}_1, \boldsymbol{t}^{I})\,\boldsymbol{u}_1 + (\boldsymbol{u}_2, \boldsymbol{t})\,\boldsymbol{u}_1 = [(\boldsymbol{x}, \boldsymbol{t}^{I}) + (\boldsymbol{x}^{I}, \boldsymbol{t})]\,\boldsymbol{u}_1 = \boldsymbol{0}\,; \end{aligned}$$

the direction of line $S + \lambda \boldsymbol{u}_2$ is $\boldsymbol{u}_2$, which agrees. For $\boldsymbol{N}_1$ we have

$$\boldsymbol{N}_1(S) = [\boldsymbol{R}_1, \boldsymbol{T}_1](S) = \boldsymbol{R}_1\boldsymbol{T}_1(S) - \boldsymbol{T}_1\boldsymbol{R}_1(S) = \boldsymbol{0}\,.$$

6.3 Theorem. *Let $g(t)$ be a motion with the fixed axoid $\mathfrak{A}(t, \lambda)$ and the moving axoid $\overline{\mathfrak{A}}(t, \lambda)$. We then have:*

The ruled surfaces $\mathfrak{A}(t, \lambda)$ and $g(t_0)\,\overline{\mathfrak{A}}(t, \lambda)$ contact along the generating line $\mathfrak{A}(t_0, \lambda) = g(t_0)\,\overline{\mathfrak{A}}(t_0, \lambda)$ for every $t = t_0$, i.e. the generating lines of the axoids gradually become one during the motion and the tangent planes coincide at the corresponding points. Further, $g(t)\,\bar{S}(t) = S(t)$, i.e. the points of the striction curve also gradually become one during the motion (for $v_0 \neq 0$ the striction curves do not roll, of course).

Proof. The theorem is a consequence of (5.1) and of Theorem 6.2.

6.4 Theorem. *For the striction curves of axoids we have:*

$$\left.\frac{dS}{d\varphi}\right|_{\varphi=\varphi_0} - g(t_0)\left(\left.\frac{d\bar{S}}{d\varphi}\right|_{\varphi=\varphi_0}\right) = v_0(\varphi_0)\,\boldsymbol{u}_1(\varphi_0)\,,$$

i.e. the striction curves contact only at those instants when the instantaneous motion is rotation.

Proof.

$$\frac{\mathrm{d}S}{\mathrm{d}\varphi} = \frac{\mathrm{d}S}{\mathrm{d}s}\frac{\mathrm{d}s}{\mathrm{d}\varphi} = \frac{\mathrm{d}S}{\mathrm{d}s}\,\kappa_1 = (q\boldsymbol{u}_1 + d\boldsymbol{u}_3)\,\kappa_1 = \mu_2\boldsymbol{u}_1 + \mu_1\boldsymbol{u}_3\,,$$

$$\frac{\mathrm{d}\bar{S}}{\mathrm{d}\varphi} = \bar{\mu}_2\bar{\boldsymbol{u}}_1 + \bar{\mu}_1\bar{\boldsymbol{u}}_3\,,\quad g(\mathrm{t})\frac{\mathrm{d}\bar{S}}{\mathrm{d}\varphi} = \bar{\mu}_2\boldsymbol{u}_1 + \bar{\mu}_1\boldsymbol{u}_3\,,$$

so that

$$\frac{\mathrm{d}S}{\mathrm{d}\varphi} - g(\mathrm{t})\frac{\mathrm{d}\bar{S}}{\mathrm{d}\varphi} = (\mu_2 - \bar{\mu}_2)\,\boldsymbol{u}_1 + (\mu_1 - \bar{\mu}_1)\,\boldsymbol{u}_3 = v_0\boldsymbol{u}_1\,.$$

6.5 Theorem. *Space motion is determined uniquely by its axoids, i.e. let two ruled surfaces $\mathfrak{A}(\mathrm{t}, \lambda)$, $\bar{\mathfrak{A}}(\mathrm{t}, \lambda)$ be given with the spherical arcs* s *and* $\bar{\mathrm{s}}$, *and let* $\bar{\mathrm{s}}(\mathrm{t}) = \mathrm{s}(\mathrm{t})$. *Let the spherical images of surfaces $\mathfrak{A}$ and $\bar{\mathfrak{A}}$ have no singular points; let $d(\mathrm{t}) = \bar{d}(\mathrm{t})$ and $k(\mathrm{t}) - \bar{k}(\mathrm{t}) \neq 0$. Further, let the ruled surfaces $\mathfrak{A}(\mathrm{t}, \lambda)$ and $\bar{\mathfrak{A}}(\mathrm{t}, \lambda)$ contact along the lines $\mathfrak{A}(0, \lambda)$ and $\bar{\mathfrak{A}}(0, \lambda)$. Then there exists just one motion $g(\mathrm{t})$ such that $g(0) = e$ and $\mathfrak{A}(\mathrm{t}, \lambda)$, $\bar{\mathfrak{A}}(\mathrm{t}, \lambda)$ are its axoids.*

Proof is similar to the proof of an analogous theorem from Chapter II; we shall only give its outline. According to the assumptions of the theorem $k - \bar{k} \neq 0$, so that the canonical parameter can be found since $k - \bar{k} = \kappa_1^{-1}(\mathrm{s})$; moreover $\kappa_1(\mathrm{s}) = \mathrm{ds}/\mathrm{d}\varphi$, so that

$$\varphi(\mathrm{s}) = \int \frac{\mathrm{ds}}{\kappa_1(\mathrm{s})}\,.$$

We now compute $\mathrm{s} = \mathrm{s}(\varphi)$, and substitute the expression for s in terms of φ; it is easy to compute $\boldsymbol{R}_1(\varphi)$ from the equations

$$\boldsymbol{R}_1 = [\boldsymbol{x}; \boldsymbol{t}]\,,\quad (\boldsymbol{x}, \boldsymbol{t}) = 0\,,\quad \mathfrak{A} = \boldsymbol{x} \times \boldsymbol{t} + \lambda\boldsymbol{x}\,;$$

from the condition $q - \bar{q} = v_0/\kappa_1$ we determine v_0 and, therefore, also $\boldsymbol{R} = \boldsymbol{R}(\varphi)$. By solving the differential equation $g' = \boldsymbol{R}g$ we now determine $g(\varphi)$. From the construction of $g(\varphi)$ it is evident that its axoids are indeed $\mathfrak{A}$ and $\bar{\mathfrak{A}}$; the uniqueness of the solution of equation $g' = \boldsymbol{R}g$ under the given initial condition $g(0) = e$ implies the uniqueness of the motion $g(\varphi)$.

6.6 Remark. In concrete cases it is not very advantageous to use Theorem 6.5 for expressing the matrix of motion. It is simpler to use the fact that the motion gradually maps the moving axoid into the fixed one. Indeed, let $\mathfrak{A}(\mathrm{t}, \lambda)$ and $\bar{\mathfrak{A}}(\mathrm{t}, \lambda)$ be axoids of the motion and denote $\mathscr{F}\{S, \boldsymbol{u}_1, \boldsymbol{u}_2, \boldsymbol{u}_3\}$

and $\mathscr{F}\{\bar{S}, \bar{u}_1, \bar{u}_2, \bar{u}_3\}$ the Frenet frames of the fixed and moving axoid. Put $\mathscr{F} = \mathscr{R}_0\gamma$, $\overline{\mathscr{F}} = \mathscr{R}_0\bar{\gamma}$. The motion transforming $\overline{\mathscr{F}}$ into $\mathscr{F}$ is then expressed as $\gamma(t)\,\bar{\gamma}(t)^{-1}$, i.e. $g(t) = \gamma(t)\,\bar{\gamma}(t)^{-1}$ [whereas $\bar{\gamma}(t)^{-1}\,\gamma(t)$ determines $\overline{\mathscr{F}}$ relative to $\mathscr{F}$].

6.7 Example. Consider the ruled surfaces $\mathfrak{A}(\alpha, \lambda)$, $\overline{\mathfrak{A}}(\alpha, \lambda)$ from Example 4.7, i.e.

$$\mathfrak{A}(\alpha, \lambda) = \frac{q}{2}\begin{pmatrix} 1 - \sin\alpha(1 + \cos^2\alpha) \\ \cos\alpha(1 + \sin^2\alpha) \\ \sin\alpha + \cos 2\alpha \end{pmatrix} + \frac{\lambda}{\sqrt{2}}\begin{pmatrix} \sin\alpha \\ \cos\alpha \\ -1 \end{pmatrix};$$

$$\overline{\mathfrak{A}}(\alpha, \lambda) = \frac{q}{2}\begin{pmatrix} \cos^2\alpha - \sin\alpha\cos 2\alpha \\ -\sin^2\alpha \\ \sin\alpha\cos\alpha(1 - 2\sin\alpha) \end{pmatrix} + \frac{\lambda}{\sqrt{2}}\begin{pmatrix} \sin\alpha \\ -1 \\ -\cos\alpha \end{pmatrix}.$$

These two ruled surfaces obviously satisfy the assumptions of Theorem 6.5 and, therefore, they determine a motion. Let us find it: put $\mathfrak{A}(\alpha, 0) = m(\alpha)$, $\overline{\mathfrak{A}}(\alpha, 0) = \bar{m}(\alpha)$. For the striction curves $S(\alpha)$ and $\bar{S}(\alpha)$ we have

$$S(\alpha) = m(\alpha) - \frac{(m^{\bullet}, u_1^{\bullet})}{(u_1^{\bullet}, u_1^{\bullet})}\, u_1\,, \quad \bar{S}(\alpha) = \bar{m}(\alpha) - \frac{(\bar{m}^{\bullet}, \bar{u}_1^{\bullet})}{(\bar{u}_1^{\bullet}, \bar{u}_1^{\bullet})}\, \bar{u}_1\,.$$

The computation yields

$$u_1^{\bullet} = \frac{\sqrt{2}}{2}\begin{pmatrix} \cos\alpha \\ -\sin\alpha \\ 0 \end{pmatrix}, \quad u_2 = \begin{pmatrix} \cos\alpha \\ -\sin\alpha \\ 0 \end{pmatrix},$$

$$u_3 = u_1 \times u_2 = \frac{\sqrt{2}}{2}\begin{pmatrix} -\sin\alpha \\ -\cos\alpha \\ -1 \end{pmatrix}.$$

Further,

$$m^{\bullet} = \frac{q}{2}\begin{pmatrix} -\cos\alpha(1 + \cos^2\alpha) + 2\cos\alpha\sin^2\alpha \\ -\sin\alpha - \sin^3\alpha + 2\cos^2\alpha\sin\alpha \\ \cos\alpha - 2\sin 2\alpha \end{pmatrix},$$

$$(m^{\bullet}, u_1^{\bullet}) = \frac{-q}{\sqrt{2}}\cos 2\alpha\,,$$

$$S(\alpha) = \frac{q}{2}\begin{pmatrix} 1 - 3\sin^3\alpha \\ 3\cos^3\alpha \\ \sin\alpha - \cos 2\alpha \end{pmatrix}.$$

Analogously,

$$\bar{\boldsymbol{u}}_1^{\bullet} = \frac{\sqrt{2}}{2}\begin{pmatrix}\cos\alpha\\ 0\\ \sin\alpha\end{pmatrix}, \quad \bar{\boldsymbol{u}}_2 = \begin{pmatrix}\cos\alpha\\ 0\\ \sin\alpha\end{pmatrix},$$

$$\bar{\boldsymbol{u}}_3 = \bar{\boldsymbol{u}}_1 \times \bar{\boldsymbol{u}}_2 = \frac{\sqrt{2}}{2}\begin{pmatrix}\sin\alpha\\ -1\\ -\cos\alpha\end{pmatrix},$$

$$\bar{m}^{\bullet} = \frac{q}{2}\begin{pmatrix}-2\cos\alpha\sin\alpha - \cos^3\alpha + 5\cos\alpha\sin^2\alpha\\ -\sin 2\alpha\\ (1 - 2\sin\alpha)\cos 2\alpha - 2\sin\alpha\cos^2\alpha\end{pmatrix},$$

$$(\bar{\boldsymbol{u}}_1^{\bullet}, \bar{m}^{\bullet}) = \frac{q}{2\sqrt{2}}(3\sin^2\alpha - 1 - \sin\alpha),$$

$$\bar{S}(\alpha) = \frac{q}{2}\begin{pmatrix}1 - \sin^3\alpha\\ 1 - 4\sin^2\alpha + \sin\alpha\\ -\cos^3\alpha\end{pmatrix}.$$

For $\gamma(\alpha)$ and $\bar{\gamma}(\alpha)$ we have thus obtained

$$\gamma(\alpha) = \begin{bmatrix}1, & 0, & 0, & 0\\ \frac{q}{2}(1 - 3\sin^3\alpha), & \frac{\sqrt{2}}{2}\sin\alpha, & \cos\alpha, & -\frac{\sqrt{2}}{2}\sin\alpha\\ \frac{q}{2}.\,3\cos^3\alpha, & \frac{\sqrt{2}}{2}\cos\alpha, & -\sin\alpha, & -\frac{\sqrt{2}}{2}\cos\alpha\\ \frac{q}{2}(\sin\alpha - \cos 2\alpha), & -\frac{\sqrt{2}}{2}, & 0, & -\frac{\sqrt{2}}{2}\end{bmatrix},$$

$$\bar{\gamma}(\alpha) = \begin{bmatrix}1, & 0, & 0, & 0\\ \frac{q}{2}(1 - \sin^3\alpha), & \frac{\sqrt{2}}{2}\sin\alpha, & \cos\alpha, & \frac{\sqrt{2}}{2}\sin\alpha\\ \frac{q}{2}(1 - 4\sin^2\alpha + \sin\alpha), & \frac{\sqrt{2}}{2}, & 0, & -\frac{\sqrt{2}}{2}\\ -\frac{q}{2}\cos^3\alpha, & -\frac{\sqrt{2}}{2}\cos\alpha, & \sin\alpha, & -\frac{\sqrt{2}}{2}\cos\alpha\end{bmatrix}.$$

Put

$$\gamma = \begin{pmatrix} 1, 0 \\ \boldsymbol{s}, \boldsymbol{a} \end{pmatrix}, \quad \bar{\gamma} = \begin{pmatrix} 1, & 0 \\ \bar{\boldsymbol{s}}, & \bar{\boldsymbol{a}} \end{pmatrix}.$$

Then

$$(\bar{\gamma})^{-1} = \begin{pmatrix} 1, & 0 \\ -\bar{\boldsymbol{a}}^{\mathrm{T}}\bar{\boldsymbol{s}}, & \bar{\boldsymbol{a}}^{\mathrm{T}} \end{pmatrix},$$

so that

$$g(\alpha) = \gamma(\bar{\gamma})^{-1} = \begin{pmatrix} 1, & 0 \\ \boldsymbol{s}, & \boldsymbol{a} \end{pmatrix} \begin{pmatrix} 1, & 0 \\ -\bar{\boldsymbol{a}}^{\mathrm{T}}\bar{\boldsymbol{s}}, & \bar{\boldsymbol{a}}^{\mathrm{T}} \end{pmatrix} = \begin{pmatrix} 1, & 0 \\ \boldsymbol{s} - \boldsymbol{a}\bar{\boldsymbol{a}}^{\mathrm{T}}\bar{\boldsymbol{s}}, & \boldsymbol{a}\bar{\boldsymbol{a}}^{\mathrm{T}} \end{pmatrix}.$$

The computation yields

$$\boldsymbol{a}\bar{\boldsymbol{a}}^{\mathrm{T}} = \begin{pmatrix} \cos^2\alpha, & \sin\alpha, & \cos\alpha\sin\alpha \\ -\cos\alpha\sin\alpha, & \cos\alpha, & -\sin^2\alpha \\ -\sin\alpha, & 0, & \cos\alpha \end{pmatrix}, \quad \boldsymbol{s} - \boldsymbol{a}\bar{\boldsymbol{a}}^{\mathrm{T}}\bar{\boldsymbol{s}} = \begin{pmatrix} 0 \\ q\cos\alpha \\ q\sin\alpha \end{pmatrix},$$

i.e.

$$g(\alpha) = \begin{pmatrix} 1, & 0, & 0, & 0 \\ 0, & \cos^2\alpha, & \sin\alpha, & \cos\alpha\sin\alpha \\ q\cos\alpha, & -\cos\alpha\sin\alpha, & \cos\alpha, & -\sin^2\alpha \\ q\sin\alpha, & -\sin\alpha, & 0, & \cos\alpha \end{pmatrix},$$

which is in agreement with the motion of Example 4.6 in which, as we know, it is necessary to put $m_1 = 0$, $m_2 = q\cos\alpha$, $m_3 = q\sin\alpha$.

Fig. 42 shows the axoids of this motion: the fixed axoid $\mathfrak{A}$ and a position of the moving axoid $\bar{\mathfrak{A}}$. Parts of the striction lines $\mathrm{s} \in \mathfrak{A}$, $\mathbf{s} \in \bar{\mathfrak{A}}$ are also shown, together with their common point S on lines $q \equiv \bar{q}$ along which the two surfaces contact. The figure is based on a computer printout.

6.8 Exercise. Determine the invariants of motion and the axoids for a motion determined by the motion of a Frenet frame along a space curve. Hint: choose the Frenet frame of the space curve as the frame in the moving space and determine the invariants of the moving directing cone with the aid of Frenet formulae, which can be obtained by differentiating the Frenet frame.

6.9 Exercise. Assume given a ruled surface which is not cylindrical. Its Frenet frame determines a space motion as in Exercise 6.8. Determine the axoids and invariants of the motion generated in this way.

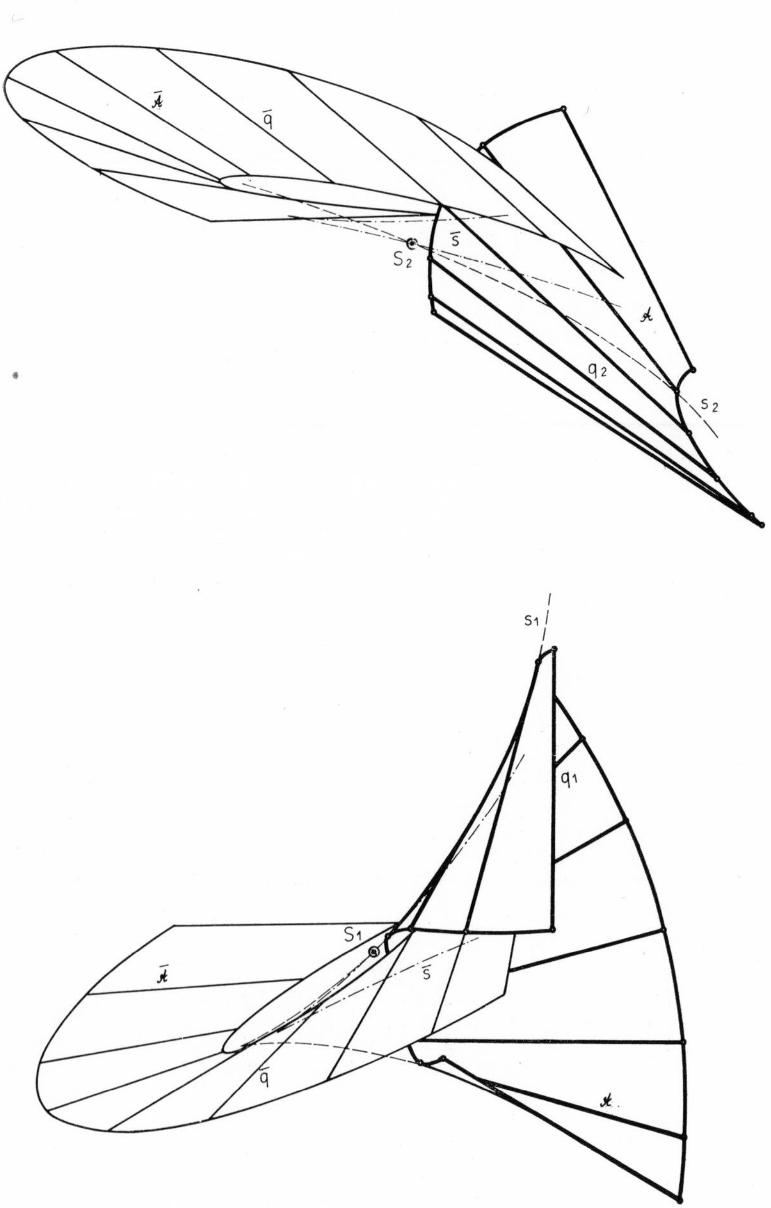

Fig. 42.

7. TRAJECTORY OF A POINT

In this section we shall deal with the trajectory of a point under general space motion. This means that we shall determine the curvature and torsion of the trajectory of the point in dependence on its position and on the invariants of motion. Let us assume, therefore, that motion $g(\varphi)$ is given as a function of the canonical parameter φ with the fixed directing cone $\boldsymbol{R}(\varphi)$, and assume that $\kappa_1 \neq 0$. Then

$$\begin{aligned} \boldsymbol{R} &= \boldsymbol{R}_1 + v_0 \boldsymbol{R}_2 , \\ \boldsymbol{R}' &= \boldsymbol{R}_1' + v_0 \boldsymbol{R}_2' + v_0' \boldsymbol{R}_2 = \kappa_1 \boldsymbol{T}_1 + \mu_1 \boldsymbol{T}_2 + \pi \boldsymbol{R}_2 + \kappa_1 v_0 \boldsymbol{T}_2 \\ &= \kappa_1 \boldsymbol{T}_1 + \psi \boldsymbol{T}_2 + \pi \boldsymbol{R}_2 , \end{aligned}$$

where we denote $v_0' = \pi$, $\mu_1 + v_0 \kappa_1 = \psi$. Further,

$$\begin{aligned} \boldsymbol{R}'' &= \kappa_1' \boldsymbol{T}_1 + \psi' \boldsymbol{T}_2 + \pi' \boldsymbol{R}_2 + \kappa_1(-\kappa_1 \boldsymbol{R}_1 + \kappa_2 \boldsymbol{N}_1 - \mu_1 \boldsymbol{R}_2 + \mu_2 \boldsymbol{N}_2) \\ &\quad + \psi(-\kappa_1 \boldsymbol{R}_2 + \kappa_2 \boldsymbol{N}_2) + \pi \kappa_1 \boldsymbol{T}_2 \\ &= -\kappa_1^2 \boldsymbol{R}_1 + a \boldsymbol{R}_2 + \kappa_1' \boldsymbol{T}_1 + b \boldsymbol{T}_2 + \kappa_1 \kappa_2 \boldsymbol{N}_1 + c \boldsymbol{N}_2 , \end{aligned}$$

where we denote

$$a = \pi' - 2\kappa_1 \mu_1 - v_0 \kappa_1^2 , \quad b = \psi' + \pi \kappa_1 , \quad c = \kappa_1 \mu_2 + \psi \kappa_2 .$$

For a certain given instant φ_0 choose an orthonormal base in E_3 to render the expression of the operators $\boldsymbol{R}, \boldsymbol{R}', \boldsymbol{R}''$ as simple as possible. This can be achieved by taking the point of the striction curve of the fixed axoid as the origin, the axis of the instantaneous helical motion as the axis $\boldsymbol{u}_1$, and the normal to the fixed axoid at the point of the striction curve as the axis $\boldsymbol{u}_2$. In matrix notation the operators $\boldsymbol{R}_1, \boldsymbol{T}_1, \ldots, \boldsymbol{N}_2$ then read

$$\boldsymbol{R}_1 = \begin{pmatrix} 0, & 0, & 0, & 0 \\ 0, & 0, & 0, & 0 \\ 0, & 0, & 0, & -1 \\ 0, & 0, & 1, & 0 \end{pmatrix} , \quad \boldsymbol{T}_1 = \begin{pmatrix} 0, & 0, & 0, & 0 \\ 0, & 0, & 0, & 1 \\ 0, & 0, & 0, & 0 \\ 0, & -1, & 0, & 0 \end{pmatrix} ,$$

$$\boldsymbol{N}_1 = \begin{pmatrix} 0, & 0, & 0, & 0 \\ 0, & 0, & -1, & 0 \\ 0, & 1, & 0, & 0 \\ 0, & 0, & 0, & 0 \end{pmatrix} , \quad \boldsymbol{R}_2 = \begin{pmatrix} 0, & 0, & 0, & 0 \\ 1, & 0, & 0, & 0 \\ 0, & 0, & 0, & 0 \\ 0, & 0, & 0, & 0 \end{pmatrix} ,$$

$$T_2 = \begin{pmatrix} 0, & 0, & 0, & 0 \\ 0, & 0, & 0, & 0 \\ 1, & 0, & 0, & 0 \\ 0, & 0, & 0, & 0 \end{pmatrix}, \quad N_2 = \begin{pmatrix} 0, & 0, & 0, & 0 \\ 0, & 0, & 0, & 0 \\ 0, & 0, & 0, & 0 \\ 1, & 0, & 0, & 0 \end{pmatrix},$$

i.e.

$$R = \begin{pmatrix} 0, & 0, & 0, & 0 \\ v_0, & 0, & 0, & 0 \\ 0, & 0, & 0, & -1 \\ 0, & 0, & 1, & 0 \end{pmatrix}, \quad R' = \begin{pmatrix} 0 & 0, & 0, & 0 \\ \pi, & 0, & 0, & \kappa_1 \\ \psi, & 0, & 0, & 0 \\ 0, & -\kappa_1, & 0, & 0 \end{pmatrix},$$

$$R'' = \begin{pmatrix} 0, & 0, & 0, & 0 \\ a, & 0, & -\kappa_1\kappa_2, & \kappa_1' \\ b, & \kappa_1\kappa_2, & 0, & \kappa_1^2 \\ c, & -\kappa_1', & -\kappa_1^2, & 0 \end{pmatrix}.$$

Further, let X be an arbitrary point from E_3. For the tangent vector of its trajectory at the point X we have $X' = \boldsymbol{R}X$. Further differentiation of this relation yields

$$X'' = R'X + RX' = (R' + R^2)\,X\,,$$

$$X''' = (R'' + RR' + R'R)\,X + (R' + R^2)\,RX$$
$$= (R'' + RR' + 2R'R + R^3)\,X\,.$$

Multiplying the appropriate matrices we easily find that

$$R^2 = \begin{pmatrix} 0, & 0, & 0, & 0 \\ 0, & 0, & 0, & 0 \\ 0, & 0, & -1, & 0 \\ 0, & 0, & 0, & -1 \end{pmatrix}, \quad R'R = \begin{pmatrix} 0, & 0, & 0, & 0 \\ 0, & 0, & \varkappa_1, & 0 \\ 0, & 0, & 0, & 0 \\ -\kappa_1 v_0, & 0, & 0, & 0 \end{pmatrix},$$

$$RR' = \begin{pmatrix} 0, & 0, & 0, & 0 \\ 0, & 0, & 0, & 0 \\ 0, & \kappa_1, & 0, & 0 \\ \psi, & 0, & 0, & 0 \end{pmatrix}, \quad R^3 = -R_1\,.$$

If we now denote the linear operators of the first, second and third derivatives by Ω_1, Ω_2, Ω_3, respectively, we have

$$\Omega_1 = \boldsymbol{R}, \quad \Omega_2 = \begin{pmatrix} 0, & 0, & 0, & 0 \\ \pi, & 0, & 0, & \kappa_1 \\ \psi, & 0, & -1. & 0 \\ 0, & -\kappa_1, & 0, & -1 \end{pmatrix},$$

$$\Omega_3 = \begin{pmatrix} 0, & 0, & 0, & 0 \\ a, & 0, & -\kappa_1\kappa_2 + 2\kappa_1, & \kappa_1' \\ b, & \varkappa_1\kappa_2 + \kappa_1, & 0, & \kappa_1^2 + 1 \\ c + \psi - 2\kappa_1 v_0, & -\kappa_1', & -\kappa_1^2 - 1, & 0 \end{pmatrix}.$$

Put $d = c + \psi - 2\kappa_1 v_0$. If a point

$$X = \begin{pmatrix} 1 \\ x^1 \\ x^2 \\ x^3 \end{pmatrix}$$

is now given, we have $X' = \Omega_1 X$, $X'' = \Omega_2 X$, $X''' = \Omega_3 X$. Upon computing we have

$$X' = \begin{pmatrix} 0 \\ v_0 \\ -x^3 \\ x^2 \end{pmatrix}, \quad X'' = \begin{pmatrix} 0 \\ \pi + \kappa_1 x^3 \\ \psi - x^2 \\ -\kappa_1 x^1 - x^3 \end{pmatrix},$$

$$X''' = \begin{pmatrix} 0 \\ a + (2\kappa_1 - \kappa_1\kappa_2)\, x^2 + \kappa_1' x^3 \\ b + (\kappa_1 + \kappa_2\kappa_1)\, x^1 + (\kappa_1^2 + 1)\, x^3 \\ d - \kappa_1' x^1 - (\kappa_1^2 + 1)\, x^2 \end{pmatrix}.$$

First, we determine the set of inflection points. An inflection point of the trajectory is a point at which the osculation plane of the trajectory is not determined, i.e. the vectors X' and X'' are linearly dependent, $X' \times X'' = 0$. Computation yields

$$X' \times X'' = \begin{pmatrix} (x^2)^2 + (x^3)^2 + \kappa_1 x^1 x^3 - \psi x^2 \\ v_0\kappa_1 x^1 + \pi x^2 + v_0 x^3 + \kappa_1 x^2 x^3 \\ v_0\psi - v_0 x^2 + \pi x^3 + \kappa_1 (x^3)^2 \end{pmatrix},$$

so that the set of inflection points is given by the equations

$$(x^2)^2 + (x^3)^2 + \kappa_1 x^1 x^3 - \psi x^2 = 0\,,$$
$$v_0 \kappa_1 x^1 + \pi x^2 + v_0 x^3 + \kappa_1 x^2 x^3 = 0\,,$$
$$v_0 \psi - v_0 x^2 + \pi x^3 + \kappa_1 (x^3)^2 = 0\,.$$

These three equations are not independent, of course: if we multiply the first by v_0, the second by $-x^3$, and the third by x^2, we arrive at zero. If we want to determine two independent equations of these three, we must eliminate the points for which $x^2 = x^3 = 0$, i.e. the points on the axis of instantaneous motion. We shall investigate these points first. Let $x^2 = x^3 = 0$. Then we have the equations $v_0 \kappa_1 x^1 = 0$ and $v_0 \psi = 0$, and we arrive at the following result: the axis of instantaneous motion belongs to the set of inflection points if and only if $v_0 = 0$. If $v_0 \neq 0$ and $\psi = 0$, it is precisely the point of the striction curve; if $v_0 \neq 0$, $\psi \neq 0$, none of the points of the axis of instantaneous motion belongs to the inflection points. Now, notice the points for which $x^2 \neq 0$ and $x^3 = 0$. For these points we have the equations $v_0 \kappa_1 x^1 + \pi x^2 = 0$, $x^2(x^2 - \psi) = 0$. If $x^3 \neq 0$, we are able to express the second equation with the aid of the first and third.

7.1 Theorem. *The set of inflection points is a curve given by equations*

$$\left.\begin{aligned} \Phi_1 &\equiv (x^2)^2 + (x^3)^2 + \kappa_1 x^1 x^3 - \psi x^2 = 0 \\ \Phi_2 &\equiv v_0 \psi - v_0 x^2 + \pi x^3 + \kappa_1 (x^3)^2 = 0 \end{aligned}\right\} \quad \textit{for} \quad x^3 \neq 0\,,$$

$$v_0 \kappa_1 x^1 + \pi x^2 = 0\,, \quad x^2 - \psi = 0 \qquad \textit{for} \quad x^3 = 0\,, \quad x^2 \neq 0\,,$$

$$v_0 \kappa_1 = 0\,, \quad v_0 \psi = 0 \qquad \textit{for} \quad x^3 = x^2 = 0\,.$$

If $v_0 \neq 0$, there is only one inflection point in the plane $x^3 = 0$ and its coordinates are

$$x^1 = -\frac{\pi \psi}{v_0 \kappa_1}\,, \quad x^2 = \psi\,, \quad x^3 = 0\,.$$

If $v_0 = 0$, a straight line of inflection points (the axis $x^2 = x^3 = 0$) lies in the plane $x^3 = 0$ for $\pi \neq 0$; if $\pi = 0$, there is also the line $x^2 = \psi$, $x^3 = 0$. (Notice that the plane $x^3 = 0$ has a geometrical meaning: it is the asymptotic plane of the fixed axoid.)

If the instantaneous motion is rotation, the inflection curve can be described very easily. Indeed, if $v_0 = 0$, the inflection curve splits into the axis $x^2 = x^3 = 0$ and the parabola

$$x^3 = -\frac{\pi}{\kappa_1}, \quad \pi x^1 + \left(\frac{\psi^2}{4} - \frac{\pi^2}{\kappa_1^2}\right) = \left(x^2 - \frac{\psi}{2}\right)^2,$$

whose axis and apex can be easily determined.

We shall now devote our attention to the general case. First, note that the curve of inflection points lies on a parabolic cylinder with its lines parallel to the axis $x^2 = x^3 = 0$ and with a base whose equation is

$$\kappa_1\left(x^3 + \frac{\pi}{2\kappa_1}\right)^2 = v_0 x^2 + \frac{\pi^2}{4\kappa_1} - v_0\psi .$$

It is again easy to determine the axis and apex of this parabola. Let us investigate whether our curve also lies on some other cone or cylinder, i.e. on some singular quadric. This means that we look for a linear combination of the equations Φ_1 and Φ_2 which would define a singular quadric. This case occurs if the determinant of the matrix formed by the coefficients of the quadric is equal to zero. Let us express the required quadric as $\Phi(x^1, x^2, x^3) \equiv \alpha\Phi_1 + \beta\Phi_2 = 0$, $\alpha, \beta \in \mathbf{R}$. Here we can put $\alpha = 1$, since if $\alpha = 0$ we obtain the familiar parabolic cylinder $\Phi_2 = 0$ as the solution. Therefore, put $\Phi = \Phi_1 + \beta\Phi_2$. The matrix of coefficients of the quadric Φ then reads

$$\begin{pmatrix} 0, & 0, & \frac{1}{2}\kappa_1, & 0 \\ 0, & 1, & 0, & -\frac{1}{2}\psi - \frac{1}{2}\beta v_0 \\ \frac{1}{2}\kappa_1, & 0, & 1 + \beta\kappa_1, & \frac{1}{2}\beta\pi \\ 0, & -\frac{1}{2}\psi - \frac{1}{2}\beta v_0, & \frac{1}{2}\beta\pi, & \beta v_0 \psi \end{pmatrix}.$$

The determinant of this matrix is $\frac{1}{4}\kappa_1^2(\psi - \beta v_0)$. This means that the equation $\kappa_1^2(\psi - \beta v_0) = 0$ for the singular quadric Φ has the only solution $\beta = \psi/v_0$. The equation of the quadric Φ now reads

$$\begin{aligned} \Phi \equiv v_0(x^2)^2 - 2v_0\psi x^2 + v_0\psi^2 + (v_0 + \kappa_1\psi)(x^3)^2 \\ + \pi\psi x^3 + v_0\kappa_1 x^1 x^3 = 0 . \end{aligned}$$

If we introduce new coordinates by the relations

$$y^2 = x^2 - \psi\,, \quad y^3 = x^3\,, \quad y^1 = x^1 - \frac{\pi\psi}{\kappa_1 v_0}\,,$$

we obtain a new equation

$$\Phi(y^1, y^2, y^3) \equiv v_0(y^2)^2 + (v_0 + \kappa_1\psi)(y^3)^2 + v_0\kappa_1 y^1 y^3 = 0\,.$$

From this equation we immediately see that Φ is a conical surface with vertex V having the coordinates $[\pi\psi\kappa_1^{-1}v_0, \psi, 0]$. In the plane $y^1 = (v_0 + \kappa_1\psi)\,\kappa_1^{-1}v_0^{-1}$ the base of the conical surface is expressed by the equation

$$\frac{v_0}{v_0 + \kappa_1\psi}(y^2)^2 + \left(y^3 + \frac{1}{2}\right)^2 = \frac{1}{4}\,.$$

Also, note that the conical surface Φ is tangent to the plane $x^3 = 0$ along the line $x^2 = \psi$, $x^3 = 0$, which lies on the parabolic cylinder $\Phi_2 = 0$ but does not belong to the curve of inflection points. On the whole we may say, therefore, that the curve of inflection points can be constructed, if $v_0 \neq 0$, as the penetration curve of a parabolic cylinder and an elliptical cone; this penetration curve splits into a third-degree curve, the set of inflection points, and the line $x^2 = \psi$, $x^3 = 0$, which does not belong to the set of inflection points.

7.2 Theorem. *The osculation plane of the trajectory of the point X is given by the equation*

$$\begin{aligned}&\xi_1[(x^2)^2 + (x^3)^2 + \kappa_1 x^1 x^3 - \psi x^2] + \\ &+ \xi_2(v_0\kappa_1 x^1 + \pi x^2 + v_0 x^3 + \kappa_1 x^2 x^3) + \\ &+ \xi_3[\kappa_1(x^3)^2 + \pi x^3 - v_0 x^2 + v_0\psi] + D = 0\,,\end{aligned}$$

where

$$\begin{aligned}D = {}&\mu_1 x^1 x^2 - \kappa_1 x^3[(x^1)^2 + (x^2)^2 + (x^3)^2] + \\ &+ (x^1 + \pi)\,[(x^2)^2 + (x^3)^2] + v_0\psi x^3\,.\end{aligned}$$

Proof. The coordinates of the vector $X' \times X''$ are the coordinates of the vector perpendicular to the osculation plane of the trajectory, i.e. the coefficients of the variables ξ_1, ξ_2, ξ_3 in its equation. If we exploit the fact that the osculation plane passes through point X, we obtain the expression for D.

7.3 Theorem. *The principal normal n_X of the trajectory of point X is given by the equation*

$$n_X = \begin{pmatrix} x^1 \\ x^2 \\ x^3 \end{pmatrix} + \lambda \begin{pmatrix} v_0(\psi x^3 + \kappa_1 x^1 x^2) + [(x^2)^2 + (x^3)^2](\pi + \kappa_1 x^3) \\ x^3(v_0\pi - \kappa_1 x^1 x^2) + [v_0^2 + (x^2)^2](\psi - x^2) + (x^3)^2(v_0\kappa_1 - x^2) \\ x^2 x^3(\psi - v_0\kappa_1 - x^2) - [v_0^2 + (x^3)^2](\kappa_1 x^1 + x^3) - v_0\pi x^2 \end{pmatrix}.$$

Proof. The principal normal of the trajectory lies in the osculation plane of the trajectory and is perpendicular to the tangent, i.e. $n_X = X + \lambda(\alpha X' + \beta X'')$, $\lambda \in \mathbf{R}$, where $(\alpha X' + \beta X'', X') = 0$, i.e. $\alpha(X')^2 + \beta(X', X'') = 0$, so that, e.g., we may put $\beta = (X')^2$, $\alpha = -(X', X'')$. The computation yields

$$(X')^2 = v_0^2 + (x^2)^2 + (x^3)^2,$$

$$(X', X'') = v_0(\pi + \kappa_1 x^3) - \psi x^3 - \kappa_1 x^1 x^2,$$

$$\alpha X' + \beta X'' = (\psi x^3 + \kappa_1 x^1 x^2 - v_0\pi - v_0\kappa_1 x^3)\begin{pmatrix} v_0 \\ -x^3 \\ x^2 \end{pmatrix}$$

$$+ [v_0^2 + (x^2)^2 + (x^3)^2]\begin{pmatrix} \pi + \kappa_1 x^3 \\ \psi - x^2 \\ -\kappa_1 x^1 - x^3 \end{pmatrix}$$

$$= \begin{pmatrix} v_0(x^3\psi + \kappa_1 x^1 x^2) + [(x^2)^2 + (x^3)^2](\pi + \kappa_1 x^3) \\ x^3(v_0\pi - \kappa_1 x^1 x^2) + [v_0^2 + (x^2)^2](\psi - x^2) + (x^3)^2(v_0\kappa_1 - x^2) \\ x^2 x^3(\psi - v_0\kappa_1 - x^2) - [v_0^2 + (x^3)^2](\kappa_1 x^1 + x^3) - v_0\pi x^2 \end{pmatrix}.$$

7.4 Theorem. *The points at which the principal normal of the trajectory intersects the axis of instantaneous motion, or is parallel with it, are given by the equation*

$$v_0\{[(x^2)^2 + (x^3)^2](\pi + \kappa_1 x^3) + v_0(\kappa_1 x^1 x^2 + \psi x^3)\} = 0.$$

Therefore, if $v_0 = 0$, each principal normal of the trajectory has the property that it intersects the instantaneous axis of rotation or is parallel with it; if $v_0 \neq 0$, the set of all points at which the principal normal of the trajectory intersects the axis of instantaneous helical motion or is parallel with it is a third-degree surface.

Proof. In the equation of the normal put the second and third coordinates equal to zero and eliminate λ.

To be brief, a line with the property that the principal normals of the trajectories at its points are perpendicular to this line is called *n*-line.

7.5 Theorem. *If $v_0 \neq 0$ and $\pi^2 - 4v_0\kappa_1 > 0$, there exist exactly three n-lines with the equations*

$$x^2 = 0\,, \quad x^3 = (2\kappa_1)^{-1}\left[-\pi \pm (\pi^2 - 4v_0\kappa_1)^{1/2}\right]; \quad x^2 = x^3 = 0\,.$$

If $v_0 \neq 0$ and $\pi^2 = 4v_0\kappa_1$, there exist exactly two n-lines with the equations

$$x^2 = 0\,, \quad x^3 = -(2\kappa_1)^{-1}\,\pi\,; \quad x^2 = x^3 = 0\,.$$

If $v_0 \neq 0$ and $\pi^2 - 4v_0\kappa_1 < 0$, there is just one n-line, namely the axis of instantaneous helical motion.

If $v_0 = 0$, all lines parallel with the axis of instantaneous rotation for which $\pi + \kappa_1 x^3 = 0$ are n-lines.

Proof. Assume given the line $p \equiv [x^1 = a^1 + tu^1,\ x^2 = a^2 + tu^2,\ x^3 = a^3 + tu^3]$, t being the parameter. Let us compute the vector $\boldsymbol{N}$ of the principal normal of the trajectory at the points of line p (up to a proportionality factor). Let the vector $\boldsymbol{N}$ have the coordinates N^1, N^2, N^3 (N^1, N^2, N^3 depend on $a^1, a^2, a^3, u^1, u^2, u^3$, and t). We obtain

$$\begin{aligned}
N^1 &= v_0[a^3 + tu^3 + \kappa_1(a^1 + tu^1)(a^2 + tu^2)] \\
&\quad + [(a^2 + tu^2)^2 + (a^3 + tu^3)^2]\,[\pi + \kappa_1(a^3 + tu^3)]\,, \\
N^2 &= (a^3 + tu^3)\,[v_0\pi - \kappa_1(a^1 + tu^1)(a^2 + tu^2)] \\
&\quad + [v_0^2 + (a^2 + tu^2)^2]\,(\psi - a^2 - tu^2) \\
&\quad + (a^3 + tu^3)^2\,(v_0\kappa_1 - a^2 - tu^2)\,, \\
N^3 &= (a^2 + tu^2)(a^3 + tu^3)\,[\psi - v_0\kappa_1 - a^2 - tu^2] \\
&\quad - [v_0^2 + (a^3 + tu^3)^2]\,[\kappa_1(a^1 + tu^1) + a^3 + tu^3] \\
&\quad - v_0\pi(a^2 + tu^2)\,.
\end{aligned}$$

The line p is an n-line if and only if $N^1u^1 + N^2u^2 + N^3u^3 = 0$ for all t. Comparing the coefficients of the highest powers we obtain the equation $[(u^2)^2 + (u^3)^2]^2 = 0$ which yields $u^2 = u^3 = 0$. If we substitute back into the equation of the n-line, we arrive at

$$v_0[a^3 + \kappa_1(a^1 + tu^1)\,a^2] + [(a^2)^2 + (a^3)^2]\,(\pi + \kappa_1 a^3) = 0\,.$$

If $v_0 = 0$, we have either $a^2 = a^3 = 0$ or $\pi + \kappa_1 a^3 + 0$. If $v_0 \neq 0$, we have $v_0\kappa_1 u^1 a^2 = 0$ for the linear term, i.e. $a^2 = 0$, and we obtain the equation $a^3[v_0 + \pi a^3 + \kappa_1(a^3)^2] = 0$ whose solution yields the desired result.

7.6 Theorem. *The curvature k_X of the trajectory at point X is given by the formula*

$$\begin{aligned}(k_X)^2 = (v_0^2 + \delta^2)^{-3}\,\{&\delta^4 \\ &+ \delta^2[2\kappa_1 x^1 x^3 - 2\psi x^2 + \pi^2 + v_0^2 + 2\pi\kappa_1 x^3 + \kappa_1^2(x^3)^2] \\ &+ (x^1)^2\,\kappa_1^2[v_0^2 + (x^3)^2] + \psi^2[v_0^2 + (x^2)^2] \\ &+ 2x^1x^2x^3\kappa_1(v_0\kappa_1 - \psi) + 2v_0\kappa_1 x^1(\pi x^2 + v_0 x^3) \\ &+ 2v_0\psi[x^3\pi + \kappa_1(x^3)^2 - v_0 x^2]\}\,,\end{aligned}$$

where $\delta^2 = (x^2)^2 + (x^3)^2$, i.e. δ is the distance of point X from the axis of instantaneous motion.

Proof. The curvature k_X is given by the formula

$$k_x^2 = \frac{(X' \times X'',\, X' \times X'')}{(X', X')^3}\,.$$

Now it suffices to substitute from the expressions for $X' \times X''$ and for X'.

The formula for the torsion τ_X of the trajectory at point X reads

$$\tau_X = \frac{|X', X'', X'''|}{(X' \times X'',\, X' \times X'')}\,.$$

Since we know the expressions for X', X'' and X''', only a mechanical computation is involved. In the general case the resulting formula is too complicated; we present it only for helical motions, i.e. for motions whose all invariants are constant. Thus we obtain

$$\psi' = 0\,, \quad \pi = v_0' = 0\,, \quad \kappa_1' = 0\,,$$

so that

$$
\begin{aligned}
&\left|X', X'', X'''\right| \\
&= \begin{vmatrix} v_0, & \kappa_1 x^3, & a + (2\kappa_1 - \kappa_1\kappa_2)\,x^2 \\ -x^3, & \psi - x^2, & (\kappa_1 + \kappa_1\kappa_2)\,x^1 + (\kappa_1^2 + 1)\,x^3 \\ x^2, & -\kappa_1 x^1 - x^3, & d - (\kappa_1^2 + 1)\,x^2 \end{vmatrix} \\
&= (\delta^2 + \kappa_1 x^1 x^3 - \psi x^2)(v_0 - 2\kappa_1 \mu_1) \\
&\quad + d[\kappa_1 (x^3)^2 + v_0(\psi - x^2)] + \kappa_1 x^2 (2 - \kappa_2)(\delta^2 - \psi x^2) \\
&\quad + v_0 \kappa_1 x^1 (\kappa_2 + 1)(\kappa_1 x^1 + x^3) + 3\kappa_1^2 x^1 x^2 x^3 \,.
\end{aligned}
$$

The formula for τ_X is now obvious. We also see that the set of points at which the trajectory has zero torsion is a third-degree surface.

Finally, let us investigate the conditions which the motion has to satisfy at a given instant so that all the points have zero torsion. If we consider the determinant $\left|X', X'', X'''\right|$ in the general case, we see that its term $x^1 x^2 x^3$ has again the coefficient $3\kappa_1^2$ as in the case of helical motion. Therefore, κ_1 must be equal to zero. The coefficient with $(x^3)^3$ is κ_1', i.e. it must also be equal to zero. Under these assumptions

$$
\left|X', X'', X'''\right| = \begin{vmatrix} v_0, & \pi, & a \\ -x^3, & \psi - x^2, & b + x^3 \\ x^2, & -x^3, & d - x^2 \end{vmatrix} = 0 \,.
$$

Writing this out we arrive at the equations

$$
v_0 + a = 0\,, \quad v_0 \psi d = 0\,, \quad v_0 b + \pi d = 0\,, \quad b\pi - v_0 d = 0\,,
$$

which have two solutions

1. $b = 0,\ d = 0,\ v_0 = -a,$

2. $\pi = 0,\ v_0 = 0,\ a = 0.$

Therefore, torsion is zero at all points if at a given instant we have

$$
\kappa_1 = \kappa_1' = \psi' = \psi(\kappa_2 + 1) = v_0 + \pi' = 0
$$

or

$$
\kappa_1 = \kappa_1' = \pi = v_0 = \pi' = 0\,.
$$

So far, we have treated only local properties of space motion, i.e. the properties of motion at a given instant. It is interesting to mention how one is able to solve problems concerning the behaviour of motion as a whole. As an example we shall determine all helical motions (i.e. motions which have constant invariants) which have at least one linear trajectory (a single point will also be considered a linear trajectory).

Let us choose a fixed point of the moving system and denote it by $\bar{A}$, also denote the Frenet frame of the moving axoid by $\bar{\mathscr{F}} = \{\bar{S}, \bar{\boldsymbol{u}}_1, \bar{\boldsymbol{u}}_2, \bar{\boldsymbol{u}}_3\}$. For all φ we have

$$\bar{A} = \bar{S}(\varphi) + x^1(\varphi)\,\bar{\boldsymbol{u}}_1(\varphi) + x^2(\varphi)\,\bar{\boldsymbol{u}}_2(\varphi) + x^3(\varphi)\,\bar{\boldsymbol{u}}_3(\varphi)\,,$$

where $x^1(\varphi), x^2(\varphi), x^3(\varphi)$ are the coordinates of point $\bar{A}$ in frame $\bar{\mathscr{F}}(\varphi)$. Point $\bar{A}$ is a fixed point of the moving system, i.e. $\bar{A}'$ must be zero, $\bar{A}' = \boldsymbol{0}$. We shall now use the Frenet formula for the fixed axoid in the form (6.1); we obtain

$$\bar{S}' = \frac{\mathrm{d}\bar{S}}{\mathrm{d}s}\frac{\mathrm{d}s}{\mathrm{d}\varphi} = \kappa_1(\bar{q}\bar{\boldsymbol{u}}_1 + \bar{d}\bar{\boldsymbol{u}}_3) = \bar{\mu}_2\bar{\boldsymbol{u}}_1 + \mu_1\bar{\boldsymbol{u}}_3\,,$$

and, similarly,

$$\bar{\boldsymbol{u}}_1' = \kappa_1\bar{\boldsymbol{u}}_2\,,\quad \bar{\boldsymbol{u}}_2' = -\kappa_1\bar{\boldsymbol{u}}_1 + \bar{\kappa}_2\bar{\boldsymbol{u}}_3\,,\quad \bar{\boldsymbol{u}}_3' = -\bar{\kappa}_2\bar{\boldsymbol{u}}_2\,.$$

We now substitute into the expression for $\bar{A}'$:

$$\begin{aligned}\bar{A}' = \boldsymbol{0} &= \bar{S}' + (x^1)'\,\bar{\boldsymbol{u}}_1 + (x^2)'\,\bar{\boldsymbol{u}}_2 + (x^3)'\,\bar{\boldsymbol{u}}_3 + x^1\kappa_1\bar{\boldsymbol{u}}_2\\ &\quad + x^2(-\kappa_1\bar{\boldsymbol{u}}_1 + \bar{\kappa}_2\bar{\boldsymbol{u}}_3) + x^3(-\bar{\kappa}_2\bar{\boldsymbol{u}}_2)\\ &= \bar{\boldsymbol{u}}_1[\bar{\mu}_2 + (x^1)' - \kappa_1x^2] + \bar{\boldsymbol{u}}_2[(x^2)' + \kappa_1x^1 - \bar{\kappa}_2x^3]\\ &\quad + \bar{\boldsymbol{u}}_3[\mu_1 + (x^3)' + \bar{\kappa}_2x^2]\,.\end{aligned}$$

7.7 Theorem. *Let $\bar{A}$ be a fixed point of the moving system; put $\bar{A} = \bar{S} + x^i\bar{\boldsymbol{u}}_i$, where $\bar{\mathscr{F}} = \{\bar{S}, \bar{\boldsymbol{u}}_1, \bar{\boldsymbol{u}}_2, \bar{\boldsymbol{u}}_3\}$ is the Frenet frame of the moving axoid. Then*

$$(x^1)' = \kappa_1x^2 - \bar{\mu}_2\,,\quad (x^2)' = \bar{\kappa}_2x^3 - \kappa_1x^1\,,\quad (x^3)' = -\bar{\kappa}_2x^2 - \mu_1\,.$$

The trajectory of point $\bar{A}$ is the curve $A(\varphi) = g(\varphi)\,\bar{A}$. Moreover,

$$A(\varphi) = g(\varphi)\,\bar{A} = g(\varphi)\,(\bar{S} + x^i\bar{\boldsymbol{u}}_i) = g(\varphi)\,\bar{S} + x^i\,g(\varphi)\,\bar{\boldsymbol{u}}_i = S + x^i\boldsymbol{u}_i\,.$$

Therefore, the point $A(\varphi)$ has the coordinates $x^1(\varphi)$, $x^2(\varphi)$, $x^3(\varphi)$ in the frame $\mathscr{F} = \{S, \boldsymbol{u}_1, \boldsymbol{u}_2, \boldsymbol{u}_3\}$ which is the Frenet frame of the fixed axoid, S being the point of the striction curve, $\boldsymbol{u}_1$ the vector of the generating line, and $\boldsymbol{u}_2$ the vector of the normal at point S. This means that if a line is to be the trajectory of point $\bar{A}$ during a motion, point $A(\varphi)$ must be an inflection point of the trajectory, i.e. the functions $x^1(\varphi), x^2(\varphi), x^3(\varphi)$ must satisfy the equations of the inflection curve for all φ. We thus have the theorem:

7.8 Theorem. *Let $g(\varphi)$ be a motion with $\kappa_1 \neq 0$; denote the Frenet frame of the moving axoid $\bar{\mathcal{F}} = \{\bar{S}, \bar{\boldsymbol{u}}_1, \bar{\boldsymbol{u}}_2, \bar{\boldsymbol{u}}_3\}$. Point $\bar{A} = \bar{S} + x^i\bar{\boldsymbol{u}}_i$ has a linear trajectory if and only if the functions x^1, x^2, x^3 satisfy the equations*

$$\text{(a)}\ (x^1)' = \kappa_1 x^2 - \bar{\mu}_2,\ (x^2)' = \bar{\kappa}_2 x^3 - \kappa_1 x^1,\ (x^3)' = -\bar{\kappa}_2 x^2 - \mu_1.$$

$$\begin{aligned}\text{(b)}\ &(x^2)^2 + (x^3)^2 + \kappa_1 x^1 x^3 - \psi x^2 = 0,\\ &v_0\kappa_1 x^1 + \pi x^2 + v_0 x^3 + \kappa_1 x^2 x^3 = 0,\\ &v_0\psi - v_0 x^2 + \pi x^3 + \kappa_1 (x^3)^2 = 0.\end{aligned}$$

Now, we shall devote our attention to helical motions. Equation (b) will become

$$\begin{aligned}\text{(b}'\text{)}\ &(x^2)^2 + (x^3)^2 + \kappa_1 x^1 x^3 - \psi x^2 &= 0,\\ &v_0(\kappa_1 x^1 + x^3) + \kappa_1 x^2 x^3 &= 0,\\ &v_0(\psi - x^2) + \kappa_1 (x^3)^2 &= 0.\end{aligned}$$

Our problem is to find the conditions under which the system (a), (b′) can be solved, not forgetting that all the invariants of motion are now constants. We distinguish two cases:

1. $x_3 = 0$.

The first of the (b′) equations yields $x^2(x^2 - \psi) = 0$, i.e. $x^2 =$ const. The second of the (a) equations yields $-\kappa_1 x^1 = 0$, i.e. $x^1 = 0$. Equations (a) and (b′) thus reduce to

$$\kappa_1 x^2 - \bar{\mu}_2 = -\bar{\kappa}_2 x^2 - \mu_1 = 0\,, \quad x^2(x^2 - \psi) = v_0(x^2 - \psi) = 0\,.$$

The solutions of these equations are

α) $\psi \neq 0$: $x^2 = \psi = \mu_1 + v_0\kappa_1$, $x^1 = x^3 = 0$, and the solvability condition is

$$\kappa_1(\mu_1 + v_0\kappa_1) - \bar{\mu}_2 = -\bar{\kappa}_2(\mu_1 + v_0\kappa_1) - \mu_1 = 0\,,$$

i.e.

$$\bar{\mu}_2 = \kappa_1(\mu_1 + v_0\kappa_1)\,, \quad \bar{\kappa}_2 = -(\mu_1 + v_0\kappa_1)^{-1}\,\mu_1\,.$$

The trajectory we are seeking is always a straight line because the point $\bar{A}$ does not lie on the axis of instantaneous motion, i.e. the tangent vector of the trajectory remains nonzero.

β) $\psi = 0$: $x^1 = x^2 = x^3 = 0$. The solvability conditions are

$$\mu_1 = \bar{\mu}_2 = 0\,, \quad \mu_1 + v_0\kappa_1 = \psi = v_0\kappa_1 = 0\,,$$

i.e.

$$v_0 = 0\,, \quad \text{so that} \quad \mu_2 = 0\,.$$

In this case spherical motion is involved and the trajectory of point $\bar{A}$ is a point, the centre of the sphere.

2. $x^3 \neq 0$ [i.e. there exists φ_0 such that $x^3(\varphi_0) \neq 0$]. The third of the (b′) equations implies $v_0 \neq 0$. We shall first prove that the solutions $x^1(\varphi)$, $x^2(\varphi)$, $x^3(\varphi)$, provided they exist, are necessarily constant. Equations (b′) must hold for all φ. Thus, let us differentiate the third of the (b′) equations:

$$v_0(x^2)' + 2\kappa_1\, x^3(x^3)' = 0\,.$$

If we substitute from (a), we obtain

$$v_0(x^3\bar{\kappa}_2 - x^1\kappa_1) + 2\kappa_1 x^3(\bar{\kappa}_2 x^2 + \mu_1) = 0\,,$$

i.e.

$$2\bar{\kappa}_2\kappa_1 x^2 x^3 + v_0 x^3\bar{\kappa}_2 - v_0 x^1\kappa_1 + 2\kappa_1 x^3\mu_1 = 0\,.$$

We substitute for $\kappa_1 x^2 x^3$ from the second of the (b′) equations and arrive at

$$2\bar{\kappa}_2(v_0\kappa_1 x^1 + v_0 x^3) - v_0 x^3\bar{\kappa}_2 + v_0 x^1\kappa_1 - 2\kappa_1 x^3\mu_1 = 0\,,$$

i.e.

$$v_0\kappa_1 x^1(1 + 2\bar{\kappa}_2) + x^3(\bar{\kappa}_2 v_0 - 2\kappa_1\mu_1) = 0\,.$$

Now, we must distinguish two cases:

α) $1 + 2\bar{\kappa}_2 \neq 0$.

Multiply the second of the (b′) equations by the number $1 + 2\bar{\kappa}_2$ and substitute for $v_0\kappa_1 x^1(1 + 2\bar{\kappa}_2)$; this yields

$$(1 + \bar{\kappa}_2)(-v_0 x^3 - \kappa_1 x^2 x^3) + x^3(\bar{\kappa}_2 v_0 - 2\kappa_1\mu_1) = 0\,,$$

i.e.

$$(1 + 2\bar{\kappa}_2)(v_0 + \kappa_1 x^2) - \bar{\kappa}_2 v_0 + 2\kappa_1\mu_1 = 0\,.$$

This is a linear equation in x^2 so that $x^2 = const.$ because the coefficient at x^2 is different from zero.

β) $1 + 2\bar{\kappa}_2 = 0$.

Then we also have $\bar{\kappa}_2 v_0 - 2\kappa_1 u_1 = 0$. Differentiating the second of the (b′) equations we obtain

$$v_0[\kappa_1(x^1)' + (x^3)'] + \kappa_1(x^2)'\, x^3 + \kappa_1 x^2 (x^3)' = 0\,.$$

Upon substitution we have

$$v_0(\kappa_1 x^2 - \kappa_1 \bar{\mu}_2 - \bar{\kappa}_2 x^2 - \mu_1) + \kappa_1 x^3(x^3 \bar{\kappa}_2 - x^1 \kappa_1)$$
$$+ \kappa_1 x^2(-\bar{\kappa}_2 x^2 - \mu_1) = 0\,,$$

i.e.

$$v_0 x^2(\kappa_1^2 - \bar{\kappa}_2) - v_0(\kappa_1 \bar{\mu}_2 + \mu_1) + \kappa_1 \bar{\kappa}_2[(x^3)^2 - (x^2)^2]$$
$$- \kappa_1^2 x^1 x^3 - \kappa_1 \mu_1 x^2 = 0\,.$$

Substituting for

$$-\kappa_1^2 x^1 x^3 = \kappa_1 (x^2)^2 + \kappa_1 (x^3)^2 - \kappa_1 x^2 \psi$$

from the first of the (b′) equations we arrive at

$$v_0 x^2(\kappa_1^2 - \bar{\kappa}_2) - v_0(\kappa_1 \bar{\mu}_2 + \mu_1) + \kappa_1 \bar{\kappa}_2[(x^3)^2 - (x^2)^2]$$
$$+ \kappa_1 (x^2)^2 + \kappa_1 (x^3)^2 - \kappa_1 x^2 \psi - \kappa_1 \mu_1 x^2 = 0\,;$$

if we substitute for $(x^3)^2$ from the third of the (b′) equations, we have

$$(\bar{\kappa}_2 + 1)\, v_0(\psi - x^2) + \kappa_1 (x^2)^2 + v_0 x^2(\kappa_1^2 - \bar{\kappa}_2) - v_0(\kappa_1 \bar{\mu}_2 + \mu_1)$$
$$- \kappa_1 x^2(\psi + \mu_1) = 0\,,$$

which is a quadratic equation in x^2 with a nonzero term at the second power; it has two solutions at the most, so that again $x^2 = \text{const}$. As in case α) we now obtain $x^1 = const.$, $x^3 = const.$

From now on we consider the two cases α) and β) together. The (a) equations read

$$\kappa_1 x^2 - \bar{\mu}_2 = -\bar{\kappa}_2 x^2 - \mu_1 = \bar{\kappa}_2 x^3 - \kappa_1 x^1 = 0\,,$$

i.e.

$$x^2 = \bar{\mu}_2 \kappa_1^{-1}\,, \quad \bar{\kappa}_2 x^3 = \kappa_1 x^1\,.$$

The second of the (b′) equations yields

$$v_0\bar{\kappa}_2 x^3 + v_0 x^3 + \kappa_1\bar{\mu}_2\kappa_1^{-1}x^3 = 0\,, \quad \text{i.e.} \quad v_0\bar{\kappa}_2 + v_0 + \bar{\mu}_2 = 0\,.$$

Since $v_0 + \bar{\mu}_2 = \mu_2$, we have $v_0\bar{\kappa}_2 + \mu_2 = 0$. Further, we have

$$0 = -\bar{\kappa}_2 x^2 - \mu_1 = -\bar{\kappa}_2\bar{\mu}_2\kappa_1^{-1} + \mu_1\,,$$

i.e.

$$\mu_1\kappa_1 + \bar{\mu}_2\bar{\kappa}_2 = 0\,.$$

The third of the (b′) equations yields

$$v_0\psi - v_0\bar{\mu}_2\kappa_1^{-1} + \kappa_1(x^3)^2 = 0\,,$$

i.e.

$$v_0\kappa_1(\mu_1 + v_0\kappa_1) - v_0\bar{\mu}_2 + \kappa_1^2(x^3)^2 = 0\,.$$

Let us show that $v_0\kappa_1(\mu_1 + v_0\kappa_1) - v_0\bar{\mu}_2 > 0$. Indeed,

$$\begin{aligned} &v_0\kappa_1(\mu_1 + v_0\kappa_1) - v_0\bar{\mu}_2 = -v_0\bar{\kappa}_2\bar{\mu}_2 + v_0^2\kappa_2^1 - v_0\bar{\mu}_2 \\ &= v_0^2\kappa_1^2 - v_0(1 + \bar{\kappa}_2)\,\bar{\mu}_2 = v_0^2\kappa_1^2 - v_0(1 + \bar{\kappa}_2)(-v_0 - v_0\bar{\kappa}_2) \\ &= v_0^2[\kappa_1^2 + (1 + \bar{\kappa}_2)^2]^2 > 0\,. \end{aligned}$$

The last inequality means that in this case the desired motion does not exist because the equation in x^3 has no real solution.

7.9 Theorem. *A helical motion has a point or linear trajectory only in these two cases*:

1. *Spherical motion. The centre of the sphere has a point trajectory.*
2. $\bar{\mu}_2 = \kappa_1(\mu_1 + v_0\kappa_1)$, $\bar{\kappa}_2 = -(\mu_1 + v_0\kappa_1)^{-1}\,\mu_1$, $\mu_1 + v_0\kappa_1 \neq 0$. *The point* $\bar{A} = \bar{S} + x^1\bar{\boldsymbol{u}}_1 + x^2\bar{\boldsymbol{u}}_2 + x^3\bar{\boldsymbol{u}}_3$, *where* $x^1 = x^3 = 0$, $x^2 = \mu_1 + v_0\kappa_1$, *has a linear trajectory.*

7.10 Example. Since the statement of Theorem 7.9 may seem rather surprising, namely the fact that there exist cases of helical motion which have a linear trajectory, let us actually mention some of them. In Example 7.10 helical motions will be found which have one linear trajectory and whose instantaneous motions are rotations, i.e. let us assume that $v_0 = 0$. Theorem 7.9 then yields $\bar{\mu}_2 = \mu_1\kappa_1$, $\bar{\kappa}_2 = -1$. To simplify the computation let us put $\kappa_1 = 1$, $\mu_1 = 1$. Now, let us determine the moving axoid. The Frenet formulae for the moving axoid read

$$\frac{\mathrm{d}\bar{S}}{\mathrm{d}\varphi} = \bar{\boldsymbol{u}}_1 + \bar{\boldsymbol{u}}_3\,, \quad \frac{\mathrm{d}\bar{\boldsymbol{u}}_1}{\mathrm{d}\varphi} = \bar{\boldsymbol{u}}_2\,, \quad \frac{\mathrm{d}\bar{\boldsymbol{u}}_2}{\mathrm{d}\varphi} = -\bar{\boldsymbol{u}}_1 - \bar{\boldsymbol{u}}_3\,, \quad \frac{\mathrm{d}\bar{\boldsymbol{u}}_3}{\mathrm{d}\varphi} = \bar{\boldsymbol{u}}_2\,.$$

Let us choose the initial condition so that

$$\bar{S}(0) = \begin{pmatrix}1\\0\\0\\0\end{pmatrix}, \quad \bar{\boldsymbol{u}}_1(0) = \begin{pmatrix}0\\1\\0\\0\end{pmatrix}, \quad \bar{\boldsymbol{u}}_2(0) = \begin{pmatrix}0\\0\\1\\0\end{pmatrix}, \quad \bar{\boldsymbol{u}}_3(0) = \begin{pmatrix}0\\0\\0\\1\end{pmatrix}.$$

Differentiating equation $\bar{\boldsymbol{u}}_2' = -\bar{\boldsymbol{u}}_1 - \bar{\boldsymbol{u}}_3$ and substituting from the other two we obtain $\bar{\boldsymbol{u}}_2'' = -2\bar{\boldsymbol{u}}_2$, i.e. $\bar{\boldsymbol{u}}_2'' + 2\bar{\boldsymbol{u}}_2 = 0$. The general solution of this equation is

$$\bar{\boldsymbol{u}}_2 = \boldsymbol{A}\cos(\varphi\sqrt{2}) + \boldsymbol{B}\sin(\varphi\sqrt{2})\,,$$

where $\boldsymbol{A}$ and $\boldsymbol{B}$ are arbitrary vectors.

The initial condition yields (we omit the first coordinate)

$$\bar{\boldsymbol{u}}_2(0) = \begin{pmatrix}0\\1\\0\end{pmatrix} = \boldsymbol{A}\,.$$

We then have

$$\bar{\boldsymbol{u}}_3 = \int \bar{\boldsymbol{u}}_2(\varphi)\,\mathrm{d}\varphi\,, \quad \text{i.e.} \quad \bar{\boldsymbol{u}}_3 = \frac{\boldsymbol{A}}{\sqrt{2}}\sin(\varphi\sqrt{2}) - \frac{\boldsymbol{B}}{\sqrt{2}}\cos(\varphi\sqrt{2}) + \boldsymbol{C}\,,$$

and, similarly,

$$\bar{\boldsymbol{u}}_1 = \frac{\boldsymbol{A}}{\sqrt{2}}\sin(\varphi\sqrt{2}) - \frac{\boldsymbol{B}}{\sqrt{2}}\cos(\varphi\sqrt{2}) + \boldsymbol{D}\,,$$

where $\boldsymbol{C}$ and $\boldsymbol{D}$ are arbitrary vectors. If we substitute into the equation $\bar{\boldsymbol{u}}_2' = -\bar{\boldsymbol{u}}_1 - \bar{\boldsymbol{u}}_3$, we obtain

$$-\sqrt{(2)}\,\boldsymbol{A}\sin(\varphi\sqrt{2}) + \sqrt{(2)}\,\boldsymbol{B}\cos(\varphi\sqrt{2})$$

$$= -\frac{\boldsymbol{A}}{\sqrt{2}}\sin(\varphi\sqrt{2}) + \frac{\boldsymbol{B}}{\sqrt{2}}\cos(\varphi\sqrt{2}) - \boldsymbol{D} - \frac{\boldsymbol{A}}{\sqrt{2}}\sin(\varphi\sqrt{2})$$

$$+ \frac{\boldsymbol{B}}{\sqrt{2}}\cos(\varphi\sqrt{2}) - \boldsymbol{C},$$

so that $\boldsymbol{C} + \boldsymbol{D} = 0$. The initial conditions imply

$$\bar{\boldsymbol{u}}_3(0) = -\frac{\boldsymbol{B}}{\sqrt{2}} + \boldsymbol{C} = \begin{pmatrix}0\\0\\1\end{pmatrix}, \quad \bar{\boldsymbol{u}}_1(0) = -\frac{\boldsymbol{B}}{\sqrt{2}} + \boldsymbol{D} = \begin{pmatrix}1\\0\\0\end{pmatrix}.$$

Addition yields

$$-\boldsymbol{B}\sqrt{2} = \begin{pmatrix}1\\0\\1\end{pmatrix},$$

i.e.

$$\boldsymbol{B} = \begin{bmatrix}-\frac{\sqrt{2}}{2}\\0\\-\frac{\sqrt{2}}{2}\end{bmatrix}, \quad \boldsymbol{C} = \begin{bmatrix}-\frac{1}{2}\\0\\\frac{1}{2}\end{bmatrix}, \quad \boldsymbol{D} = \begin{bmatrix}\frac{1}{2}\\0\\-\frac{1}{2}\end{bmatrix}.$$

Thus, the solution is

$$\bar{\boldsymbol{u}}_1(\varphi) = \begin{bmatrix}\frac{\cos(\varphi\sqrt{2}) + 1}{2}\\\frac{\sin(\varphi\sqrt{2})}{\sqrt{2}}\\\frac{\cos(\varphi\sqrt{2}) - 1}{2}\end{bmatrix}, \quad \bar{\boldsymbol{u}}_2(\varphi) = \begin{bmatrix}\frac{-\sin(\varphi\sqrt{2})}{\sqrt{2}}\\\cos(\varphi\sqrt{2})\\\frac{-\sin(\varphi\sqrt{2})}{\sqrt{2}}\end{bmatrix},$$

$$\bar{\boldsymbol{u}}_3(\varphi) = \begin{bmatrix}\frac{\cos(\varphi\sqrt{2}) - 1}{2}\\\frac{\sin(\varphi\sqrt{2})}{\sqrt{2}}\\\frac{\cos(\varphi\sqrt{2}) + 1}{2}\end{bmatrix}.$$

Further,

$$\bar{S}' = \begin{bmatrix} \cos(\varphi\sqrt{2}) \\ \sqrt{(2)}\sin(\varphi\sqrt{2}) \\ \cos(\varphi\sqrt{2}) \end{bmatrix}, \quad \text{i.e.} \quad S = \begin{bmatrix} \dfrac{\sin(\varphi\sqrt{2})}{\sqrt{2}} + C_1 \\ -\cos(\varphi\sqrt{2}) + C_2 \\ \dfrac{\sin(\varphi\sqrt{2})}{\sqrt{2}} + C_3 \end{bmatrix}.$$

The initial condition yields $C_1 = C_3 = 0$, $C_2 = 1$, so that

$$\bar{S} = \begin{bmatrix} \dfrac{\sin(\varphi\sqrt{2})}{\sqrt{2}} \\ 1 - \cos(\varphi\sqrt{2}) \\ \dfrac{\sin(\varphi\sqrt{2})}{\sqrt{2}} \end{bmatrix},$$

and the moving axoid is

$$\overline{\mathfrak{A}}(\varphi, \lambda) = \begin{bmatrix} \dfrac{\sin(\varphi\sqrt{2})}{\sqrt{2}} \\ 1 - \cos(\varphi\sqrt{2}) \\ \dfrac{\sin(\varphi\sqrt{2})}{\sqrt{2}} \end{bmatrix} + \lambda \begin{bmatrix} \dfrac{\cos(\varphi\sqrt{2}) + 1}{2} \\ \dfrac{\sin(\varphi\sqrt{2})}{\sqrt{2}} \\ \dfrac{\cos(\varphi\sqrt{2}) - 1}{2} \end{bmatrix}.$$

If we eliminate the parameter φ, we find that the striction curve of the moving axoid has the equations $x^1 = x^3$, $(x^1)^2 + (x^3)^2 + (x^2 - 1)^2 = 1$, thus it is a circle whose centre has the coordinates $x^1 = x^3 = 0$, $x^2 = 1$. The spherical image of the moving axoid is described by the equation $x^1 - x^3 = 1$ [the spherical image is the set of terminal points of the vector $\boldsymbol{u}_1(\varphi)$, located at the same origin]. Consequently, the spherical image is also a circle, namely a circle in a plane parallel with the plane of the striction curve. This means that the moving axoid is a one-sheet hyperboloid of revolution. Also, note that the point which is to have a linear trajectory is the point with coordinates $x^1 = x^3 = 0$, $x^2 = 1$, relative to the frame $\overline{\mathscr{F}}$.

But since this point should have constant coordinates, and according to the chosen initial conditions one of the moving frames is equal to the fundamental frame in E_3 (for $\varphi = 0$), the considered point is at the same time the center of the mentioned one-sheet hyperboloid of revolution.

We now construct the fixed axoid. The relations $\mu_2 - \bar{\mu}_2 = v_0$, $\kappa_2 - \bar{\kappa}_2 = 1$ yield $\mu_2 = \bar{\mu}_2 = \mu_1 = 1$, $\kappa_2 = 0$. The Frenet formulae for the fixed axoid read

$$S' = \boldsymbol{u}_1 + \boldsymbol{u}_3\,, \quad \boldsymbol{u}_1' = \boldsymbol{u}_2\,, \quad \boldsymbol{u}_2' = -\boldsymbol{u}_1\,, \quad \boldsymbol{u}_3' = \boldsymbol{0}\,.$$

Let us choose the same initial condition as for the moving axoid. The solution of the differential equations is

$$\boldsymbol{u}_1 = \begin{pmatrix} \cos\varphi \\ \sin\varphi \\ 0 \end{pmatrix}, \quad \boldsymbol{u}_2 = \begin{pmatrix} -\sin\varphi \\ \cos\varphi \\ 0 \end{pmatrix}, \quad \boldsymbol{u}_3 = \begin{pmatrix} 0 \\ 0 \\ 1 \end{pmatrix}, \quad S = \begin{pmatrix} \sin\varphi \\ 1 - \cos\varphi \\ \varphi \end{pmatrix}.$$

The striction curve is a helix lying on the cylinder $(x^1)^2 + (x^2 - 1)^2 = 1$, i.e. on a cylinder with the axis $x^1 = 0$, $x^2 = 1$, the generating lines of the fixed axoid are perpendicular to this axis, i.e. a rectangular open helical surface is involved.

The desired motion has to move the moving axoid (and its Frenet frame) into the fixed axoid (and its Frenet frame). If we put $\overline{\mathscr{F}} = \mathscr{R}_0\bar{\gamma}$, $\mathscr{F} = \mathscr{R}_0\gamma$, where $\mathscr{R}_0$ is the fundamental frame in E_3, the resulting motion has the form $g(\varphi) = \gamma(\varphi)\,\bar{\gamma}^{-1}(\varphi)$. [To be precise $\overline{\mathscr{R}}_0(\varphi) = \mathscr{R}_0\, g(\varphi)$.] Here

$$\bar{\gamma}(\varphi) = \begin{bmatrix} 1, & 0, & 0, & 0, \\ \dfrac{\sin(\varphi\sqrt{2})}{\sqrt{2}}, & \dfrac{\cos(\varphi\sqrt{2}) + 1}{2}, & \dfrac{-\sin(\varphi\sqrt{2})}{\sqrt{2}}, & \dfrac{\cos(\varphi\sqrt{2}) - 1}{2} \\ 1 - \cos(\varphi\sqrt{2}), & \dfrac{\sin(\varphi\sqrt{2})}{\sqrt{2}}, & \cos(\varphi\sqrt{2}), & \dfrac{\sin(\varphi\sqrt{2})}{\sqrt{2}} \\ \dfrac{\sin(\varphi\sqrt{2})}{\sqrt{2}}, & \dfrac{\cos(\varphi\sqrt{2}) - 1}{2}, & \dfrac{-\sin(\varphi\sqrt{2})}{\sqrt{2}}, & \dfrac{\cos(\varphi\sqrt{2}) + 1}{2} \end{bmatrix},$$

$$\gamma(\varphi) = \begin{pmatrix} 1, & 0, & 0, & 0 \\ \sin\varphi, & \cos\varphi, & -\sin\varphi, & 0 \\ 1-\cos\varphi, & \sin\varphi, & \cos\varphi, & 0 \\ \varphi, & 0, & 0, & 1 \end{pmatrix}.$$

Now, consider the point

$$\bar{A} = \begin{pmatrix} 1 \\ 0 \\ 1 \\ 0 \end{pmatrix}.$$

The trajectory of point $\bar{A}$ is $A(\varphi) = \gamma(\varphi)\,\bar{\gamma}^{-1}(\varphi)\,\bar{A}$, where $\gamma^{-1}(\varphi)\,\bar{A} = \bar{A}$, so that the trajectory $A(\varphi)$ has in the fundamental frame $\mathscr{R}_0$ the parametric equations

$$x^1(\varphi) = 0\,, \quad x^2(\varphi) = 1\,, \quad x^3(\varphi) = \varphi\,,$$

this is the axis of the fixed axoid. We may say, therefore, that in our case the motion is a rolling ($v_0 = 0$) of a one-sheet hyperboloid of revolution along a rectangular open helical surface, where the centre of the moving axoid moves along the axis of the fixed axoid.

7.11 Example. Let us consider briefly the case from Theorem 7.9 with developable surfaces as axoids, i.e. the case $\mu_1 = 0$. According to Theorem 7.9 we then have $\bar{\mu}_2 = v_0\kappa_1^2$, $\bar{\kappa}_2 = 0$. For the sake of simplicity, put $\kappa_1 = 1$. Similarly as in the case of the fixed axoid in Example 7.10, we obtain the following relations for the moving axoid:

$$\bar{\boldsymbol{u}}_1 = \begin{pmatrix} \cos\varphi \\ \sin\varphi \\ 0 \end{pmatrix}, \quad \bar{\boldsymbol{u}}_2 = \begin{pmatrix} -\sin\varphi \\ \cos\varphi \\ 0 \end{pmatrix}, \quad \bar{\boldsymbol{u}}_3 = \begin{pmatrix} 0 \\ 0 \\ 1 \end{pmatrix}, \quad \bar{S} = \begin{pmatrix} -v_0\sin\varphi \\ v_0(1-\cos\varphi) \\ 0 \end{pmatrix}.$$

The striction curve of the moving axoid is a circle given by the equations

$$x^3 = 0\,, \quad (x^1)^2 + (x^2 - v_0)^2 = v_0^2\,,$$

i.e. with centre at the point $x^1 = 0$, $x^2 = v_0$, $x^3 = 0$. The moving axoid is the set of tangents to this circle.

If we use the formulae $\kappa_2 - \bar{\kappa}_2 = 1$, $\mu_2 - \bar{\mu}_2 = v_0$, we obtain $\kappa_2 = 1$, $\mu_2 = 2v_0$ for the fixed axoid, so that the Frenet formulae are

$$S' = 2v_0\boldsymbol{u}_1\,, \quad \boldsymbol{u}_1' = \boldsymbol{u}_2\,, \quad \boldsymbol{u}_2' = -\boldsymbol{u}_1 + \boldsymbol{u}_3\,, \quad \boldsymbol{u}_3' = -\boldsymbol{u}_2\,.$$

The solution is

$$\boldsymbol{u}_1(\varphi) = \begin{bmatrix} \dfrac{\cos(\varphi\sqrt{2}) + 1}{2} \\ \dfrac{\sin(\varphi\sqrt{2})}{\sqrt{2}} \\ -\dfrac{\cos(\varphi\sqrt{2}) + 1}{2} \end{bmatrix}, \quad \boldsymbol{u}_2(\varphi) = \begin{bmatrix} \dfrac{-\sin(\varphi\sqrt{2})}{\sqrt{2}} \\ \cos(\varphi\sqrt{2}) \\ \dfrac{\sin(\varphi\sqrt{2})}{\sqrt{2}} \end{bmatrix},$$

$$\boldsymbol{u}_3(\varphi) = \begin{bmatrix} \dfrac{-\cos(\varphi\sqrt{2}) + 1}{2} \\ \dfrac{-\sin(\varphi\sqrt{2})}{\sqrt{2}} \\ \dfrac{\cos(\varphi\sqrt{2}) + 1}{2} \end{bmatrix},$$

$$S(\varphi) = v_0 \begin{bmatrix} \varphi + \dfrac{\sin(\varphi\sqrt{2})}{\sqrt{2}} \\ 1 - \cos(\varphi\sqrt{2}) \\ \varphi - \dfrac{\sin(\varphi\sqrt{2})}{\sqrt{2}} \end{bmatrix}.$$

The fixed axoid has the equation

$$\mathfrak{A}(\varphi, \lambda) = v_0 \begin{bmatrix} \varphi + \dfrac{\sin(\varphi\sqrt{2})}{\sqrt{2}} \\ 1 - \cos(\varphi\sqrt{2}) \\ \varphi - \dfrac{\sin(\varphi\sqrt{2})}{\sqrt{2}} \end{bmatrix} + \lambda \begin{bmatrix} \dfrac{\cos(\varphi\sqrt{2}) + 1}{2} \\ \dfrac{\sin(\varphi\sqrt{2})}{\sqrt{2}} \\ \dfrac{-\cos(\varphi\sqrt{2}) + 1}{2} \end{bmatrix}.$$

In the fundamental frame $\mathscr{R}_0$ the striction curve has the parametric equations

$$x^1 = v_0\left[\varphi + \frac{\sin(\varphi\sqrt{2})}{\sqrt{2}}\right], \quad x^2 = v_0[1 - \cos(\varphi\sqrt{2})],$$

$$x^3 = v_0\left[\varphi - \frac{\sin(\varphi\sqrt{2})}{\sqrt{2}}\right].$$

Therefore, the striction curve is a helix and for its tangent vectors we have

$$(x^1)' = v_0[1 + \cos(\varphi\sqrt{2})], \quad (x^2)' = v_0\sqrt{(2)}\sin(\varphi\sqrt{2}),$$
$$(x^3)' = v_0[1 - \cos(\varphi\sqrt{2})].$$

This means that the tangent vector of the striction curve is a multiple of the vector $\boldsymbol{u}_1$. Thus, the fixed axoid is the surface of tangents to the helix. If we introduce new coordinates by the relations

$$y^1 = \frac{x^1 - x^3}{\sqrt{2}}, \quad y^2 = \frac{x^1 + x^3}{\sqrt{2}}, \quad y^3 = x^2,$$

we obtain

$$y^1(\varphi) = v_0 \sin(\varphi\sqrt{2}), \quad y^2 = v_0\varphi\sqrt{2}, \quad y^3 = v_0[1 - \cos(\varphi\sqrt{2})],$$

i.e. the striction curve has the axis $y^1 = 0$, $y^3 = v_0$, radius v_0, and reduced thread pitch also v_0. The motion is such that the plane of the moving axoid gradually becomes the osculation plane of the fixed helix, with the circle with radius v_0 in the moving plane identifying successively with the points of the fixed helix (this does not involve the development of a developable helical surface into a plane, because in that case the circle in the moving plane would have the radius $2v_0$). Under this motion the centre of the circle in the moving plane has a linear trajectory.

7.12 Exercise. Find the representation of the motion from Example 7.11 and prove that the described point really has a linear trajectory.

7.13 Exercise. Find the matrix representation of all helical motions which have a linear trajectory.

Below, we shall deal with cylindrical motions and show that the formulae for general space motion can be applied to them if we put $\kappa_1 = 0$ in these formulae. We shall thus consider cylindrical motions as a special case of space motion, i.e. as a motion in group $\mathscr{E}_3$ and not as a motion in group $Z(\boldsymbol{R}_2)$ as was the case in Theorem 5.18. The difference is that we must now define the Frenet frame in $\mathscr{E}_3$ which consists of the vectors $\boldsymbol{R}_1, \boldsymbol{T}_1, \boldsymbol{N}_1, \boldsymbol{R}_2, \boldsymbol{T}_2, \boldsymbol{N}_2$. For this purpose it is sufficient to take into account that the axoids of cylindrical motion are cylindrical surfaces. This means that there is no special point on the generating line of the axoid. As the origin of the Frenet frame in the fixed system we thus choose an arbitrary point S of the straight line of the fixed axoid. This defines the vector $\boldsymbol{T}_1(\boldsymbol{N}_1)$ uniquely – it is defined by the line passing through point S whose direction is determined by the vector $\boldsymbol{T}_2(\boldsymbol{N}_2)$. Then the following theorem holds:

7.14 Theorem. *Let $g(\varphi)$ be a cylindrical motion with canonical parameter φ and fixed directing cone $\boldsymbol{R}$ for which $\mathbf{K}(\boldsymbol{R}, \boldsymbol{R}) \neq 0$, $\boldsymbol{R}' \neq 0$. Then the Frenet formulae from Theorem* 5.1 *are true if we put $\kappa_1 = 0$, $\mu_1 > 0$. μ_2 is an arbitrary function (for the sake of simplicity we put $\mu_2 = v_0$).*

Proof. We use Theorem 5.18 and define the vectors $\boldsymbol{T}_1$ and $\boldsymbol{N}_1$ as outlined above. For simplicity, let us adjust the parameter of motion so that $\mathbf{T}(\boldsymbol{T}_2', \boldsymbol{T}_2') = 1$. Then $\boldsymbol{R}_1 = [\boldsymbol{x}; \boldsymbol{t}]$, where $(\boldsymbol{x}, \boldsymbol{t}) = 0$ and the vectors $\boldsymbol{x}, \boldsymbol{t}', \boldsymbol{x} \times \boldsymbol{t}'$ constitute an orthonormal base. Further, $\mu_1 = 1$, $\boldsymbol{t}'' = \kappa_2 \boldsymbol{x} \times \boldsymbol{t}'$. If also $\boldsymbol{t} = a_1 \boldsymbol{t}' + a_2 \boldsymbol{x} \times \boldsymbol{t}'$, differentiation of this equation yields $a_1 \kappa_2 + a_2' = 0$, $a_1' = -a_2 \kappa_2 = 1$.

For $\boldsymbol{T}_1$ we obtain the expression $\boldsymbol{T}_1 = [\boldsymbol{t}'; a\boldsymbol{x} + b\boldsymbol{x} \times \boldsymbol{t}']$, where the condition $\mathscr{K}(\boldsymbol{R}_1, \boldsymbol{T}_1) = 0$ yields $a = -a_1$. For $\boldsymbol{N}_1$ we have $\boldsymbol{N}_1 = [\boldsymbol{R}_1, \boldsymbol{T}_1] = [\boldsymbol{x} \times \boldsymbol{t}'; -b\boldsymbol{t}' - a_2 \boldsymbol{x}]$. Altogether, $\boldsymbol{T}_1' - \kappa_2 \boldsymbol{N}_1 = [\boldsymbol{0}; -\boldsymbol{x} + b'\boldsymbol{x} \times \boldsymbol{t}'] = -\boldsymbol{R}_2 + \mu_2 \boldsymbol{N}_2$, $\mu_2 = b'$ is an arbitrary function, $\boldsymbol{N}_1' + \kappa_2 \boldsymbol{T}_1 = [\boldsymbol{0}; -b'\boldsymbol{t}'] = -\mu_2 \boldsymbol{T}_2$. This concludes the proof.

7.15 Theorem. *Theorems* 5.12 *and* 5.13 *are valid for cylindrical motions also if we put* $\overline{\boldsymbol{T}}_1 = \operatorname{ad} g^{-1} \boldsymbol{T}_1$, $\overline{\boldsymbol{N}}_1 = \operatorname{ad} g^{-1} \boldsymbol{N}_1$, $\kappa_1 = 0$ ($\mu_2 = v_0$, $\bar{\mu}_2 = 0$).

7.16 Remark. Theorems 7.14 and 7.15 mean that the preceding considerations of this section are valid also for cylindrical motions, if we put $\kappa_1 = 0$ everywhere (and possibly also $\mu_2 = v_0$, $\bar{\mu}_2 = 0$).

7.17 Example. Let us show that only cylindrical motion may have an infinite number of linear trajectories. If $g(\varphi)$ is such a motion, then the equations of Theorem 7.7 have an infinite number of solutions. Let $v_0 \neq 0$. From equations (b) of Theorem 7.7 we then compute:

$$v_0 x^2 = v_0 \psi + \pi x^3 + \kappa_1 (x^3)^2 \tag{7.1}$$

$$v_0 \kappa_1 x^1 = -\pi x^2 - v_0 x^3 - \kappa_1 x^2 x^3 \,. \tag{7.2}$$

In this way x^1 and x^2 are represented in terms of x^3. If we now differentiate equations (7.1) and (7.2) and substitute from Theorem 7.7, we obtain two algebraic equations for x^3 which should have an infinite number of solutions. This means that the equations must be identically equal to zero. Let us consider only the terms of the highest order in x^3. In (7.1), $v_0(-\kappa_1 x^1) = 2\kappa_1 x^3 - (\bar{\kappa}_2 x^2)$ up to terms of degree $\leqq 2$, so that $\kappa_1(1 + 2\bar{\kappa}_2) = 0$.

In (7.2), $0 = -\kappa_1(-\kappa_1 x^1)\, x^3 - \kappa_1 x^2(-\bar{\kappa}_2 x^2)$ up to terms of degree $\leqq 3$, so that $\kappa_1(1 - \bar{\kappa}_2) = 0$. Necessarily we then have $\kappa_1 = 0$. If $v_0 = 0$ and $\kappa_1 \neq 0$, then $x^3 = 0$ and $x^2(x^2 - \mu_1) = 0$. Further, $x^1 = 0$ and we have at most two solutions. Therefore, $\kappa_1 = 0$. The motion involved is thus cylindrical.

7.18 Exercise. Prove that if cylindrical motion has one linear trajectory, then it has an infinite number of linear trajectories. The similar holds for a plane trajectory.

7.19 Exercise. Determine all space motions whose all trajectories of points are plane curves. Hint: the equation $|X', X'', X'''| = 0$ must be satisfied identically. This equation yields the conditions for the invariants of motion and, integrating the Frenet formulae, we obtain the explicit expressions for these motions. See also [24].

In the formula of Theorem 7.6 we see that investigation of the properties of the trajectory of a point is very difficult, because the formula for the trajectory curvature is complicated. Now, let us show that a certain line can be defined which is associated with the curvature of the trajectory and whose representation is simpler.

7.20 Definition. The line which passes through the centre of the osculation circle of the trajectory of point X and is perpendicular to the osculation plane of the trajectory of point X is called the *axis of curvature* c_X of point X.

7.21 Lemma. Let $X(t)$ be a curve in E_3. Then the Plücker coordinates of its axis of curvature are

$$c_X = |X' \times X''|^{-1} \left[X' \times X''; \; X \times (X' \times X'') + |X'|^2 X'\right].$$

Proof. The parametric representation $A(\lambda)$ of the axis of curvature X is given by the relation $A(\lambda) = X + (k_X)^{-1}\, \boldsymbol{n}_X + \lambda \boldsymbol{b}_X$. Thus, we have

$$c_X = \left[\boldsymbol{b}_X; X \times \boldsymbol{b}_X + k_X^{-1} \boldsymbol{t}_X\right].$$

Upon substitution the desired formula is obtained.

7.22 Lemma. Plücker's coordinates $p^1, \ldots, p^6$ of the axis of curvature of point $X = (x^1, x^2, x^3)$ are given, up to a proportionality factor, by the equations

$$p^1 = (x^2)^2 + (x^3)^2 + \kappa_1 x^1 x^3 - \psi x^2 ,$$
$$p^2 = v_0 \kappa_1 x^1 + \pi x^2 + v_0 x^3 + \kappa_1 x^2 x^3 ,$$
$$p^3 = v_0 \psi - v_0 x^2 + \pi x^3 + \kappa_1 (x^3)^2 ,$$
$$p^4 = v_0 \psi x^2 + v_0^3 - v_0 \kappa_1 x^1 x^3 ,$$
$$p^5 = -\pi x^1 x^3 - \psi x^2 x^3 - v_0^2 x^3 - v_0 \psi x^1 + v_0 x^1 x^2 ,$$
$$p^6 = \pi x^1 x^2 + (x^2)^2 \psi + v_0^2 x^2 + v_0 \kappa_1 (x^1)^2 + v_0 x^1 x^3 .$$

The proof is performed by substitution and is left to the reader.

7.23 Remark. The axis of curvature of a curve is the set of all centres of all spheres which have at least second-order contact with the curve, i.e. in particular the centre of the osculation circle and the centre of the osculation sphere lie on it. Of course, in the formula from 7.22 we must omit the inflection points for which the axis of curvature does not exist.

8. SPECIAL MOTIONS

Consider two general helical motions denoted by $g_1(t)$ and $g_2(t)$ and given as follows: $g_1(t)$ is a general helical motion about line p_1 which is determined by its Plücker coordinates $X = [x; t]$. Motion $g_1(t)$ is determined by rotation through angle $\varphi_1(t)$ and by translation $a_1(t)$ in the direction of the axis of the helical motion. Moreover, we shall assume that a_1 is defined as a composite function of the angle of rotation φ_1 which is given as a function of time, i.e. $a_1(t) \equiv a_1(\varphi_1(t))$. Let us write motion g_1 in the following form: $g_1 = g_1(\varphi_1(t), a_1(\varphi_1(t)))$. Motion $g_2(t)$ is defined analogously as a general helical motion about line p_2 determined by its Plücker coordinates $Y = [y; s]$; the angle of rotation is $\varphi_2(t)$ and the translation $a_2(t) \equiv a_2(\varphi_2(t))$. Note that by definition g_1 and g_2 are general helical motions, i.e. the displacements a_1 and a_2 may be arbitrary (differentiable) functions of the angles φ_1 and φ_2, respectively. Let us consider as special the case in which a_1 and a_2 are linear functions of the angles φ_1 and φ_2, respectively. Further, let us note that for the angle ϑ of lines p_1 and p_2 and for their distance d we have $(\boldsymbol{x}, \boldsymbol{y}) = \cos\vartheta$, $(\boldsymbol{x}, \boldsymbol{s}) + (\boldsymbol{y}, \boldsymbol{t}) = d \sin\vartheta$ (the distance and angle are oriented, otherwise see assertion 2.6).

8.1 Definition. The motion $g(t) = g_1(\varphi_1(t), a_1(\varphi_1(t)))\, g_2(\varphi_2(t), a_2(\varphi_2(t)))$ is called a *general double-helical motion* defined by the angle ϑ, the distance d of the axis of the two helical motions, the angles of rotation $\varphi_1(t)$ and $\varphi_2(t)$, and by the translations $a_1(\varphi_1(t))$ and $a_2(\varphi_2(t))$.

8.2 Remark. A general double-helical motion is generated by performing first helical motion g_2 about the axis p_2 and then helical motion g_1 about the axis p_1, or also by first performing helical motion g_1 about the axis p_1 and then helical motion g_2 about the axis $g_1 p_1$, as was the case of rotations in Chapter II.

8.3 Definition. Let $\omega_1 = d\varphi_1/dt$, $\omega_2 = d\varphi_2/dt$ denote the *instantaneous angular velocities* of motions g_1 and g_2, $v_1 = da_1/d\varphi_1$, $v_2 = da_2/d\varphi_2$ the *instantaneous parameters* (instantaneous reduced heights of threads) of motions g_1 and g_2, respectively.

For the directing functions $\boldsymbol{r}_1(t)$ and $\boldsymbol{r}_2(t)$ of motions $g_1(t)$ and $g_2(t)$ we have

$$\boldsymbol{r}_1(t) = \omega_1[\boldsymbol{x}; \boldsymbol{t} + v_1\boldsymbol{x}]\,, \quad \boldsymbol{r}_2(t) = \omega_2[\boldsymbol{y}; \boldsymbol{s} + v_2\boldsymbol{y}]\,.$$

For the directing function $\boldsymbol{r}(t)$ of motion $g(t)$ we obtain

$$\begin{aligned}
\boldsymbol{r}(t) &= g'(t)\, g^{-1}(t) = (g_1 g_2)'\,(g_1 g_2)^{-1} = (g_1' g_2 + g_1 g_2')\, g_2^{-1} g_1^{-1} \\
&= g_1' g_1^{-1} + g_1 g_2' g_2^{-1} g_1^{-1} = \boldsymbol{r}_1(t) + \operatorname{ad} g_1\, \boldsymbol{r}_2(t) \\
&= \operatorname{ad} g_1[\boldsymbol{r}_1(t) + \boldsymbol{r}_2(t)] \\
&= \operatorname{ad} g_1[\omega_1 \boldsymbol{x} + \omega_2 \boldsymbol{y};\ \omega_1 \boldsymbol{t} + \omega_2 \boldsymbol{s} + v_1\omega_1\boldsymbol{x} + v_2\omega_2\boldsymbol{y}]\,,
\end{aligned}$$

because $\operatorname{ad} g_1\, \boldsymbol{r}_1(t) = \boldsymbol{r}_1(t)$ and helical motions about a common axis mutually commute (they constitute a commutative group as we know from Section 1; besides, it is easy to verify this directly).

8.4 Theorem. *For the double-helical motion* $g(t)$ *we have*

$$\begin{aligned}
\boldsymbol{r}(t) &= \operatorname{ad} g_1[\boldsymbol{r}_1(t) + \boldsymbol{r}_2(t)] \\
&= \operatorname{ad} g_1[\omega_1 \boldsymbol{x} + \omega_2 \boldsymbol{y};\ \omega_1 \boldsymbol{t} + \omega_2 \boldsymbol{s} + v_1\omega_1\boldsymbol{x} + v_2\omega_2\boldsymbol{y}]\,,
\end{aligned}$$

$$\omega \;= \frac{d\varphi}{dt} = [\omega_1^2 + \omega_2^2 + 2\omega_1\omega_2 \cos \vartheta)]^{1/2} ,$$

$$v_0 \;= \omega^{-2}\{v_1\omega_1^2 + v_2\omega_2^2 + \omega_1\omega_2[d \sin \vartheta + (v_1 + v_2) \cos \vartheta]\} ,$$

where φ is the canonical parameter, ω the angular velocity and v_0 the parameter of the instantaneous helical motion

Proof. We have

$$\omega^2 = \mathbf{K}(\boldsymbol{r}, \boldsymbol{r}) = (\omega_1 \boldsymbol{x} + \omega_2 \boldsymbol{y},\ \omega_1 \boldsymbol{x} + \omega_2 \boldsymbol{y})$$
$$= \omega_1^2 + \omega_2^2 + 2\omega_1\omega_2 \cos \vartheta .$$

Further,

$$v_0 = \mathscr{K}(\boldsymbol{r}, \boldsymbol{r})\, \omega^{-2}$$
$$= (\omega_1 \boldsymbol{x} + \omega_2 \boldsymbol{y},\ \omega_1 \boldsymbol{t} + \omega_2 \boldsymbol{s} + v_1\omega_1 \boldsymbol{x} + v_2\omega_2 \boldsymbol{y})\, \omega^{-2}$$
$$= \omega^{-2}[\omega_1\omega_2(\boldsymbol{x}, \boldsymbol{s}) + v_1\omega_1^2 + \omega_1\omega_2 v_2(\boldsymbol{x}, \boldsymbol{y}) + \omega_2\omega_1(\boldsymbol{y}, \boldsymbol{t})$$
$$+ v_1\omega_1\omega_2(\boldsymbol{x}, \boldsymbol{y}) + v_2\omega_2^2]$$
$$= \omega^{-2}\{v_1\omega_1^2 + v_2\omega_2^2 + \omega_1\omega_2[d \sin \vartheta + (v_1 + v_2) \cos \vartheta]\} ,$$

where we have used the relations $(\boldsymbol{x}, \boldsymbol{t}) = 0$, $(\boldsymbol{y}, \boldsymbol{s}) = 0$, which hold for Plücker coordinates.

8.5 Definition. Let $\varepsilon_1 = d\omega_1/dt$, $\varepsilon_2 = d\omega_2/dt$, $\varepsilon = d\omega/dt$ denote the *angular acceleration* of motions g_1, g_2 and g, let $\pi_1 = dv_1/d\varphi_1$, $\pi_2 = dv_2/d\varphi_2$, $\pi = dv_0/d\varphi$ be their *helical accelerations*.

Let us now compute the functions ε and π. Differentiating equation

$$\omega^2 = \omega_1^2 + \omega_2^2 + 2\omega_1\omega_2 \cos \vartheta$$

with respect to time t we arrive at

$$2\omega\varepsilon = 2\omega_1\varepsilon_1 + 2\omega_2\varepsilon_2 + 2(\omega_1\varepsilon_2 + \varepsilon_1\omega_2) \cos \vartheta ,$$

i.e.

$$\varepsilon = \omega^{-1}[\omega_1\varepsilon_1 + \omega_2\varepsilon_2 + (\omega_1\varepsilon_2 + \omega_2\varepsilon_1) \cos \vartheta] .$$

To compute π we use a special procedure which will also prove convenient below. Therefore, we explain it in some detail. Assume that the resulting double-helical motion is defined as a function of the parameter φ_1. Under this assumption we shall compute π and change the parameter to t in the resulting formula. The values corresponding to parameter φ_1 will be marked with an asterisk. We thus have $\omega_1^* = 1$, so that

$$(\omega^*)^2 = 1 + (\omega_2^*)^2 + 2\omega_2^* \cos \vartheta \,;$$

further,

$$\omega_2^* = \frac{d\varphi_2}{d\varphi_1} = \frac{d\varphi_2}{dt}\frac{dt}{d\varphi_1} = \frac{\omega_2}{\omega_1}\,;$$

analogously, we also have

$$\frac{d\varphi}{d\varphi_1} = \omega^* = \frac{\omega}{\omega_1}\,,$$

and

$$\varepsilon_2^* = \frac{d\omega_2^*}{d\varphi_1} = \frac{d\omega_2^*}{dt}\frac{dt}{d\varphi_1} = \omega_1^{-1}\frac{d}{dt}\left(\frac{\omega_2}{\omega_1}\right) = \omega_1^{-3}(\varepsilon_2\omega_1 - \omega_2\varepsilon_1)\,.$$

Now, compute π with the aid of φ_1:

$$\begin{aligned}\pi &= \frac{dv_0}{d\varphi} = \frac{dv_0}{d\varphi_1}\frac{d\varphi_1}{d\varphi} = (\omega^*)^{-1}\frac{d}{d\varphi_1}\,.\\ &\quad .\{(\omega^*)^{-2}\left[v_1 + v_2(\omega_2^*)^2 + \omega_2^*(d \sin \vartheta + (v_1 + v_2)\cos \vartheta)\right]\}\\ &= (\omega^*)^{-5}\{[\pi_1(1 + \omega_2^* \cos \vartheta) + \pi_2(\omega_2^*)^2(\omega_2^* + \cos \vartheta)](\omega_2^*)^2\\ &\quad + \varepsilon_2^*[\![2\omega_2^*(v_2 - v_1) + d[1 - (\omega_2^*)^2]\sin \vartheta\\ &\quad + (v_2 - v_1)[1 + (\omega_2^*)^2]\cos \vartheta]\!]\}\,.\end{aligned}$$

If we substitute the functions of time back again, we obtain

$$\begin{aligned}\pi &= \omega^{-5}\omega_1^5\{[\pi_1(1 + \omega_2\omega_1^{-1} \cos \vartheta)\\ &\quad + \pi_2\omega_2^2\omega_1^{-2}(\omega_2\omega_1^{-1} + \cos \vartheta)]\,\omega^2\omega_1^{-2}\\ &\quad + \omega_1^{-3}(\varepsilon_2\omega_1 - \omega_2\varepsilon_1)[2\omega_2\omega_1^{-1}(v_2 - v_1) + d(1 - \omega_2^2\omega_1^{-2})\sin\\ &\quad + (v_2 - v_1)(1 + \omega_2^2\omega_1^{-2})\cos \vartheta]\}\,,\end{aligned}$$

which, after some manipulation, yields the desired result.

8.6 Theorem. *For double-helical motion we have*

$$\varepsilon = \omega^{-1}[\omega_1\varepsilon_1 + \omega_2\varepsilon_2 + (\omega_1\varepsilon_2 + \omega_2\varepsilon_1)\cos\vartheta],$$

$$\begin{aligned}\pi = \omega^{-5}\{&\omega^2[\pi_1\omega_1^2(\omega_1 + \omega_2\cos\vartheta) + \pi_2\omega_2^2(\omega_2 + \omega_1\cos\vartheta)]\\ &+ (\varepsilon_2\omega_1 - \varepsilon_1\omega_2)[2\omega_1\omega_2(v_2 - v_1) + d(\omega_1^2 - \omega_2^2)\sin\vartheta\\ &+ (v_2 - v_1)(\omega_1^2 + \omega_2^2)\cos\vartheta]\}.\end{aligned}$$

The ruling line of the fixed axoid of motion $g(t)$ is defined by the vector $\boldsymbol{r}(t)$ of the directing function with

$$\boldsymbol{r}(t) = \mathrm{ad}g_1[\omega_1\boldsymbol{x} + \omega_2\boldsymbol{y};\ \omega_1\boldsymbol{t} + \omega_2\boldsymbol{s} + v_1\omega_1\boldsymbol{x} + v_2\omega_2\boldsymbol{y}].$$

Therefore, let us put

$$\boldsymbol{\sigma}(t) = [\omega_1\boldsymbol{x} + \omega_2\boldsymbol{y};\ \omega_1\boldsymbol{t} + \omega_2\boldsymbol{s} + r_1\omega_1\boldsymbol{x} + r_2\omega_2\boldsymbol{y}],$$

denote by $q(t)$ the line defined by the vector $\boldsymbol{\sigma}(t) \in \mathfrak{E}$ and by $p(t)$ the ruling line of the fixed axoid.

8.7 Theorem. *The fixed axoid of a double-helical motion arises from the action of helical motion $g_1(t)$ on line $q(t)$, the moving axoid arises from the action of motion $g_2^{-1}(t)$ on the same line [of course, the position of line $q(t)$ changes in general with time].*

Proof. The first part of the assertion follows from the relation

$$\boldsymbol{r}(t) = \mathrm{ad}g_1(t)\,\boldsymbol{\sigma}(t), \quad \text{i.e.} \quad p(t) = g_1(t)\,q(t),$$

for the second we have

$$(g_1g_2)[g_2^{-1}\,q(t)] = g_1\,q(t) = p(t).$$

Theorem 8.7 implies that line $q(t)$ is of principal importance for the resulting motion. Let us deal, therefore, in detail with its position with respect to the axes p_1 and p_2. The unit vector of the same direction as line q is $\omega^{-1}(\omega_1\boldsymbol{x} + \omega_2\boldsymbol{y})$. The following holds for the angles ψ_1 and ψ_2 between the lines $q(t)$ and p_1 and p_2, respectively:

$$\cos\psi_1 = (\boldsymbol{x}, \omega^{-1}[\omega_1\boldsymbol{x} + \omega_2\boldsymbol{y}]) = \omega^{-1}(\omega_1 + \omega_2\cos\vartheta),$$

$$\sin\psi_1 = \omega^{-1}\omega_2\sin\vartheta,$$

$$\cos \psi_2 = (\boldsymbol{y}, \omega^{-1}[\omega_1 \boldsymbol{x} + \omega_2 \boldsymbol{y}]) = \omega^{-1}(\omega_2 + \omega_1 \cos \vartheta),$$
$$\sin \psi_2 = \omega^{-1}\omega_1 \sin \vartheta .$$

We also see that the directions of lines p_1, p_2, and q are determined by the vectors $\boldsymbol{x}, \boldsymbol{y}, \omega^{-1}(\omega_1 \boldsymbol{x} + \omega_2 \boldsymbol{y})$. Moreover, if $\vartheta \neq 0$, $\vartheta \neq \pi$. These three vectors lie in the plane defined by the vectors $\boldsymbol{x}$ and $\boldsymbol{y}$. This means that the lines p_1, p_2, and q determine a hyperbolic paraboloid.

8.8 Theorem. *The lines p_1, p_2 and $q(\mathrm{t})$ determine a hyperbolic paraboloid (which may also be a plane) for every* t.

Denote by $\boldsymbol{S}(\mathrm{t})$ a vector from $\mathfrak{E}$ which determines the Plücker coordinates of line $q(\mathrm{t})$. Then

$$\begin{aligned} \boldsymbol{S}(\mathrm{t}) &= \omega^{-1}[\omega_1 \boldsymbol{x} + \omega_2 \boldsymbol{y};\ \omega_1 \boldsymbol{t} + \omega_2 \boldsymbol{s} + v_1 \omega_1 \boldsymbol{x} + v_2 \omega_2 \boldsymbol{y}] \\ &\quad - v_0 \omega^{-1}[\boldsymbol{0};\ \omega_1 \boldsymbol{x} + \omega_2 \boldsymbol{y}] \\ &= \omega^{-1}[\omega_1 \boldsymbol{x} + \omega_2 \boldsymbol{y};\ \omega_1 \boldsymbol{t} + \omega_2 \boldsymbol{s} + [v_1 - v_0]\, \omega_1 \boldsymbol{x} \\ &\quad + [v_2 - v_0]\, \omega_2 \boldsymbol{y}] . \end{aligned}$$

8.9 Theorem. *Denote by d_1, d_2 the distances of line q from lines p_1 and p_2. Then*

$$d_1 = \omega_2^{-1}[\omega_2 d \sin \vartheta + (v_1 - v_0)\, \omega_1 + (v_2 - v_0)\, \omega_2 \cos \vartheta] \sin^{-1} \vartheta,$$
$$d_2 = \omega_1^{-1}[\omega_1 d \sin \vartheta + (v_2 - v_0)\, \omega_2 + (v_1 - v_0)\, \omega_1 \cos \vartheta] \sin^{-1} \vartheta .$$

Proof. We use the expression for $\boldsymbol{S}(\mathrm{t})$ and for $\sin \psi_1$ and $\sin \psi_2$.

8.10 Theorem. *We have $d_1 + d_2 = d$.*

Proof. Theorem 8.9 yields

$$d_1 - d = \omega_2^{-1}[(v_1 - v_0)\, \omega_1 + (v_2 - v_0)\, \omega_2 \cos \vartheta] \sin^{-1} \vartheta ,$$

and

$$d_2 - d = \omega_1^{-1}[(v_2 - v_0)\, \omega_2 + (v_1 - v_0)\, \omega_1 \cos \vartheta] \sin^{-1} \vartheta ,$$

with

$$\begin{aligned} & v_1 - v_0 \\ &= \omega^{-2}\{v_1 \omega_1^2 + 2 v_1 \omega_1 \omega_2 \cos \vartheta - v_1 \omega_1^2 - v_2 \omega_2^2 + v_1 \omega_2^2 \\ &\quad - \omega_1 \omega_2 [d \sin \vartheta + (v_1 + v_2) \cos \vartheta]\} \\ &= \omega^{-2}\{(v_1 - v_2)\, \omega_2^2 + \omega_1 \omega_2 [-d \sin \vartheta + (v_1 - v_2) \cos \vartheta]\} . \end{aligned}$$

Similarly,

$$v_2 - v_0 = \omega^{-2}\{(v_2 - v_1)\,\omega_1^2 + \omega_1\omega_2[-d\sin\vartheta + (v_2 - v_1)\cos\vartheta]\}\,.$$

Upon substitution and some manipulations we obtain the formulae

$$d_1 - d = \omega_1\omega^{-2}[(v_1 - v_2)\,\omega_2\sin\vartheta - d(\omega_1 + \omega_2\cos\vartheta)]\,,$$
$$d_2 - d = \omega_2\omega^{-2}[(v_2 - v_1)\,\omega_1\sin\vartheta - d(\omega_2 + \omega_1\cos\vartheta)]\,.$$

If we add these two formulae we arrive at

$$d_1 + d_2 - 2d$$
$$= \omega^{-2}(-d)\,(\omega_1^2 + \omega_1\omega_2\cos\vartheta + \omega_2^2 + \omega_1\omega_2\cos\vartheta) = -d\,,$$

i.e.

$$d_1 + d_2 = d\,.$$

As a consequence of Theorem 8.10 we have the following two theorems:

8.11 Theorem. *The lines p_1, p_2, and $q(\mathrm{t})$ have a common shortest transversal, i.e. the line $q(\mathrm{t})$ changes only so that it intersects at right angles the shortest transversal of the lines p_1 and p_2 all the time (if p_1 and p_2 are concurrent, we again consider the perpendicular to both, constructed at the point where they intersect, to be their shortest transversal).*

The case of parallel lines p_1 and p_2 is treated separately; Theorem 8.11 is true in this case as well. Further, let $d\sin\vartheta \neq 0$.

8.12 Theorem. *The hyperbolic paraboloid determined by lines p_1, p_2, and $q(\mathrm{t})$ has perpendicular directing planes.*

Proof. The shortest transversal of the skew lines p_1 and p_2 is perpendicular to both and belongs to the mentioned hyperbolic paraboloid according to Theorem 8.11. Its direction is thus perpendicular to the directing plane of one system and belongs to the directing plane of the other system of lines of the hyperbolic paraboloid.

An important special case occurs if $d_1 = const.$; then we also have $d_2 = const.$ In this case the ruling lines of the fixed axoid are tangents to the cylindrical surface with axis p_1 and radius d_1, and the ruling lines of the moving axoid are tangents to the cylindrical surface with radius d_2 and axis p_2 (where $d_1 + d_2 = d$, of course).

8.13 Theorem. *To given* ω_1, ω_2, p_1, and p_2, *an infinite number of motions can be found for which the ruling lines of the fixed axoid are tangent to a cylindrical surface of given radius* d_1. *For this purpose it is sufficient to choose* v_1 *and* v_2 *so that*

$$v_1 - v_2 = (\omega_1 \omega_2 \cos \vartheta)^{-1} \left[d_1 \omega^2 - d\omega_2(\omega_2 + \omega_1 \cos \vartheta)\right].$$

Now, we shall turn our attention to the case where the hyperbolic paraboloid, determined by lines p_1, p_2, and $q(t)$, is independent of time, i.e. it is the same for all t. The mentioned hyperbolic paraboloid is defined by the vectors $\boldsymbol{X}$, $\boldsymbol{Y}$, $\boldsymbol{S}(t)$ and will be independent of time if and only if the vector space generated by the vectors $\boldsymbol{X}$, $\boldsymbol{Y}$, $\boldsymbol{S}(t)$ is always the same, i.e. if

$$\boldsymbol{S}(t) = \alpha(t)\,\boldsymbol{X} + \beta(t)\,\boldsymbol{Y} + \gamma(t)\,\boldsymbol{S}(0),$$

α, β, γ are real functions. Let us express $\boldsymbol{S}(t)$ as

$$\begin{aligned}\boldsymbol{S}(t) &= \omega^{-1}[\omega_1 \boldsymbol{x} + \omega_2 \boldsymbol{y};\ \omega_1 \boldsymbol{t} + \omega_2 \boldsymbol{s} + [v_1 - v_0]\,\omega_1 \boldsymbol{x} \\ &\quad + [v_2 - v_0]\,\omega_2 \boldsymbol{y}] \\ &= \omega^{-1}\omega_1 \boldsymbol{X} + \omega^{-1}\omega_2 \boldsymbol{Y} + \omega^{-1}[\boldsymbol{0};\ [v_1 - v_0]\,\omega_1 \boldsymbol{x} \\ &\quad + [v_2 - v_0]\,\omega_2 \boldsymbol{y}].\end{aligned}$$

Thus, if we put

$$\boldsymbol{V}(t) = (\boldsymbol{0};\ [v_1 - v_0]\,\omega_1 \boldsymbol{x} + [v_2 - v_0]\,\omega_2 \boldsymbol{y}),$$

we have

$$\begin{aligned}&\omega^{-1}\omega_1 \boldsymbol{X} + \omega^{-1}\omega_2 \boldsymbol{Y} + \boldsymbol{V}(t) \\ &= \alpha \boldsymbol{X} + \beta \boldsymbol{Y} + \gamma[\omega^{-1}(0)\,\boldsymbol{X} + \omega^{-1}(0)\,\omega_2(0)\,\boldsymbol{Y} + \boldsymbol{V}(0)],\end{aligned}$$

i.e.

$$\boldsymbol{V}(t) = \alpha_1 \boldsymbol{X} + \beta_1 \boldsymbol{Y} + \gamma_1\,\boldsymbol{V}(0).$$

This equation for the first components means that $\boldsymbol{0} = \alpha_1 \boldsymbol{x} + \beta_1 \boldsymbol{y}$, i.e. with respect to the assumption that $\sin \vartheta \neq 0$, α_1 and β_1 must be equal to $\boldsymbol{0}$. This implies that $\boldsymbol{V}(t) = \gamma_1(t)\,\boldsymbol{V}(0)$, which occurs if and only if the vectors $\boldsymbol{V}(t)$ and $\boldsymbol{V}^{\bullet}(t)$ are linearly dependent. We have thus obtained the following condition: the hyperbolic paraboloid, determined by the lines p_1, p_2 and $q(t)$, is independent of time if and only if $\boldsymbol{V}(t) \times \boldsymbol{V}^{\bullet}(t) = 0$.

To solve this differential equation we shall use the trick from the proof of Theorem 8.6; for the sake of clarity we shall now omit the asterisk. Put $v_1 - v_0 = f$, $v_2 - v_0 = g$, and differentiate with respect to φ_1 under the assumption that the motion is given as a function of the angle φ_1. Then

$$V(\varphi_1) = \boldsymbol{x}f + \omega_2 \boldsymbol{y}g\,, \quad V^{\bullet}(\varphi_1) = \boldsymbol{x}\dot{f} + \varepsilon_2 \boldsymbol{y}g + \omega_2 \boldsymbol{y}\dot{g}\,,$$

so that

$$\begin{aligned} V(\varphi_1) \times V^{\bullet}(\varphi_1) &= (\boldsymbol{x}f + \omega_2 \boldsymbol{y}g) \times (\boldsymbol{x}\dot{f} + \varepsilon_2 \boldsymbol{y}g + \omega_2 \boldsymbol{y}\dot{g}) \\ &= \boldsymbol{x} \times \boldsymbol{y}(\varepsilon_2 fg + \omega_2 f\dot{g} - \omega_2 \dot{f}g) = \boldsymbol{0}\,. \end{aligned}$$

Upon manipulation we have

$$\frac{\varepsilon_2}{\omega_2} = \frac{\dot{f}g - f\dot{g}}{fg}\,, \quad \text{i.e.} \quad \frac{\dot{\omega}_2}{\omega_2} = \left(\frac{f}{g}\right)^{\bullet}\left(\frac{f}{g}\right)^{-1}.$$

Integration yields

$$\ln \omega_2 = \ln\left(\frac{f}{g}\right) + \ln C\,, \quad \text{i.e.} \quad \omega_2^* = C\frac{f}{g}\,,$$

where the asterisk indicates differentiation with respect to φ_1. If we substitute for ω_2^*, we obtain $\omega_2\omega_1^{-1} = Cfg^{-1}$, i.e. $\omega_2 g = C\omega_1 f$, i.e. $\omega_2(v_2 - v_0) = C\omega_1(v_1 - v_0)$. Of course, in the computation it was assumed that $f \neq 0$, $g \neq 0$. However, a closer look at the vectors $\boldsymbol{V}$ and $\boldsymbol{V}^{\bullet}$ will disclose that $f = 0$, g arbitrary, and f arbitrary, $g = 0$, are also solutions. On the whole, we thus have:

8.14 Theorem. *The hyperbolic paraboloid determined by lines p_1, p_2, and $q(\mathrm{t})$, is independent of time* t *if and only if there exist constants C_1 and C_2 such that*

$$C_1\omega_2(v_2 - v_0) + C_2\omega_1(v_1 - v_0) = 0\,.$$

Let us give another formulation of Theorem 8.14:

8.15 Theorem. *The hyperbolic paraboloid determined by lines p_1, p_2, and $q(\mathrm{t})$, is independent of time if and only if there exist constants C_1 and C_2 such that*

$$v_2 - v_1 = d\,\frac{C_1\omega_2 + C_2\omega_1}{C_1(\omega_1 + \omega_2\cos\vartheta) - C_2(\omega_2 + \omega_1\cos\vartheta)}\sin\vartheta\,.$$

The last special case we consider occurs if line $q(t)$ is fixed, i.e. if it does not change with time. When does this occur?

If $q(t)$ is a constant line, then $\sin\psi_1 = const.$, $\sin\psi_2 = const.$, $d_1 = const.$, $d_2 = const.$, and the satifaction of these conditions in itself quarantees that $q(t)$ is really constant. The first two conditions yield $\omega_2 = C_1\omega_1$, $C_1 \neq 0$, the second two conditions yield $v_1 - v_2 = C_2$, where C_1 and C_2 are constants.

8.16 Theorem. *Line $q(t)$ is constant if and only if there exist constants $C_1 \neq 0$ and C_2 such that $\omega_2 = C_1\omega_1$, $v_2 - v_1 = C_2$.*

The motions which satisfy Theorem 8.16 are all the motions under which the fixed and moving axoids are created by a helical motion of one fixed line (in general a helical motion with a variable parameter v_0 is involved, of course).

We shall now determine the invariants of general double-helical motion. For this purpose the first two derivatives of the directing function $\boldsymbol{r}(t)$ are required. Again assume, as in the proof of Theorem 8.6, that motion is given as a function of the angle φ_1; we again denote the appropriate expressions by an asterisk. The invariants of motion do not depend on the choice of the parameter and, therefore, it is of no consequence whether they are computed with the aid of parameter t or φ_1. Finally, we again substitute the parameter t into the resulting expressions. Thus, we have

$$\boldsymbol{r}(\varphi_1) = \mathrm{ad}g_1[\boldsymbol{x} + \overset{*}{\omega}_2\boldsymbol{y};\ \boldsymbol{t} + \overset{*}{\omega}_2\boldsymbol{s} + v_1\boldsymbol{x} + v_2\overset{*}{\omega}_2\boldsymbol{y}]\,.$$

Put

$$\boldsymbol{r}(\varphi_1) = \mathrm{ad}g_1(\varphi_1)\, s(\varphi_1)\,, \quad \text{i.e.} \quad \boldsymbol{r}(\varphi_1) = g_1(\varphi_1)\, s(\varphi_1)\, g_1^{-1}(\varphi_1)$$

if we use matrix notation for the moment. Further, we have

$$\begin{aligned} \boldsymbol{r}^{\boldsymbol{\cdot}}(\varphi_1) &= \mathrm{ad}g_1\dot{s} + \dot{g}_1 s g_1^{-1} + g_1\, s(g_1^{-1})^{\boldsymbol{\cdot}} \\ &= \mathrm{ad}g_1\dot{s} + \dot{g}_1 g_1^{-1}\,\mathrm{ad}g_1 s - \mathrm{ad}g_1 s\dot{g}_1 g_1^{-1} \\ &= \mathrm{ad}g_1\dot{s} + [\boldsymbol{r}_1, \mathrm{ad}g_1 s] = \mathrm{ad}g_1\dot{s} + [\mathrm{ad}g_1\boldsymbol{r}_1, \mathrm{ad}g_1 s] \\ &= \mathrm{ad}g_1\dot{s} + \mathrm{ad}g_1[\boldsymbol{r}_1, s] = \mathrm{ad}g_1(\dot{s} + [\boldsymbol{r}_1, s])\,. \end{aligned}$$

If we put $\dot{s} + [\boldsymbol{r}_1, s] = v$, we similarly have $\boldsymbol{r}^{\boldsymbol{\cdot\cdot}} = \mathrm{ad}g_1(\dot{v} + [\boldsymbol{r}_1, v])$. The computation yields

$$\boldsymbol{r}^{\boldsymbol{\cdot}} = \mathrm{ad}g_1\{[\overset{*}{\varepsilon}_2\boldsymbol{y};\ \overset{*}{\varepsilon}_2\boldsymbol{s} + \pi_1\boldsymbol{x} + \pi_2(\overset{*}{\omega}_2)^2\,\boldsymbol{y} + v_2\overset{*}{\varepsilon}_2\boldsymbol{y}]$$

$$+ [[x; t + v_1 x], [x + \omega_2^* y; t + \omega_2^* s + v_1 x + v_2 \omega_2^* y]]\}$$

$$= \mathrm{ad} g_1 [\varepsilon_2^* y + \omega_2^* x \times y;\ \varepsilon_2^*(s + v_2 y) + \pi_1 x + \pi_2 (\omega_2^*)^2 y$$

$$+ \omega_2^* \{x \times s + t \times y + (v_1 + v_2) x \times y\}],$$

$$r^{\cdot\cdot} = \mathrm{ad} g_1 [\dot{\varepsilon}_2^* y + 2\varepsilon_2^* x \times y + \omega_2^*(x \cos \vartheta - y);\ \dot{\varepsilon}_2^*(s + v_2 y)$$

$$+ \dot{\pi}_1 x + \dot{\pi}_2 (\omega_2^*)^2 y + 3\pi_2 \omega_2^* \varepsilon_2^* y + \omega_2^* (\pi_1 + 2\pi_2 \omega_2^*) x \times y$$

$$+ 2\varepsilon_2^* \{x \times s + t \times y + (v_1 + v_2) x \times y\}$$

$$+ \omega_2^* \{dx \sin \vartheta + t \cos \vartheta - s + (2v_1 + v_2)(x \cos \vartheta - y)\}].$$

If we put $r(\mathrm{t}) = [z; u]$, by the formulae of Theorem 5.9 we have

$$V = (\dot{\varphi})^2 (z^{\cdot}, z^{\cdot}) - (z, z^{\cdot})^2 = [(\omega_2^*)^2 (\omega^*)^2 + (\varepsilon_2^*)^2] \sin^2 \vartheta,$$

$$|z, z^{\cdot}, z^{\cdot\cdot}| = |x + \omega_2^* y, \varepsilon_2^* y + \omega_2^* x \times y, \dot{\varepsilon}_2^* y + 2\varepsilon_2^* x \times y$$

$$+ \omega_2^*(x \cos \vartheta - y)|$$

$$= (\varepsilon_2^* x \times y + \omega_2^* [x \cos \vartheta - y]$$

$$+ [\omega_2^*]^2 [x - y \cos \vartheta], \dot{\varepsilon}_2^* y + 2\varepsilon_2^* x \times y$$

$$+ \omega_2^* [x \cos \vartheta - y])$$

$$= 2(\varepsilon_2^*)^2 \sin^2 \vartheta - \omega_2^* \dot{\varepsilon}_2^* \sin^2 \vartheta + (\omega_2^*)^2 \sin^2 \vartheta$$

$$+ (\omega_2^*)^3 \cos \vartheta \sin^2 \vartheta$$

$$= [2(\varepsilon_2^*)^2 - \omega_2^* \dot{\varepsilon}_2^* + (\omega_2^*)^2 + (\omega_2^*)^3 \cos \vartheta] \sin^2 \vartheta.$$

To compute the expression Z, let

$$\alpha = -d \sin \vartheta \cos \vartheta + (v_1 + v_2) \sin^2 \vartheta, \quad \Delta = d \sin \vartheta + (v_1 + v_2) \cos \vartheta;$$

for the sake of simplicity we omit the asterisks. Thus

$$(z^{\cdot}, u^{\cdot}) = \varepsilon_2(\pi_1 \cos \vartheta + \pi_2 \omega_2^2 + \varepsilon_2 v_2) + \omega_2^2 \alpha,$$

$$(z, u) = v_1 + v_2 \omega_2^2 + \omega_2 \Delta,$$

$$(z, u)^{\cdot} = \pi_1 + \pi_2 \omega_2^3 + 2v_2 \omega_2 \varepsilon_2 + \varepsilon_2 \Delta + \omega_2(\pi_1 + \pi_2 \omega_2) \cos \vartheta.$$

$$Z = (z^{\cdot}, u^{\cdot}) \omega^4 - (z, u)^{\cdot} \omega^2 (z, z^{\cdot}) + (z, u)(z, z^{\cdot})^2$$

$$= \varepsilon_2 \omega^4 (\pi_1 \cos \vartheta + \pi_2 \omega_2^2 + \varepsilon_2 v_2) + \omega^4 \omega_2^2 \alpha$$

$$- \omega^2 [\pi_1 + \pi_2 \omega_2^3 + 2v_2 \omega_2 \varepsilon_2 + \varepsilon_2 \Delta + \omega_2 (\pi_1 + \pi_2 \omega_2) \cos \vartheta].$$

$$\varepsilon_2(\omega_2 + \cos\vartheta) + (v_1 + v_2\omega_2^2 + \omega_2\Delta)\,\varepsilon_2^2(\omega_2 + \cos\vartheta)^2$$
$$= \omega^4\omega_2^2\alpha + \omega^2\omega_2\varepsilon_2(\omega_2\pi_2 - \pi_1)\sin^2\vartheta + \varepsilon_2^2\big[v_1(\omega_2 + \cos\vartheta)^2$$
$$+ v_2(1 + \omega_2\cos\vartheta)^2 - \Delta(\omega_2 + \cos\vartheta)(1 + \omega_2\cos\vartheta)\big].$$

Further,

$$Z - (\boldsymbol{z}, \boldsymbol{u})\,V = \omega^4\omega_2^2\alpha + \omega^2\omega_2\varepsilon_2(\omega_2\pi_2 - \pi_1)\sin^2\vartheta$$
$$+ \varepsilon_2^2[v_1(\omega_2 + \cos\vartheta)^2 + v_2(1 + \omega_2\cos\vartheta)^2$$
$$- \Delta(\omega_2 + \cos\vartheta)(1 + \omega_2\cos\vartheta)]$$
$$- (v_1 + v_2\omega_2^2 + \omega_2\Delta)(\omega^2\omega_2^2 + \varepsilon_2^2)\sin^2\vartheta.$$

The formulae for the invariants are

$$\kappa_1 = (\omega^*)^{-3}\,V^{1/2}, \quad \kappa_2 = V^{-1}\left|\boldsymbol{z}, \boldsymbol{z}^{\bullet}, \boldsymbol{z}^{\bullet\bullet}\right|,$$
$$\mu_1 = (\omega^*)^{-5}\,V^{-1/2}\big[Z - (\boldsymbol{z}, \boldsymbol{u})\,V\big].$$

If we substitute for the asterisk values from the proof of Theorem 8.6, we obtain

$$\kappa_1 = \omega^{-3}\big[\omega^2\omega_1^2\omega_2^2 + (\omega_1\varepsilon_2 - \omega_2\varepsilon_1)^2\big]^{1/2}\sin\vartheta,$$
$$\kappa_2 = \big[\omega^2\omega_1^2\omega_2^2 + (\omega_1\varepsilon_2 - \omega_2\varepsilon_1)^2\big]^{-1}\{2\omega_1^2\varepsilon_2^2 - \omega_2^2\varepsilon_1^2$$
$$+ \omega_1\omega_2[\omega_2\dot{\varepsilon}_1 - \omega_1\dot{\varepsilon}_2 - \varepsilon_1\varepsilon_2 + \omega_2\omega_1^2(\omega_1 + \omega_2\cos\vartheta)]\},$$
$$\mu_1 = \omega^{-5}\big[\omega^2\omega_1^2\omega_2^2 + (\omega_1\varepsilon_2 - \omega_2\varepsilon_1)^2\big]^{-1/2}\{\omega^4\omega_1^2\omega_2^2\alpha$$
$$+ \omega^2\omega^1\omega_2(\omega_1\varepsilon_2 - \omega_2\varepsilon_1)(\omega_2\pi_2 - \omega_1\pi_1)\sin^2\vartheta$$
$$+ (\omega_1\varepsilon_2 - \omega_2\varepsilon_1)^2\,[v_1(\omega_2 + \omega_1\cos\vartheta)^2$$
$$+ v_2(\omega_1 + \omega_2\cos\vartheta)^2 - \Delta(\omega_2 + \omega_1\cos\vartheta)(\omega_1 + \omega_2\cos\vartheta)]$$
$$- (v_1\omega_1^2 + v_2\omega_2^2 + \omega_1\omega_2\Delta)\,[\omega^2\omega_2^2\omega_1^2$$
$$+ (\omega_1\varepsilon_2 - \omega_2\varepsilon_1)^2]\sin^2\vartheta\}\sin^{-1}\vartheta.$$

The general expression for μ_2 is too complicated; we give it only for $v_1, v_2, \omega_1, \omega_2$ constant, i.e. for the case of helical motion. Then $\varepsilon_2^* = \pi_1 = \pi_2 = 0$. Once again compute the derivatives with respect to φ_1 and, for the sake of simplicity, omit the asterisk:

$$\boldsymbol{r} = \mathrm{ad}g_1[\boldsymbol{x} + \omega_2\boldsymbol{y};\ \boldsymbol{t} + \omega_2\boldsymbol{s} + v_1\boldsymbol{x} + v_2\omega_2\boldsymbol{y}],$$

$$\begin{aligned}
r^{\cdot} &= \operatorname{ad} g_1[\omega_2 x \times y;\ \omega_2\{x \times s + t \times y + (v_1 + v_2)\, x \times y\}]\,,\\
r^{\cdot\cdot} &= \operatorname{ad} g_1[\omega_2(x \cos \vartheta - y);\ \omega_2\{dx \sin \vartheta + t \cos \vartheta - s\\
&\qquad + (2v_1 + v_2)(\cos \vartheta - y)\}]\,.
\end{aligned}$$

$$V = \omega^2\omega_2^2 \sin^2 \vartheta\,, \quad |z, z^{\cdot}, z^{\cdot\cdot}| = \omega_2^2(1 + \omega_2 \cos \vartheta) \sin^2 \vartheta\,,$$

$$Z = \omega^4\omega_2^2\alpha\,,$$

$$(z, z^{\cdot\cdot}) = -\omega_2^2 \sin^2 \vartheta\,,$$

$$z \times z^{\cdot} = \omega_2[(\cos \vartheta + \omega_2)\, x - (1 + \omega_2 \cos \vartheta)\, y]\,,$$

$$\dot{V} = 0\,.$$

$$\begin{aligned}
&u^{\cdot\cdot}\omega^2 - (z, z^{\cdot\cdot})\, u\\
&= \omega_2[\omega_2 t(1 + \cos^2 \vartheta) + (\omega_2^2 + 1)\, t \cos \vartheta - s - \omega_2^2 s \cos^2 \vartheta\\
&\quad - 2s\omega_2 \cos \vartheta + \omega^2 dx \sin \vartheta + v_1\omega_2 x \sin^2 \vartheta + v_2\omega_2^2 y \sin^2 \vartheta\\
&\quad + \omega^2(2v_1 + v_2)(x \cos \vartheta - y)]\,,
\end{aligned}$$

$$\begin{aligned}
&V|z, z^{\cdot}, u^{\cdot\cdot}\omega^2 - (z, z^{\cdot\cdot})\, u| - Z|z, z^{\cdot}, z^{\cdot\cdot}|\\
&= \omega^2\omega_2^4[\omega_2^2(v_1 - v_2) \sin^2 \vartheta + v_1\omega^2(1 + \omega_2 \cos \vartheta) \sin^2 \vartheta\\
&\quad + d\omega_2^2(\cos \vartheta + \omega_2) \sin^3 \vartheta] \sin^2 \vartheta\,,
\end{aligned}$$

and thus, after removing the fictitious asterisks, we obtain

$$\begin{aligned}
\mu_2 &= \omega^{-4}[\omega_1^2\omega_2^2(v_1 - v_2) \sin^2 \vartheta + v_1\omega^2\omega_1(\omega_1 + \omega_2 \cos \vartheta) \sin^2 \vartheta\\
&\qquad + d\omega_2^2(\cos \vartheta + \omega_2) \sin^3 \vartheta] \sin^2 \vartheta\,,
\end{aligned}$$

The results can be summarized in the following theorem:

8.17 Theorem. *Put*

$$\begin{aligned}
\alpha &= -d \sin \vartheta \cos \vartheta + (v_1 + v_2) \sin^2 \vartheta\,,\\
\Delta &= d \sin \vartheta + (v_1 + v_2) \cos \vartheta\,,\\
\omega_1 + \omega_2 \cos \vartheta &= \omega \cos \psi_1\,,\\
\omega_2 + \omega_1 \cos \vartheta &= \omega \cos \psi_2\,,\\
M &= \omega^2\omega_1^2\omega_2^2 + (\omega_1\varepsilon_2 - \omega_2\varepsilon_1)^2\,.
\end{aligned}$$

For the invariants of general double-helical motion we have

$$\frac{d\varphi}{dt} = \omega = (\omega_1^2 + \omega_2^2 + 2\omega_1\omega_2 \cos \vartheta)^{1/2},$$

$$v_0 = \omega^{-2}(v_1\omega_1^2 + v_2\omega_2^2 + \omega_1\omega_2\Delta),$$

$$\kappa_1 = \omega^{-3}M^{1/2} \sin \vartheta,$$

$$\kappa_2 = M^{-1}[2\omega_1^2\varepsilon_2^2 - \omega_2^2\varepsilon_1^2 + \omega_1\omega_2(\omega_2\dot{\varepsilon}_1 - \omega_1\dot{\varepsilon}_2 - \varepsilon_1\varepsilon_2) + \omega_2\omega_1^2\omega \cos \psi_1],$$

$$\mu_1 = \omega^{-3}M^{-1/2}[\omega^2\omega_1^2\omega_2^2\alpha + \omega_1\omega_2(\omega_1\varepsilon_2 - \omega_2\varepsilon_1) . (\omega_2\pi_2 - \omega_1\pi_1)\sin^2\vartheta + (\omega_1\varepsilon_2 - \omega_2\varepsilon_1)^2 . (v_1 \cos^2\psi_2 + v_2\cos^2\psi_1 - \Delta \cos\psi_1 \cos\psi_2)] \sin^{-1}\vartheta - \omega^{-3}v_0M^{1/2}\sin\vartheta .$$

If $\omega_1 : \omega_2 =$ *const.*, $v_1 =$ *const.*, $v_2 =$ *const.*, *we also have*

$$\mu_2 = \omega^{-4}[\omega_1^2\omega_2^2(v_1 - v_2) + v_1\omega^3\omega_1 \cos\psi_1 + d\omega\omega_1\omega_2^2 \sin\vartheta \cos\psi_2].$$

Let us also discuss the invariants of the moving directing cone of double-helical motion. If $g(t)$ is a motion, the motion $g^{-1}(-t)$ is the inverse motion of the given motion, $g^{in}(\tau) = g^{-1}(-t)$. For the directing cone of the given and inverse motions it is $\boldsymbol{R}^{in} = \bar{\boldsymbol{R}}$, $\bar{\boldsymbol{R}}^{in} = \boldsymbol{R}$, as in Chapter II.

We also have

$$\frac{d\bar{\boldsymbol{R}}^{in}(\tau)}{d\tau} = \frac{d\boldsymbol{R}(\tau)}{d\tau} = \frac{d\boldsymbol{R}(-t)}{d(-t)} = -\frac{d\boldsymbol{R}(-t)}{dt}, \quad \text{i.e.} \quad \frac{d\bar{\boldsymbol{R}}^{in}}{d\tau} = -\frac{d\boldsymbol{R}}{dt},$$

similarly

$$\frac{d^2\bar{\boldsymbol{R}}^{in}}{d\tau^2} = \frac{d^2\boldsymbol{R}}{dt^2}.$$

8.18 Theorem. *Let $g(t)$ be a space motion, $g^{in}(\tau) = g^{-1}(-t)$ its inverse motion. For their invariants we have*

$$\omega = \omega^{in}, \quad v_0 = v_0^{in}, \quad \kappa_1 = \kappa_1^{in}, \quad \kappa_2 = -\kappa_2^{in}, \quad \mu_1 = \mu_1^{in},$$

$$\mu_2 = -\bar{\mu}_2^{in},$$

and, of course,

$$\bar{\kappa}_2 = -\kappa_2^{in}, \quad \bar{\mu}_2 = -\mu_2^{in}.$$

8.19 Theorem. *Let* $g(t) = g_1(\varphi_1(t), a_1(\varphi_1(t)))\, g_2(\varphi_2(t), a_2(\varphi_2(t)))$ *be a general double-helical motion. Then its inverse motion is*

$$g^{in}(\tau) = g_2(-\varphi_2(-t), -a_2(\varphi_2)(-t)))\, g_1(-\varphi_1(-t), -a_1(\varphi_1(-t))).$$

Proof. It must be $g(t)\, g^{in}(-\tau) = e$. Then $g(t)\, g^{in}(-\tau) = g_1(\varphi_1(t), a_1(\varphi_1(t)))\, g_2(\varphi_2(t), a_2(\varphi_2(t)))\,.$

$$\cdot g_2(-\varphi_2(t), -a_2(\varphi_2(t))\, g_1(-\varphi_1(t), -a_1(\varphi_1(t))) =$$
$$= g_1(\varphi_1, a_1(\varphi_1))\, g_2(\varphi_2, a_2(\varphi_2))\, g_2(-\varphi_2, -a_2(\varphi_2))\,.$$
$$\cdot\, g_1(-\varphi_1, -a_1(\varphi_1)) = g_1(\varphi_1, a_1(\varphi_1))\, g_1(-\varphi_1, -a_1(\varphi_1)) = e\,.$$

8.20 Theorem. *If a double-helical motion is determined by the functions* $\omega_1, \omega_2, v_1, v_2$, *the inverse motion is determined by the functions* $\omega_2, \omega_1, v_2, v_1$.

If we wish to determine the invariants of a motion which is inverse to a given double-helical motion, we must substitute in the formulae of Theorem 8.17 according to the following table:

ω_1	ω_2	v_1	v_2	ε_1	ε_2	$\dot{\varepsilon}_1$	$\dot{\varepsilon}_2$	π_1	π_2
ω_2	ω_1	v_2	v_1	$-\varepsilon_2$	$-\varepsilon_1$	$\dot{\varepsilon}_2$	$\dot{\varepsilon}_1$	$-\pi_2$	$-\pi_1$

8.21 Theorem. *For the invariants of the moving directing cone of a double-helical motion we have*

$$\bar{\kappa}_2 = M^{-1}[\omega_1^2\varepsilon_2^2 - 2\omega_2^2\varepsilon_1^2$$
$$+ \omega_1\omega_2(\omega_2\dot{\varepsilon}_1 - \omega_1\dot{\varepsilon}_2 + \varepsilon_1\varepsilon_2 - \omega\omega_1\omega_2^2 \cos\psi_2)],$$

and for $\omega_1 : \omega_2 = const.$, $v_1 = const.$, $v_2 = const.$

$$\bar{\mu}_2 = \omega^{-4}[\omega_1^2\omega_2^2(v_2 - v_1) - v_2\omega^3\omega_2\cos\psi_2 - d\omega\omega_2\omega_1^2 \sin\vartheta\cos\psi_1].$$

8.22 Exercise. Prove that the following holds for a double-helical motion:

$$\bar{v}_0 = v_0, \quad \bar{\kappa}_1 = \kappa_1, \quad \kappa_2 - \bar{\kappa}_2 = 1, \quad \mu_2 - \bar{\mu}_2 = v_0, \quad \mu_1 = \bar{\mu}_1.$$

Finally, we shall find the points of the striction curve of the fixed axoid of a double-helical motion. To avoid complications, we consider only the case $\omega_1 : \omega_2 = const.$; in the general case the computational procedure is quite analogous and there are no new manipulations introduced, only the corresponding formulae are slightly more complicated.

Denote the striction curve of the fixed axoid $A(t)$ and assume that $A(t) = g_1(t)\,X(t)$. We would like to determine the set $X(t)$, i.e. the points on lines $q(t)$ which gradually become points of the striction curve under motion $g_1(t)$. Again, assume that the motion is given as a function of the angle φ_1 and omit the asterisks. We know that we can obtain the point $A(t)$ of the striction curve as the point of intersection of the normal to the axoid at point $A(t)$ and of the ruling line of the axoid. We also know that the ruling line of the axoid is determined by the vector $\boldsymbol{R}_1$ of the fixed directing cone, and that the normal to the axoid at the point of the striction curve is determined by the vector $\boldsymbol{R}_1'$. We thus have

$$\begin{aligned} \boldsymbol{r} &= \mathrm{ad}g_1[\boldsymbol{x} + \omega_2\boldsymbol{y};\, \boldsymbol{t} + \omega_2\boldsymbol{s} + v_1\boldsymbol{x} + v_2\omega_2\boldsymbol{y}]\,, \\ \boldsymbol{R}_1 &= \mathrm{ad}g_1\omega^{-1}[\boldsymbol{x} + \omega_2\boldsymbol{y};\, \boldsymbol{t} + \omega_2\boldsymbol{s} + (v_1 - v_0)\,\boldsymbol{x} + (v_2 - v_0)\,\omega_2\boldsymbol{y}]\,. \end{aligned}$$

If we differentiate with respect to φ_1 ($\omega_2 = const.$), we obtain

$$\begin{aligned} \boldsymbol{R}_1^{\bullet} &= \mathrm{ad}g_1\omega^{-1}[\boldsymbol{0};\, (\pi_1 - \pi\omega)\,\boldsymbol{x} + (\pi_2\omega_2 - \pi\omega)\,\omega_2\boldsymbol{y}] + \mathrm{ad}g_1\omega^{-1}\,. \\ &\quad .\,[[\boldsymbol{x};\, \boldsymbol{t} + v_1\boldsymbol{x}],\, [\boldsymbol{x} + \omega_2\boldsymbol{y};\, \boldsymbol{t} + \omega_2\boldsymbol{s} + (v_1 - v_0)\,\boldsymbol{x} \\ &\quad + (v_2 - v_0)\,\omega_2\boldsymbol{y}]] \\ &= \omega^{-1}\,\mathrm{ad}g_1[\omega_2\boldsymbol{x} \times \boldsymbol{y};\, (\pi_1 - \pi\omega)\,\boldsymbol{x} + (\pi_2\omega_2 - \pi\omega)\,\omega_2\boldsymbol{y} \\ &\quad + \omega_2\{\boldsymbol{x} \times \boldsymbol{s} + \boldsymbol{t} \times \boldsymbol{y} + (v_1 + v_2 - v_0)\,\boldsymbol{x} \times \boldsymbol{y}\}]\,, \end{aligned}$$

which, upon substitution for π, yields

$$\begin{aligned} \boldsymbol{R}_1^{\bullet} &= \omega^{-1}\omega_2\,\mathrm{ad}g_1[\boldsymbol{x} \times \boldsymbol{y};\, \omega^2(\pi_1 - \pi_2\omega_2)\,\{\boldsymbol{x}(\omega_2 + \cos\vartheta) \\ &\quad - \boldsymbol{y}(1 + \omega_2\cos\vartheta)\} \\ &\quad + \boldsymbol{x} \times \boldsymbol{s} + \boldsymbol{t} \times \boldsymbol{y} + (v_1 + v_2 - v_0)\,\boldsymbol{x} \times ,\boldsymbol{y}]\,; \end{aligned}$$

upon normalization we have

$$\boldsymbol{R}_1^{\bullet} \cong \sin^{-1}\vartheta\,\mathrm{ad}g_1[\boldsymbol{x} \times \boldsymbol{y};\, \omega^{-2}(\pi_1 - \pi_2\omega_2)\,\{\boldsymbol{x}(\omega_2 + \cos\vartheta)$$

$$- \boldsymbol{y}(1 + \omega_2 \cos \vartheta)\}$$
$$+ \boldsymbol{x} \times \boldsymbol{s} + \boldsymbol{t} \times \boldsymbol{y} + (v_1 + v_2 - v_0)\, \boldsymbol{x} \times \boldsymbol{y}] .$$

Now, it is easy to find the equation of line q from the expression for the vector $\boldsymbol{R}_1$:

$$q = \omega^{-2}[\boldsymbol{x} \times \boldsymbol{t} + \omega_2^2 \boldsymbol{y} \times \boldsymbol{s} + \omega_2(\boldsymbol{x} \times \boldsymbol{s} + \boldsymbol{y} \times \boldsymbol{t})$$
$$+ \omega_2(v_2 - v_1)\, \boldsymbol{x} \times \boldsymbol{y}] + \nu(\boldsymbol{x} + \omega_2 \boldsymbol{y}), \ \nu \in \mathbf{R} .$$

The equation of the normal can be obtained in the same way, but before writing it down we should realize one more fact: the equation of the normal will include the quantities $|\boldsymbol{x}, \boldsymbol{y}, \boldsymbol{t}|$ and $|\boldsymbol{x}, \boldsymbol{y}, \boldsymbol{s}|$. These two determinants are constant and it is easy to prove that a fundamental frame in the space E_3 can be selected so that both are zero. Assume, therefore, that $|\boldsymbol{x}, \boldsymbol{y}, \boldsymbol{t}| = 0$ and $|\boldsymbol{x}, \boldsymbol{y}, \boldsymbol{s}| = 0$. Denote by n the normal we are looking for, and write $n = g_1 m$. It is evident that it suffices to determine the equation of line m. We immediately find that

$$m = \omega^{-2}(\pi_1 - \pi_2\omega_2)\,(\boldsymbol{x} + \boldsymbol{y}\omega_2) + \lambda \boldsymbol{x} \times \boldsymbol{y}, \quad \lambda \in \mathbf{R} .$$

The point of intersection of lines m and q can be found as follows:

$$X = \omega^{-2}[\boldsymbol{x} \times \boldsymbol{t} + \omega_2^2 \boldsymbol{y} \times \boldsymbol{s} + \omega_2(\boldsymbol{x} \times \boldsymbol{s} + \boldsymbol{y} \times \boldsymbol{t})$$
$$+ \omega_2(v_2 - v_1)\, \boldsymbol{x} \times \boldsymbol{y}] + \nu_1(\boldsymbol{x} + \omega_2 \boldsymbol{y})$$
$$= \omega^{-2}(\pi_1 - \pi_2\omega_2)\,(\boldsymbol{x} + \omega_2 \boldsymbol{y}) + \lambda_1 \boldsymbol{x} \times \boldsymbol{y} .$$

Find the inner product of this equation with the vector $\boldsymbol{x}$. We have

$$\nu_1(1 + \omega_2 \cos \vartheta) = \omega^{-2}(\pi_1 - \pi_2\omega_2)\,(1 + \omega_2 \cos \vartheta) ,$$

i.e.

$$\nu_1 = \omega^{-2}(\pi_1 - \pi_2\omega_2) .$$

If we revert to parameter t, we have

$$X(\mathrm{t}) = \omega^{-2}[\omega_1^2 \boldsymbol{x} \times \boldsymbol{t} + \omega_2^2 \boldsymbol{y} \times \boldsymbol{s} + \omega_1\omega_2(\boldsymbol{x} \times \boldsymbol{s} + \boldsymbol{y} \times \boldsymbol{t})$$
$$+ \omega_1\omega_2(v_2 - v_1)\, \boldsymbol{x} \times \boldsymbol{y} + (\pi_1\omega_1 - \pi_2\omega_2)\,(\omega_1 \boldsymbol{x} + \omega_2 \boldsymbol{y})] .$$

If we wish to determine the position of point X, it is better to determine its distance from the shortest transversal of lines p_1 and p_2; denote this shortest transversal by l. The line l is determined by the vector

$$[[x; t], [y; s]] = [x \times y; x \times s + t \times y],$$

i.e. upon normalization by the vector

$$[x \times y;\ x \times s + t \times y] \sin^{-1} \vartheta,$$

and its equation is, therefore,

$$l = (|x, y, s|\, x - |x, y, t|\, y) \sin^{-2} \vartheta + \eta x \times y \equiv \eta x \times y, \quad \eta \in \mathbf{R}.$$

The point of intersection Y of lines q and l is

$$\begin{aligned} Y &= \omega^{-2}[\omega_1^2 x \times t + \omega_2^2 y \times s + \omega_1\omega_2(x \times s + y \times t) \\ &\quad + \omega_1\omega_2(v_2 - v_1)\, x \times y] + v_2(\omega_1 x + \omega_2 y) \\ &= \eta_2 x \times y, \end{aligned}$$

if we proceed with the originally adopted notation. The inner product of the equation with the vector x yields

$$v_2(\omega_1 + \omega_2 \cos \vartheta) = 0, \quad \text{i.e.} \quad v_2 = 0.$$

Therefore,

$$X(t) - Y(t) = \omega^{-2}(\pi_1\omega_1 - \pi_2\omega_2)(\omega_1 x + \omega_2 y).$$

Putting $D(t)$ for the distance of points $X(t)$ and $Y(t)$, we get

$$\begin{aligned} D^2 &= \omega^{-4}(\pi_1\omega_1 - \pi_2\omega_2)^2 (\omega_1^2 + \omega_2^2 + 2\omega_1\omega_2 \cos \vartheta) \\ &= \omega^{-2}(\pi_1\omega_1 - \pi_2\omega_2)^2. \end{aligned}$$

8.23 Theorem. *Let $A(t)$ be the striction curve of the fixed axoid of a double-helical motion with the constant ratio $\omega_1 : \omega_2$. Put $A(t) = g_1(t)\,X(t)$ and denote by $D(t)$ the distance of point $X(t)$ from the shortest transversal of lines p_1 and p_2. Then*

$$D(t) = \omega^{-1}(\pi_1\omega_1 - \pi_2\omega_2)$$

8.24 Theorem. *If $\omega_1 : \omega_2 = const.$, then $D(t) = const.$ if and only if $v_1 - v_2$ is a linear function of the angle φ_1 (and therefore also of φ_2).*

The proof is evident.

We shall now devote our attention to special cases of double-helical motion:

1. $v_1 = 0$, $v_2 = 0$, i.e. we are combining two rotations with variable angular velocities. In this case $v_0 = \omega^{-2}\omega_1\omega_2 d \sin \vartheta$, so that by the composition of two rotations a rotation can be generated only if the axes are parallel or concurrent. Inspection of the formulae will immediately tell us that if $d \sin \vartheta = 0$, the axoids are developable surfaces, i.e. $\mu_1 = 0$.

2. $v_1 = const.$, $v_2 = const.$, $\omega_1 : \omega_2 = const.$ In this case helical motions are involved, i.e. motions which have all invariants constant. It can be proved (even though not quite trivially) that all motions whose all invariants are constant are double-helical motions with $v_1, v_2, \omega_1 : \omega_2$ constant. Let us note that helical motions are the three-dimensional analogy of cyclical motions in a plane. Further, note that the spherical motion associated to a helical motion is a cyclical spherical motion, i.e. the rolling of two circles on a sphere. Assume that $\omega_1 \neq 0$, $\omega_2 \neq 0$, $\sin \vartheta \neq 0$. If we are interested only in the geometrical aspect of the problem, we may choose the parameter so that $\omega_1 = const.$, $\omega_2 = const.$, which is what will be assumed. We obtain the following expression for μ_1:

$$\mu_1 = \omega^{-4}\omega_1\omega_2\{[v_1\omega_2^2 + v_2\omega_1^2 + (v_1 + v_2)\,\omega_1\omega_2 \cos \vartheta] \sin \vartheta - \\ - d(\omega^2 \cos \vartheta + \omega_1\omega_2 \sin^2 \vartheta)\} ;$$

the other invariants are in Theorem 8.17. For the angles of lines p_1q and p_2q we have

$$\cos \psi_1 = \omega^{-1}(\omega_1 + \omega_2 \cos \vartheta) , \quad \cos \psi_2 = \omega^{-1}(\omega_2 + \omega_1 \cos \vartheta) ,$$

while the distances of lines p_1q and p_2q are

$$d_1 = \omega_2^{-1} \sin^{-1} \vartheta[\omega_2 d \sin \vartheta + (v_1 - v_0)\,\omega_1 + (v_2 - v_0)\,\omega_2 \cos \vartheta] ,$$
$$d_2 = \omega_1^{-1} \sin^{-1} \vartheta[\omega_1 d \sin \vartheta + (v_2 - v_0)\,\omega_2 + (v_1 - v_0)\,\omega_1 \cos \vartheta] .$$

Obviously, the line $q(t)$ is constant as is the point $X(t)$ from Theorem 8.23 – it is the point intersection of line q with the shortest transversal l of lines p_1 and p_2. This implies that:

a) $v_1 \neq 0$, $v_2 \neq 0$. The axoids are ruled helical surfaces (in general oblique and open). The fixed axoid is a rectangular helical surface if and only if $\omega_1 + \omega_2 \cos \vartheta = 0$; the moving axoid is a rectangular helical surface if and only if $\omega_1 + \omega_2 \cos \vartheta = 0$. The fixed axoid is a closed helical surface if and only if

$$\omega_2 d \sin \vartheta + (v_1 - v_0)\,\omega_1 + (v_2 - v_0)\,\omega_2 \cos \vartheta = 0 ,$$

which yields the condition $\omega_1(v_2 - v_1) \sin \vartheta = d(\omega_2 + \omega_1 \cos \vartheta)$; the moving axoid is a closed helical surface if and only if

$$\omega_2(v_1 - v_2) \sin \vartheta = d(\omega_1 + \omega_2 \cos \vartheta) .$$

For example, the fixed axoid is a rectangular closed helical surface if and only if $\omega_1 : \omega_2 = -\cos \vartheta$, $(v_2 - v_1) + d \tan \vartheta = 0$, and similarly for other combinations. Both axoids are closed helical surfaces if $d = 0$, $v_1 = v_2$. Both axoids cannot be rectangular helical surfaces simultaneously. The axoids are developable if $\mu_1 = 0$, i.e. if

$$\begin{aligned}&[v_1\omega_2^2 + v_2\omega_1^2 + (v_1 + v_2)\,\omega_1\omega_2 \cos \vartheta] \sin \vartheta \\ &= d(\omega^2 \cos \vartheta + \omega_1\omega_2 \sin^2 \vartheta) .\end{aligned}$$

It is interesting to note when the fixed axoid is a developable closed helical surface, i.e. when the equations

$$\omega_1(v_2 - v_1) \sin \vartheta = d(\omega_2 + \omega_1 \cos \vartheta)$$

and

$$\begin{aligned}&[v_1\omega_2^2 + v_2\omega_1^2 + (v_1 + v_2)\,\omega_1\omega_2 \cos \vartheta] \sin \vartheta \\ &= d(\omega^2 \cos \vartheta + \omega_1\omega_2 \sin^2 \vartheta) ,\end{aligned}$$

hold simultaneously. After some manipulation we obtain $\omega^2 v_1 \sin \vartheta = 0$, i.e. $v_1 = 0$. For v_2 it must then be $\omega_1 v_2 \sin \vartheta = d(\omega_2 + \omega_1 \cos \vartheta)$, from which it is easy to derive $d_1 = 0$. We have expected this result, because a developable closed helical surface is necessarily a conical surface of revolution, and this can only be generated by rotation of a line concurrent with the axis. The moving axoid is then a developable helical surface with radius d. An analogous consideration applies if the moving axoid is a conical surface of revolution.

b) $v_1 = 0$, $v_2 \neq 0$. The fixed axoid is a hyperboloid of revolution with jugular circle radius $d_1 = \omega^{-2}\omega_2[d(\omega_2 + \omega_1 \cos \vartheta) - v_2\omega_1 \,.\, \sin \vartheta]$; it may be a right circular cone as we have seen. If $\omega_1 + \omega_2 \cos \vartheta = 0$, the fixed axoid is the set of all tangents to the circle. The moving axoid is a developable helical surface since then $\mu_1 = 0$. This motion is the developing of a developable helical surface into the plane if $v_0 = 0$. For v_0 we have

$$\begin{aligned}v_0 &= \omega^{-2}[v_2\omega_2^2 + \omega_1\omega_2(d \sin \vartheta + v_2 \cos \vartheta)] \\ &= \omega_2\omega^{-2}[v_2(\omega_2 + \omega_1 \cos \vartheta) + \omega_1 d \sin \vartheta]\end{aligned}$$

$$= \omega_2 \omega^{-2}(v_2 \omega_2 \sin^2 \vartheta - \omega_2 d \sin \vartheta \cos \vartheta) .$$

Thus, $v_0 = 0$ if $v_2 \sin \vartheta - d \cos \vartheta = 0$. This means that the developing of a developable helical surface into a plane is characterized by the conditions

$$v_1 = 0 , \quad \omega_1 + \omega_2 \cos \vartheta = 0 , \quad v_2 = d \cot \vartheta .$$

In case (b) the moving axoid is, in general, an oblique open helical surface; special cases can be analysed in the same way as case (a).

The case $v_1 \neq 0$, $v_2 = 0$ can be solved in the same way as (b).

(c) $v_1 = 0$, $v_2 = 0$. The fixed axoid is a hyperboloid of revolution with jugular circle radius $d_1 = \omega^{-2}\omega_2 d(\omega_2 + \omega_1 \cos \vartheta)$, the moving axoid is a hyperboloid of revolution with jugular circle radius $d_2 = \omega^{-2}\omega_1 d(\omega_1 + \omega_2 \cos \vartheta)$. The fixed axoid is a right circular cone if and only if either $d = 0$ or $\omega_2 + \omega_1 \cos \vartheta = 0$. In the former case the moving axoid is also a right circular cone, in the latter it is a pencil of lines in a plane. In both cases we arrive at spherical motions. If the axes of rotations are mutually perpendicular, and if $d \neq 0$, the axoids are hyperboloids of revolution with the radii $d_1 = \omega^{-2}\omega_2^2 d$ and $d_2 = \omega^{-2}\omega_1^2 d$. Let us also note that $v_0 = \omega^{-2}\omega_1\omega_2 d \sin \vartheta$, so that if $d \sin \vartheta \neq 0$, the motion is not a rolling.

3. Now, let us consider the case of parallel axes which was originally excluded. Therefore, let $\sin \vartheta = 0$, $\cos \vartheta = 1$. In this case $\omega = \omega_1 + \omega_2$, $\kappa_1 = 0$. The line $q(t)$ is parallel with p_1 and p_2. The axoids are thus cylindrical surfaces, if $\omega_1 : \omega_2 = const.$ they are surfaces of revolution. The projection of the motion into the plane perpendicular to p_1 and p_2 is a plane motion. Together with this motion, translation takes place in the direction of p_1, with $v_0 = (\omega_1 + \omega_2)^{-1} (v_1\omega_1 + v_2\omega_2)$. We arrive at a singular case of space motion. The axoids exist but they do not determine the motion. This case, of course, belongs to plane kinematics and it is being left to the reader to analyse.

4. Now, consider a motion composed of a general helical motion, $g_1(\varphi_1(t), a_1(\varphi_1(t)))$, and of motion $g_2(\varphi_2(t), a_2(\varphi_2(t)))$, where $\varphi_2 = const.$, i.e. g_2 is a general translation. Here, it is not possible to express the translation a_2 as a function of the constant φ_2; assume, therefore, that a_2 is a function of φ_1; put $\varphi_2 = 0$. Then $g_2 = g_2(0, a_2(\varphi_1(t)))$. Clearly, we cannot use anything of what we have so far derived in this section and we must consider this case separately. Let $g_1(\varphi_1(t), a_1(\varphi_1(t)))$ be a general helical motion with rotation $\varphi_1 = \varphi_1(t)$ and translation $a_1(\varphi_1(t))$; assume the axis of this

helical motion to be determined by the Plücker coordinates $[\boldsymbol{x}; \boldsymbol{t}]$. Let g_2 be the translation in the direction of the unit vector $\boldsymbol{y}$, let the translation be given by the function $a_2(\varphi_1(\mathrm{t}))$. Put $\tau_2 = \mathrm{d}a_2/\mathrm{d}\varphi$, $(\boldsymbol{x}, \boldsymbol{y}) = \cos\alpha$. Then

$$\boldsymbol{r}(\mathrm{t}) = \mathrm{ad}g_1\omega_1[\boldsymbol{x}; \boldsymbol{t} + v_1\boldsymbol{x} + \tau_2\boldsymbol{y}],$$

so that

$$\omega = \omega_1, \quad v_0 = v_1 + \tau\cos\alpha.$$

Also

$$\boldsymbol{q}(\mathrm{t}) = \boldsymbol{x} \times \boldsymbol{t} + \tau_2\boldsymbol{x} \times \boldsymbol{y} + \lambda\boldsymbol{x}, \quad \lambda \in \mathbf{R},$$

i.e. for $\boldsymbol{x} \neq \boldsymbol{y}$ line $q(\mathrm{t})$ translates parallel and is parallel with the axis p_1, so that the fixed axoid is a general cylindrical surface; only if $\tau_2 = const.$, i.e. if $a_2 = m\varphi_1 + n$, is it a right circular cylinder. The moving axoid is created by the parallel translation of line $q(\mathrm{t})$, i.e. it is also a cylindrical surface. If $\tau_2 = const.$, we arrive at a plane. We again obtain a singular case of motion such as in 3. If $\boldsymbol{x} = \boldsymbol{y}$, $q = p$; we obtain a general helical motion with the axis p_1 and $v_0 = v_1 + \tau_2$. The axoids are formed by line p_1. An analogous result is obtained by combining translation g_1 with helical motion $g_2(\varphi_2, a_2)$.

In cases 3 and 4 the motion is fully determined by plane motion and by the function v_0 which, in case 3, reads

$$v_0 = (\omega_1 + \omega_2)^{-1}(v_1\omega_1 + v_2\omega_2)$$

and, in case 4, is

$$v_0 = v_1 + \tau_2\cos\alpha.$$

8.25 Exercise. Find the invariants of Bennet's mechanism, i.e. of the motion in Example 3.5.

Hint: Bennet's mechanism is given as a double-rotation motion, the distance and angle of the axes being given and φ_2 being a function of φ_1. It is therefore sufficient to substitute into the formulae of Theorem 8.17 (with the exception of the invariant μ_2 for which the corresponding formula must first be derived; this is not very difficult, since $v_1 = v_2 = 0$).

8.26 Exercise. Determine the equations of both axoids for Bennet's mechanism using the results of Section 8, i.e. first find the line $q(\mathrm{t})$ and then subject this line to motion $g_1(t)$ and $g_2^{-1}(t)$. Notice that this is a geometrical problem which means that the angle φ_1 may be chosen directly as the parameter of the motion.

To conclude this section, let us derive the matrices of all double-helical motions. Let it be noted that the matrices of these motions are not defined uniquely and that their expressions depend on the choice of the base.

Take $p_1 = \lambda \boldsymbol{u}_1$, $p_2 = d\boldsymbol{u}_3 + \mu(\boldsymbol{u}_1 \cos \vartheta + \boldsymbol{u}_2 \sin \vartheta)$; the Plücker coordinates of lines p_1 and p_2 are

$$\boldsymbol{x} = \begin{pmatrix} 1 \\ 0 \\ 0 \end{pmatrix}, \quad \boldsymbol{t} = \begin{pmatrix} 0 \\ 0 \\ 0 \end{pmatrix}, \quad \boldsymbol{y} = \begin{pmatrix} \cos \vartheta \\ \sin \vartheta \\ 0 \end{pmatrix}, \quad \boldsymbol{s} = \begin{pmatrix} -d \sin \vartheta \\ d \cos \vartheta \\ 0 \end{pmatrix}.$$

Denote by $\gamma(\varphi_i, a_i)$ rotation about axis $\boldsymbol{u}_1$ through angle φ_i and translation along $\boldsymbol{u}_1$ by a_i, let h be the rotation about $\boldsymbol{u}_3$ through angle ϑ and the translation along $\boldsymbol{u}_3$ be d. The general helical motion can then be expressed as

$$g = g_1(\varphi_1, a_1)\, g_2(\varphi_2, a_2) = \gamma(\varphi_1, a_1)\, h\gamma(\varphi_2, a_2)\, h^{-1},$$

with

$$h = \begin{pmatrix} 1, & 0, & 0, & 0 \\ 0, & \cos \vartheta, & -\sin \vartheta, & 0 \\ 0, & \sin \vartheta, & \cos \vartheta, & 0 \\ d, & 0, & 0, & 0 \end{pmatrix}, \quad \gamma(\varphi_i, a_i) = \begin{pmatrix} 1, & 0, & 0, & 0 \\ a_i, & 1, & 0, & 0 \\ 0, & 0, & \cos \varphi_i, & -\sin \varphi_i \\ 0, & 0, & \sin \varphi_i, & \cos \varphi_i \end{pmatrix}.$$

Upon multiplication we arrive at

$$g(t) = \begin{pmatrix} 1, & 0, & 0, & 0 \\ a_1, & 1, & 0, & 0 \\ 0, & 0, & \cos \varphi_1, & -\sin \varphi_1 \\ 0, & 0, & \sin \varphi_1, & \cos \varphi_1 \end{pmatrix}.$$

$$\begin{pmatrix} 1, & 0, & 0, & 0 \\ a_2 \cos \vartheta - d \sin \vartheta \sin \varphi_2, & \cos^2 \vartheta + \sin^2 \vartheta \cos \varphi_2, & \cos \vartheta \sin \vartheta(1 - \cos \varphi_2) & \sin \vartheta \sin \varphi_2 \\ a_2 \sin \vartheta + d \sin \varphi_2 \cos \vartheta, & \sin \vartheta \cos \vartheta(1 - \cos \varphi_2), & \sin^2 \vartheta + \cos^2 \vartheta \cos \varphi_2, & -\cos \vartheta \sin \varphi_2 \\ d(1 - \cos \varphi_2), & -\sin \vartheta \sin \varphi_2 & \cos \vartheta \sin \varphi_2, & \cos \varphi_2 \end{pmatrix}$$

where it was assumed that $a_1, a_2, \varphi_1, \varphi_2$ are given functions of time. Helical motion is obtained by putting $a_1 = v_1\varphi_1$, $a_2 = v_2\varphi_2$, where v_1 and v_2 are constants. We obtain a more lucid result if the two axes are mutually perpendicular. Then $\sin \vartheta = 1$, $\cos \vartheta = 0$ and, upon multiplication,

$$g(t) = \begin{pmatrix} 1, & 0, & 0, & 0 \\ a_1 - d \sin \varphi_2, & \cos \varphi_2, & 0, & \sin \varphi_2 \\ a_2 \cos \varphi_1 - d \sin \varphi_1(1 - \cos \varphi_2). & \sin \varphi_1 \sin \varphi_2, & \cos \varphi_1, & -\sin \varphi_1 \cos \varphi_2 \\ a_2 \sin \varphi_1 + d \cos \varphi_1(1 - \cos \varphi_2), & -\cos \varphi_1 \sin \varphi_2, & \sin \varphi_1, & \cos \varphi_1 \cos \varphi_2 \end{pmatrix}.$$

Now, the case remains in which general helical motion is combined with translation. Here, $g(t) = \gamma(\varphi_1, a_1)\, T$, where T is the translation in the direction $\boldsymbol{y} = \boldsymbol{u}_1 \cos \alpha + \boldsymbol{u}_2 \sin \alpha$ by a_2. We have

$$T = \begin{pmatrix} 1, & 0, & 0, & 0 \\ a_2 \cos \alpha, & 1, & 0, & 0 \\ a_2 \sin \alpha, & 0, & 1, & 0 \\ 0, & 0, & 0, & 1 \end{pmatrix},$$

so that

$$g(t) = \begin{pmatrix} 1, & 0, & 0, & 0 \\ a_1 + a_2 \cos \alpha, & 1, & 0, & 0 \\ a_2 \cos \varphi_1 \sin \alpha, & 0, & \cos \varphi_1, & -\sin \varphi_1 \\ a_2 \sin \varphi_1 \sin \alpha, & 0, & \sin \varphi_1, & \cos \varphi_1 \end{pmatrix}.$$

8.27 Example. Let us investigate whether the combination of helical motion and translation can generate a motion whose trajectories will be plane curves. Let us consider the simplest case in which translation along the axis of the helical motion is involved, i.e. let $\alpha = 0$. Therefore, put $a_1 + a_2 = b$. Point $X = 0 + x_0^i \boldsymbol{u}_i$ has the trajectory

$$x^1(\varphi) = b + x_0^1\,,$$

$$x^2(\varphi) = x_0^2 \cos \varphi - x_0^3 \sin \varphi\,,$$

$$x^3(\varphi) = x_0^2 \sin \varphi + x_0^3 \cos \varphi\,.$$

This trajectory is a plane curve if and only if $|\boldsymbol{x}^{\cdot}, \boldsymbol{x}^{\cdot\cdot}, \boldsymbol{x}^{\cdot\cdot\cdot}| = 0$. Computation yields the equation $(\dddot{b} + \dot{b})\,[(x_0^2)^2 + (x_0^3)^2] = 0$ which implies the following: if point X is on the axis, its trajectory is always plane (it is a line). If a point which is not on the axis has a plane trajectory, all points have plane trajectories. This case occurs if and only if $a_1 + a_2 = A \sin(\varphi + \varphi_0) + C$, where A, φ_0, C are constants. In this case the trajectories are ellipses.

8.28 Exercise. Solve the problem of the preceding example for $\alpha \neq 0$.

9. DUAL VECTORS AND TRAJECTORIES OF LINES

In space kinematics so-called dual numbers and dual vectors are sometimes used. We shall briefly acquaint the reader with this apparatus and mention the associations between our approach to notation and the approach which uses dual vectors. As an example of using dual vectors in kinematics, we shall derive the relations for the invariants of the trajectory of a line with the aid of this formalism. First, let us introduce dual numbers and the algebra of dual vectors.

9.1 Definition. The *ring D of dual numbers* is the set of all pairs of real numbers (we shall write these pairs as $\boldsymbol{\alpha} = a_1 + \mathbf{e}a_2$, $a_1 a_2 \in \mathbf{R}$, a_1 is called the *real*, a_2 the *dual part of the dual number*, $\mathbf{e}$ is the so-called *dual unit*), in which the operations of addition and multiplication are defined as follows: If $\boldsymbol{\alpha} = a_1 + \mathbf{e}a_2$, $\boldsymbol{\beta} = b_1 + \mathbf{e}b_2 \in D$, then

$$\boldsymbol{\alpha} + \boldsymbol{\beta} = a_1 + b_1 + \mathbf{e}(a_2 + b_2)\,,$$
$$\boldsymbol{\alpha} \circ \boldsymbol{\beta} = (a_1 + \mathbf{e}a_2)(b_1 + \mathbf{e}b_2) = a_1 b_1 + \mathbf{e}(a_1 b_2 + a_2 b_1)\,.$$

Remark. It is immediately seen from the definition that we may formally operate with dual numbers as with binomials if we put $\mathbf{e}^2 = 0$.

9.2 Exercise. Prove that D is really a ring, i.e. that it is a commutative group under addition, that the associative law holds for multiplication, that it contains a unit element with respect to multiplication, and that both distributive laws hold. Also prove that D is a commutative ring, i.e. that the commutative law for multiplication holds in it.

9.3 Exercise. Show that one cannot divide without restriction in D, and find all the dual numbers to which there exists the inverse element and determine it.

9.4 Definition. Let V_3 be a three-dimensional vector space over the field of real numbers **R** with the cross product $\times$. Denote by $\mathscr{D}$ the set of all pairs of vectors in V_3 (the elements of $\mathscr{D}$ will be expressed as $\mathbf{A} = \boldsymbol{u}_1 + \mathbf{e}\boldsymbol{u}_2$, where $\boldsymbol{u}_2, \boldsymbol{u}_2 \in V_3$, **e** is as yet only a symbol, but we shall soon see that it is the dual unit) in which the following operations are defined: let $\mathbf{A} = \boldsymbol{u}_1 + \mathbf{e}\boldsymbol{u}_2 \in \mathscr{D}$, $\mathbf{B} = \boldsymbol{v}_1 + \mathbf{e}\boldsymbol{v}_2 \in \mathscr{D}$, $\boldsymbol{\alpha} = a_1 + \mathbf{e}a_2 \in D$. Then

$$\mathbf{A} + \mathbf{B} = \boldsymbol{u}_1 + \boldsymbol{v}_1 + \mathbf{e}(\boldsymbol{u}_2 + \boldsymbol{v}_2),$$

$$\boldsymbol{\alpha}\mathbf{A} = (a_1 + \mathbf{e}a_2)(\boldsymbol{u}_1 + \mathbf{e}\boldsymbol{u}_2) = a_1\boldsymbol{u}_1 + \mathbf{e}(a_1\boldsymbol{u}_2 + a_2\boldsymbol{u}_1),$$

$$\begin{aligned}\mathbf{A} \times \mathbf{B} &= (\boldsymbol{u}_1 + \mathbf{e}\boldsymbol{u}_2) \times (\boldsymbol{v}_1 + \mathbf{e}\boldsymbol{v}_2) \\ &= \boldsymbol{u}_1 \times \boldsymbol{v}_1 + \mathbf{e}(\boldsymbol{u}_2 \times \boldsymbol{v}_1 + \boldsymbol{u}_1 \times \boldsymbol{v}_2).\end{aligned}$$

Remark. If we put $\mathbf{e}^2 = 0$, we can treat the elements of $\mathscr{D}$, or rather the elements of $\mathscr{D}$ and the dual numbers, as binomials.

The set $\mathscr{D}$ with the given operations is called the *algebra of dual vectors*. If $\mathscr{R} = \{\boldsymbol{f}_1, \boldsymbol{f}_2, \boldsymbol{f}_3\}$ is a base in V_3 (not necessarily orthonormal), we may write for $\mathbf{A} = \boldsymbol{u}_1 + \mathbf{e}\boldsymbol{u}_2$ that $\boldsymbol{u}_1 = a_1^i\boldsymbol{f}_i$, $\boldsymbol{u}_2 = a_2^i\boldsymbol{f}_i$, i.e. $\mathbf{A} = \boldsymbol{\alpha}^i\boldsymbol{f}_i$, where $\boldsymbol{\alpha}^i = a_1^i + \mathbf{e}a_2^i$, so that every element of $\mathscr{D}$ can be represented as a linear combination with dual coefficients of the vectors $\boldsymbol{f}_1, \boldsymbol{f}_2, \boldsymbol{f}_3$. It seems, therefore, that we may consider the set $\mathscr{D}$ as a vector space of dimension three (with the domain of coefficients D). Of course, the situation is slightly more complicated. The set $\mathscr{D}$ satisfies all the axioms of vector spaces, but its domain D is only a ring and not a field, i.e. we cannot divide without restriction in D, not even if we omit the zero element. For example, the expression $1/\mathbf{e}$ has no meaning because no dual number multiplied by **e** can yield unity. Nevertheless, we shall call the elements of $\mathscr{D}$ vectors. However, we must pay attention when computing with them. In dividing a dual vector by a dual number, we must always exclude the case in which this division has no meaning. Let us give a simple example. In the original vector space the following is true: if two vectors $\boldsymbol{u}, \boldsymbol{v}$ are linearly dependent, at least one of them is a multiple of the other. Let $\mathbf{eA} + \mathbf{eB} = 0$, $\mathbf{A}, \mathbf{B} \neq 0$, hold for vectors $\mathbf{A}, \mathbf{B} \in \mathscr{D}$. $\mathbf{A}, \mathbf{B}$ are linearly dependent (with coefficients from D), but they can be chosen so that neither is a multiple of the other. For example, it is sufficient to take $\mathbf{A} = \mathbf{e}\boldsymbol{u}$, $\mathbf{B} = \mathbf{e}\boldsymbol{v}$, where $\boldsymbol{u}, \boldsymbol{v}$ are independent vectors of V_3. Then we indeed have $\mathbf{eA} + \mathbf{eB} = \mathbf{e}^2\boldsymbol{u} + \mathbf{e}^2\boldsymbol{v} = 0$, but $(a_1 + \mathbf{e}a_2)\mathbf{A} = a_1\mathbf{e}\boldsymbol{u} \neq \mathbf{e}\boldsymbol{v}$ and $(a_1 + \mathbf{e}a_2)\mathbf{B} = a_1\mathbf{e}\boldsymbol{v} \neq \mathbf{e}\boldsymbol{u}$ for all $a_1 + \mathbf{e}a_2 \in \mathscr{D}$.

9.5 Definition. Let $\mathscr{R} = \{f_1, f_2, f_3\}$ be an orthonormal base in V_3, and let $\mathbf{A} = \alpha^i f_i$, $\mathbf{B} = \beta^i f_i \in \mathscr{D}$. Define the *inner product* $(\mathbf{A}, \mathbf{B})$ by the formula $(\mathbf{A}, \mathbf{B}) = \alpha^i \beta^i$.

9.6 Exercise. Let $\mathbf{A}, \mathbf{B}, \mathbf{C} \in \mathscr{D}$, $\alpha, \beta \in D$. Show that

$$(\mathbf{A}, \mathbf{B}) = (\mathbf{B}, \mathbf{A}), \quad (\alpha\mathbf{A} + \beta\mathbf{B}, \mathbf{C}) = \alpha(\mathbf{A}, \mathbf{C}) + \beta(\mathbf{B}, \mathbf{C}).$$

Thus the inner product in $\mathscr{D}$ has been formally introduced in the same way as the inner product in V_3. Of course, the inner product of two dual vectors is a general dual number. The inner product which was just introduced could depend on which base was chosen for its definition. That this is not so will be seen later.

9.7 Exercise. Let $\mathbf{A}, \mathbf{B}, \mathbf{C}, \mathbf{D} \in \mathscr{D}$. Then we have

$$\mathbf{A} \times (\mathbf{B} \times \mathbf{C}) = (\mathbf{A}, \mathbf{C})\,\mathbf{B} - (\mathbf{A}, \mathbf{B})\,\mathbf{C},$$

$$(\mathbf{A}, \mathbf{B} \times \mathbf{C}) = (\mathbf{A} \times \mathbf{B}, \mathbf{C}),$$

$$\mathbf{A} \times \mathbf{B} = -\mathbf{B} \times \mathbf{A},$$

$$(\alpha\mathbf{A} + \beta\mathbf{B}) \times \mathbf{C} = \alpha\mathbf{A} \times \mathbf{C} + \beta\mathbf{B} \times \mathbf{C},$$

$$(\mathbf{A} \times \mathbf{B}, \mathbf{C} \times \mathbf{D}) = (\mathbf{A}, \mathbf{C})(\mathbf{B}, \mathbf{D}) - (\mathbf{A}, \mathbf{D})(\mathbf{B}, \mathbf{C}),$$

$$\mathbf{A} \times (\mathbf{B} \times \mathbf{C}) + \mathbf{B} \times (\mathbf{C} \times \mathbf{A}) + \mathbf{C} \times (\mathbf{A} \times \mathbf{B}) = \mathbf{0},$$

where $\alpha, \beta \in D$.

The algebra of dual vectors is a real vector space over the field of real numbers, of course, and as such its dimension is 6. Indeed, if $\{f_1, f_2, f_3\}$ is a base in V_3, $\{f_1, f_2, f_3\}$ is also a base in $\mathscr{D}$ considered over the domain of coefficients D, and $\{f_1, f_2, f_3, \mathbf{e}f_1, \mathbf{e}f_2, \mathbf{e}f_3\}$ is a base in $\mathscr{D}$ considered over the field $\mathbf{R}$. Recall now that the Lie algebra $\mathfrak{E}$ of the group of congruences of the Euclidean space is a vector space of dimension 6 over $\mathbf{R}$, in which the bracket $[\cdot, \cdot]$ of two vectors is defined. Also, let us notice the connection between the vector product in $\mathscr{D}$ and the bracket in $\mathfrak{E}$. Assign the vector $[\boldsymbol{u}_1; \boldsymbol{u}_2] \in \mathfrak{E}$ to the dual vector $\mathbf{A} = \boldsymbol{u}_1 + \mathbf{e}\boldsymbol{u}_2$; denote this correspondence by $\bar{\imath}$. Thus, we define $\bar{\imath} : \mathscr{D} \to \mathfrak{E}$, $\boldsymbol{u}_1 + \mathbf{e}\boldsymbol{u}_2 \to [\boldsymbol{u}_1; \boldsymbol{u}_2]$. The correspondence $\bar{\imath}$

is a one-to-one linear (over $\mathbf{R}$) mapping of $\mathscr{D}$ onto $\mathfrak{E}$; the desired connection is described in the following theorem.

9.8 Theorem. *Let* $\mathbf{A}, \mathbf{B} \in \mathscr{D}$. *Then* $\bar{\imath}(\mathbf{A} \times \mathbf{B}) = [\bar{\imath}\mathbf{A}, \bar{\imath}\mathbf{B}]$.

Proof. The theorem is a consequence of Lemma 1.1 and of the definition of the cross product in $\mathscr{D}$.

9.9 Theorem. *Let* $\mathbf{A}, \mathbf{B} \in \mathscr{D}$. *Then* $(\mathbf{A}, \mathbf{B}) = \mathbf{K}(\bar{\imath}\mathbf{A}, \bar{\imath}\mathbf{B}) + \mathbf{e}\mathscr{K}(\bar{\imath}\mathbf{A}, \bar{\imath}\mathbf{B})$.

Proof. The assertion follows immediately from Section 1 and Definition 9.5.

Now, identify the sets $\mathscr{D}$ and $\mathfrak{E}$ with the aid of $\bar{\imath}$. Then we may say that the dual vector is a Klein image of a line if and only if it is unit, i.e. if $(\mathbf{A}, \mathbf{A}) = 1$, so that in $\mathscr{D}$ the Klein quadric is the unit sphere. If we wish to find a more profound analogy between spherical and space kinematics, we must also express the matrix of space motion in terms of dual numbers. This is possible as we shall see further on.

9.10 Definition. Denote by $O(n, D)$ the set of all $n \times n$ matrices g whose elements are dual numbers and which satisfy the equation $g^T g = e$. (e is the unit $n \times n$ matrix.)

9.11 Theorem. *$O(n, D)$ constitutes a Lie group (over $\mathbf{R}$) of dimension $n(n-1)$ which contains $O(n, \mathbf{R})$ as a subgroup.*

Proof. $O(n, D)$ constitutes a group because, if $g, h \in O(n, D)$,

$$(gh)(gh)^{\mathrm{T}} = ghh^{\mathrm{T}}g^{\mathrm{T}} = gg^{\mathrm{T}} = e\,, \quad \text{i.e.} \quad gh \in O(n.\,D)\,,$$

also $e \in O(n, D)$, the transposed matrix is the inverse element, and $gg^T = e$ follows from $g^{\mathrm{T}}g = e$ since the multiplication of matrices with dual elements is associative. If we split the equation $gg^{\mathrm{T}} = e$ into components, we obtain: let $g = (\alpha^i_j) = (\gamma^i_j + \mathbf{e}\beta^i_j)$. Then

$$gg^{\mathrm{T}} = (\gamma^i_j + \mathbf{e}\beta^i_j)(\gamma^i_k + \mathbf{e}\beta^i_k) = \gamma^i_j\gamma^i_k + \mathbf{e}(\beta^i_j\gamma^i_k + \gamma^i_j\beta^i_k) = \delta_{jk}\,,$$

i.e. we obtain the equations $\gamma^i_j\gamma^i_k = \delta^i_k$, $\beta^i_j\gamma^i_k + \gamma^i_j\beta^i_k = 0$. If we take the matrix from $O(n, D)$ as a point of $(2n^2)$-dimensional Euclidean space, $O(n, D)$ is determined by quadratic equations and, therefore, it is a Lie group. If we write the matrix $g \in O(n, D)$ again in the form $g = (\alpha^i_j) = (\gamma^i_j + \mathbf{e}\beta^i_j)$, we see that γ^i_j belongs to $O(n, \mathbf{R})$, i.e. that it is an ordinary orthogonal matrix, and conversely, every orthogonal matrix belongs to $O(n, D)$.

If we wish to determine the dimension of the group $O(n, D)$, we must find the expressions for its elements. Symbolically, let us write $g = \gamma + \mathbf{e}a\gamma$, where γ and a are real matrices. This is possible. Indeed, if $g = \gamma + \mathbf{e}\beta$, it is sufficient to put $a = \beta\gamma^{\mathrm{T}}$. Then

$$gg^{\mathrm{T}} = (\gamma + \mathbf{e}a\gamma)(\gamma^{\mathrm{T}} + \mathbf{e}\gamma^{\mathrm{T}}a^{\mathrm{T}}) = e + \mathbf{e}(\gamma\gamma^{\mathrm{T}}a^{\mathrm{T}} + a\gamma\gamma^{\mathrm{T}})$$
$$= e + \mathbf{e}(a + a^{\mathrm{T}}).$$

This means that $g = \gamma + \mathbf{e}a\gamma$ belongs to $O(n, D)$ if and only if $\gamma \in O(n, \mathbf{R})$ and a is skew-symmetric. Further, $O(n)$ has dimension $\frac{1}{2}n(n-1)$, the set of all skew-symmetric matrices also has dimension $\frac{1}{2}n(n-1)$, so that $O(n, D)$ has dimension $n(n-1)$.

9.12 Theorem. *$g = (e + \mathbf{e}a)\gamma$ belongs to $O(n, D)$ if and only if $\gamma \in O(n, \mathbf{R})$ and $a \in \mathfrak{O}(n, \mathbf{R})$.*

9.13 Theorem. *The Lie algebra of the group $O(3, D)$ is the algebra $\mathscr{D}$ of dual vectors.*

Proof. Consider a curve $g(t) = \gamma(t) + \mathbf{e}a(t)\gamma(t)$ on the group $O(3, D)$ which passes through the origin, i.e. let $g(0) = e$, so that $\gamma(0) = e$, $a(0) = 0$. For the tangent vector $g'(0)$ we have

$$g'(0) = \gamma'(0) + \mathbf{e}[a'(0)\gamma(0) + a(0)\gamma'(0)] = \gamma'(0) + \mathbf{e}a'(0),$$

which is a 3×3 skew-symmetric matrix with dual elements. Using the method from Section 3 of Chapter II we are now able to assign to every such matrix a dual vector such that the bracket becomes the cross product. This proves the theorem.

Incidentally, we have also proved that the algebra of dual vectors is, in fact, the algebra of skew-symmetric matrices with dual coefficients.

Below we use again the notation from Section 3 of Chapter II: If X is a skew-symmetric matrix, we shall denote its corresponding vector by X^v. Also, recall that we have $(gXg^{\mathrm{T}})^v = gX^v$ for $g \in O(3, \mathbf{R})$.

9.14 Theorem. *Groups $O(3, D)$ and $\mathscr{E}$ are isomorphic.*

Proof. Take $g = (e + \mathbf{e}a)\gamma \in O(3, D)$ and assign to it the matrix

$$\bar{\imath}'(g) = \begin{pmatrix} 1, & 0 \\ a^v, & \gamma \end{pmatrix}.$$

The mapping $\bar{\imath}' : O(3, D) \to \mathscr{E}$ is an isomorphism. Indeed, we immediately see that $\bar{\imath}'$ is one-to-one. Further, put $h = (e + \mathbf{e}b)\delta \in O(3, D)$. Then

$$\begin{aligned} gh &= (e + \mathbf{e}a)\gamma(e + \mathbf{e}b)\delta = (\gamma + \mathbf{e}a\gamma)(\delta + \mathbf{e}b\delta) \\ &= \gamma\delta + \mathbf{e}(a\gamma\delta + \gamma b\delta) \\ &= \gamma\delta + \mathbf{e}(a + \gamma b\gamma^{-1})\gamma\delta = [e + \mathbf{e}(a + \gamma b\gamma^{-1})]\gamma\delta, \end{aligned}$$

so that

$$\bar{\imath}'(gh) = \begin{pmatrix} 1, & 0 \\ a^v + \gamma b^v, & \gamma\delta \end{pmatrix},$$

in virtue of $(\gamma b\gamma^{-1})^v = \gamma b^v$. On the other hand, of course,

$$\bar{\imath}'(g)\,\bar{\imath}'th) = \begin{pmatrix} 1, & 0 \\ a^v, & \gamma \end{pmatrix}\begin{pmatrix} 1, & 0 \\ b^v, & \delta \end{pmatrix} = \begin{pmatrix} 1, & 0 \\ a^v + \gamma b^v, & \gamma\delta \end{pmatrix}.$$

And this proves the theorem.

Now we need to find out how group $O(3, D)$ acts on the dual vectors. Therefore, let $\mathbf{A} \in \mathscr{D}$, $g \in O(3, D)$; put $\mathbf{A} = \boldsymbol{u}_1 + \mathbf{e}\boldsymbol{u}_2$, $g = (e + \mathbf{e}a)\gamma$; then

$$\bar{\imath}\mathbf{A} = [\boldsymbol{u}_1; \boldsymbol{u}_2], \quad \bar{\imath}'g = \begin{pmatrix} 1, & 0 \\ a^v, & \gamma \end{pmatrix}, \quad \bar{\imath}'g(\bar{\imath}\mathbf{A}) = [\gamma\boldsymbol{u}_1; \gamma\boldsymbol{u}_2 + a^v \times \gamma\boldsymbol{u}_1]$$

according to (1.2).

On the other side, let us compute the formal product $g\mathbf{A}$:

$$\begin{aligned} g\mathbf{A} &= (e + \mathbf{e}a)\gamma(\boldsymbol{u}_1 + \mathbf{e}\boldsymbol{u}_2) = (\gamma + \mathbf{e}a\gamma)(\boldsymbol{u}_1 + \mathbf{e}\boldsymbol{u}_2) \\ &= \gamma\boldsymbol{u}_1 + \mathbf{e}(a\gamma\boldsymbol{u}_1 + \gamma\boldsymbol{u}_2) = \gamma\boldsymbol{u}_1 + \mathbf{e}(a^v \times \gamma\boldsymbol{u}_1 + \gamma\boldsymbol{u}_2), \end{aligned}$$

which is in agreement with the above expression.

9.15 Theorem. *Group $O(3, D)$ acts naturally in $\mathscr{D}$, i.e. as the product of a matrix and a column.*

It remains to introduce the action of group $O(3, D)$ on the points of E_3 and on the vectors from the vector space of all vectors from E_3. Therefore, to the point $M \in E_3$ determined by the radius-vector X (i.e. $M = O + X$) we assign the matrix $m = e + \mathbf{e}x$, where $x^v = X$ (we assign the skew-symmetric matrix x to the coordinates of point M). Then we have

$$(e + \mathbf{e}a)\,\gamma(e + \mathbf{e}x)\,\gamma^T = (\gamma + \mathbf{e}a\gamma)\,(\gamma^T + \mathbf{e}x\gamma^T) = e + \mathbf{e}(a + \gamma x \gamma^T)$$
$$= e + \mathbf{e}(a + \gamma x^v)\,.$$

On the other hand,

$$\bar{\imath}'(g)\,M = \begin{pmatrix} 1, & 0 \\ a, & \gamma \end{pmatrix} \begin{pmatrix} 1 \\ x^v \end{pmatrix} = \begin{pmatrix} 1 \\ a + \gamma x^v \end{pmatrix},$$

which leads us to the following definition:

$$g(m) = gm(\operatorname{Re} g)^T\,,$$

where Re g represents the real part of matrix g.

9.16 Example. Let us elucidate the above considerations on a simple example. Refer to Example 1.4 in which we chose $\boldsymbol{X}, \boldsymbol{Y} \in \mathfrak{E}$, $g \in \mathscr{E}$. According to our definitions it is

$$\bar{\imath}^{-1}(\boldsymbol{X}) = \begin{pmatrix} 1 + 2\mathbf{e} \\ \mathbf{e} \\ -2 \end{pmatrix}, \quad \bar{\imath}^{-1}(\boldsymbol{Y}) = \begin{pmatrix} 0 \\ 1 - \mathbf{e} \\ -3 + \mathbf{e} \end{pmatrix};$$

according to the computation in 1.4 we have

$$\bar{\imath}^{-1}[\boldsymbol{X}, \boldsymbol{Y}] = \begin{pmatrix} 2 - 5\mathbf{e} \\ 3 + 5\mathbf{e} \\ 1 + \mathbf{e} \end{pmatrix}.$$

Direct computation yields

$$\bar{\imath}^{-1}(\boldsymbol{X}) \times \bar{\imath}^{-1}(\boldsymbol{Y}) = \begin{pmatrix} 2 - 5\mathbf{e} \\ 3 + 5\mathbf{e} \\ 1 + \mathbf{e} \end{pmatrix},$$

which agrees. Further,

$$(\bar{\imath}')^{-1}(g) = \begin{bmatrix} 1, & \mathbf{e}\sqrt{3}, & -\mathbf{e} \\ 0 & \frac{1}{2}, & \frac{\sqrt{3}}{2} \\ 2\mathbf{e}, & -\frac{\sqrt{3}}{2}, & \frac{1}{2} \end{bmatrix},$$

$$(\bar{\imath}')^{-1}(g)\,\bar{\imath}^{-1}(X) = \begin{bmatrix} 1 + 4\mathbf{e} \\ \frac{1}{2}\mathbf{e} - \sqrt{3} \\ 2\mathbf{e} - \frac{\sqrt{3}}{2}\mathbf{e} - 1 \end{bmatrix};$$

on the other hand, we obtain

$$\bar{\imath}'(\mathrm{ad}g(X)) = \begin{bmatrix} 1 + 4\mathbf{e} \\ -\sqrt{3} + \frac{1}{2}\mathbf{e} \\ -1 + \mathbf{e}\left(2 - \frac{\sqrt{3}}{2}\right) \end{bmatrix}.$$

Now, let us elucidate how group $O(3, D)$ acts on the points of E_3. Choose g as above and let

$$M = \begin{pmatrix} 1 \\ 2 \\ 3 \\ 7 \end{pmatrix}.$$

Then

$$gM = \begin{bmatrix} 1 \\ 2 \\ -\frac{1}{2} + \frac{7\sqrt{3}}{2} \\ -\frac{3\sqrt{3}}{2} + \frac{7}{2} \end{bmatrix}.$$

On the other hand, to point M we assign the matrix

$$m = \begin{pmatrix} 1, & -7\mathrm{e}, & 3\mathrm{e} \\ 7\mathrm{e}, & 1, & -2\mathrm{e} \\ -3\mathrm{e}, & 2\mathrm{e}, & 1 \end{pmatrix},$$

so that

$$\begin{aligned} &((\bar{\imath}')^{-1}(g))(m) \\ &= \begin{bmatrix} 1, & \sqrt{3}\mathrm{e}, & -\mathrm{e} \\ 0, & \frac{1}{2}, & \frac{\sqrt{3}}{2} \\ 2\mathrm{e}, & -\frac{\sqrt{3}}{2}, & \frac{1}{2} \end{bmatrix} \begin{bmatrix} 1, & -7\mathrm{e}, & 3\mathrm{e} \\ 7\mathrm{e}, & 1, & -2\mathrm{e} \\ -3\mathrm{e}, & 2\mathrm{e}, & 1 \end{bmatrix} \begin{bmatrix} 1, & 0, & 0 \\ 0, & \frac{1}{2}, & -\frac{\sqrt{3}}{2} \\ 0, & \frac{\sqrt{3}}{2}, & \frac{1}{2} \end{bmatrix} \\ &= \begin{bmatrix} 1, & \left(\frac{3\sqrt{3}}{2} - \frac{7}{2}\right)\mathrm{e}, & \left(\frac{7\sqrt{3}}{2} - \frac{1}{2}\right)\mathrm{e} \\ \left(\frac{7}{2} - \frac{3\sqrt{3}}{2}\right)\mathrm{e}, & 1, & -2\mathrm{e} \\ \left(\frac{1}{2} - \frac{7\sqrt{3}}{2}\right)\mathrm{e}, & 2\mathrm{e}, & 1 \end{bmatrix}, \end{aligned}$$

which again agrees. The last expression, due to its relative complexity, is of course more of theoretical significance. The symbolism used is suitable if lines are involved, because the action of the group is then simple multiplication of a matrix and a column. However, we shall rather take a look at the profit one could gain by using dual numbers in kinematics, in the light of what has been already formulated.

9.17 Theorem. *Let $g(\mathrm{t}) \in \mathscr{E}$ be a motion, $\boldsymbol{r}(\mathrm{t}), \bar{\boldsymbol{r}}(\mathrm{t})$ its directing functions. Put $g^* = \bar{\imath}^{-1}(g)$, $\boldsymbol{r}^* = \bar{\imath}^{-1}(\boldsymbol{r})$, $\bar{\boldsymbol{r}}^* = \bar{\imath}^{-1}(\bar{\boldsymbol{r}})$. Then $\boldsymbol{r}^*(\mathrm{t}) = (g^*)^{\boldsymbol{\cdot}}(g^*)^{\mathrm{T}}$, $\bar{\boldsymbol{r}}^*(\mathrm{t}) = (g^*)^{\mathrm{T}}(g^*)^{\boldsymbol{\cdot}}$, i.e. the directing functions can be computed with the aid of dual numbers in the same way as in spherical kinematics with the aid of real numbers.*

Proof. Let $g(\mathrm{t}) = \begin{pmatrix} 1, & 0 \\ a^v(\mathrm{t}), & \gamma(\mathrm{t}) \end{pmatrix}$. Then

$$r(t) = \begin{pmatrix} 0, & 0 \\ (\dot{a})^v - \dot{\gamma}\gamma^T a^v, & \dot{\gamma}\gamma^T \end{pmatrix}, \quad \bar{r}(t) = \begin{pmatrix} 0, & 0 \\ \gamma^T(\dot{a})^v, & \gamma^T\dot{\gamma} \end{pmatrix}$$

according to Theorem 4.2. Further,

$$g^*(t) = (e + \mathbf{e}a)\,\gamma\,,$$

so that

$$\begin{aligned}(g^*)^\cdot\,(g^*)^T &= (\gamma + \mathbf{e}a\gamma)^\cdot\,\gamma^T(e - \mathbf{e}a) = (\dot{\gamma} + \mathbf{e}\dot{\sigma}\gamma + \mathbf{e}a\dot{\gamma})\,\gamma^T(e - \mathbf{e}a) \\ &= (\dot{\gamma}\gamma^T + \mathbf{e}\dot{a} + \mathbf{e}a\dot{\gamma}\gamma^T)\,(e - \mathbf{e}a) \\ &= \dot{\gamma}\gamma^T + \mathbf{e}\dot{a} + \mathbf{e}a\dot{\gamma}\gamma^T - \mathbf{e}\dot{\gamma}\gamma^T a \\ &= \gamma^\cdot\gamma^T + \mathbf{e}(\dot{a} - [\dot{\gamma}\gamma^T, a]) = \dot{\gamma}\gamma^T + \mathbf{e}(\dot{a} - \dot{\gamma}\gamma^T a^v) = r^*(t)\end{aligned}$$

and, analogously,

$$\begin{aligned}(g^*)^T\,(g^*)^\cdot &= \gamma^T(e - \mathbf{e}a)\,(\dot{\gamma} + \mathbf{e}\dot{a}\gamma + \mathbf{e}a\dot{\gamma}) \\ &= \gamma^T(\dot{\gamma} + \mathbf{e}\dot{a}\gamma + \mathbf{e}a\dot{\gamma} - \mathbf{e}a\dot{\gamma}) \\ &= \gamma^T(\dot{\gamma} + \mathbf{e}\dot{a}\gamma) = \gamma^T\dot{\gamma} + \mathbf{e}\gamma^T\dot{a}\gamma = \gamma^T\dot{\gamma} + \mathbf{e}\,\mathrm{ad}(\gamma^T)\,\dot{a} \\ &= \gamma^T\dot{\gamma} + \mathbf{e}\gamma^T(\dot{a})^v = \bar{r}^*(t)\,.\end{aligned}$$

Here the fact was exploited that the bracket of two skew-symmetric matrices corresponds to the action of the first matrix on the vector which corresponds to the second matrix, and that the group ad$O(3)$ acts on skew-symmetric matrices in the same way as $O(3)$ acts on the respective vectors-see Chapter II, Section 10.

9.18 Example. Theorem 9.17 deserves to be demonstrated on an actual example. Consider the motion from Example 4.6 with the specialization from 4.7, i.e. $m_1 = 0$, $m_2 = q\cos\alpha$, $m_3 = q\sin\alpha$. The matrix of motion reads

$$g(\alpha) = \begin{pmatrix} 1, & 0, & 0, & 0 \\ 0, & \cos^2\alpha, & \sin\alpha, & \sin\alpha\cos\alpha \\ q\cos\alpha, & -\sin\alpha\cos\alpha, & \cos\alpha, & -\sin^2\alpha \\ q\sin\alpha, & -\sin\alpha, & 0, & \cos\alpha \end{pmatrix},$$

so that

$$a = \begin{pmatrix} 0, & -q\sin\alpha, & q\cos\alpha \\ q\sin\alpha, & 0, & 0 \\ -q\cos\alpha, & 0, & 0 \end{pmatrix},$$

$$\mathbf{e}a\gamma = \mathbf{e}q \begin{pmatrix} \sin^2\alpha\cos\alpha - \sin\alpha\cos\alpha, & -\sin\alpha\cos\alpha, & \sin^3\alpha + \cos^2\alpha \\ \sin\alpha\cos^2\alpha, & \sin^2\alpha, & \sin^2\alpha\cos\alpha \\ -\cos^3\alpha, & -\sin\alpha\cos\alpha, & -\sin\alpha\cos^2\alpha \end{pmatrix},$$

$$g^*(\alpha) = \begin{pmatrix} \cos^2\alpha + \mathbf{e}q\sin\alpha\cos\alpha(\sin\alpha - 1), & \sin\alpha(1 - \mathbf{e}q\cos\alpha), & \sin\alpha\cos\alpha + \mathbf{e}q\sin^3\alpha + \mathbf{e}q\cos^3\alpha \\ \sin\alpha\cos\alpha(-1 + \mathbf{e}q\cos\alpha), & \cos\alpha + \mathbf{e}q\sin^2\alpha, & \sin^2\alpha(-1 + \mathbf{e}q\cos\alpha) \\ -\sin\alpha - \mathbf{e}q\cos^3\alpha, & -\mathbf{e}q\sin\alpha\cos\alpha, & \cos\alpha(1 - \mathbf{e}q\sin\alpha\cos\alpha) \end{pmatrix}.$$

Differentiation yields

$$[g^*(\alpha)]^\cdot = \begin{pmatrix} -2\cos\alpha\sin\alpha + \mathbf{e}q(2\cos^2\alpha\sin\alpha - \cos^2\alpha - \sin^3\alpha + \sin^2\alpha), & \cos\alpha - \mathbf{e}q\cos 2\alpha, & \cos 2\alpha + \mathbf{e}q\sin 2\alpha(3/2\sin\alpha - 1) \\ \sin^2\alpha - \cos^2\alpha + \mathbf{e}q\cos\alpha(\cos^2\alpha - 2\sin^2\alpha), & -\sin\alpha + \mathbf{e}q\sin 2\alpha, & -\sin 2\alpha + 2\mathbf{e}q\sin\alpha(2 - 3\sin^2\alpha) \\ -\cos\alpha + 3\mathbf{e}q\cos^2\alpha\sin\alpha, & -\mathbf{e}q\cos 2\alpha, & -\sin\alpha + \mathbf{e}q\cos\alpha(2 - 3\cos^2\alpha) \end{pmatrix}$$

After a rather tedious and unpleasant computation one obtains

$$(g^*)^\cdot(g^*)^{\mathrm{T}} = \begin{pmatrix} 0 & 1 + \mathbf{e}q\cos\alpha(\sin\alpha - 1), & \cos\alpha + \mathbf{e}q\sin\alpha(\sin\alpha - 1) \\ -1 + \mathbf{e}q\cos\alpha(1 - \sin\alpha), & 0, & -\sin\alpha + \mathbf{e}q\cos\alpha(1 + \sin\alpha) \\ -\cos\alpha - \mathbf{e}q\sin\alpha(\sin\alpha - 1), & \sin\alpha - \mathbf{e}q\cos\alpha(1 + \sin\alpha), & 0 \end{pmatrix},$$

which agrees with the value obtained in Example 4.7.

Example 9.18 demonstrated that Theorem 9.17 does not provide us with a particularly ideal method of computing the directing functions. The next theorem will be more interesting in this connection.

9.19 Theorem. *Consider motion $g(\varphi)$ with the directing cone $\boldsymbol{R}(\varphi)$, expressed as a dual vector function. Then we may write the Frenet formulae in the form*

$$\boldsymbol{R} = \sigma_0 \boldsymbol{R}_1,$$

$$\begin{aligned} R_1' &= \sigma_1 T_1 , \\ T_1' &= -\sigma_1 R_1 + \sigma_2 N_1 , \\ N_1' &= -\sigma_2 T_1 , \end{aligned}$$

where R_1, T_1, N_1 are dual notations of these vectors from Theorem 5.1, and further $\sigma_0 = 1 + \mathrm{e}v_0$, $\sigma_1 = \kappa_1 + \mathrm{e}\mu_1$, $\sigma_2 = \kappa_2 + \mathrm{e}\mu_2$. Moreover, R_1, T_1, N_1 constitute an orthonormal base in $\mathcal{D}$.

Proof. Put $R = [x; t]$, $R_1 = [x; t - v_0 x]$, or in dual vector notation $R = x + \mathrm{e}t$, $R_1 = x + \mathrm{e}(t - v_0 x)$; further,

$$\sigma_0 R_1 = (1 + \mathrm{e}v_0)\left[x + \mathrm{e}(t - v_0 x)\right] = x + \mathrm{e}v_0 x + \mathrm{e}t - \mathrm{e}v_0 x = R .$$

In the further course of the proof let us again introduce the arc of the spherical image as the parameter, denoting it by s, with the derivative with respect to s being indicated by a Roman numeral. Now, let $R_1 = x + \mathrm{e}t$. Then $R_1^{\mathrm{I}} = x^{\mathrm{I}} + \mathrm{e}t^{\mathrm{I}}$. If we refer to the proof of Theorem 5.1, we obtain

$$T_1 = x^{\mathrm{I}} + \mathrm{e}\left[t^{\mathrm{I}} - (x^{\mathrm{I}}, t^{\mathrm{I}})\, x^{\mathrm{I}}\right] .$$

Further,

$$\begin{aligned} \sigma_1 \kappa_1^{-1} T_1 &= (1 + \mathrm{e}\mu_1 \kappa_1^{-1})\, T_1 \\ &= \left[1 + \mathrm{e}(x^{\mathrm{I}}, t^{\mathrm{I}})\right] \{x^{\mathrm{I}} + \mathrm{e}[t^{\mathrm{I}} - (x^{\mathrm{I}}, t^{\mathrm{I}})\, x^{\mathrm{I}}]\} = x^{\mathrm{I}} + \mathrm{e}t^{\mathrm{I}} = R_1^{\mathrm{I}} . \end{aligned}$$

In the proof of Theorem 5.1 the equation

$$T_1^{\mathrm{I}} + R_1 - kN_1 = \mu_1 \kappa_1^{-1} R_2 - \mu_2 \kappa_1^{-1} N_2$$

was proved. If we adopt dual vector notation,

$$T_1^{\mathrm{I}} + R_1 - kN_1 = \mathrm{e}\mu_1 \kappa_1^{-1} R_1 - \mathrm{e}\mu_2 \kappa_1^{-1} N_1 ,$$

i.e.

$$T_1^{\mathrm{I}} = -R_1 + \kappa_2 \kappa_1^{-1} N_1 - \mu_1 \kappa_1^{-1} \mathrm{e} R_1 - \mu_2 \kappa_1^{-1} \mathrm{e} N_1 ,$$

i.e.

$$\begin{aligned} T_1^{\mathrm{I}} &= \kappa_1^{-1}\left[-R_1(\kappa_1 + \mathrm{e}\mu_1) + N_1(\kappa_2 + \mathrm{e}\mu_2)\right] \\ &= \kappa_1^{-1}(-\sigma_1 R_1 + \sigma_2 N_1) , \end{aligned}$$

so that also

$$T_1' = -\sigma_1 R_1 + \sigma_2 N_1 .$$

The last equation can be proved similarly from the relation $N_1^I + kT_1 = -\mu_2\kappa_1^{-1}T_2$, derived also in the proof of Theorem 5.1.

Differentiating the relation $(R_1, R_1) = 1$ we obtain

$$(R_1', R_1) = 0\,,$$

i.e.

$$(R_1, T_1) = 0\,.$$

Further, we have

$$\begin{aligned}
(R_1, N_1) &= (R_1, R_1 \times \dot{T}_1) = (R_1 \times R_1, T_1) = 0\,,\\
(N_1, T_1) &= (R_1 \times T_1, T_1) = (R_1, T_1 \times T_1) = 0\,,\\
(N_1, N_1) &= (R_1 \times T_1, R_1 \times T_1) = (R_1, R_1)(T_1, T_1) - (R_1, T_1)^2 = 1\,,
\end{aligned}$$

so that R_1, T_1, N_1 constitute indeed an orthonormal base in $\mathscr{D}$.

9.20 Theorem. *Let motion $g(t)$ be a motion with the directing function $r(t)$, written as a dual vector function. Put $r = \sigma R_1$. For the invariants $\sigma, \sigma_1, \sigma_2$ we then have*

$$\sigma = \omega\sigma_0\,, \quad \textit{where} \quad \omega = \frac{d\varphi}{dt}\,,$$

$$\begin{aligned}
&(r, r) = \sigma^2\,,\\
&(r, r)(r^\bullet, r^\bullet) - (r, r^\bullet)^2 = \sigma^4\sigma_1^2\omega^2\,,\\
&|r, r^\bullet, r^{\bullet\bullet}| = \sigma^3\sigma_1^2\sigma_2\omega^3\,.
\end{aligned}$$

Proof. $R = \omega r = \sigma_0 R_1 = \omega\sigma R_1$, so that $\sigma_0 = \omega\sigma$. Further,

$$(r\omega, r\omega) = (r, r)\,\omega^2 = (\omega\sigma R_1, \omega\sigma R_1) = \omega^2\sigma^2(R_1, R_1) = \omega^2\sigma^2\,.$$

Theorem 9.19 yields

$$R_1^\bullet = R_1'\omega = \omega\sigma T_1\,,$$

so that

$$r^\bullet = \sigma^\bullet R_1 + \sigma R_1^\bullet = \sigma^\bullet R_1 + \omega\sigma\sigma_1 T_1\,,$$

i.e.

$$\begin{aligned}
&(r,r)(r^\bullet, r^\bullet) - (r, r^\bullet)^2\\
&= (\sigma R_1, \sigma R_1)(\sigma^\bullet R_1 + \omega\sigma\sigma_1 T_1, \dot{\sigma} R_1 + \omega\sigma\sigma_1 T_1)
\end{aligned}$$

$$- (\sigma \boldsymbol{R}_1, \dot{\sigma} \boldsymbol{R}_1 + \omega\sigma\sigma_1 \boldsymbol{T}_1)^2$$
$$= \sigma^2[(\dot{\sigma})^2 + \omega^2\sigma_1^2\sigma^2] - (\sigma\dot{\sigma})^2 = \omega^2\sigma^4\sigma_1^2 .$$

Further, we have

$$\boldsymbol{r}^{\cdot\cdot} = \ddot{\sigma}\boldsymbol{R}_1 + \dot{\sigma}\omega\sigma_1\boldsymbol{T}_1 + (\omega\sigma\sigma_1)^{\cdot}\,\boldsymbol{T}_1 + \omega^2\sigma\sigma_1(-\sigma_1\boldsymbol{R}_1 + \sigma_2\boldsymbol{N}_1)\,,$$

$$|\boldsymbol{r}, \boldsymbol{r}^{\cdot}, \boldsymbol{r}^{\cdot\cdot}| = |\sigma\boldsymbol{R}_1, \dot{\sigma}\boldsymbol{R}_1 + \omega\sigma\sigma_1\boldsymbol{T}_1, \ddot{\sigma}\boldsymbol{R}_1 + \dot{\sigma}\omega\sigma_1\boldsymbol{T}_1 + (\omega\sigma\sigma_1)^{\cdot}\,\boldsymbol{T}_1$$
$$- \omega^2\sigma\sigma_1^2\boldsymbol{R}_1 + \omega^2\sigma\sigma_1\sigma_2\boldsymbol{N}_1|$$
$$= |\sigma\boldsymbol{R}_1, \omega\sigma\sigma_1\boldsymbol{T}_1, \omega^2\sigma\sigma_1\sigma_2\boldsymbol{N}_1|$$
$$= \omega^3\sigma^3\sigma_1^2\sigma_2|\boldsymbol{R}_1, \boldsymbol{T}_1, \boldsymbol{N}_1| = \omega^3\sigma^3\sigma_1^2\sigma_2 .$$

9.21 Example. Let us prove that Theorems 9.20 and 5.9 yield the same expression for μ_2. We first compute $\omega^3\sigma^3\sigma_1^2\sigma_2$:

$$\omega^3\sigma^3\sigma_1^2\sigma_2 = \omega^6(1 + \mathrm{e}v_0)^3\,(\kappa_1 + \mathrm{e}\mu_1)^2\,(\kappa_2 + \mathrm{e}\mu_2)$$
$$= \omega^6(1 + 3\mathrm{e}v_0)\,(\kappa_1^2 + 2\mathrm{e}\kappa_1\mu_1)\,(\kappa_2 + \mathrm{e}\mu_2)$$
$$= \omega^6\kappa_1^2\kappa_2 + \mathrm{e}\omega^6(3v_0\kappa_1^2\kappa_2 + 2\kappa_1\mu_1\kappa_2 + \mu_2\kappa_1^2)\,.$$

Put $\boldsymbol{r} = \boldsymbol{x} + \mathrm{e}\boldsymbol{t}$. The dual part of the determinant

$$|\boldsymbol{r}, \boldsymbol{r}^{\cdot}, \boldsymbol{r}^{\cdot\cdot}| \equiv |\boldsymbol{x} + \mathrm{e}\boldsymbol{t},\ \boldsymbol{x}^{\cdot} + \mathrm{e}\boldsymbol{t}^{\cdot},\ \boldsymbol{x}^{\cdot\cdot} + \mathrm{e}\boldsymbol{t}^{\cdot\cdot}|$$

is equal to

$$|\boldsymbol{t}, \boldsymbol{x}^{\cdot}, \boldsymbol{x}^{\cdot\cdot}| + |\boldsymbol{x}, \boldsymbol{t}^{\cdot}, \boldsymbol{x}^{\cdot\cdot}| + |\boldsymbol{x}, \boldsymbol{x}^{\cdot}, \boldsymbol{t}^{\cdot\cdot}|\,,$$

so that

$$|\boldsymbol{t}, \boldsymbol{x}^{\cdot}, \boldsymbol{x}^{\cdot\cdot}| + |\boldsymbol{x}, \boldsymbol{t}^{\cdot}, \boldsymbol{x}^{\cdot\cdot}| + |\boldsymbol{x}, \boldsymbol{x}^{\cdot}, \boldsymbol{t}^{\cdot\cdot}|$$
$$= 3\omega^6 v_0\kappa_1^2\kappa_2 + 2\omega^6\kappa_1\mu_1\kappa_2 + \omega^6\mu_2\kappa_1^2 .$$

If we employ the notation from Theorem 5.9, we may write

$$3\omega^6 v_0\kappa_1^2\kappa_2 + 2\omega^6\kappa_1\mu_1\kappa_2 = \omega^6\kappa_1\kappa_2(3v_0\kappa_1 + 2\mu_1)$$
$$= \omega^6\omega^{-3}V^{1/2}V^{-1}|\boldsymbol{x}, \boldsymbol{x}^{\cdot}, \boldsymbol{x}^{\cdot\cdot}|\,\{3(\boldsymbol{x}, \boldsymbol{t})\,\omega^{-2}\omega^{-3}V^{1/2}$$
$$+ 2\omega^{-5}V^{-1/2}[Z - (\boldsymbol{x}, \boldsymbol{t})\,V]\} = \omega^{-2}|\boldsymbol{x}, \boldsymbol{x}^{\cdot}, \boldsymbol{x}^{\cdot\cdot}|\,[(\boldsymbol{x}, \boldsymbol{t}) + 2V^{-1}Z]\,,$$

and upon substitution we obtain

$$\begin{aligned}\omega^8\mu_2\kappa_1^2 V &= \omega^8\mu_2\omega^{-6}(V^{1/2})^2\, V = \mu_2\omega^2 V^2 \\ &= \omega^2 V[|\boldsymbol{t}, \boldsymbol{x}^\cdot, \boldsymbol{x}^{\cdot\cdot}| + |\boldsymbol{x}, \boldsymbol{t}^\cdot, \boldsymbol{x}^{\cdot\cdot}| + |\boldsymbol{x}, \boldsymbol{x}^\cdot, \boldsymbol{t}^{\cdot\cdot}|] \\ &\quad - V(\boldsymbol{x}, \boldsymbol{t})\,|\boldsymbol{x}, \boldsymbol{x}^\cdot, \boldsymbol{x}^{\cdot\cdot}| - 2Z|\boldsymbol{x}, \boldsymbol{x}^\cdot, \boldsymbol{x}^{\cdot\cdot}|\,,\end{aligned}$$

i.e.

$$\begin{aligned}\mu_2 = \omega^{-2}V^{-2}\{&\omega^2 V[|\boldsymbol{t}, \boldsymbol{x}^\cdot, \boldsymbol{x}^{\cdot\cdot}| + |\boldsymbol{x}, \boldsymbol{t}^\cdot, \boldsymbol{x}^{\cdot\cdot}| + |\boldsymbol{x}, \boldsymbol{x}^\cdot, \boldsymbol{t}^{\cdot\cdot}|] \\ &- V|\boldsymbol{x}, \boldsymbol{x}^\cdot, \boldsymbol{x}^{\cdot\cdot}|\,(\boldsymbol{x}, \boldsymbol{t}) - 2Z|\boldsymbol{x}, \boldsymbol{x}^\cdot, \boldsymbol{x}^{\cdot\cdot}|\}\,.\end{aligned}$$

Now, we have also obtained a new formula for computing the invariant μ_2. However, we have to show first that the two formulas actually mean the same. To achieve this we use the following trick:

Express the vector $\boldsymbol{x}^{\cdot\cdot}$ in the base $\boldsymbol{x}, \boldsymbol{x}^\cdot, \boldsymbol{x} \times \boldsymbol{x}^\cdot$ (which is not orthonormal in general), i.e. we write

$$\boldsymbol{x}^{\cdot\cdot} = \alpha\boldsymbol{x} + \beta\boldsymbol{x}^\cdot + \gamma\boldsymbol{x} \times \boldsymbol{x}^\cdot\,, \quad \alpha, \beta, \gamma \in \mathbf{R}\,.$$

Successive inner multiplication of this equation by the vectors $\boldsymbol{x}, \boldsymbol{x}^\cdot, \boldsymbol{x} \times \boldsymbol{x}^\cdot$ yields

$$\begin{aligned}(\boldsymbol{x}, \boldsymbol{x}^{\cdot\cdot}) &= \alpha(\boldsymbol{x}, \boldsymbol{x}) + \beta(\boldsymbol{x}, \boldsymbol{x}^\cdot)\,, \\ (\boldsymbol{x}^\cdot, \boldsymbol{x}^{\cdot\cdot}) &= \alpha(\boldsymbol{x}, \boldsymbol{x}^\cdot) + \beta(\boldsymbol{x}^\cdot, \boldsymbol{x}^\cdot)\,, \\ |\boldsymbol{x}, \boldsymbol{x}^\cdot, \boldsymbol{x}^{\cdot\cdot}| &= \gamma(\boldsymbol{x} \times \boldsymbol{x}^\cdot, \boldsymbol{x} \times \boldsymbol{x}^\cdot) = \gamma[(\boldsymbol{x}, \boldsymbol{x})(\boldsymbol{x}^\cdot, \boldsymbol{x}^\cdot) - (\boldsymbol{x}, \boldsymbol{x}^\cdot)^2] = \gamma V.\end{aligned}$$

Therefore, we have $\gamma = V^{-1}|\boldsymbol{x}, \boldsymbol{x}^\cdot, \boldsymbol{x}^{\cdot\cdot}|$ and the determinant of the system for α and β reads

$$(\boldsymbol{x}, \boldsymbol{x})(\boldsymbol{x}^\cdot, \boldsymbol{x}^\cdot) - (\boldsymbol{x}, \boldsymbol{x}^\cdot)^2 = V,$$

so that

$$\alpha = V^{-1}\begin{vmatrix}(\boldsymbol{x}, \boldsymbol{x}^{\cdot\cdot}), & (\boldsymbol{x}, \boldsymbol{x}^\cdot) \\ (\boldsymbol{x}^\cdot, \boldsymbol{x}^{\cdot\cdot}), & (\boldsymbol{x}^\cdot, \boldsymbol{x}^\cdot)\end{vmatrix} = V^{-1}[(\boldsymbol{x}, \boldsymbol{x}^{\cdot\cdot})(\boldsymbol{x}^\cdot, \boldsymbol{x}^\cdot) - (\boldsymbol{x}, \boldsymbol{x}^\cdot)(\boldsymbol{x}^\cdot, \boldsymbol{x}^{\cdot\cdot})]\,;$$

write

$$\begin{aligned}\alpha &= V^{-1}W, \\ \beta &= V^{-1}\begin{vmatrix}(\boldsymbol{x}, \boldsymbol{x}), & (\boldsymbol{x}, \boldsymbol{x}^{\cdot\cdot}) \\ (\boldsymbol{x}, \boldsymbol{x}^\cdot), & (\boldsymbol{x}^\cdot, \boldsymbol{x}^{\cdot\cdot})\end{vmatrix} \\ &= V^{-1}[(\boldsymbol{x}, \boldsymbol{x})(\boldsymbol{x}^\cdot, \boldsymbol{x}^{\cdot\cdot}) - (\boldsymbol{x}, \boldsymbol{x}^\cdot)(\boldsymbol{x}, \boldsymbol{x}^{\cdot\cdot})] = \tfrac{1}{2}\dot{V}V^{-1}\,.\end{aligned}$$

Multiply now the two expressions for μ_2 by the factor $\omega^2 V^2$ and subtract them from one another. This yields

$$
\begin{aligned}
&\{\omega^2 V|\boldsymbol{t}, \boldsymbol{x}^{\boldsymbol{\cdot}}, \boldsymbol{x}^{\boldsymbol{\cdot\cdot}}| + \omega^2 V|\boldsymbol{x}, \boldsymbol{t}^{\boldsymbol{\cdot}}, \boldsymbol{x}^{\boldsymbol{\cdot\cdot}}| + \omega^2 V|\boldsymbol{x}, \boldsymbol{x}^{\boldsymbol{\cdot}}, \boldsymbol{t}^{\boldsymbol{\cdot\cdot}}| - \\
&\quad (\boldsymbol{x}, \boldsymbol{t})\, V|\boldsymbol{x}, \boldsymbol{x}^{\boldsymbol{\cdot}}, \boldsymbol{x}^{\boldsymbol{\cdot\cdot}}| - 2Z|\boldsymbol{x}, \boldsymbol{x}^{\boldsymbol{\cdot}}, \boldsymbol{x}^{\boldsymbol{\cdot\cdot}}|\} - \\
&\quad \{V|\boldsymbol{x}, \boldsymbol{x}^{\boldsymbol{\cdot}}, \boldsymbol{t}^{\boldsymbol{\cdot\cdot}}|\, \omega^2 - V(\boldsymbol{x}, \boldsymbol{x}^{\boldsymbol{\cdot\cdot}})\, |\boldsymbol{x}, \boldsymbol{x}^{\boldsymbol{\cdot}}, \boldsymbol{t}| - \tfrac{1}{2}\dot{V}|\boldsymbol{x}, \boldsymbol{x}^{\boldsymbol{\cdot}}, \boldsymbol{t}^{\boldsymbol{\cdot}}| + \\
&\quad \tfrac{1}{2}\dot{V}|\boldsymbol{x}, \boldsymbol{x}^{\boldsymbol{\cdot}}, \boldsymbol{t}|\, (\boldsymbol{x}, \boldsymbol{x}^{\boldsymbol{\cdot}}) - Z|\boldsymbol{x}, \boldsymbol{x}^{\boldsymbol{\cdot}}, \boldsymbol{x}^{\boldsymbol{\cdot\cdot}}|\} \\
&= \omega^2|\boldsymbol{t}, \boldsymbol{x}^{\boldsymbol{\cdot}}, W\boldsymbol{x} + \tfrac{1}{2}\dot{V}\boldsymbol{x}^{\boldsymbol{\cdot}} + |\boldsymbol{x}, \boldsymbol{x}^{\boldsymbol{\cdot}}, \boldsymbol{x}^{\boldsymbol{\cdot\cdot}}|\, \boldsymbol{x} \times \boldsymbol{x}^{\boldsymbol{\cdot}}| + \\
&\quad \omega^2|\boldsymbol{x}, \boldsymbol{t}^{\boldsymbol{\cdot}}, W\boldsymbol{x} + \tfrac{1}{2}\dot{V}\boldsymbol{x}^{\boldsymbol{\cdot}} + |\boldsymbol{x}, \boldsymbol{x}^{\boldsymbol{\cdot}}, \boldsymbol{x}^{\boldsymbol{\cdot\cdot}}|\, \boldsymbol{x} \times \boldsymbol{x}^{\boldsymbol{\cdot}}| - \\
&\quad (\boldsymbol{x}, \boldsymbol{t})\, V|\boldsymbol{x}, \boldsymbol{x}^{\boldsymbol{\cdot}}, \boldsymbol{x}^{\boldsymbol{\cdot\cdot}}| - [(\boldsymbol{x}^{\boldsymbol{\cdot}}, \boldsymbol{t}^{\boldsymbol{\cdot}})\, \omega^4 - (\boldsymbol{x}, \boldsymbol{t})^{\boldsymbol{\cdot}}\, \omega^2(\boldsymbol{x}, \boldsymbol{x}^{\boldsymbol{\cdot}}) + \\
&\quad (\boldsymbol{x}, \boldsymbol{t})\, (\boldsymbol{x}, \boldsymbol{x}^{\boldsymbol{\cdot}})^2]\, |\boldsymbol{x}, \boldsymbol{x}^{\boldsymbol{\cdot}}, \boldsymbol{x}^{\boldsymbol{\cdot\cdot}}| + V(\boldsymbol{x}, \boldsymbol{x}^{\boldsymbol{\cdot\cdot}})\, |\boldsymbol{x}, \boldsymbol{x}^{\boldsymbol{\cdot}}, \boldsymbol{t}| + \\
&\quad \tfrac{1}{2}\dot{V}|\boldsymbol{x}, \boldsymbol{x}^{\boldsymbol{\cdot}}, \boldsymbol{t}^{\boldsymbol{\cdot}}|\, \omega^2 - \tfrac{1}{2}\dot{V}|\boldsymbol{x}, \boldsymbol{x}^{\boldsymbol{\cdot}}, \boldsymbol{t}|\, (\boldsymbol{x}, \boldsymbol{x}^{\boldsymbol{\cdot}}) \\
&= |\boldsymbol{x}, \boldsymbol{x}^{\boldsymbol{\cdot}}, \boldsymbol{t}|\, [-W\omega^2 - \tfrac{1}{2}\dot{V}(\boldsymbol{x}, \boldsymbol{x}^{\boldsymbol{\cdot}}) + V(\boldsymbol{x}, \boldsymbol{x}^{\boldsymbol{\cdot\cdot}})] + \\
&\quad |\boldsymbol{x}, \boldsymbol{x}^{\boldsymbol{\cdot}}, \boldsymbol{x}^{\boldsymbol{\cdot\cdot}}|\, \{\omega^2(\boldsymbol{t}, \boldsymbol{x})\, (\boldsymbol{x}^{\boldsymbol{\cdot}}, \boldsymbol{x}^{\boldsymbol{\cdot}}) - (\boldsymbol{t}, \boldsymbol{x}^{\boldsymbol{\cdot}})\, (\boldsymbol{x}, \boldsymbol{x}^{\boldsymbol{\cdot}}) + \\
&\quad \omega^2(\boldsymbol{x}^{\boldsymbol{\cdot}}, \boldsymbol{t}^{\boldsymbol{\cdot}}) - (\boldsymbol{x}, \boldsymbol{x}^{\boldsymbol{\cdot}})\, (\boldsymbol{t}^{\boldsymbol{\cdot}}, \boldsymbol{x})] - (\boldsymbol{x}, \boldsymbol{t})\, [\omega^2(\boldsymbol{x}^{\boldsymbol{\cdot}}, \boldsymbol{x}^{\boldsymbol{\cdot}}) - (\boldsymbol{x}, \boldsymbol{x}^{\boldsymbol{\cdot}})^2] - \\
&\quad (\boldsymbol{x}^{\boldsymbol{\cdot}}, \boldsymbol{t}^{\boldsymbol{\cdot}})\, \omega^4 + (\boldsymbol{x}, \boldsymbol{t})^{\boldsymbol{\cdot}}\, \omega^2(\boldsymbol{x}, \boldsymbol{x}^{\boldsymbol{\cdot}}) - (\boldsymbol{x}, \boldsymbol{t})\, (\boldsymbol{x}, \boldsymbol{x}^{\boldsymbol{\cdot}})^2\} \\
&= |\boldsymbol{x}, \boldsymbol{x}^{\boldsymbol{\cdot}}, \boldsymbol{t}|\, [-\omega^2(\boldsymbol{x}^{\boldsymbol{\cdot}}, \boldsymbol{x}^{\boldsymbol{\cdot}})\, (\boldsymbol{x}, \boldsymbol{x}^{\boldsymbol{\cdot\cdot}}) + \omega^2(\boldsymbol{x}, \boldsymbol{x}^{\boldsymbol{\cdot}})\, (\boldsymbol{x}, \boldsymbol{x}^{\boldsymbol{\cdot\cdot}}) \\
&\quad - \omega^2(\boldsymbol{x}, \boldsymbol{x}^{\boldsymbol{\cdot}})\, (\boldsymbol{x}^{\boldsymbol{\cdot}}, \boldsymbol{x}^{\boldsymbol{\cdot\cdot}}) + (\boldsymbol{x}, \boldsymbol{x}^{\boldsymbol{\cdot}})^2\, (\boldsymbol{x}, \boldsymbol{x}^{\boldsymbol{\cdot\cdot}}) \\
&\quad + \omega^2(\boldsymbol{x}, \boldsymbol{x}^{\boldsymbol{\cdot\cdot}})\, (\boldsymbol{x}^{\boldsymbol{\cdot}}, \boldsymbol{x}^{\boldsymbol{\cdot}}) - (\boldsymbol{x}, \boldsymbol{x}^{\boldsymbol{\cdot\cdot}})\, (\boldsymbol{x}, \boldsymbol{x}^{\boldsymbol{\cdot}})^2] = 0\,.
\end{aligned}
$$

9.22 Exercise. Verify that the expressions for the other invariants, obtained by Theorem 9.20, are the same as the expressions in Theorem 5.9.

9.23 Example. Let us actually compute the invariants for the motion from Example 4.7 with the aid of dual vectors. For the fixed directing function we have

$$
\boldsymbol{r}(\alpha) = \begin{pmatrix} \sin\alpha - eq\cos\alpha(1 + \sin\alpha) \\ \cos\alpha + eq\sin\alpha(\sin\alpha - 1) \\ -1 \quad + eq\cos\alpha(1 - \sin\alpha) \end{pmatrix},
$$

$$
\boldsymbol{r}^{\boldsymbol{\cdot}}(\alpha) = \begin{pmatrix} \cos\alpha - eq[-\sin\alpha(1 + \sin\alpha) + \cos^2\alpha] \\ -\sin\alpha + eq[\cos\alpha(\sin\alpha - 1) + \sin\alpha\cos\alpha] \\ eq[-\sin\alpha(1 - \sin\alpha) - \cos^2\alpha] \end{pmatrix},
$$

$$r^{\bullet\bullet}(\alpha) = \begin{pmatrix} -\sin\alpha - \mathbf{e}q[-\cos\alpha(1+\sin\alpha) - \sin\alpha\cos\alpha - 2\cos\alpha\sin\alpha] \\ -\cos\alpha + \mathbf{e}q[-\sin\alpha(\sin\alpha - 1) + 2\cos^2\alpha - \sin^2\alpha] \\ \mathbf{e}q[-\cos\alpha(1-\sin\alpha) + \sin\alpha\cos\alpha + 2\cos\alpha\sin\alpha] \end{pmatrix}.$$

Therefore,

$$\begin{aligned}&(r(\alpha), r(\alpha)) \\ &= 2 + 2\mathbf{e}q[-\sin\alpha\cos\alpha(1+\sin\alpha) + \sin\alpha\cos\alpha(\sin\alpha - 1) - \\ &\quad \cos\alpha(1-\sin\alpha)] = 2 + 2\mathbf{e}q(-\cos\alpha - \sin\alpha\cos\alpha),\end{aligned}$$

i.e.

$$r^2 = 2[1 - \mathbf{e}q\cos\alpha(1+\sin\alpha)].$$

Further,

$$\begin{aligned}(r, r^{\bullet}) &= \sin\alpha\cos\alpha - \cos\alpha\sin\alpha + \mathbf{e}q[\sin^2\alpha(1+\sin\alpha) - \\ &\quad \sin\alpha\cos^2\alpha - \cos^2\alpha(1+\sin\alpha) + \cos^2\alpha(\sin\alpha - 1) + \\ &\quad \sin\alpha\cos^2\alpha - \sin^2\alpha(\sin\alpha - 1) + \sin\alpha(1-\sin\alpha) + \\ &\quad \cos^2\alpha] = \mathbf{e}q(\sin^2\alpha - \cos^2\alpha + \sin\alpha),\end{aligned}$$

$$(r, r^{\bullet})^2 = 0;$$

i.e.

$$\begin{aligned}(r^{\bullet}, r^{\bullet}) &= 1 + 2\mathbf{e}q[\cos\alpha\sin\alpha(1+\sin\alpha) - \cos^3\alpha - \\ &\quad \sin\alpha\cos\alpha(\sin\alpha - 1) - \sin^2\alpha\cos\alpha] \\ &= 1 + 2\mathbf{e}q(-\cos\alpha + 2\cos\alpha\sin\alpha),\end{aligned}$$

$$\begin{aligned}&(r, r)(r^{\bullet}, r^{\bullet}) - (r, r^{\bullet})^2 \\ &= 2[1 - \mathbf{e}q\cos\alpha(1+\sin\alpha)][1 + 2\mathbf{e}q\cos\alpha(2\sin\alpha - 1)] \\ &= 2[1 + 3\mathbf{e}q\cos\alpha(\sin\alpha - 1)].\end{aligned}$$

Further, we have

$$\sigma^4\sigma_1^2\omega^2 = \omega^6\kappa_1^2 + \mathbf{e}\omega^6(2\kappa_1\mu_1 + 4v_0\kappa_1^2),$$

so that

$$\sigma^2 = \omega^2(1 + 2\mathbf{e}v_0) = 2[1 - \mathbf{e}q\cos\alpha(1+\sin\alpha)],$$

$$\omega^6\kappa_1^2 + \mathbf{e}\omega^6(2\kappa_1\mu_1 + 4v_0\kappa_1^2) = 2 + 6\mathbf{e}q\cos\alpha(\sin\alpha - 1),$$

so that

$$\omega = \sqrt{2}\,, \quad v_0 = -\tfrac{1}{2}q \cos\alpha(1 + \sin\alpha)\,, \quad \kappa_1^2 = \tfrac{1}{4}\,, \quad \kappa_1 = \tfrac{1}{2}\,,$$
$$8(\mu_1 + v_0) = 6\mathbf{e}q \cos\alpha(\sin\alpha - 1)\,,$$

i.e.

$$4\mu_1 = 2q \cos\alpha(1 + \sin\alpha) + 3q \cos\alpha(\sin\alpha - 1)$$
$$= q \cos\alpha(5 \sin\alpha - 1)\,,$$

which agrees.

According to Example 9.21 we have

$$\omega^3\sigma^3\sigma_1^2\sigma_2 = \omega^6\kappa_1^2\kappa_2 + \mathbf{e}\omega^6(3v_0\kappa_1^2\kappa_2 + 2\kappa_1\mu_1\kappa_2 + \mu_2\kappa_1^2)$$
$$= 2\kappa_2 + 2\mathbf{e}(3v_0\kappa_0 + 4\mu_1\kappa_2 + \mu_2)\,.$$

If we compute the determinant, we obtain

$$|\boldsymbol{r}, \boldsymbol{r}^{\bullet}, \boldsymbol{r}^{\bullet\bullet}| = 1 - 3\mathbf{e}q \cos\alpha(1 + \sin\alpha)\,,$$

so that $\kappa_2 = \frac{1}{2}$, and further

$$2\mu_2 = |\boldsymbol{r}, \boldsymbol{r}^{\bullet}, \boldsymbol{r}^{\bullet\bullet}| - 3v_0 - 4\mu_1$$
$$= -\tfrac{1}{2}q(6\cos\alpha + 6\cos\alpha \sin\alpha - 3\cos\alpha - 3\cos\alpha\sin\alpha +$$
$$10\cos\alpha\sin\alpha - 2\cos\alpha)\,,$$

so that

$$\mu_2 = -\tfrac{1}{4}q\cos\alpha(1 + 13\sin\alpha)\,,$$

which again agrees.

In the next part of this section we shall deal with the determination of the invariants of the trajectory of a line. Therefore, assume the motion $g(\varphi)$ to be given by the Frenet frame of the fixed directing cone $\boldsymbol{R}_1, \boldsymbol{T}_1, \boldsymbol{N}_1$; we express the vectors $\boldsymbol{R}_1, \boldsymbol{T}_1, \boldsymbol{N}_1 \in \mathfrak{E}$ as dual vector functions. Let us also assume that with the moving axoid (with the moving system) a straight line $\bar{p}$ is bound fixed. Its Klein image is denoted by $\overline{\boldsymbol{X}}$, i.e. $\overline{\boldsymbol{X}}$ is a unit dual vector. According to Theorem 9.15 the trajectory $\boldsymbol{X}(\varphi)$ of vector $\overline{\boldsymbol{X}}$ is equal to $g(\varphi)\,\overline{\boldsymbol{X}}$, where $g(\varphi) \in O(3, D)$. In virtue of Theorem 9.17, the derivatives of the vector function $\boldsymbol{X}(\varphi)$ are

$$\boldsymbol{X}' = \boldsymbol{R} \times \boldsymbol{X}\,, \quad \boldsymbol{X}'' = \boldsymbol{R}' \times \boldsymbol{X} + \boldsymbol{R} \times (\boldsymbol{R} \times \boldsymbol{X})\,,$$

where $\boldsymbol{R}$ is the fixed directing cone, also in dual notation. The invariants of the ruled surface which originates from the motion of line $\bar{p}$ can be computed finding first the invariants $(\kappa_1)_X, (\kappa_2)_X, (\mu_1)_X, (\mu_2)_X$ of the vector function $\boldsymbol{X}(\varphi)$, and then using the fact that for the invariants of the corresponding ruled surface we have

$$k_p = (\kappa_2)_X (\kappa_1)_X^{-1}, \quad d_p = (\mu_1)_X (\kappa_1)_X^{-1}, \quad q_p = (\mu_2)_X (\kappa_1)_X^{-1}.$$

Denote the dual invariants of the vector function $\boldsymbol{X}(\varphi)$ by τ, τ_1, τ_2, with $\tau = 1$ because $(\boldsymbol{X}, \boldsymbol{X}) = 1$ [$\boldsymbol{X}$ is the Klein image of a line so that $(v_0)_X = 0$]. According to Theorem 9.20 we have the following expressions for τ_1 and τ_2:

$$\tau_1^2 = (\boldsymbol{X}', \boldsymbol{X}')\,\omega^2, \quad |\boldsymbol{X}, \boldsymbol{X}', \boldsymbol{X}''| = \tau_1^2 \tau_2 \omega^3,$$

where ω is some unknown angular velocity, the derivative of the canonical parameter corresponding to $\boldsymbol{X}(\varphi)$ with respect to φ. It will not appear in the resultant formulae because we shall be forming quotients of the invariants; therefore, put $\omega = 1$. Further, we have

$$\boldsymbol{X}' = \boldsymbol{R} \times \boldsymbol{X} = \sigma_0 \boldsymbol{R}_1 \times \boldsymbol{X},$$

$$\begin{aligned}\boldsymbol{X}'' &= \boldsymbol{R}' \times \boldsymbol{X} + \boldsymbol{R} \times (\boldsymbol{R} \times \boldsymbol{X}) = (\sigma_0 \boldsymbol{R}_1)' \times \boldsymbol{X} + \boldsymbol{R} \times (\boldsymbol{R} \times \boldsymbol{X}) \\ &= \sigma_0' \boldsymbol{R}_1 \times \boldsymbol{X} + \sigma_0 \sigma_1 \boldsymbol{T}_1 \times \boldsymbol{X} + \sigma_0^2 [\boldsymbol{R}_1 \times (\boldsymbol{R}_1 \times \boldsymbol{X})],\end{aligned}$$

so that

$$\tau_1^2 = \sigma_0^2 (\boldsymbol{R}_1 \times \boldsymbol{X}, \boldsymbol{R}_1 \times \boldsymbol{X}) = \sigma_0^2 [1 - (\boldsymbol{R}_1, \boldsymbol{X})^2],$$

where $\sigma_0^2 = 1 + 2\mathbf{e} v_0$.

$$\begin{aligned}\tau_1^2 \tau_2 &= |\boldsymbol{X}, \sigma_0 \boldsymbol{R}_1 \times \boldsymbol{X}, \sigma_0' \boldsymbol{R}_1 \times \boldsymbol{X} + \sigma_0 \sigma_1 \boldsymbol{T}_1 \times \boldsymbol{X} + \\ &\qquad \sigma_0^2 (\boldsymbol{R}_1, \boldsymbol{X}) \boldsymbol{R}_1 - \sigma_0^2 \boldsymbol{X}| \\ &= \sigma_0^2 |\boldsymbol{X}, \boldsymbol{R}_1 \times \boldsymbol{X}, \sigma_1 \boldsymbol{T}_1 \times \boldsymbol{X} + \sigma_0 (\boldsymbol{R}_1, \boldsymbol{X}) \boldsymbol{R}_1| \\ &= \sigma_0^2 (\boldsymbol{X} \times (\boldsymbol{R}_1 \times \boldsymbol{X}), \sigma_1 \boldsymbol{T}_1 \times \boldsymbol{X} + \sigma_0 (\boldsymbol{R}_1, \boldsymbol{X}) \boldsymbol{R}_1) \\ &= \sigma_0^2 (\boldsymbol{R}_1 - (\boldsymbol{X}, \boldsymbol{R}_1) \boldsymbol{X}, \sigma_1 \boldsymbol{T}_1 \times \boldsymbol{X} + \sigma_0 (\boldsymbol{R}_1 \times \boldsymbol{X}) \boldsymbol{R}_1) \\ &= \sigma_0^2 [\sigma_1 (\boldsymbol{N}_1, \boldsymbol{X}) + \sigma_0 (\boldsymbol{R}_1, \boldsymbol{X}) - \sigma_0 (\boldsymbol{R}_1 \boldsymbol{X})^3].\end{aligned}$$

On the other hand, of course,

$$\tau_1^2 = (\kappa_1)_X^2 + 2\mathbf{e} (\kappa_1)_X (\mu_1)_X,$$

$$\tau_1^2\tau_2 = [(\kappa_1)_X^2 + 2\mathbf{e}(\kappa_1)_X(\mu_1)_X][(\kappa_2)_X + \mathbf{e}(\mu_2)_X]$$
$$= (\kappa_1)_X^2(\kappa_2)_X + \mathbf{e}[(\kappa_1)_X^2(\mu_2)_X + 2(\kappa_1)_X(\mu_1)_X(\kappa_2)_X].$$

Now, consider a certain instant φ_0, and at this instant let $\boldsymbol{X}$ have the dual coordinates ξ_1, ξ_2, ξ_3 in the Frenet frame of the fixed axoid, i.e. let

$$\boldsymbol{X} = \xi_1\boldsymbol{R}_1 + \xi_2\boldsymbol{T}_1 + \xi_3\boldsymbol{N}_1,$$

and put

$$\xi_1 = p_1 + \mathbf{e}p_4, \quad \xi_2 = p_2 + \mathbf{e}p_5, \quad \xi_3 = p_3 + \mathbf{e}p_6,$$

which means that $(p_1, p_2, p_3; p_4, p_5, p_6)$ are the Plücker coordinates of the line $p = g(\varphi_0)\,\bar{p}$.

For the invariants we obtain

$$(\kappa_1)_X^2 + 2\mathbf{e}(\kappa_1)_X(\mu_1)_X = (1 + 2\mathbf{e}v_0)(1 - \xi_1^2)$$
$$= (1 + 2\mathbf{e}v_0)(1 - p_1^2 - 2\mathbf{e}p_1p_4)$$
$$= 1 - p_1^2 + 2\mathbf{e}[v_0(1 - p_1^2) - p_1p_4].$$

If we denote the angle between line p and the axis of the instantaneous motion by ψ, then

$$p_1 = \cos\psi, \quad \sin^2\psi = 1 - p_1^2 = p_2^2 + p_3^2.$$

Further, we have $(\kappa_1)_X^2 = \sin^2\psi$, i.e.

$$(\kappa_1)_X = \sin\psi, \quad (\mu_1)_X\sin\psi = v_0\sin^2\psi - p_1p_4,$$

so that

$$(\mu_1)_X = v_0\sin\psi - p_1p_4\sin^{-1}\psi.$$

Further, assume that $\sin\psi \neq 0$; if $\sin\psi = 0$, the spherical image of the trajectory of line $\bar{p}$ has a singular point.

Further,

$$\tau_1^2\tau_2$$
$$= (1 + 2\mathbf{e}v_0)[(\kappa_1 + \mathbf{e}\mu_1)(p_3 + \mathbf{e}p_6) +$$
$$(1 + \mathbf{e}v_0)(p_1 + \mathbf{e}p_4 - p_1^3 - 3\mathbf{e}p_1^2p_4)]$$
$$= (1 + 2\mathbf{e}v_0)[\kappa_1p_3 + p_1\sin^2\psi$$

$$+ \mathbf{e}(\mu_1 p_3 + \kappa_1 p_6 + v_0 p_1 \sin^2\psi + p_4 - 3p_1^2 p_4)]$$
$$= \kappa_1 p_3 + p_1 \sin^2\psi + \mathbf{e}(2v_0\kappa_1 p_3 + 2v_0 p_1 \sin^2\psi + \mu_1 p_3 + \kappa_1 p_6$$
$$+ v_0 p_1 \sin^2\psi + p_4 - 3p_1^2 p_4)$$
$$= (\kappa_1)_X^2 (\kappa_2)_X + \mathbf{e}[(\kappa_1)_X^2 (\mu_2)_X + 2(\kappa_1)_X (\mu_1)_X (\kappa_2)_X]\,.$$

Therefore, we have

$$(\kappa_2)_X = \sin^{-2}\psi(\kappa_1 p_3 + p_1 \sin^2\psi)\,,$$
$$\sin^2\psi(\mu_2)_X + 2\sin\psi(v_0 \sin\psi - p_1 p_4 \sin^{-1}\psi)\,.$$
$$\sin^{-2}\psi(\kappa_1 p_3 + p_1 \sin^2\psi)$$
$$= \sin^2\psi(\mu_2)_X + 2\sin^{-2}\psi(v_0 \sin^2\psi - p_1 p_4)(\kappa_1 p_3 + p_1 \sin^2\psi)$$
$$= 2v_0\kappa_1 p_3 + 3v_0 p_1 \sin^2\psi + \mu_1 p_3 + \kappa_1 p_6 + p_4 - 3p_1^2 p_4\,.$$

This implies that

$$(\mu_2)_X \sin^2\psi$$
$$= -\,2v_0 p_1 \sin^2\psi + 2p_1 p_4 p_3 \kappa_1 \sin^{-2}\psi + 2p_1^2 p_4 + 3v_0 p_1 \sin^2\psi$$
$$+ \mu_1 p_3 + \kappa_1 p_6 + p_4 - 3p_1^2 p_4$$
$$= 2p_1 p_3 p_4 \kappa_1 \sin^{-2}\psi + v_0 r_1 \sin^2\psi + \mu_1 p_3 + \kappa_1 p_6 + p_4 \sin^2\psi\,.$$

9.24 Theorem. *For the invariants of the trajectory of line $\bar{p}$ in position p with the Plücker coordinates $p_1, \ldots, p_6$ in the Frenet frame of the fixed axoid it is*

$$d_p = \sin^{-2}\psi(v_0 \sin^2\psi - p_1 p_4)\,,$$
$$q_p = \sin^{-5}\psi[2p_1 p_3 p_4 \kappa_1 + (\mu_1 p_3 + \kappa_1 p_6)\sin^2\psi +$$
$$(v_0 p_1 + p_4)\sin^4\psi]\,,$$
$$k_p = \sin^{-3}\psi(\kappa_1 p_3 + p_1 \sin^2\psi)\,.$$

If we look at Theorem 10.15, we see that $d_p = 0$ if and only if p belongs to the quadratic complex of tangents.

9.25 Theorem. *The trajectory of a line has torsal rulings just at those instants at which it belongs to the quadratic complex of tangents (see Section 10).*

The point of the striction curve of the trajectory of line $\bar{p}$ lies on the shortest transversal of the axis of instantaneous motion and of line p. This shortest transversal is the normal to the trajectory of line $\bar{p}$ at the point of the striction curve.

Proof. Denote the vector which determines the normal of the trajectory at the point of the striction curve by $\boldsymbol{N}_X$. Then $\boldsymbol{X}' = \alpha \boldsymbol{N}_X$, $\boldsymbol{R}_1 = \beta \boldsymbol{R}$, $\alpha, \beta \in D$, $\boldsymbol{X}' = \boldsymbol{R} \times \boldsymbol{X}$, so that $\boldsymbol{N}_X = \gamma \boldsymbol{R}_1 \times \boldsymbol{X}$, where $\gamma \in D$. This implies that $(\boldsymbol{N}_X, \boldsymbol{X}) = 0$, $(\boldsymbol{N}_X, \boldsymbol{R}_1) = 0$, and this is the required assertion.

9.26 Example. As an application of Theorem 9.24 we shall find all lines with the property that the tangent vector of the striction curve of their trajectory is colinear with the tangent vector of the trajectory of the point of the striction curve. Therefore, let p be a line with the Plücker coordinates $p_1, \ldots, p_6$ in the Frenet frame $\mathscr{F} = \{S, \boldsymbol{u}_1, \boldsymbol{u}_2, \boldsymbol{u}_3\}$ of the fixed axoid; put

$$\boldsymbol{X} = \xi_1 \boldsymbol{R}_1 + \xi_2 \boldsymbol{T}_1 + \xi_3 \boldsymbol{N}_1 ,$$

where

$$\xi_1 = p_1 + \mathbf{e} p_4 , \quad \xi_2 = p_2 + \mathbf{e} p_5 , \quad \xi_3 = p_3 + \mathbf{e} p_6 .$$

Also, put $\boldsymbol{X} = \boldsymbol{x}_1 + e\boldsymbol{x}_2$. The point M of the striction curve on line p is the point of intersection of line p and of the normal to the trajectory of line p at the point of the striction curve, i.e. the point of intersection of line p and of the shortest transversal l of the axis of instantaneous motion and of line p. Let us find it. The line p has the vector equation $p = \boldsymbol{x}_1 \times \boldsymbol{x}_2 + \lambda \boldsymbol{x}_1$, $\lambda \in \mathbf{R}$, the line l is determined by the dual vector

$$\boldsymbol{R}_1 \times \boldsymbol{X} = \boldsymbol{u}_1 \times \boldsymbol{X} = \boldsymbol{u}_1 \times (\boldsymbol{x}_1 + \mathbf{e}\boldsymbol{x}_2) = \boldsymbol{u}_1 \times \boldsymbol{x}_1 + \mathbf{e}\boldsymbol{u}_1 \times \boldsymbol{x}_2 ;$$

its vector equation is

$$\begin{aligned} l &= (\boldsymbol{u}_1 \times \boldsymbol{x}_1) \times (\boldsymbol{u}_1 \times \boldsymbol{x}_2) + \mu \boldsymbol{u}_1 \times \boldsymbol{x}_1 \\ &= |\boldsymbol{u}_1, \boldsymbol{x}_1, \boldsymbol{x}_2| + \mu \boldsymbol{u}_1 \times \boldsymbol{x}_1 , \quad \mu \in \mathbf{R} . \end{aligned}$$

For the point of intersection M of lines p and l it must be

$$\boldsymbol{x}_1 \times \boldsymbol{x}_2 + \lambda \boldsymbol{x}_1 = |\boldsymbol{u}_1, \boldsymbol{x}_1, \boldsymbol{x}_2| \, \boldsymbol{u}_1 + \mu \boldsymbol{u}_1 \times \boldsymbol{x}_1$$

for certain $\lambda, \mu \in \mathbf{R}$. Perform the inner product of this equation and the vector $\boldsymbol{u}_1 \times \boldsymbol{x}_1$. Then

$$(\boldsymbol{x}_1 \times \boldsymbol{x}_2, \ \boldsymbol{u}_1 \times \boldsymbol{x}_1) = -(\boldsymbol{u}_1, \boldsymbol{x}_2) = \mu \sin^2 \psi ,$$

where ψ is again the angle between the axis of instantaneous motion and line p.

Thus, $\mu \sin^2 \psi = -p_2$, i.e. $\mu = -p_2 \sin^{-2} \psi$. Point M is then determined by the relation

$$M = |\boldsymbol{u}_1, \boldsymbol{x}_1, \boldsymbol{x}_2| \, \boldsymbol{u}_1 - p_2 \sin^{-2} \psi \boldsymbol{u}_1 \times \boldsymbol{x}_1 .$$

If we denote by $\boldsymbol{t}_M$ the tangent vector to the trajectory of point M, we obtain

$$\begin{aligned} \boldsymbol{t}_M &= \boldsymbol{u}_1 \times (|\boldsymbol{u}_1, \boldsymbol{x}_1, \boldsymbol{x}_2| \, \boldsymbol{u}_1 - p_2 \sin^{-2} \psi \boldsymbol{u}_1 \times \boldsymbol{x}_1) + v_0 \boldsymbol{u}_1 \\ &= -p_2 \sin^{-2} \psi (p_1 \boldsymbol{u}_1 - \boldsymbol{x}_1) + v_0 \boldsymbol{u}_1 \\ &= v_0 \boldsymbol{u}_1 + p_2 \sin^{-2} \psi (\boldsymbol{x}_1 - p_1 \boldsymbol{u}_1) . \end{aligned}$$

If we wish to determine the tangent vector $M^{\bullet}$ of the striction curve, we must first determine the vectors of the Frenet frame of the trajectory of line p; denote these by $\bar{\boldsymbol{u}}_1, \bar{\boldsymbol{u}}_2, \bar{\boldsymbol{u}}_3$. But this is simple, since $\bar{\boldsymbol{u}}_1 = \boldsymbol{x}_1$, $\bar{\boldsymbol{u}}_2$ has the direction of the shortest transversal l, i.e.

$$\bar{\boldsymbol{u}}_2 = \sin^{-1} \psi \boldsymbol{u}_1 \times \boldsymbol{x}_1 ,$$

and, finally,

$$\bar{\boldsymbol{u}}_3 = \bar{\boldsymbol{u}}_1 \times \bar{\boldsymbol{u}}_2 = \sin^{-1} \psi \boldsymbol{x}_1 \times (\boldsymbol{u}_1 \times \boldsymbol{x}_1) = \sin^{-1} \psi (\boldsymbol{u}_1 - p_1 \boldsymbol{x}_1) .$$

According to the form of the Frenet formulae for a ruled surface [see, e.g., (6.1)],

$$\frac{\mathrm{d}M}{\mathrm{d}s} = q_p \bar{\boldsymbol{u}}_1 + d_p \bar{\boldsymbol{u}}_3 = q_p \boldsymbol{x}_1 + d_p \sin^{-1} \psi (\boldsymbol{u}_1 - p_1 \boldsymbol{x}_1) ,$$

where q_p and d_p are from Theorem 9.24. The vectors $\boldsymbol{t}_M$ and $\mathrm{d}M/\mathrm{d}s$ are colinear if and only if their cross product is equal to zero. Therefore,

$$\begin{aligned} M^{\bullet} \times \boldsymbol{t}_M &= [q_p \boldsymbol{x}_1 + d_p \sin^{-1} \psi (\boldsymbol{u}_1 - p_1 \boldsymbol{x}_1)] \\ &\quad \times [v_0 \boldsymbol{u}_1 + p_2 \sin^{-2} \psi (\boldsymbol{x}_1 - p_1 \boldsymbol{u}_1)] \\ &= v_0 q_p \boldsymbol{x}_1 \times \boldsymbol{u}_1 - q_p p_2 \sin^{-2} \psi p_1 \boldsymbol{x}_1 \times \boldsymbol{u}_1 + \end{aligned}$$

$$d_p \sin^{-1}\psi(p_4 \sin^{-2}\psi \boldsymbol{u}_1 \times \boldsymbol{x}_1) -$$
$$p_1 v_0 \boldsymbol{x}_1 \times \boldsymbol{u}_1 + p_4 \sin^{-2}\psi p_1^2 \boldsymbol{x}_1 \times \boldsymbol{u}_1)$$
$$= \left[v_0 q_p - q_p p_4 \sin^{-2}\psi p_1 + d_p \sin^{-1}\psi(-p_4 - p_1 v_0)\right] \cdot$$
$$\boldsymbol{x}_1 \times \boldsymbol{u}_1 \,.$$

Thus, for the lines with the desired property the following equation must hold:

$$v_0 q_p - q_p p_1 p_4 \sin^{-2}\psi - d_p(p_1 v_0 + p_4)\sin^{-1}\psi = 0\,,$$

or

$$q_p(v_0 \sin^2\psi - p_1 p_4) - d_p(p_1 v_0 + p_4)\sin\psi = 0\,.$$

Upon substitution from Theorem 9.24 we arrive at

$$(v_0 \sin^{-2}\psi - p_1 p_4)\left[2p_1 p_3 p_4 \kappa_1 + (\mu_1 p_3 + \kappa_1 p_6)\sin^2\psi\right] = 0\,.$$

We may therefore state the following. Lines for which the tangent of the striction curve is colinear with the tangent to the trajectory of the point of the striction curve constitute two complexes: the quadratic complex $v_0[(p_2)^2 + (p_3)^2] - p_1 p_4 = 0$, which is the complex of tangents (see Section 10), and the third-degree complex with the equation

$$2p_1 p_3 p_4 \kappa_1 + (\mu_1 p_3 + \kappa_1 p_6)\left[(p_2)^2 + (p_3)^2\right] = 0\,.$$

Finally, we shall show that the connection between spherical and space kinematics is much deeper than would seem from what was said so far. We first introduce functions of the dual variable as an extension of the function of the real variable.

9.27 Definition. Let $f(x)$ be a real function which has the derivative, and let $\boldsymbol{\alpha} = a_1 + \mathbf{e}a_2$ be a dual number. We then define $f(\boldsymbol{\alpha}) = f(a_1) + \mathbf{e}a_2 f'(a_1)$.

In particular we obtain

$$\cos\boldsymbol{\alpha} = \cos a_1 - \mathbf{e}a_2 \sin a_1\,, \quad \sin\boldsymbol{\alpha} = \sin a_1 + \mathbf{e}a_2 \cos a_1\,.$$

9.28 Exercise. Prove that formulae analogous to those for ordinary circular functions are also true for circular functions of a dual angle. For example, $\cos(\boldsymbol{\alpha} + \boldsymbol{\beta}) = \cos\boldsymbol{\alpha}\cos\boldsymbol{\beta} - \sin\boldsymbol{\alpha}\sin\boldsymbol{\beta}$, etc. If p_1 and p_2 are two lines, **A** and **B** the dual vectors corresponding to them, we may write

$(\mathbf{A}, \mathbf{B}) = \cos \boldsymbol{\alpha}$, where $\boldsymbol{\alpha}$ is the dual angle between the lines p_1 and p_2, i.e. $\boldsymbol{\alpha} = \varphi + \mathbf{e}d$, where φ is the ordinary angle between p_1 and p_2 and d is the distance between them.

9.29 Exercise. Let a be the orthogonal matrix of rotation through angle φ about the axis determined by the unit vector $\boldsymbol{x}$. If we substitute into this matrix for φ the dual angle $\varphi_1 + \mathbf{e}\varphi_2$, and for the vector $\boldsymbol{x}$ the dual vector $\boldsymbol{x}_1 + \mathbf{e}\boldsymbol{x}_2$, we obtain a new matrix $\tilde{a}$ whose elements are dual numbers. Prove that $\tilde{a} \in (3, D)$ and that this matrix expresses rotation about the axis determined by the dual vector $\boldsymbol{x}_1 + \mathbf{e}\boldsymbol{x}_2$ through angle φ_1 and translation in the direction of this axis by φ_2.

9.30 Exercise. Prove that there exists a dual analogy to Euler angles, i.e. that any congruence in space can be expressed as a product of three dual rotations through the dual Euler angles ψ_1, ψ_2, ψ_3 as in spherical kinematics.

By way of illustration, we shall derive a ruled analogy of the concept of the osculation circle of a trajectory by using the apparatus just derived. Since the necessary considerations slightly exceed the scope of this book, we shall only give the principal ideas. A circle on a sphere is described by the equation $(\boldsymbol{X}, \boldsymbol{Y}) = \cos \varphi$, where φ is a given angle (spherical radius) and $\boldsymbol{Y}$ is a fixed unit vector which determines the circle's centre. On the dual unit sphere this equation describes the set of all lines $\boldsymbol{X}$ which forms the given dual angle with line $\boldsymbol{Y}$ (i.e. a real angle and distance). Such a set of lines depends on two parameters and is called *linear congruence of lines* (it is determined by two linear equations for the Plücker coordinates). Now, if $\boldsymbol{X}(\mathrm{t})$ is a curve on the unit sphere, the osculation circle of this curve at point $\boldsymbol{X}(\mathrm{t}_0)$ is determined by the equations

$$(\boldsymbol{X}(\mathrm{t}_0), \boldsymbol{Y}) = \cos \varphi\,, \quad (\boldsymbol{X}'(\mathrm{t}_0), \boldsymbol{Y}) = 0\,, \quad (\boldsymbol{X}''(\mathrm{t}_0), \boldsymbol{Y}) = 0\,,$$

which are obtained from the condition that the osculation circle should have contact of at least second order with the curve. The last two equations determine the coordinates of the centre of this osculation circle. Now, if we apply the same consideration to the dual sphere, we arrive at the equation for the osculation linear congruence of the trajectory of a line and its axis $\boldsymbol{Y}$ is determined by the last two equations. This axis $\boldsymbol{Y}$ is called the axis of curvature of line $\boldsymbol{X}$ and its Plücker coordinates can be obtained by substituting dual values for α, β, γ, κ_1, $\boldsymbol{R}$ and $\boldsymbol{X}$ into the formula in Theorem II, 4.2. Of course, the result must be normalized to obtain a unit dual vector. The significance of the axis of curvature is, e.g., that it is the axis of the osculation helical surface of the trajectory of a line because the ruled helical surface is obtained by the helical motion of a line, i.e. the ruling line is always at the same angle with the axis and always at the same distance from it. For details see [44], [47] and [131].

This means that upon substitution we have for the axis of curvature $\boldsymbol{Y}_X = \mathbf{a}\boldsymbol{R}_1 + \mathbf{b}\boldsymbol{T}_1 + \mathbf{c}\boldsymbol{N}_1$ of line $\boldsymbol{X} = \boldsymbol{\alpha}\boldsymbol{R}_1 + \boldsymbol{\beta}\boldsymbol{T}_1 + \gamma\boldsymbol{N}_1$, where $\boldsymbol{R}_1, \boldsymbol{T}_1, \boldsymbol{N}_1$ is the Frenet frame of the fixed axoid in dual vector notation,

$$\mathbf{a} \doteq \sigma_1\gamma\boldsymbol{\alpha} + \sigma_0(\boldsymbol{\beta}^2 + \gamma^2)\,, \quad \mathbf{b} \doteq \sigma_1\boldsymbol{\beta}\gamma\,, \quad \mathbf{c} \doteq \sigma_1\gamma^2\,,$$

where σ_0 and σ_1 are given in Theorem 9.19 and equality holds up to a dual factor.

In what follows we identify lines with unit dual vectors; recall that the cross product of dual vectors represents the shortest transversal of the corresponding lines. If $g(t)$ is a motion, then $\boldsymbol{R}_1$ and $\bar{\boldsymbol{R}}_1$ represent the ruling lines of the fixed and moving axoid, respectively, and the formula from Theorem 9.19 holds. The axis of curvature of the generating straight line of the axoid is defined in the same way as for the trajectory of a straight line: it is the straight line $\boldsymbol{o}$ for the vector of which we have $(\boldsymbol{R}_0', \boldsymbol{o}) = (\boldsymbol{R}_0'', \boldsymbol{o}) = 0$ for the fixed and similarly for the moving axoid.

9.31 Theorem. *Let $g(t)$ be a space motion for which the formulae from Theorem 9.19 are true. Denote by $\boldsymbol{o}$ and $\bar{\boldsymbol{o}}$ the axis of curvature of the fixed and moving axoid respectively, let $\boldsymbol{X}$ be an arbitrary line and let $\boldsymbol{Y}_X$ be the axis of curvature of its trajectory. If we put $\boldsymbol{n} = \boldsymbol{R}_1 \times \boldsymbol{X}$, $\boldsymbol{n}' = \boldsymbol{R}_1 \times \boldsymbol{n}$, $\bar{\mathrm{v}} = \boldsymbol{X} \times \bar{\boldsymbol{o}}$, $\boldsymbol{p} = \bar{\mathrm{v}} \times \boldsymbol{n}'$, $\mathrm{v} = \boldsymbol{p} \times \boldsymbol{o}$, then $\boldsymbol{Y}_X = \mathrm{v} \times \boldsymbol{n}$* (Disteli).

Proof. From Theorem 9.19 we obtain the dual coordinates of the axis of curvature of the axoid $\boldsymbol{o} = (\sigma_2, 0, \sigma_1)$ and $\bar{\boldsymbol{o}} = (\bar{\sigma}_2, 0, \sigma_1)$, equality holding up to a dual factor because $(\boldsymbol{T}_1, \boldsymbol{o}) = 0$ and $(-\sigma_1\boldsymbol{R}_1 + \sigma_2\boldsymbol{N}_1, \boldsymbol{o}) = 0$ must hold for the fixed axoid, and analogously for the moving axoid. We shall determine the vector $\boldsymbol{Y}_X$ up to a proportionality factor. Computation yields

$$\begin{aligned}\boldsymbol{Y}_X &= \mathrm{v} \times \boldsymbol{n} = (\boldsymbol{p} \times \boldsymbol{o}) \times \boldsymbol{n} = [(\bar{\mathrm{v}} \times \boldsymbol{n}') \times \boldsymbol{o}] \times \boldsymbol{n} \\ &= \{[(\boldsymbol{x} \times \bar{\boldsymbol{o}}) \times \boldsymbol{n}'] \times \boldsymbol{o}\} \times \boldsymbol{n}\end{aligned}$$

$$= \{[(X, n')\bar{o} - (\bar{o}, n')X] \times o\} \times n$$
$$= [(X, n')\bar{o} \times o - (\bar{o}, n')X \times o] \times (R_1 \times X)$$
$$= (X, n')|\bar{o}, o, X|R_1 - (X, n')|\bar{o}, o, R_1|X$$
$$(\bar{o}, n')|X, o, X|R_1 + (\bar{o}, n')|X, o, R_1|X$$
$$= (X, n')|\bar{o}, o, X|R_1 + (\bar{o}, n')|X, o, R_1|X$$

because $|\bar{o}, o, R_1| = |X, o, X| = 0$. If $Y_X = \alpha R_1 + \beta T_1 + \gamma N_1$, we obtain $n' = \alpha R_1 - X$, so that $(X, n') = \alpha^2 - 1$, $(\bar{o}, n') = \sigma_1 \gamma$, $|\bar{o}, o, X| = -\beta\sigma_1(\bar{\sigma}_2 - \sigma_2) = \beta\sigma_1\sigma_0$, $|X, o, R_1| = \alpha\beta\sigma_1$, because $\sigma_2 - \bar{\sigma}_2 = \sigma_0$. Substitution now yields the required result.

Theorem 9.31 is valid only in the general case, of course, but we shall not discuss this point in detail. Just note that the Disteli Theorem transforms the construction of the axis of curvature into a multiple construction of the axis of two skew or concurrent lines.

9.32 Exercise. Assume a linear helical surface to be given by the angle α between the axis and the ruling line, by the radius a and the parameter v_0. Prove that the following holds for its invariants k, q, d:

$$k = \cot\alpha\,, \quad d = v_0 - a\cot\alpha\,, \quad q = a + v_0\cot\alpha$$

and conversely,

$$v_0 = (1 + k^2)^{-1}(d + kq)\,, \quad a = (1 + k^2)^{-1}(q - dk)\,.$$

9.33 Exercise. Let a ruled surface $X(\mathrm{t})$ be given. Prove that the ruled helical surface, which contacts the surface $X(\mathrm{t})$ along the ruling $X(\mathrm{t}_0)$ and whose invariants are $k(\mathrm{t}_0)$, $d(\mathrm{t}_0)$, $q(\mathrm{t}_0)$, has its axis in the axis of curvature of the ruling $X(\mathrm{t})$.

9.34 Remark. The surface in Exercise 9.33 is called the osculation helical surface of a ruled surface.

9.35 Exercise. Assume a space motion is given. Find all the lines which intersect the axis of curvature of their trajectory. *Hint*: It must be $q = dk$. We now substitute from Theorem 9.24. The following is obtained:

$$(p_2^2 + p_3^2)(p_3\psi + \kappa_1 p_6 - p_4) - 3p_1p_3p_4\kappa_1 = 0\,.$$

10. FIELDS OF VELOCITIES AND ACCELERATION

Assume that motion $g(\mathrm{t})$ is given as a function of time t with the directing function $\boldsymbol{r}(\mathrm{t})$. Denote by $\omega = \mathrm{d}\varphi/\mathrm{dt}$ the angular velocity, and let $\omega(\mathrm{t}) \neq 0$ for all values of t. For the directing cone $\boldsymbol{R}(\mathrm{t})$ we have $\boldsymbol{R} = \omega^{-1}\boldsymbol{r}$. For the velocity $\boldsymbol{v}_A$ at point A we obtain

$$A(\mathrm{t}) = g(\mathrm{t})\,\bar{A}\,, \quad \boldsymbol{v}_{A(\mathrm{t})} = \frac{\mathrm{d}A(\mathrm{t})}{\mathrm{dt}} = \frac{\mathrm{d}[g(\mathrm{t})\,\bar{A}]}{\mathrm{dt}} = \frac{\mathrm{d}g(\mathrm{t})}{\mathrm{dt}}\,g^{-1}(\mathrm{t})\,g(\mathrm{t})\,\bar{A}$$

$$= \boldsymbol{r}(\mathrm{t})\,A(\mathrm{t}) = \omega(\mathrm{t})\,\boldsymbol{R}(\mathrm{t})\,A(\mathrm{t})\,,$$

i.e., as in Chapter II we have

$$\boldsymbol{v}_A = \omega\boldsymbol{R}A\,, \quad \text{or} \quad \dot{A} = \omega\boldsymbol{R}A\,.$$

By further differentiation we obtain for the acceleration $\boldsymbol{a}_A$ at point A

$$\boldsymbol{a}_A = \ddot{A} = \frac{\mathrm{d}\omega}{\mathrm{dt}}\boldsymbol{R}A + \omega\frac{\mathrm{d}\boldsymbol{R}}{\mathrm{dt}}A + \omega\boldsymbol{R}\frac{\mathrm{d}A}{\mathrm{dt}}$$

$$= \frac{\mathrm{d}\omega}{\mathrm{dt}}\boldsymbol{R}A + \omega^2\frac{\mathrm{d}\boldsymbol{R}}{\mathrm{d}\varphi}A + \omega^2\boldsymbol{R}^2A\,.$$

10.1 Theorem. *Denote by Ω and Θ the operators of velocity and acceleration, respectively, of motion $g(\mathrm{t})$ with the directing cone $\boldsymbol{R}$, i.e. $\boldsymbol{v}_A = \Omega A$, $\boldsymbol{a}_A = \Theta A$. Then Ω and Θ are linear operators and their representation is:*

$$\Omega = \omega\boldsymbol{R}\,, \quad \Theta = \varepsilon\boldsymbol{R} + \omega^2\frac{\mathrm{d}\boldsymbol{R}}{\mathrm{d}\varphi} + \omega^2\boldsymbol{R}^2\,,$$

where φ is the canonical parameter, $\omega = \mathrm{d}\varphi/\mathrm{dt}$ is angular velocity, $\varepsilon = \mathrm{d}\omega/\mathrm{dt}$ is angular acceleration.

Considering that $\boldsymbol{R} = \boldsymbol{R}_1 + v_0\boldsymbol{R}_2$, we have

$$\frac{\mathrm{d}\boldsymbol{R}}{\mathrm{d}\varphi} = \boldsymbol{R}' = \boldsymbol{R}_1' + v_0'\boldsymbol{R}_2 + v_0\boldsymbol{R}_2' = \kappa_1\boldsymbol{T}_1 + \mu_1\boldsymbol{T}_2 + v_0'R_2 + v_0\kappa_1\boldsymbol{T}_2\,.$$

Denote by $v_0' = \mathrm{d}v_0/\mathrm{d}\varphi = \pi$ the helical acceleration and note that π is a kinematico-geometrical quantity as opposed to the angular acceleration ε which is a kinematic quantity.

10.2 Theorem.

$$\Omega = \omega(\boldsymbol{R}_1 + v_0\boldsymbol{R}_2),$$
$$\Theta = \varepsilon(\boldsymbol{R}_1 + v_0\boldsymbol{R}_2) + \omega^2[\kappa_1\boldsymbol{T}_1 + \pi\boldsymbol{R}_2 + (\mu_1 + v_0\kappa_1)\boldsymbol{T}_2 + \boldsymbol{R}_1^2].$$

Proof. It suffices to prove that $\boldsymbol{R}^2 = \boldsymbol{R}_1^2$, which can be done by direct computation.

Choose the frame $\mathscr{F}_0 = \{S(t_0), \boldsymbol{u}_1(t_0), \boldsymbol{u}_2(t_0)\, \boldsymbol{u}_3(t_0)\}$ from Section 6; it is the Frenet frame of the fixed axoid at time instant t_0. At this instant the vectors $\boldsymbol{R}_1, \boldsymbol{T}_1, \ldots, \boldsymbol{N}_2$ have the same representation as at the beginning of Section 7.

10.3 Theorem. *The operators* $\Omega(t_0)$ *and* $\Theta(t_0)$ *have the following representation in* $\mathscr{F}_0$:

$$\Omega = \begin{pmatrix} 0, & 0, 0, & 0 \\ \omega v_0, & 0, 0, & 0 \\ 0, & 0, 0, & -\omega \\ 0, & 0, \omega, & 0 \end{pmatrix},$$

$$\Theta = \begin{pmatrix} 0, & 0, & 0, & 0 \\ \varepsilon v_0 + \omega^2\pi, & 0, & 0, & \omega^2\kappa_1 \\ \omega^2(\mu_1 + v_0\kappa_1), & 0, & -\omega^2, & -\varepsilon \\ 0, & -\omega^2\kappa_1, & \varepsilon, & -\omega^2 \end{pmatrix}.$$

10.4 Theorem. *Let* $\boldsymbol{R} = [\boldsymbol{x}; \boldsymbol{t}]$ *be the directing cone of motion* $g(t)$. *For the velocity* $\boldsymbol{v}_A$ *and acceleration* $\boldsymbol{a}_A$ *of point A we have*

$$\boldsymbol{v}_A = \omega(\boldsymbol{x} \times A + \boldsymbol{t}),$$
$$\boldsymbol{a}_A = \varepsilon(\boldsymbol{x} \times A + \boldsymbol{t}) + \omega(\boldsymbol{x}^{\bullet} \times A + \boldsymbol{t}^{\bullet}) + \omega^2\boldsymbol{x} \times (\boldsymbol{x} \times A + \boldsymbol{t}).$$

Proof. The theorem is a direct consequence of the definition of the action of the vectors from $\mathfrak{E}$ on the points and vectors from E_3, and of Theorem 10.1.

10.5 Theorem. *Let* $\boldsymbol{r}(t)$ *be the directing function of motion* $g(t)$. *For the velocity* $\boldsymbol{v}_A$ *and acceleration* $\boldsymbol{a}_A$ *of point A we have*

$$\boldsymbol{v}_A = \boldsymbol{r}A, \quad \boldsymbol{a}_A = (\boldsymbol{r}^{\bullet} + \boldsymbol{r}^2)A,$$

or, if we put $\boldsymbol{r} = [\boldsymbol{x}_0; \boldsymbol{t}_0]$,

$$\boldsymbol{v}_A = \boldsymbol{x}_0 \times A + \boldsymbol{t}_0\,, \quad \boldsymbol{a}_A = \dot{\boldsymbol{x}}_0 \times A + \boldsymbol{x}_0 \times (\boldsymbol{x}_0 \times A + \boldsymbol{t}_0)\,.$$

Proof. We know the relation $\boldsymbol{v}_A = \boldsymbol{r}A$; differentiation yields the other relation.

Now, let us devote our attention to the analysis of the velocity field. First, note that for $A = S + a^i \boldsymbol{u}_i$ we have $\boldsymbol{v}_A = \omega(v_0 \boldsymbol{u}_1 - a^3 \boldsymbol{u}_2 - a^2 \boldsymbol{u}_3)$ because

$$\boldsymbol{v}_A = \begin{pmatrix} 0, & 0,\ 0, & 0 \\ \omega v_0, & 0,\ 0, & 0 \\ 0, & 0,\ 0, & -\omega \\ 0, & 0,\ \omega, & 0 \end{pmatrix} \begin{pmatrix} 1 \\ a^1 \\ a^2 \\ a^3 \end{pmatrix} = \begin{pmatrix} 0 \\ \omega v_0 \\ -\omega a^3 \\ \omega a^2 \end{pmatrix}.$$

Further, if $\boldsymbol{X} \in \mathfrak{E}$, we may write

$$\boldsymbol{X} = p_1 \boldsymbol{R}_1 + p_2 \boldsymbol{T}_1 + p_3 \boldsymbol{N}_1 + p_4 \boldsymbol{R}_2 + p_5 \boldsymbol{T}_2 + p_6 \boldsymbol{N}_2\,.$$

Then $\mathscr{K}(\boldsymbol{X}, \boldsymbol{X}) = 0$ if and only if $p_1 p_4 + p_2 p_5 + p_3 p_6 = 0$. Moreover, we assume that

$$(p_1)^2 + (p_2)^2 + (p_3)^2 = \mathbf{K}(\boldsymbol{X}, \boldsymbol{X}) = 1$$

holds for the Klein image of a line. Also, let us recall the relation between the Plücker coordinates of a line and the line's parametric representation. Therefore, consider line p with the following Plücker coordinates in frame $\boldsymbol{R}_1, \ldots, \boldsymbol{N}_2$:

$$\boldsymbol{X} = [\boldsymbol{x}; \boldsymbol{t}] = \begin{bmatrix} p_1; & p_4 \\ p_2; & p_5 \\ p_3; & p_6 \end{bmatrix}.$$

The parametric representation of line p in frame $\mathscr{F}_0$ then reads $p = \boldsymbol{x} \times \boldsymbol{t} + \lambda \boldsymbol{x}$, $\lambda \in \mathbf{R}$. Conversely, if p is given by the equation $p = A + \lambda \boldsymbol{x}$, its Plücker coordinates read $\boldsymbol{X} = [\boldsymbol{x}; A \times \boldsymbol{x}]$.

From Chapters I and II we know that the velocity field is a field of tangent vectors to the trajectories of instantaneous motion, which are helices with the same axis, the ruling line of the axoid, and with the same parameter v_0 (if $v_0 = 0$, they are circles). Moreover, these helices are the integral curves of the velocity vector field. Hence, or directly from the coordinate notation, we can see that $\boldsymbol{v}_A = 0$ may occur only if $v_0 = 0$. Then the points on the instantaneous axis of rotation have zero velocity. Further, the only line which has the property that the velocities of its points lie on this line is the axis of instantaneous motion (sometime also called the axis of skew motion).

Indeed, we have

$$\boldsymbol{v}_A = \omega \begin{pmatrix} v_0 \\ -a^3 \\ a^2 \end{pmatrix},$$

for $A = X + \lambda \boldsymbol{u}$ we have

$$\boldsymbol{v}_A = \begin{pmatrix} v_0 \\ -x^3 - \lambda u^3 \\ x^2 + \lambda u^2 \end{pmatrix}.$$

If $\boldsymbol{v}_A = \mu(\lambda)\, \boldsymbol{u}$ is to hold for all λ, we must have

$$\omega v_0 = \mu(\lambda)\, u^1, \quad -x^3 - \lambda u^3 = \mu(\lambda)\, u^2, \quad x^2 + \lambda u^2 = \mu(\lambda)\, u^3.$$

From the second and third equations we eliminate $\mu(\lambda)$ and obtain

$$x^2 u^2 + x^3 u^3 = -\lambda[(u^2)^2 + (u^3)^2],$$

so that $u^2 = u^3 = 0$ and thus also $x^2 = x^3 = 0$.

10.6 Theorem. *For two points A, B we have $(\boldsymbol{v}_B - \boldsymbol{v}_A, B - A) = 0$. This is the so-called condition of line-segment rigidity. (The projection of the velocity of the points of a line segment on to this line segment is constant.) The vector $\boldsymbol{v}_B - \boldsymbol{v}_A$ is perpendicular to the axis of instantaneous motion.*

Proof. Let $\boldsymbol{r} = [\boldsymbol{x}; \boldsymbol{t}]$. Then

$$\boldsymbol{v}_A = \boldsymbol{x} \times A + \boldsymbol{t}, \quad v_B = \boldsymbol{x} \times B + \boldsymbol{t},$$

$$\boldsymbol{v}_B - \boldsymbol{v}_A = \boldsymbol{x} \times B - \boldsymbol{x} \times A = \boldsymbol{x} \times (B - A),$$

so that

$$(\boldsymbol{v}_B - \boldsymbol{v}_A, B - A) = (\boldsymbol{x} \times [B - A], B - A) = 0.$$

Further, we have

$$(v_B - v_A, x) = (x \times (B - A), x) = 0\,.$$

Theorem 10.6 can be used to construct the magnitude of the velocity at other points if its direction is known, and to construct the direction of the line of the axoid. A point on the line of the axoid can be constructed with the aid of Theorem 10.4. For details see, e.g., [18].

A line which contains the velocity of one of its points is called a *tangent line.*

10.7 Theorem. *Tangent lines belong to the quadratic complex with the equation*

$$p_1 p_4 - (p_3^2 + p_2^2)\, v_0 = 0$$

and they are determined in this complex by the condition $(p_2)^2 + (p_3)^2 \neq 0$. *If* $v_0 = 0$, *the complex decomposes into two linear complexes, into the complex of all lines intersecting the axis of skew motion or parallel with it, and into the complex of all lines perpendicular to this axis.*

Remark. The lines parallel with the axis of skew motion ($p_2 = p_3 = 0$) belong to the mentioned quadratic complex. However, they are not tangent with the exception of the axis of skew motion.

Proof. Let A be a point with the coordinates a^1, a^2, a^3; v_A then has the coordinates $v_0, -a^3, a^2$. The tangent T_A at point A has the equation

$$T_A = A + \lambda\gamma^{-1} v_A\,, \quad \text{where} \quad \lambda \in \mathcal{R}\,, \quad \gamma = [(v_0)^2 + (a^3)^2 + (a^2)^2]^{1/2}\,.$$

The Plücker coordinates of the tangent T_A are $(v_A\gamma^{-1};\ A \times v_A\gamma^{-1})$. Computation yields

$$p_1 = \gamma^{-1} v_0\,, \quad p_2 = -\gamma^{-1} a^3\,, \quad p_3 = \gamma^{-1} a^2\,,$$

$$p_4 = \gamma^{-1}[(a^2)^2 + (a^3)^2]\,, \quad p_5 = \gamma^{-1}(-a^1 a^2 + v_0 a^3)\,,$$

$$p_6 = \gamma^{-1}(-a^1 a^3 - v_0 a^2)\,,$$

$$p_1 p_4 = \gamma^{-2} v_0[(a^2)^2 + (a^3)^2] = [(p_2)^2 + (p_3)^2]\, v_0\,.$$

What remains is evident.

10.8 Theorem. *Let* p *be a line which does not belong to the complex* $p_1 p_4 - v_0[(p_2)^2 + p_3)^2] = 0$. *The tangents of the trajectories constructed* *at the points of line p constitute one system of lines of a hyperbolic paraboloid. The joins of the endpoints of the velocity vectors belong to the second system of lines of this hyperbolic paraboloid (for any ω).*

Proof. Let $p \equiv X(\lambda) = A + \lambda \boldsymbol{u}, \lambda \in \mathbf{R}$. Then $\boldsymbol{v}_X = \omega(\boldsymbol{u}_1 \times X + v_0 \boldsymbol{u}_1)$, so that the ruled surface of all tangents has the equation

$$\mathfrak{A}(\lambda, \mu) = X + \mu\omega^{-1}\boldsymbol{v}_X = A + \lambda\boldsymbol{u} + \mu[\boldsymbol{u}_1 \times (A + \lambda\boldsymbol{u}) + v_0\boldsymbol{u}_1],$$
$$\lambda, \mu \in \mathbf{R}.$$

If $\lambda = const. = \lambda_0$, $\mathfrak{A}(\lambda_0, \mu)$ is a line lying on the surface, if $\mu = \mu_0$, then

$$\mathfrak{A}(\lambda, \mu_0) = A + \mu_0(\boldsymbol{u}_1 \times A + v_0\boldsymbol{u}_1) + \lambda(\boldsymbol{u} + \mu_0\boldsymbol{u}_1 \times \boldsymbol{u})$$

is also a line. This means that $\mathfrak{A}(\lambda, \mu)$ is a warped (ruled) quadric. The lines of the first system have their direction given by the vector $\boldsymbol{u}_1 \times (A + \lambda\boldsymbol{u}) + v_0\boldsymbol{u}_1$. If this is to be a hyperbolic paraboloid these directions must lie in a single plane, which is evident; this plane is determined by the vectors $\boldsymbol{u}_1 \times A + v_0\boldsymbol{u}_1$ and $\boldsymbol{u}_1 \times \boldsymbol{u}$. For the second system this plane is determined by the vectors $\boldsymbol{u}$ and $\boldsymbol{u}_1 \times \boldsymbol{u}$. If the lines $\mathfrak{A}(\lambda_0, \mu)$ are skew lines, the surface $\mathfrak{A}(\lambda, \mu)$ is a hyperbolic paraboloid. Let us also investigate when the lines $\mathfrak{A}(\lambda_0, \mu)$ lie in a single plane. This occurs if and only if the vectors $\boldsymbol{u}_1 \times (A + \lambda\boldsymbol{u}) + v_0\boldsymbol{u}_1$ lie in a plane which contains the vector $\boldsymbol{u}$. This occurs if and only if $|\boldsymbol{u}, \boldsymbol{u}_1 \times A + v_0\boldsymbol{u}_1, \boldsymbol{u}_1 \times \boldsymbol{u}| = 0$.

Assume that $A = \boldsymbol{u} \times \boldsymbol{t}$, where $(\boldsymbol{u}, \boldsymbol{t}) = 0$, so that $[\boldsymbol{u}; \boldsymbol{t}]$ are the Plücker coordinates of line p. Then

$$\boldsymbol{u}_1 \times A = \boldsymbol{u}_1 \times (\boldsymbol{u} \times \boldsymbol{t}) = t^1\boldsymbol{u} - u^1\boldsymbol{t}.$$

Our condition then reads

$$|\boldsymbol{u}, t^1\boldsymbol{u} - u^1\boldsymbol{t} + v_0\boldsymbol{u}_1, \boldsymbol{u}_1 \times \boldsymbol{u}| = 0,$$

i.e.

$$(\boldsymbol{u}, [v_0\boldsymbol{u}_1 - u^1\boldsymbol{t}] \times [\boldsymbol{u}_1 \times \boldsymbol{u}]) = 0,$$

i.e.

$$(\boldsymbol{u}, v_0u^1\boldsymbol{u}_1 - v_0\boldsymbol{u} + u^1t^1\boldsymbol{u}) = v_0(u^1)^2 - v_0 + u^1t^1 = 0,$$

which when written in terms of Plücker coordinates really yields

$$p_1p_4 - v_0[(p_2)^2 + (p_3)^2] = 0.$$

10.9 Theorem. *Let p be a line of the complex* $p_1 p_4 - v_0[(p_2)^2 + (p_3)^2] = 0$. *Then, if p is a tangent line at point A, the tangents T_B of the trajectories constructed at all points B of line p are located in a plane α. Plane α is the osculation plane of the helix S_A of point A under instantaneous motion. The tangents T_B in plane α envelop a parabola with vertex at A and with principal axis along the principal normal of the helix S_A at point A. If we denote the focus of this parabola by F, we have*

$$\overline{FA} = [v_0^2 + (a^2)^2 + (a^3)^2][(a^2)^2 + (a^3)^2]^{-1/2},$$

where a^i are the coordinates of point A. The join of the endpoints of the velocity vectors is a tangent to this parabola for any value of ω. If $v_0 = 0$, the focus F lies on the axis of instantaneous rotation.

If line p is parallel to the axis of instantaneous motion, the tangents at its points are mutually parallel.

Proof. Substitute

$$\boldsymbol{u} = \boldsymbol{u}_1 \times A + v_0 \boldsymbol{u}_1$$

into equation

$$A + \lambda \boldsymbol{u} + \mu[\boldsymbol{u}_1 \times (A + \lambda \boldsymbol{u}) + v_0 \boldsymbol{u}_1] = \mathfrak{A}(\lambda, \mu).$$

We obtain

$$\begin{aligned}
&\mathfrak{A}(\lambda, \mu) = \\
&= A + \lambda(\boldsymbol{u}_1 \times A + v_0\boldsymbol{u}_1) + \mu\{\boldsymbol{u}_1 \times [A + \lambda(\boldsymbol{u}_1 \times A + v_0\boldsymbol{u}_1)] + v_0\boldsymbol{u}_1\} \\
&= A + \lambda(\boldsymbol{u}_1 \times A + v_0\boldsymbol{u}_1) + \mu(\boldsymbol{u}_1 \times A + v_0\boldsymbol{u}_1) + \lambda\mu\boldsymbol{u}_1 \times (\boldsymbol{u}_1 \times A) \\
&= A + (\lambda + \mu)(\boldsymbol{u}_1 \times A + v_0\boldsymbol{u}_1) + \lambda\mu\boldsymbol{u}_1 \times (\boldsymbol{u}_1 \times A).
\end{aligned}$$

First, this implies that the lines T_B lie in one plane. Note also that $(\boldsymbol{u}_1, \boldsymbol{u}_1 \times [\boldsymbol{u}_1 \times A]) = 0$, so that $\mathfrak{A}(\lambda, \mu)$ contains a line perpendicular to the axis of instantaneous motion; this line is

$$A + \nu\boldsymbol{u}_1 \times (\boldsymbol{u}_1 \times A) = A + \nu[(\boldsymbol{u}_1, A)\,\boldsymbol{u}_1 - A], \quad \nu \in \mathbf{R},$$

and it intersects the axis of instantaneous motion. It is thus the principal normal of the helix S_A of point A, and the plane α is indeed its osculation plane. Also, note that the vectors $\boldsymbol{u}_1 \times A + v_0\boldsymbol{u}_1$ and $\boldsymbol{u}_1 \times (\boldsymbol{u}_1 \times A)$ constitute an orthogonal base in plane α because $(\boldsymbol{u}_1 \times A + v_0\boldsymbol{u}_1, \boldsymbol{u}_1$

$\times [\boldsymbol{u}_1 \times A]) = 0$. Denote by γ, δ the magnitudes of these vectors. Then $\{A, \gamma^{-1}(\boldsymbol{u}_1 \times A + v_0\boldsymbol{u}_1), \delta^{-1}\boldsymbol{u}_1 \times (\boldsymbol{u}_1 \times A)\}$ constitutes an orthonormal frame in plane α. Our system of lines in plane α has parametric equations $x^1 = \gamma(\lambda + \mu)$, $x^2 = \delta\lambda\mu$ in this frame. We obtain the implicit equation by eliminating the parameter μ:

$$x^1\gamma^{-1} = \lambda + \mu\,, \quad x^2\delta^{-1} = \lambda\mu\,,$$

i.e.

$$\mu = x^1\gamma^{-1} - \lambda\,, \quad x^2\delta^{-1} = \lambda(x^1\gamma^{-1} - \lambda)\,,$$

so that

$$\lambda^2 - \lambda x^1\gamma^{-1} + x^2\delta^{-1} = 0\,.$$

If we are to determine the envelope of this system of lines, we must partially differentiate the equation with respect to λ and eliminate λ from both equations, i.e. $2\lambda - x^1\gamma^{-1} = 0$, i.e. $\lambda = \frac{1}{2}x^1\gamma^{-1}$, and after substitution $\frac{1}{4}(x^1)^2\gamma^{-2} - \frac{1}{2}(x^1)^2\gamma^{-2} + x^2\delta^{-1} = 0$, i.e. $\frac{1}{4}(x^1)^2\gamma^{-2} = x^2\delta^{-1}$, i.e. $x^2 = \frac{1}{4}(x^1)^2\delta\gamma^{-2}$. For the focal distance $\overline{FA}$ we have $\overline{FA} = \gamma^2\delta^{-1}$; it is easy to compute that

$$\gamma^2 = v_0^2 + (a^2)^2 + (a^3)^2\,, \quad \delta^2 = (a^2)^2 + (a^3)^2\,.$$

Consider an arbitrary plane α. Every point of plane α at which the velocity is perpendicular to plane α is called the *focus of plane* α.

10.10 Theorem. *A plane not parallel with the axis of instantaneous motion has exactly one focus. If* $v_0 = 0$, *then this focus lies on the axis of instantaneous motion. Let further* α *be a plane parallel with the axis of instantaneous motion. Then if* $v_0 \neq 0$, α *has no foci. Let* $v_0 = 0$. *Then either* α *contains the axis of instantaneous rotation and all points of* α *are foci or there is no focus.*

Proof. Let $\alpha \equiv X = A + \lambda\boldsymbol{u} + \mu\boldsymbol{w}$, $\lambda, \mu \in \mathbf{R}$, where $\boldsymbol{u}, \boldsymbol{w}$ constitute an orthonormal base in the plane α. For the velocity $\boldsymbol{v}_X$ at the point X we have

$$\boldsymbol{v}_X = \boldsymbol{u}_1 \times X + v_0\boldsymbol{u}_1 = \boldsymbol{u}_1 \times (A + \lambda\boldsymbol{u} + \mu\boldsymbol{w}) + v_0\boldsymbol{u}_1\,.$$

The condition for focus F reads $(\boldsymbol{v}_F, \boldsymbol{u}) = (\boldsymbol{v}_F, \boldsymbol{w}) = 0$, i.e.

$$\begin{aligned}&(\boldsymbol{u}_1 \times [A + \lambda\boldsymbol{u} + \mu\boldsymbol{w}] + v_0\boldsymbol{u}_1, \boldsymbol{u})\\ &= (\boldsymbol{u}_1 \times [A + \lambda\boldsymbol{u} + \mu\boldsymbol{w}] + v_0\boldsymbol{u}_1, \boldsymbol{w}) = 0\,.\end{aligned}$$

We obtain equations

$$\left|u_1\, A, u\right| + \mu\left|u_1, w, u\right| + v_0 u^1 = \left|u_1, A, w\right| + \lambda\left|u_1, u, w\right| + v_0 w^1 = 0 \,.$$

These equations have precisely one solution if $\left|u_1, u, w\right| \neq 0$, i.e. if α is not parallel with the axis of instantaneous motion. If it is parallel, the focus can exist only if $\left|u_1, A, u\right| + v_0 u^1 = \left|u_1, A, w\right| + v_0 w^1 = 0$. In the given case we can assume that $u = u_1$; then $(w, u) = (w, u_1) = 0$, i.e. $w^1 = 0$. Therefore, it must be $\left|u_1, A, u_1\right| + v_0 = \left|u_1, A, w\right| = v_0 = 0$. If $v_0 \neq 0$, the focus does not exist; if $v_0 = 0$, u_1, A, w must lie in a single plane, i.e. the plane contains the axis of rotation and the focal equation has a solution for all λ and μ. The uniqueness of the solution of the focal equation implies that for $v_0 = 0$ the focus is located on the axis of rotation.

10.11 Theorem. *If two tangent lines are parallel, the join of the tangent points lies in the same plane as the axis of instantaneous motion and, moreover, if $v_0 \neq 0$, it is also parallel with this axis.*

Proof. $v_A = u_1 \times A + v_0 u_1$, $v_B = u_1 \times B + v_0 u_1$, and if they are to be parallel $v_A \times v_B = 0$ must hold. Put $B = A + u$. Then it must be

$$\begin{aligned} 0 &= (u_1 \times A + v_0 u_1) \times (u_1 \times A + u_1 \times u + v_0 u_1) \\ &= (u_1 \times A) \times (u_1 \times u) + v_0 (u_1 \times A) \times u_1 + v_0 u_1 \\ &\quad \times (u_1 \times A) + v_0 u_1 \times (u_1 \times u) \\ &= (u_1 \times A) \times (u_1 \times u) + v_0 u_1 \times (u_1 \times u)\,, \end{aligned}$$

i.e.

$$\left|u_1, A, u\right| u_1 + v_0 (u^1 u_1 - u) = 0 \,.$$

If we take the inner product of this equation with the vector u_1, we obtain $\left|u_1, A, u\right| = 0$, i.e. u_1, A, u lie in the same plane, which is the first assertion. If $v_0 \neq 0$, then $u^2 u_2 + u^3 u_3 = 0$, i.e. $u^2 = u^3 = 0$.

Consequences. The foci of a system of parallel planes, concurrent with the axis of instantaneous motion, lie on a line parallel with the axis of instantaneous motion. Indeed: for $v_0 \neq 0$ this follows from Theorem 10.11, if $v_0 = 0$, this follows from Theorem 10.10.

The focus of a plane perpendicular to the axis of instantaneous motion lies on the axis of instantaneous motion.

A straight line is called a *zero line* if the velocities at its points are perpendicular to it. The condition of rigidity of a segment implies that it is sufficient to satisfy the mentioned condition at a single point of the line. This means that if we pass a plane through a zero line, the zero line must pass through the focus of this plane and, conversely, every line which passes through the focus of a plane is a zero line.

10.12 Theorem. *Line p is a zero line if and only if there exists a plane α containing p so that the focus of α lies on p.*

10.13 Theorem. *Zero lines constitute a linear complex which has the equation $v_0 p_1 + p_4 = 0$.*

Proof. Let $p = \boldsymbol{u} \times \boldsymbol{t} + \lambda \boldsymbol{u}$, $\lambda \in \mathbf{R}$, be a line, $(\boldsymbol{u}, \boldsymbol{t}) = 0$. Line p is a zero line if

$$(\boldsymbol{v}_{\boldsymbol{u} \times \boldsymbol{t}}, \boldsymbol{u}) = 0\,,$$

i.e.

$$(\boldsymbol{u}_1 \times [\boldsymbol{u} \times \boldsymbol{t}] + v_0 \boldsymbol{u}_1, \boldsymbol{v}) = 0\,,$$

i.e.

$$(t^1 \boldsymbol{u} - u^1 \boldsymbol{t} + v_0 \boldsymbol{u}_1, \boldsymbol{u}) = t^1 + v_0 u^1 = 0\,;$$

which, in terms of Plücker coordinates, reads $v_0 p_1 + p_4 = 0$.

10.14 Theorem. *Let $\boldsymbol{R}$ be a vector of the directing cone. Then the zero complex has the equation $\mathscr{K}(\boldsymbol{R}, \boldsymbol{X}) = 0$.*

Proof. The equation is correct under our choice of coordinate system; indeed

$$\mathscr{K}(\boldsymbol{R}, \boldsymbol{X}) = \mathscr{K}\left(\begin{bmatrix} 1;\ v_0 \\ 0;\ 0 \\ 0;\ 0 \end{bmatrix}, \begin{bmatrix} p_1;\ p_4 \\ p_2;\ p_5 \\ p_3;\ p_6 \end{bmatrix}\right) = p_4 + v_0 p_1\,.$$

The invariance of the form $\mathscr{K}$ implies that the equation is valid in the general case as well.

10.15 Theorem. *Let $\boldsymbol{R}$ be a vector of the directing cone. Then the tangent quadratic complex has the equation $\mathscr{K}([\boldsymbol{R}, \boldsymbol{X}], [\boldsymbol{R}, \boldsymbol{X}]) = 0$.*

Proof. Let us prove that the equation holds under our choice of coordinate system. Then the invariance of the form $\mathscr{K}$ implies its validity in any system of coordinates.

$$R = [u_1; v_0 u_1], \quad X = [u; t].$$

Then

$$[R, X] = [[u_1; v_0 u_1], [u; t]] = [u_1 \times u; u_1 \times t + v_0 u_1 \times u],$$

$$\mathscr{K}([R, X], [R, X]) = (u_1 \times u, u_1 \times t + v_0 u_1 \times u)$$

$$= -u^1 t^1 + v_0 - v_0 (u^1)^2 = v_0[(u^2)^2 + (u^3)^2] - u^1 t^1,$$

which in terms of Plücker coordinates reads $p_1 p_4 - v_0[(p_2)^2 + (p_3)^2]$.

10.16 Theorem. *Let p be a line with Plücker coordinates Y. Then the set of tangents constructed at the points of line p has the equation*

$$\mathscr{K}([R, X], [R, X]) = 0, \quad \mathscr{K}(Y, X) = 0.$$

We may now devote our attention to the analysis of the vector field of acceleration, i.e. to the analysis of the linear operator Θ. For the purposes of our analysis, put

$$\omega^2(\mu_1 + v_0 \kappa_1) = \nu_2, \quad \varepsilon v_0 + \omega^2 \pi = \nu_1.$$

Then

$$\Theta = \begin{pmatrix} 0, & 0, & 0, & 0 \\ \nu_1, & 0, & 0, & \omega^2 \kappa_1 \\ \nu_2, & 0, & -\omega^2, & -\varepsilon \\ 0, & -\omega^2 \kappa_1, & \varepsilon, & -\omega^2 \end{pmatrix}.$$

First, note that

$$\begin{vmatrix} 0, & 0, & \omega^2 \kappa_1 \\ 0, & -\omega^2, & -\varepsilon \\ -\omega^2 \kappa_1, & \varepsilon, & -\omega^2 \end{vmatrix} = \omega^6 \kappa_1^2.$$

Thus, we have to distinguish two cases:

1. $\kappa_1 \neq 0$. According to the theorem on the solution of linear equations, the equation $\Theta M = 0$ has precisely one solution. Let us call this solution the *centre of acceleration*; it is the only point at which acceleration is zero. For the coordinates m^1, m^2, m^3 of the centre of acceleration we have

$$\nu_1 + \omega^2 \kappa_1 m^3 = \nu_2 - \omega^2 m^2 - \varepsilon m^3 = -\omega^2 \kappa_1 m^1 + \varepsilon m^2 - \omega^2 m^3 = 0.$$

The solution yields

$$m^1 = \omega^{-6}\kappa_1^{-2}[\varepsilon v_2\omega^2\kappa_1 + v_1(\varepsilon^2 + \omega^2\kappa_1)],$$
$$m^2 = \omega^{-4}\kappa_1^{-2}(v_2\omega^2\kappa_1 + \varepsilon v_1),$$
$$m^3 = -\omega^{-2}\kappa_1^{-1}v_1 .$$

From this expression we obtain, e.g., the following: the centre of acceleration lies in the asymptotic plane of the axoid if and only if $\varepsilon v_0 + \omega^2\pi = 0$. If the centre of acceleration is on the axis of instantaneous motion, it will merge with the point of the striction curve. If $v_0 = const.$, $\omega = const.$, we have $\varepsilon = \pi = 0$ and the centre of acceleration lies on the normal to the axoid at the point of the striction curve and its distance from the point of the striction curve is $\mu_1 + v_0\kappa_1$.

2. $\kappa_1 = 0$. Then

$$\Theta = \begin{pmatrix} 0, & 0, & 0, & 0 \\ v_1, & 0, & 0, & 0 \\ v_2, & 0, & -\omega^2, & -\varepsilon \\ 0, & 0, & \varepsilon, & -\omega^2 \end{pmatrix}.$$

If $v_1 \neq 0$, there exists no point at which acceleration would be zero. Thus let us look for the line at whose points acceleration would have the direction of this line – the axis of acceleration. Denote it by $p = A + \lambda \boldsymbol{u}$, $\lambda \in \mathbf{R}$. Put

$$\Theta = \begin{pmatrix} 0, & 0 \\ \boldsymbol{t}, & \boldsymbol{x} \end{pmatrix}.$$

Then

$$\boldsymbol{a}_{A+\lambda\boldsymbol{u}} = \begin{pmatrix} 0, & 0 \\ \boldsymbol{t}, & \boldsymbol{x} \end{pmatrix}\begin{pmatrix} 1 \\ A + \lambda\boldsymbol{u} \end{pmatrix} = \begin{pmatrix} 0 \\ \boldsymbol{t} + \boldsymbol{x}A + \lambda\boldsymbol{x}\boldsymbol{u} \end{pmatrix}.$$

If $\boldsymbol{t} + \boldsymbol{x}A + \lambda\boldsymbol{x}\boldsymbol{u} = \alpha(\lambda)\,\boldsymbol{u}$ is to hold, then $\boldsymbol{x}\boldsymbol{u} = \delta\boldsymbol{u}$, i.e. we must have

$$\begin{pmatrix} 0, & 0, & 0 \\ 0, & -\omega^2, & -\varepsilon \\ 0, & \varepsilon, & -\omega^2 \end{pmatrix}\begin{pmatrix} u^1 \\ u^2 \\ u^3 \end{pmatrix} = \begin{pmatrix} 0 \\ -\omega^2u^2 - \varepsilon u^3 \\ \varepsilon u^2 - \omega^2u^3 \end{pmatrix} = \begin{pmatrix} \delta u^1 \\ \delta u^2 \\ \delta u^3 \end{pmatrix}.$$

a) $\delta = 0$. Then $\omega^2u^2 - \varepsilon u^3 = 0$, $\varepsilon u^2 - \omega^2u^3 = 0$. The determinant of the system is $\omega^4 + \varepsilon^2 \neq 0$, i.e. $u^2 = u^3 = 0$.

b) $u^1 = 0$. Then we must have

$$\begin{vmatrix} -\omega^2 - \delta, & -\varepsilon \\ \varepsilon, & -\omega^2 - \delta \end{vmatrix} = (\omega^2 + \delta)^2 + \varepsilon^2 = 0\,,$$

which has no real solution. Therefore, $p = A + \lambda \boldsymbol{u}_1$ and $\boldsymbol{a}_A = \alpha \boldsymbol{u}_1$ must hold, which yields the following equations:

$$\begin{pmatrix} 0, & 0, & 0, & 0 \\ v_1, & 0, & 0, & 0 \\ v_2, & 0, & -\omega^2, & -\varepsilon \\ 0, & 0, & \varepsilon, & -\omega^2 \end{pmatrix} \begin{pmatrix} 1 \\ a^1 \\ a^2 \\ a^3 \end{pmatrix} = \begin{pmatrix} 0 \\ v_1 \\ v_2 - \omega^2 a^2 - \varepsilon a^3 \\ \varepsilon a^2 - \omega^2 a^3 \end{pmatrix} = \begin{pmatrix} 0 \\ \alpha \\ 0 \\ 0 \end{pmatrix}.$$

The solution is

$$a^2 = (\omega^4 + \varepsilon^2)^{-1}\, v_2 \omega^2\,, \quad a^3 = (\omega^4 + \varepsilon^2)^{-1}\, v_2 \varepsilon\,.$$

In case 2 there exists precisely one axis of acceleration, namely the line with the equations

$$x^2 = (\omega^4 + \varepsilon^2)^{-1}\, v_2 \omega^2\,, \quad x^3 = (\omega^4 + \varepsilon^2)^{-1}\, v_2 \varepsilon\,.$$

Remark. The axis of acceleration always exists, i.e. even if $\kappa_1 \neq 0$; however, to find this line we have to solve an equation of the 3rd degree, for this reason we do not treat the general case. The reader is asked to find examples in which the mentioned equation can be solved. Let us just note that if $\kappa_1 \neq 0$, there may be one, two or three axes of acceleration. For the sake of illustration, let us give an example in which three axes of acceleration exist. Put $v_1 = v_2 = \varepsilon = 0$, $\kappa_1 = \sqrt{3}/4$. The axes of acceleration have the equations

$$p_1 = \lambda \boldsymbol{u}_2\,, \quad p_2 = \lambda(\boldsymbol{u}_1 \sqrt{3} - \boldsymbol{u}_3)\,, \quad p_3 = \lambda[\boldsymbol{u}_1 - \sqrt{(3)}\, \boldsymbol{u}_3]\,.$$

Indeed, in our case

$$\Theta = \begin{bmatrix} 0, & 0, & 0, & 0 \\ 0, & 0, & 0, & \dfrac{\omega^2 \sqrt{3}}{4} \\ 0, & 0, & -\omega^2, & 0 \\ 0, & -\dfrac{\omega^2 \sqrt{3}}{4}, & 0, & -\omega^2 \end{bmatrix},$$

$$\begin{bmatrix} 0, & 0, & \dfrac{\omega^2\sqrt{3}}{4} \\ 0, & -\omega^2, & 0 \\ -\dfrac{\omega^2\sqrt{3}}{4}, & 0, & -\omega^2 \end{bmatrix} \begin{bmatrix} 0 \\ 1 \\ 0 \end{bmatrix} = \begin{bmatrix} 0 \\ -\omega^2 \\ 0 \end{bmatrix},$$

$$\begin{bmatrix} 0, & 0, & \dfrac{\omega^2\sqrt{3}}{4} \\ 0, & -\omega^2, & 0 \\ -\dfrac{\omega^2\sqrt{3}}{4}, & 0, & -\omega^2 \end{bmatrix} \begin{bmatrix} \sqrt{3} \\ 0 \\ -1 \end{bmatrix} =$$

$$= \begin{bmatrix} -\dfrac{\omega^2\sqrt{3}}{4} \\ 0 \\ \dfrac{\omega^2}{4} \end{bmatrix} = -\frac{\omega^2}{4} \begin{bmatrix} \sqrt{3} \\ 0 \\ -1 \end{bmatrix},$$

$$\begin{bmatrix} 0, & 0, & \dfrac{\omega^2\sqrt{3}}{4} \\ 0, & -\omega^2, & 0 \\ -\dfrac{\omega^2\sqrt{3}}{4}, & 0, & -\omega^2 \end{bmatrix} \begin{bmatrix} 1 \\ 0 \\ -\sqrt{3} \end{bmatrix} =$$

$$= \begin{bmatrix} 3\dfrac{\omega^2}{4} \\ 0 \\ -\dfrac{\omega^2 . 3\sqrt{3}}{4} \end{bmatrix} = 3\frac{\omega^2}{4} \begin{bmatrix} 1 \\ 0 \\ -\sqrt{3} \end{bmatrix}.$$

A simpler situation will arise if we decompose acceleration into two components: the first is called centripetal acceleration while the second is called tangent acceleration. Let us define the operators $\Theta_c = \omega^2 \boldsymbol{R}^2$ of centripetal acceleration, and $\Theta_r = \Theta - \Theta_c$ of tangent acceleration, thus

$\Theta = \Theta_c + \Theta_r$. Denote $\Theta_c A = \boldsymbol{c}_A$, $\Theta_r A = \boldsymbol{r}_A$ the centripetal and tangent accelerations of point A, respectively. The meaning of the operator Θ_c is clear, $\boldsymbol{c}_A$ lies on the perpendicular dropped from point A on to the axis of instantaneous motion; its length can be constructed easily. If we denote by d_A the distance of point A from the axis of instantaneous motion, by $\|\boldsymbol{v}'_A\|$ the length of the projection of the velocity vector of point $\boldsymbol{A}$ into the plane perpendicular to the axis, then $\|\boldsymbol{c}_A\| : \|\boldsymbol{v}'_A\| = \|\boldsymbol{v}'_A\| : d_A$. Indeed, if point A has the coordinates a^1, a^2, a^3, we have $\boldsymbol{c}_A = -\omega^2(a^2\boldsymbol{u}_2 + a^3\boldsymbol{u}_3)$, which is evident from the representation of the operator Θ_c. Then

$$d_A = [(a^2)^2 + (a^3)^2]^{1/2}, \quad \boldsymbol{v}_A = \omega(v_0\boldsymbol{u}_1 - a^3\boldsymbol{u}_2 + a^2\boldsymbol{u}_3),$$
$$\|\boldsymbol{v}'_A\| = \omega[(a^2)^2 + (a^3)^2]^{1/2}, \quad \|\boldsymbol{c}_A\| = \omega^2[(a^2)^2 + (a^3)^2]^{1/2}.$$

Now, we shall treat the operator Θ_r. Assume that $\kappa_1 \neq 0$, the case $\kappa_1 = 0$ is easy. First of all, let us express Θ_r in matrix notation,

$$\Theta_r = \begin{pmatrix} 0, & 0, & 0, & 0 \\ v_1, & 0, & 0, & \omega^2\kappa_1 \\ v_2, & 0, & 0, & -\varepsilon \\ 0, & -\omega^2\kappa_1, & \varepsilon, & 0 \end{pmatrix}, \quad \text{further}$$

$$\begin{vmatrix} 0, & 0, & \omega^2\kappa_1 \\ 0, & 0, & -\varepsilon \\ -\omega^2\kappa_1, & \varepsilon, & 0 \end{vmatrix} = 0,$$

so that, in general, there exists no point at which the acceleration $\boldsymbol{r}_A$ is zero. Therefore, let us find the axis of acceleration in the same way as above. Put

$$\Theta_r = \begin{pmatrix} 0, & 0 \\ \boldsymbol{t}, & \boldsymbol{x} \end{pmatrix}, \quad p = A + \lambda\boldsymbol{u}, \quad \lambda \in \mathbf{R}.$$

For the direction of the axis the following must hold: $\boldsymbol{x}\boldsymbol{u} = \alpha\boldsymbol{u}$, where $\alpha \in \mathbf{R}$. This condition leads to the equation

$$\begin{vmatrix} -\alpha, & 0, & \omega^2\kappa_1 \\ 0, & -\alpha, & -\varepsilon \\ -\omega^2\kappa_1, & \varepsilon, & -\alpha \end{vmatrix} = 0, \quad \text{i.e.} \quad \alpha(\alpha^2 + \omega^4\kappa_1^2 + \varepsilon^2) = 0.$$

This equation has a single solution $\alpha = 0$. Therefore, there exists precisely one axis of tangent acceleration and for its direction we have $\boldsymbol{x}\boldsymbol{u} = 0$. The last condition yields $\omega^2\kappa_1 u^3 = 0$, $-\varepsilon u^3 = 0$, $-\omega^2\kappa_1 u^1 + \varepsilon u^2 = 0$, e.g., $u^3 = 0$, $u^1 = \varepsilon$, $u^2 = \omega^2\kappa_1$. If we put $A = a^1\boldsymbol{u}_1 + a^3\boldsymbol{u}_3$, which is possible since $u^2 = \omega^2\kappa_1 \neq 0$, we obtain the condition $\boldsymbol{r}_A = \mu\boldsymbol{u}$ in the form

$$\begin{pmatrix} 0, & 0, & 0, & 0 \\ v_1, & 0, & 0, & \omega^2\kappa_1 \\ v_2, & 0, & 0, & -\varepsilon \\ 0, & -\omega^2\kappa_1, & \varepsilon, & 0 \end{pmatrix} \begin{pmatrix} 1 \\ a^1 \\ 0 \\ a^3 \end{pmatrix} = \begin{pmatrix} 0 \\ v_1 + \omega^2\kappa_1 a^3 \\ v_2 - \varepsilon a^3 \\ -\omega^2\kappa_1 a^1 \end{pmatrix} = \mu \begin{pmatrix} 0 \\ \varepsilon \\ \omega^2\kappa_1 \\ 0 \end{pmatrix},$$

i.e.

$$-\omega^2\kappa_1 a^1 = 0\,, \quad v_1 + \omega^2\kappa_1 a^3 = \mu\varepsilon\,, \quad v_2 - \varepsilon a^3 = \mu\omega^2\kappa_1\,.$$

We obtain

$$a^1 = a^2 = 0\,, \quad a^3 = \left(\varepsilon^2 + \omega^4\kappa_1^2\right)^{-1}\left(\varepsilon v_2 - v_1\omega^2\kappa_1\right),$$

where

$$v_1 = \varepsilon v_0 + \omega^2\pi\,, \quad v_2 = \omega^2(\mu_1 + v_0\kappa_1)\,.$$

Thus, the axis of tangent acceleration has the equation

$$p = \left(\varepsilon^2 + \omega^4\kappa_1^2\right)^{-1}\left(\varepsilon\mu_1 - \omega^2\pi\kappa_1\right)\boldsymbol{u}_3 + \lambda\left(\varepsilon\boldsymbol{u}_1 + \omega^2\kappa_1\boldsymbol{u}_2\right).$$

For the sake of simplicity, put $C = \left(\varepsilon^2 + \omega^4\kappa_1^2\right)^{-1}\left(\varepsilon\mu_1 - \omega^2\pi\kappa_1\right)\boldsymbol{u}_3$.

It can be said that the axis of tangent acceleration is parallel with the asymptotic plane of the axoid and that it intersects the tangent plane along axis $\boldsymbol{u}_3$. The axis of tangent acceleration passes through the point of the striction curve if and only if $\omega^2\pi\kappa_1 = \varepsilon\mu_1$. This happens, e.g., always when $v_0 = const.$, $\omega = const.$ It is also easy to see that the axis of tangent acceleration will merge with the normal to the axoid at the point of the striction curve if and only if $\varepsilon = 0$, $\pi = 0$.

Now, let us consider the expression for the operator $\Theta_{\boldsymbol{r}}$ if the axis of acceleration is taken as one of the coordinate axes. Thus, let us consider the frame $\mathscr{R}' = \{C, \boldsymbol{u}_1', \boldsymbol{u}_2', \boldsymbol{u}_3'\}$ where we put

$$\boldsymbol{u}_1' = \left(\varepsilon^2 + \omega^4\kappa_1^2\right)^{-1/2}\left(\varepsilon\boldsymbol{u}_1 + \omega^2\kappa_1\boldsymbol{u}_2\right), \quad \boldsymbol{u}_3' = \boldsymbol{u}_3\,,$$

so that

$$\boldsymbol{u}_2' = \left(\varepsilon^2 + \omega^4\kappa_1^2\right)^{-1/2}\left(-\omega^2\kappa_1\boldsymbol{u}_1 + \varepsilon\boldsymbol{u}_2\right).$$

Direct computation yields

$$\Theta_r(C) = (\varepsilon^2 + \omega^4\kappa_1^2)^{-1/2}\,[\omega^4\kappa_1(\mu_1 + v_0\kappa_1) + \varepsilon(\varepsilon v_0 + \omega^2\pi]\,\boldsymbol{u}'_1\,,$$
$$\Theta_r(\boldsymbol{u}'_1) = 0\,,$$

as we know from computing the axis of acceleration; finally

$$\Theta_r(\boldsymbol{u}'_3) = -(\varepsilon^2 + \omega^4\kappa_1^2)^{1/2}\,\boldsymbol{u}'_2\,,\quad \Theta_r(\boldsymbol{u}'_2) = (\varepsilon^2 + \omega^4\kappa_1^2)^{1/2}\,\boldsymbol{u}'_3\,.$$

If we put

$$(\varepsilon^2 + \omega^4\kappa_1^2)^{1/2} = \omega_r\,,\quad V_0 = (\varepsilon^2 + \omega^4\kappa_1^2)^{-1}\,(\omega^2\kappa_1 v_2 + \varepsilon v_1)\,,$$

we obtain the following expression for Θ_r in frame $\mathscr{R}'$:

$$(\Theta_r)_{\mathscr{R}'} = \begin{pmatrix} 0, & 0, & 0, & 0 \\ \omega_r V_0, & 0, & 0, & 0 \\ 0, & 0, & 0, & -\omega_r \\ 0, & 0, & \omega_r, & 0 \end{pmatrix}$$

This means that Θ_r is the velocity operator if the axis of acceleration is the instantaneous axis, ω_r the angular velocity and V_0 the parameter. The operator of tangent acceleration is thus fully described.

If we write the operator Θ_r as $\Theta_r = \omega_r[\boldsymbol{x}; \boldsymbol{t}]$, we obtain

$$\boldsymbol{x} = (\varepsilon^2 + \omega^4\kappa_1^2)^{-1/2} \begin{pmatrix} \varepsilon \\ \kappa_1\omega^2 \\ 0 \end{pmatrix},\quad \boldsymbol{t} = (\varepsilon^2 + \omega^4\kappa_1^2)^{-1/2} \begin{pmatrix} v_1 \\ v_2 \\ 0 \end{pmatrix}.$$

All what was said about the velocity vector field can now be also said on the vector field of tangent accelerations. In particular, we can speak of the zero complex of tangent acceleration and of the quadratic complex of tangent acceleration.

10.17 Theorem. *The zero complex of tangent acceleration has the equation* $\mathscr{K}(\Theta_r, X) = 0$, the quadratic complex of the carriers of tangent acceleration has the equation

$$\mathscr{K}([\Theta_r, X], [\Theta_r, X]) = 0\,.$$

Let us determine the representation of these complexes in coordinate notation.

10.18 Theorem. *The zero complex of tangent acceleration has the equation*

$$\nu_1 p_1 + \nu_2 p_2 + \varepsilon p_4 + \omega^2 \kappa_1 p_5 = 0 ,$$

where

$$\nu_1 = \varepsilon v_0 + \omega^2 \pi , \quad \nu_2 = \omega^2(\mu_1 + v_0 \kappa_1) ,$$

the quadratic complex of the carriers of tangent acceleration has the equation

$$\omega_r^2 V_0 - (\varepsilon p_1 + \omega^2 \kappa_1 p_2)(\nu_1 p_1 + \nu_2 p_2 + \varepsilon p_4 + \omega^2 \kappa_1 p_5) = 0 .$$

Proof. Put $\Theta_r = [\boldsymbol{x}; \boldsymbol{t}]$, $\boldsymbol{X} = [\boldsymbol{r}; \boldsymbol{s}]$. Then $(\boldsymbol{r}, \boldsymbol{s}) = 0$ and

$$\mathscr{K}(\Theta_r, \boldsymbol{X}) = (\boldsymbol{x}, \boldsymbol{s}) + (\boldsymbol{t}, \boldsymbol{r}) ; \quad [\Theta_r, \boldsymbol{X}] = (\boldsymbol{x} \times \boldsymbol{r}; \boldsymbol{x} \times \boldsymbol{s} + \boldsymbol{t} \times \boldsymbol{r}) ,$$

$$\begin{aligned}\mathscr{K}([\Theta_r, \boldsymbol{X}], [\Theta_r, \boldsymbol{X}]) &= (\boldsymbol{x} \times \boldsymbol{r}, \boldsymbol{x} \times \boldsymbol{s} + \boldsymbol{t} \times \boldsymbol{r}) \\ &= -(\boldsymbol{x}, \boldsymbol{r})(\boldsymbol{x}, \boldsymbol{s}) + (\boldsymbol{x}, \boldsymbol{t}) - (\boldsymbol{x}, \boldsymbol{r})(\boldsymbol{t}, \boldsymbol{r}) \\ &= (\boldsymbol{x}, \boldsymbol{t}) - (\boldsymbol{x}, \boldsymbol{r})[(\boldsymbol{x}, \boldsymbol{s}) + (\boldsymbol{t}, \boldsymbol{r})] .\end{aligned}$$

If we substitute

$$\boldsymbol{x} = \begin{pmatrix} \varepsilon \\ \omega^2 \kappa_1 \\ 0 \end{pmatrix}, \quad \boldsymbol{t} = \begin{pmatrix} \nu_1 \\ \nu_2 \\ 0 \end{pmatrix}, \quad \boldsymbol{r} = \begin{pmatrix} p_1 \\ p_2 \\ p_3 \end{pmatrix}, \quad \boldsymbol{s} = \begin{pmatrix} p_4 \\ p_5 \\ p_6 \end{pmatrix},$$

we arrive at the desired equations.

Finally, let us point out several of the properties of the vector field of acceleration. First, the decomposition $\boldsymbol{a}_A = \boldsymbol{c}_A + \boldsymbol{r}_A$ is not a decomposition into mutually perpendicular components. Let us therefore find the points for which $(\boldsymbol{c}_A, \boldsymbol{r}_A) = 0$. If

$$A = a^1 \boldsymbol{u}_1 + a^2 \boldsymbol{u}_2 + a^3 \boldsymbol{u}_3 , \quad \text{then} \quad \boldsymbol{c}_A = -\omega^2(a^2 \boldsymbol{u}_2 + a^3 \boldsymbol{u}_3) ,$$

$$\boldsymbol{r}_A = (\nu_1 + \omega^2 \kappa_1 a^3)\, \boldsymbol{u}_1 + (\nu_2 - \varepsilon a^3)\, \boldsymbol{u}_2 + (-\omega^2 \kappa_1 a^1 + \varepsilon a^2)\, \boldsymbol{u}_3 .$$

The condition $(\boldsymbol{c}_A, \boldsymbol{r}_A) = 0$ yields the equation

$$(\nu_2 - \varepsilon a^3)\, a^2 + (-\omega^2 \kappa_1 a^1 + \varepsilon a^2)\, a^3 = 0 ,$$

i.e.

$$\nu_2 x^2 - \omega^2 \kappa_1 x^1 x^3 = 0 ,$$

which is the equation of a hyperbolic paraboloid with the directing planes $\boldsymbol{u}_1$, $\boldsymbol{u}_2$ and $\boldsymbol{u}_2$, $\boldsymbol{u}_3$. Note also that the principle of line-segment rigidity holds for $\boldsymbol{r}_A$, i.e. that the perpendicular projections of the tangent acceleration at the endpoints of the line segment onto this line segment are the same. This follows from the fact that $\Theta_{\mathbf{r}}$ is the velocity operator for a suitable axis and parameter. This means that if the tangent acceleration is perpendicular to a line at one of its points, the same applies to all its points. The lines with this property are the lines of the zero complex of tangent acceleration. Further: lines which display the property that both velocity and tangent acceleration at their points are perpendicular to these lines constitute a linear congruence with the equations

$$v_0 p_1 + p_4 = 0\,, \quad \nu_1 p_1 + \nu_2 p_2 + \varepsilon p_4 + \omega^2 \kappa_1 p_5 = 0\,.$$

Lines which are simultaneously carriers of velocity and tangent acceleration constitute a 4th-degree congruence. Similarly, the tangent lines to which tangent acceleration is perpendicular constitute a quadratic congruence with the equations

$$\nu_1 p_1 + \nu_2 p_2 + \varepsilon p_4 + \omega^2 \kappa_1 p_5 = 0\,, \quad p_1 p_4 - v_0\left[(p_2)^2 + (p_3)^2\right] = 0\,.$$

The set of points at which tangent acceleration is parallel with the axis of instantaneous motion is a line with the equations

$$\nu_2 - \varepsilon x^3 = 0\,, \quad \varepsilon x^2 - \omega^2 \kappa_1 x^1 = 0\,.$$

This is one of the ruling lines of the hyperbolic paraboloid

$$\nu_2 x^2 - \omega^2 \kappa_1 x^1 x^3 = 0\,.$$

Further, note the acceleration at the points of the axis of instantaneous motion. If $A = a^1 \boldsymbol{u}_1$, then clearly $\boldsymbol{c}_A = 0$, so that $\boldsymbol{r}_A = \boldsymbol{a}_A$, with

$$\boldsymbol{r}_A = \nu_1 \boldsymbol{u}_1 + \nu_2 \boldsymbol{u}_2 - \omega^2 \kappa_1 a^1 \boldsymbol{u}_3\,.$$

The hyperbolic paraboloid of the carriers of acceleration along the axis of instantaneous motion has the equation

$$(\nu_2)^2\, x^3 + x^2 \omega^2 \kappa_1 (x^1 \nu_2 - x^2 \nu_1) = 0\,.$$

Finally, we shall determine the so-called *translational velocity* and *translational acceleration* of the points on the axis of instantaneous motion.

The point of the striction curve and the axis of instantaneous motion change with t, i.e. $S = S(\mathrm{t})$, $\boldsymbol{u}_1 = \boldsymbol{u}_1(\mathrm{t})$. Therefore, we may determine how the point

$$X(\mathrm{t}) = S(\mathrm{t}) + \lambda\, \boldsymbol{u}_1(\mathrm{t})$$

changes, λ being constant. Note that $X(\mathrm{t})$ is not a trajectory of a point; point $X(\mathrm{t})$ is defined by constructing the instantaneous axis at every instant t, on this axis the point of the striction curve, and by plotting the distance λ from this point in the direction of the axis. To be able to determine $\mathrm{d}X/\mathrm{dt}$, $\mathrm{d}^2X/\mathrm{dt}^2$, we must return to the Frenet formulae for the fixed axoid. They read

$$\frac{\mathrm{d}S}{\mathrm{ds}} = q\boldsymbol{u}_1 + d\boldsymbol{u}_3\,, \quad \frac{\mathrm{d}\boldsymbol{u}_1}{\mathrm{ds}} = \boldsymbol{u}_2\,, \quad \frac{\mathrm{d}\boldsymbol{u}_2}{\mathrm{ds}} = -\boldsymbol{u}_1 + k\boldsymbol{u}_3\,, \quad \frac{\mathrm{d}\boldsymbol{u}_3}{\mathrm{ds}} = -k\boldsymbol{u}_2\,;$$

if we differentiate with respect to time t we obtain

$$\frac{\mathrm{d}S}{\mathrm{dt}} = \frac{\mathrm{d}S}{\mathrm{ds}}\frac{\mathrm{ds}}{\mathrm{d}\varphi}\frac{\mathrm{d}\varphi}{\mathrm{dt}} = \frac{\mathrm{d}S}{\mathrm{ds}}\kappa_1\omega = \omega(\mu_2\boldsymbol{u}_1 + \mu_1\boldsymbol{u}_3)\,;$$

similarly,

$$\dot{\boldsymbol{u}}_1 = \omega\kappa_1\boldsymbol{u}_2\,, \quad \dot{\boldsymbol{u}}_2 = \omega(-\boldsymbol{u}_1\kappa_1 + \kappa_2\boldsymbol{u}_3)\,, \quad \dot{\boldsymbol{u}}_3 = -\omega\kappa_2\boldsymbol{u}_2\,.$$

Further,

$$\begin{aligned}\ddot{S} &= \varepsilon(\mu_2\boldsymbol{u}_1 + \mu_1\boldsymbol{u}_3) + \omega^2(\mu_2'\boldsymbol{u}_1 + \mu_1'\boldsymbol{u}_3) + \omega\mu_2(\omega\kappa_1\boldsymbol{u}_2) + \\ &\quad \omega\mu_1(-\omega\kappa_2\boldsymbol{u}_2) \\ &= \boldsymbol{u}_1(\varepsilon\mu_2 + \omega^2\mu_2') + \boldsymbol{u}_2(\omega^2\mu_2\kappa_1 - \omega^2\mu_1\kappa_2) + \boldsymbol{u}_3(\varepsilon\mu_1 + \omega^2\mu_1')\,,\end{aligned}$$

$$\begin{aligned}\ddot{\boldsymbol{u}}_1 &= \varepsilon\kappa_1\boldsymbol{u}_2 + \omega^2\kappa_1'\boldsymbol{u}_2 + \omega^2\kappa_1(-\kappa_1\boldsymbol{u}_1 + \kappa_2\boldsymbol{u}_3) \\ &= -\omega^2\kappa_1^2\boldsymbol{u}_1 + (\varepsilon\kappa_1 + \omega^2\kappa_1')\,\boldsymbol{u}_2 + \omega^2\kappa_1\kappa_2\boldsymbol{u}_3\,.\end{aligned}$$

For $X(\mathrm{t}) = S(\mathrm{t}) + \lambda\,\boldsymbol{u}_1(\mathrm{t})$ we now have

$$\frac{\mathrm{d}X}{\mathrm{dt}} = \omega(\mu_1\boldsymbol{u}_1 + \mu_1\boldsymbol{u}_3) + \lambda\omega\kappa_1\boldsymbol{u}_2\,,$$

$$\frac{\mathrm{d}^2X}{\mathrm{dt}^2} = \left(\varepsilon\mu_2 + \omega^2\frac{\mathrm{d}\mu_2}{\mathrm{d}\varphi}\right)\boldsymbol{u}_1 + (\mu_2\kappa_1 - \mu_1\kappa_2)\,\omega^2\boldsymbol{u}_2 +$$

$$+ \left(\varepsilon\mu_1 + \omega^2 \frac{\mathrm{d}\mu_1}{\mathrm{d}\varphi} \right) \boldsymbol{u}_3 +$$

$$+ \lambda \left[-\omega^2 \kappa_1^2 \boldsymbol{u}_1 + \left(\varepsilon\kappa_1 + \omega^2 \frac{\mathrm{d}\kappa_1}{\mathrm{d}\varphi} \right) \boldsymbol{u}_2 + \omega^2 \kappa_1 \kappa_2 \boldsymbol{u}_3 \right].$$

11. KINEMATIC GENERATION OF AVERAGE ENVELOPES

Let $\mathfrak{F}$ be a surface connected fixed with the moving system $\bar{E}_3$. If $g(\mathrm{t})$ is a motion, $g(\mathrm{t})\,\mathfrak{F}$ is a one-parametric system of surfaces in the fixed system E_3. The surface Φ in the fixed system is called the *average envelope of surface* $\mathfrak{F}$ under motion $g(\mathrm{t})$ if surface Φ contacts each surface $g(\mathrm{t})\,\mathfrak{F}$ along some curve. We shall call this curve the *characteristic* of the average envelope at time t and denote it by $c(\mathrm{t})$. Note that the curve $c(\mathrm{t})$ lies on the average envelope Φ as well as on the surface $g(\mathrm{t})\,\mathfrak{F}$.

Now, let us derive the conditions for determining the characteristic $c(\mathrm{t})$. Let $\boldsymbol{r}(\mathrm{t})$ be the fixed directing function of motion $g(\mathrm{t})$. Assume that surface $\mathfrak{F}$ and surface Φ are given by the vector equations $\mathfrak{F} = \mathfrak{F}(u, v)$ and $\Phi = \Phi(u', v')$, respectively. The tangent plane of surface $\mathfrak{F}$ at point $\mathfrak{F}(u, v)$ is then given by the vector equation

$$\tau(\lambda, \mu) = \mathfrak{F} + \lambda \frac{\partial \mathfrak{F}}{\partial u} + \mu \frac{\partial \mathfrak{F}}{\partial v}, \quad \lambda, \mu \in \mathbf{R},$$

and, similarly, the tangent plane of surface Φ at point $\Phi(u', v')$ is

$$\tau'(\lambda', \mu') = \Phi + \lambda' \frac{\partial \Phi}{\partial u'} + \mu' \frac{\partial \Phi}{\partial v'}, \quad \lambda', \mu' \in \mathbf{R}.$$

Now, assume that there exists a curve $c(\mathrm{t})$ on surfaces $g(\mathrm{t})\,\mathfrak{F}$ and Φ for every t such that the surfaces $g(\mathrm{t})\,\mathfrak{F}$ and Φ contact along this curve. Then there exist functions $u(\sigma, \mathrm{t})$, $v(\sigma, \mathrm{t})$, $u'(\sigma, \mathrm{t})$, $v'(\sigma, \mathrm{t})$ such that

$$g(\mathrm{t})\,\mathfrak{F}(u(\sigma, \mathrm{t}), v(\sigma, \mathrm{t})) = \Phi(u'(\sigma, \mathrm{t}), v'(\sigma, \mathrm{t})). \tag{11.1}$$

If the surfaces $g(\mathrm{t})\,\mathfrak{F}$ and Φ contact along curve $c(\mathrm{t})$, they have common tangent planes at the points of $c(\mathrm{t})$:

$$g(\mathrm{t}) \left[\mathfrak{F}(u(\sigma, \mathrm{t}), v(\sigma, \mathrm{t})) + \lambda \frac{\partial \mathfrak{F}}{\partial u} + \mu \frac{\partial \mathfrak{F}}{\partial v} \right] = \Phi + \lambda' \frac{\partial \Phi}{\partial u'} + \mu' \frac{\partial \Phi}{\partial v'},$$

and $\partial\Phi/\partial u'$, $\partial\Phi/\partial v'$ are thus linear combinations of $\partial\mathfrak{F}/\partial u$ and $\partial\mathfrak{F}/\partial v$. If we differentiate equation (11.1) we obtain

$$g'(t)\,\mathfrak{F}(u(\sigma, t), v(\sigma, t)) + g(t)\left(\frac{\partial\mathfrak{F}}{\partial u}\frac{\partial u}{\partial t} + \frac{\partial\mathfrak{F}}{\partial v}\frac{\partial v}{\partial t}\right)$$
$$= \frac{\partial\Phi}{\partial u'}\frac{\partial u'}{\partial t} + \frac{\partial\Phi}{\partial v'}\frac{\partial v'}{\partial t},$$

i.e.

$$\boldsymbol{r}g\mathfrak{F} + g\left(\frac{\partial\mathfrak{F}}{\partial u}\frac{\partial u}{\partial t} + \frac{\partial\mathfrak{F}}{\partial v}\frac{\partial v}{\partial t}\right) = \frac{\partial\Phi}{\partial u'}\frac{\partial u'}{\partial t} + \frac{\partial\Phi}{\partial v'}\frac{\partial v'}{\partial t}.$$

At the points of the characteristic $c(t)$ we then have

$$\boldsymbol{r}g\mathfrak{F} = g\left(a\frac{\partial\mathfrak{F}}{\partial u} + b\frac{\partial\mathfrak{F}}{\partial v}\right).$$

Conversely, let $c(t)$ be a curve on $g(t)\,\mathfrak{F}(u, v)$ such that $\boldsymbol{r}(t)\,c(t)$ lies in the tangent plane of the surface. Let $c(t) = [u(\sigma, t), v(\sigma, t)]$ represent its parametric equations on $g(t)\,\mathfrak{F}(u, v)$; then the surface

$$\Phi(t, \sigma) = g(t)\,\mathfrak{F}(u(\sigma, t), v(\sigma, t))$$

is the average envelope of the system $g(t)\,\mathfrak{F}$. Indeed, if $t = t_0$, the surfaces $g(t_0)\,\mathfrak{F}$ and $\Phi(t, \sigma)$ have a common curve

$$\Phi(t_0, \sigma) = g(t_0)\,\mathfrak{F}(u(\sigma, t_0), v(\sigma, t_0))$$

and the tangent plane of the surface $\Phi(t, \sigma)$ at the points of this curve is determined by the vectors

$$\frac{\partial\Phi}{\partial t} = g'(t)\,\mathfrak{F} = g(t)\left[\frac{\partial\mathfrak{F}}{\partial u}\frac{\partial u}{\partial t} + \frac{\partial\mathfrak{F}}{\partial v}\frac{\partial v}{\partial t}\right],$$

$$\frac{\partial\Phi}{\partial\sigma} = g(t)\left[\frac{\partial\mathfrak{F}}{\partial u}\frac{\partial u}{\partial\sigma} + \frac{\partial\mathfrak{F}}{\partial v}\frac{\partial v}{\partial\sigma}\right].$$

11.1 Theorem. *Point C of surface $g(t)\,\mathfrak{F}$ is a point of the characteristic $c(t)$ if and only if the tangent plane of surface $g(t)\,\mathfrak{F}$ at point C contains the tangent of the trajectory of point C, or if it is parallel with the velocity*

$\boldsymbol{v}_C$ *of point C. Therefore, if* $\mathscr{X}(u, v)$ *is the expression for the surface* $g(\mathrm{t})\,\mathfrak{F}$, *the characteristic* $c(\mathrm{t})$ *is given by the equation*

$$\left|\boldsymbol{R}\mathscr{X}, \frac{\partial \mathscr{X}}{\partial u}, \frac{\partial \mathscr{X}}{\partial v}\right| = 0\,. \tag{11.2}$$

If we bear in mind that the vector $(\partial\mathscr{X}/\partial u) \times (\partial\mathscr{X}/\partial v)$ has the direction of the normal n of the surface, and if we rewrite this equation to read

$$\left(\boldsymbol{R}\mathscr{X}, \frac{\partial \mathscr{X}}{\partial u} \times \frac{\partial \mathscr{X}}{\partial v}\right) = 0\,,$$

we also see that the normal to the surface is perpendicular to the tangent of the trajectory at the points of the characteristic (see Fig. 43).

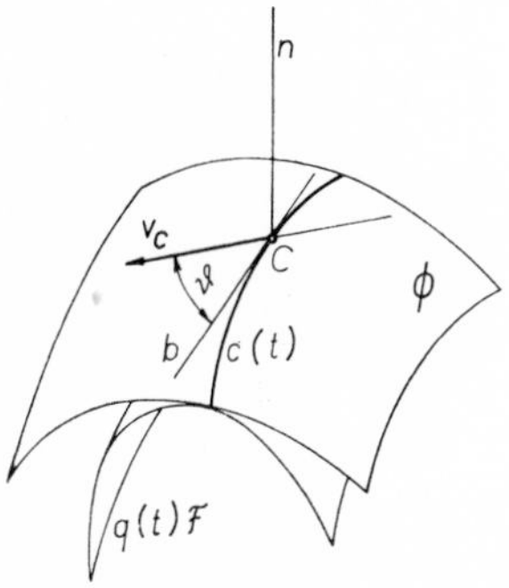

Fig. 43.

Sometimes a different method of determining the characteristic is more suitable under which we do not have to change the position of surface $\mathfrak{F}$. Let us describe this method. Denote by $\bar{c}(\mathrm{t}) = g^{-1}(\mathrm{t})\,c(\mathrm{t})$ the inverse of the characteristic $c(\mathrm{t})$ under motion $g(\mathrm{t})$. Then, of course, $\bar{c}(\mathrm{t})$ lies on surface $\mathfrak{F}$ because $g(\mathrm{t})\,\bar{c}(\mathrm{t}) = c(\mathrm{t})$ lies on $g(\mathrm{t})\,\mathfrak{F}$. Let us write equation (11.2) in the original form:

$$\left|g'\mathfrak{F}, g\frac{\partial \mathfrak{F}}{\partial u}, g\frac{\partial \mathfrak{F}}{\partial v}\right| = 0\,.$$

If we apply the mapping g^{-1} to all three terms, we obtain

$$\left|g^{-1}g'\mathfrak{F}, \frac{\partial\mathfrak{F}}{\partial u}, \frac{\partial\mathfrak{F}}{\partial v}\right| = 0\,.$$

Curve $\bar{c}(\mathrm{t})$ is thus characterized by the equation

$$\left|\bar{\boldsymbol{R}}\mathfrak{F}, \frac{\partial\mathfrak{F}}{\partial u}, \frac{\partial\mathfrak{F}}{\partial v}\right| = 0\,;$$

its solution is a curve which becomes the characteristic under motion $g(\mathrm{t})$. If the set of curves $\bar{c}(\mathrm{t})$ forms the surface $\mathfrak{F}$, the surfaces $\mathfrak{F}$ and Φ are called *conjugate surfaces*. The average envelope of surface Φ under inverse motion $g^{-1}(\mathrm{t})$ lies necessarily on surface $\mathfrak{F}$. Indeed, multiplying equation (11.1) by $g^{-1}(\mathrm{t})$ we obtain

$$g^{-1}(\mathrm{t})\,\Phi(u'(\sigma, \mathrm{t}), v'(\sigma, \mathrm{t})) = \mathfrak{F}(u(\sigma, \mathrm{t}), v(\sigma, \mathrm{t}))\,.$$

Differentiating with respect to t, we verify that these surfaces contact along the characteristic.

If $g(\mathrm{t})$ is an elementary motion, $\bar{\boldsymbol{R}}$ is constant, so that $\bar{c}(\mathrm{t}) \equiv \bar{c}$ is also a constant curve independent of t. Under elementary motion, therefore, the average envelope is formed as the trajectory of curve $\bar{c}$. Therefore, it is a surface of the same type as is the type of the elementary motion $g(\mathrm{t})$.

Further, if $\exp(\varphi\boldsymbol{R})\,g(\mathrm{t}_0)$ is the instantaneous motion of motion $g(\mathrm{t})$ at time $\mathrm{t} = \mathrm{t}_0$, the characteristic $c(\mathrm{t})$ is simultaneously the characteristic under instantaneous motion and the average envelope of surface $\mathfrak{F}$ under instantaneous motion is $\exp(\varphi\boldsymbol{R})\,c(\mathrm{t}_0)$.

As already mentioned, a point of the characteristic is also determined by the condition that at this point the normal to the surface is perpendicular to the tangent of the trajectory, i.e. the normal to the surface belongs to the complex of all trajectory normals. This complex was called the zero complex and its equation reads $\mathscr{K}(\boldsymbol{R}, \boldsymbol{X}) = 0$. Let us denote it by $\mathscr{N}(\mathrm{t})$ or, rather, by $\mathscr{N}$.

11.2 Theorem. *The characteristic $c(\mathrm{t})$ is the set of feet of the normals of surface $g(\mathrm{t})\,\mathfrak{F}$ which belong to $\mathscr{N}(\mathrm{t})$. The inverse of the characteristic $\bar{c}(\mathrm{t})$ is the set of feet of the normals to surface $\mathfrak{F}$ which belong to the complex $\overline{\mathscr{N}}(\mathrm{t})$ with the equation $\mathscr{K}(\bar{\boldsymbol{R}}, \boldsymbol{X}) = 0$.*

Recall that, for $\bar{\boldsymbol{R}} = [\boldsymbol{u}_1; v_0\boldsymbol{u}_1]$, $\overline{\mathcal{N}}(\mathrm{t})$ is given by the equation $v_0 p_1 + p_4 = 0$.

Now, let us introduce the concept of conjugate chords of contact of a linear complex.

11.3 Definition. Let p and q be different lines with the Plücker coordinates $\boldsymbol{P}$ and $\boldsymbol{Q}$, which are interpreted as vectors from $\mathfrak{E}$. Further, let $\mathcal{N}$ be a linear complex given by the equation $\mathcal{K}(\boldsymbol{R}, \boldsymbol{X}) = 0$. The lines p and q are called *conjugate chords of contact of complex* $\mathcal{N}$ if vectors $\boldsymbol{P}, \boldsymbol{Q}, \boldsymbol{R}$ lie in a single plane.

11.4 Theorem. *Different lines p and q are conjugate chords of contact of complex $\mathcal{N}$ if and only if each of their transversals belongs to $\mathcal{N}$.*

Proof. Let $p \neq q$, and let p and q be conjugate chords of contact of the linear complex $\mathcal{N}$. Then $\boldsymbol{R} = \alpha\boldsymbol{P} + \beta\boldsymbol{Q}$. Let $\mathcal{K}(\boldsymbol{P}, \boldsymbol{X}) = \mathcal{K}(\boldsymbol{Q}, \boldsymbol{X}) = 0$ for $\boldsymbol{X}$, i.e. $\boldsymbol{X}$ determines a transversal of $\boldsymbol{P}$ and $\boldsymbol{Q}$. Then

$$\mathcal{K}(\boldsymbol{R}, \boldsymbol{X}) = \mathcal{K}(\alpha\boldsymbol{P} + \beta\boldsymbol{Q}, \boldsymbol{X}) = \alpha\, \mathcal{K}(\boldsymbol{P}, \boldsymbol{X}) + \beta\, \mathcal{K}(\boldsymbol{Q}, \boldsymbol{X}) = 0\,,$$

i.e. this transversal belongs to $\mathcal{N}$. Conversely, let $\mathcal{K}(\boldsymbol{P}, \boldsymbol{X}) = \mathcal{K}(\boldsymbol{Q}, \boldsymbol{X}) = 0$ imply that $\mathcal{K}(\boldsymbol{R}, \boldsymbol{X}) = 0$. Then $\mathcal{K}(\boldsymbol{P}, \boldsymbol{X}) = 0$ and $\mathcal{K}(\boldsymbol{Q}, \boldsymbol{X}) = 0$ is a system of two homogeneous linear equations, and each of its solutions is also a solution of the equation $\mathcal{K}(\boldsymbol{R}, \boldsymbol{X}) = 0$. Therefore, the latter equation is a linear combination of the first two, i.e. $\boldsymbol{R} = \alpha\boldsymbol{P} + \beta\boldsymbol{Q}$.

Now, let us find the line q which is the conjugate chord of contact to line p with respect to the complex $\mathcal{N}$. The same notation as in Theorem 11.4 is used. Put $\boldsymbol{Q} = \mu\boldsymbol{R} + \nu\boldsymbol{P}$. It should be $\mathcal{K}(\boldsymbol{Q}, \boldsymbol{Q}) = 0$, with $\mathcal{K}(\boldsymbol{P}, \boldsymbol{P}) = 0$. Also, we have $\mathcal{K}(\boldsymbol{R}, \boldsymbol{R}) = v_0$ and, therefore,

$$\mathcal{K}(\boldsymbol{Q}, \boldsymbol{Q}) = \mathcal{K}(\mu\boldsymbol{R} + \nu\boldsymbol{P},\ \mu\boldsymbol{R} + \nu\boldsymbol{P}) =$$
$$= \mu^2\, \mathcal{K}(\boldsymbol{R}, \boldsymbol{R}) + 2\mu\nu\, \mathcal{K}(\boldsymbol{P}, \boldsymbol{R}) = 0\,.$$

If $\mu = 0$, then $\boldsymbol{Q} = \nu\boldsymbol{P}$, i.e. $p = q$, which case was excluded. Thus we have the equation

$$\mu v_0 + \nu\,.\,2\,\mathcal{K}(\boldsymbol{P}, \boldsymbol{R}) = 0\,.$$

Note that for $v_0 = 0$ and $\mathcal{K}(\boldsymbol{P}, \boldsymbol{R}) \neq 0$ we obtain $\nu = 0$, i.e. in the case of instantaneous rotation the conjugate chord of contact of line p different from the axis of rotation is the axis of rotation itself. If $v_0 \neq 0$ and $\mathcal{K}(\boldsymbol{P}, \boldsymbol{R}) = 0$, i.e. if the line belongs to $\mathcal{N}$, we obtain $p = q$, i.e. there is no conjugate chord of contact. If $v_0 \neq 0$ and $\mathcal{K}(\boldsymbol{P}, \boldsymbol{R}) \neq 0$, put $\mu = -2\mathcal{K}(\boldsymbol{P}, \boldsymbol{R})$, $\nu = v_0$, and we obtain $\boldsymbol{Q} = -2\mathcal{K}(\boldsymbol{P}, \boldsymbol{R})\boldsymbol{R} + v_0\boldsymbol{P}$, up to a proportionality factor.

11.5 Theorem. *If $v_0 \neq 0$, then just one conjugate chord of contact exists to every line which does not belong to $\mathcal{N}$; there are no conjugate chords of contact to the lines from $\mathcal{N}$. If $v_0 = 0$, then every line (different from the instantaneous axis) is a conjugate chord of contact of the instantaneous axis.*

Now, we demonstrate the construction of the conjugate chord of contact q to line p. First, we prove that

$$2\mathcal{K}(\boldsymbol{P}, \boldsymbol{R}) = v_0 \cos \varphi_1 - d_1 \sin \varphi_1$$

where φ_1 is the oriented angle of the axis of skew motion o and of the line p while d_1 is their oriented distance.

Proof. Choose a frame $\{O, \boldsymbol{u}_1, \boldsymbol{u}_2, \boldsymbol{u}_3\}$ so that $\boldsymbol{R} = [\boldsymbol{u}_1; v_0\boldsymbol{u}_1]$, and assume $\boldsymbol{u}_2$ to be such that the vector of line p lies in plane $(\boldsymbol{u}_1\boldsymbol{u}_2)$. Then

$$p = d_1\boldsymbol{u}_3 + \lambda(\boldsymbol{u}_1 \cos \varphi_1 + \boldsymbol{u}_2 \sin \varphi_1) .$$

The Plücker coordinates of line p are

$$\boldsymbol{P} = (\boldsymbol{u}_1 \cos \varphi_1 + \boldsymbol{u}_2 \sin \varphi_1;\ d_1\boldsymbol{u}_3 \times (\boldsymbol{u}_1 \cos \varphi_1 + \boldsymbol{u}_2 \sin \varphi_1)) ,$$

i.e.

$$\boldsymbol{P} = (\boldsymbol{u}_1 \cos \varphi_1 + \boldsymbol{u}_2 \sin \varphi_1;\ d_1\boldsymbol{u}_2 \cos \varphi_1 - \boldsymbol{u}_1 d_1 \sin \varphi_1) ;$$

consequently, we have

$$\boldsymbol{R} = \begin{bmatrix} 1;\ v_0 \\ 0;\ 0 \\ 0;\ 0 \end{bmatrix}, \quad \boldsymbol{P} = \begin{bmatrix} \cos \varphi_1; & -d_1 \sin \varphi_1 \\ \sin \varphi_1; & d_1 \cos \varphi_1 \\ 0; & 0 \end{bmatrix},$$

so that $2\mathcal{K}(\boldsymbol{P}, \boldsymbol{R}) = -d_1 \sin \varphi_1 + v_0 \cos \varphi_1$, which was to be proved. The form $\mathcal{K}$ is invariant and the relation thus holds in every orthonormal right-handed frame.

Let $\boldsymbol{R}$ and the line p with the Plücker coordinates $\boldsymbol{P}$ be given. For $\boldsymbol{Q}$ we then obtain

$$\boldsymbol{Q} = \lambda[(d_1 \sin \varphi_1 - v_0 \cos \varphi_1)\boldsymbol{R} + v_0\boldsymbol{P}], \quad \text{where} \quad \lambda \in \mathbf{R}.$$

$\mathbf{K}(\boldsymbol{Q}, \boldsymbol{Q})$ must equal 1, i.e.

$$\begin{aligned}&\mathbf{K}(\boldsymbol{Q}, \boldsymbol{Q})\\ &= \lambda^2[(d_1 \sin \varphi_1 - v_0 \cos \varphi_1)^2 + v_0^2\\ &\quad + 2(d_1 \sin \varphi_1 - v_0 \cos \varphi_1)\, v_0 \cos \varphi_1]\\ &= \lambda^2[(d_1^2 + v_0^2) \sin^2 \varphi_1].\end{aligned}$$

Put $D_1 = [d_1^2 + v_0^2]^{1/2}$, then $\lambda^2 D_1^2 \sin^2 \varphi_1 = 1$, so that

$$\boldsymbol{Q} = \frac{d_1 \sin \varphi_1 - v_0 \cos \varphi_1}{D_1 \sin \varphi_1} \boldsymbol{R} + \frac{v_0}{D_1 \sin \varphi_1} \boldsymbol{P}.$$

For the angle φ_2 between line q and the axis of skew motion o we have

$$\cos \varphi_2 = \mathbf{K}(\boldsymbol{Q}, \boldsymbol{R}) = \frac{d_1 \sin \varphi_1 - v_0 \cos \varphi_1}{D_1 \sin \varphi_1} + \frac{v_0 \cos \varphi_1}{D_1 \sin \varphi_1} = \frac{d_1}{D_1},$$

then $\sin \varphi_2 = v_0 D_1^{-1}$, so that

$$\frac{\sin \varphi_2}{\cos \varphi_2} = \frac{v_0}{d_1}, \quad \text{i.e.} \quad d_1 \tan \varphi_2 = v_0 .$$

Analogously, if we start with line p we arrive at the relation

$$d_2 \tan \varphi_1 = v_0 .$$

Thus the following relations hold for the conjugate chords of contact:

$$d_1 \tan \varphi_2 = d_2 \tan \varphi_1 = v_0 . \tag{11.3}$$

Further, note that the relation $\boldsymbol{R} = \alpha\boldsymbol{P} + \beta\boldsymbol{Q}$ implies that the lines o, p, q have a common shortest transversal. Indeed, the shortest transversal of lines p and o has its maximal commutative algebra determined by the vector $[\boldsymbol{P}, \boldsymbol{R}] = [\boldsymbol{P}, \alpha\boldsymbol{P} + \beta\boldsymbol{Q}] = \alpha[\boldsymbol{P}, \boldsymbol{Q}]$, which is the vector determining the maximal commutative algebra of the shortest transversal of lines p and q. It is then easy to find the direction and distance of the conjugate chord of contact from the axis of skew motion from relations (11.3).

Fig. 44 shows the construction of the conjugate chord of contact q to a given line p under a given instantaneous helical motion which is left-handed, has the axis o, and reduced thread height v_0. The numbers indicate the order in which the construction should proceed.

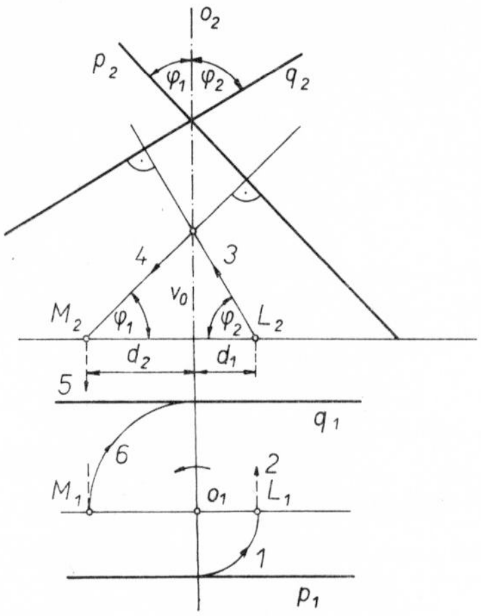

Fig. 44.

12. APPLICATION OF AVERAGE ENVELOPES

Kinematically generated average envelopes can be widely applied for practical purposes and understanding them is a necessary part of the theoretical training required in studying the problems of gearing and in designing cutting tools. The relatively complicated geometrical relations encountered in these fields can be explained with the aid of average envelopes.

A. *Conjugate surface of the tool and workpiece*

The geometrical interpretation of the relation between the cutting tool and the workpiece leads to a pair of surfaces, where the first generates a one-parametric system of surfaces, under relative motion with respect to the other, while the other surface is the envelope of this system. Such a pair of surfaces is formed, e.g., by the surface $\mathfrak{F}$ generated by the cutting edge of a disk cutter when it rotates round its axis, or by the surface of a grinding disk, and the surface Φ of the respective workpiece.

The investigation of surface $\mathfrak{F}$, which is called the *tool surface* or the *initial tool surface*, and of the surface Φ of the *workpiece* (*workpiece surface*) represents the contents of two fundamental problems, the solution of which is discussed in this section in detail. The construction of the average envelope in Section 11 infers the solution of the first fundamental problem: to determine the workpiece surface Φ if the surface $\mathfrak{F}$ of the tool is given together with its motion $g(\mathrm{t})$ relative to the workpiece.

The set of all inverses $\bar{c}(\mathrm{t})$ of the characteristics $c(\mathrm{t})$ on surfaces $g(\mathrm{t})\,\mathfrak{F}$, i.e. $\bar{c}(\mathrm{t}) = g^{-1}(\mathrm{t})\,c(\mathrm{t})$, forms its so-called *active part* $\bar{\mathfrak{F}}$ on the surface $\mathfrak{F}$. Since the surfaces $\mathfrak{F}$ and Φ are conjugate surfaces by Section 11, i.e. surfaces which constitute envelopes of one another, the surface $\bar{\mathfrak{F}}$ is the envelope of surface Φ under motion $g^{-1}(\mathrm{t})$. This then infers the solution of the second fundamental problem: to determine the surface $\mathfrak{F}$ of the tool if its motion $g(\mathrm{t})$ relative to the workpiece is given together with the workpiece surface Φ.

To start with, let us solve the first fundamental problem under the assumption that the surface $\mathfrak{F}$ of the tool is a surface of revolution and that its motion $g(\mathrm{t})$ relative to the workpiece consists of the elementary motion $g_1(\mathrm{t})$ and of the cutting rotation $g_2(\mathrm{t})$ about the tool axis, i.e. $g(\mathrm{t}) = g_1(\mathrm{t})\,g_2(\mathrm{t})$. However, rotation $g_2(\mathrm{t})$ which is the generating motion of surface $\mathfrak{F}$ has no effect on the surface Φ because the surfaces $g_1(\mathrm{t})\,\mathfrak{F}$ and $g(\mathrm{t})\,\mathfrak{F}$ are identical for every t and, therefore, their average envelopes are also identical. It is thus sufficient to investigate the average envelope of surface $\mathfrak{F}$ under the elementary motion $g_1(\mathrm{t})$. As proved in Section 11, the average envelope Φ is then a helical or translational surface or surface of revolution and its generating curve is the characteristic c on surface $\mathfrak{F}$, i.e. $\Phi \equiv g_1(\mathrm{t})\,c$. By rotating the characteristic c we obtain the active part $\bar{\mathfrak{F}}$ of surface $\mathfrak{F}$.

Thus, our problem is solved in two steps:

1. the construction of characteristic c on surface $\mathfrak{F}$,

2. the construction of the meridian d_0 of the helical surface or the surface of revolution Φ of the workpiece.

Let a surface of revolution $\mathfrak{F}$ be given by the meridian m and the axis o', and take a helical motion $g_1(t)$ with a constant reduced height of thread v_0 and with the axis $o = A + \lambda \boldsymbol{u}$, where $A = d\boldsymbol{u}_3$ and $\boldsymbol{u} = \boldsymbol{u}_1 \cos \alpha + \boldsymbol{u}_2 \sin \alpha$. The corresponding directing cone is $\boldsymbol{R} = [\boldsymbol{x}; \boldsymbol{t}]$, where $\boldsymbol{x} = \boldsymbol{u}$, $\boldsymbol{t} = A \times \boldsymbol{u} + v_0 \boldsymbol{u}$. Computation yields

$$\boldsymbol{R} = \begin{bmatrix} \cos \alpha; & v_0 \cos \alpha - d \sin \alpha \\ \sin \alpha; & v_0 \sin \alpha + d \cos \alpha \\ 0; & 0 \end{bmatrix}.$$

Further, let a surface of revolution be given with the axis of rotation $o' = \lambda \boldsymbol{u}_1$, $\lambda \in \mathbf{R}$ and the meridian m. Let m be given by parametric equations

$$x^1 = m_1(a), \quad x^2 = m_2(a), \quad x^3 = 0.$$

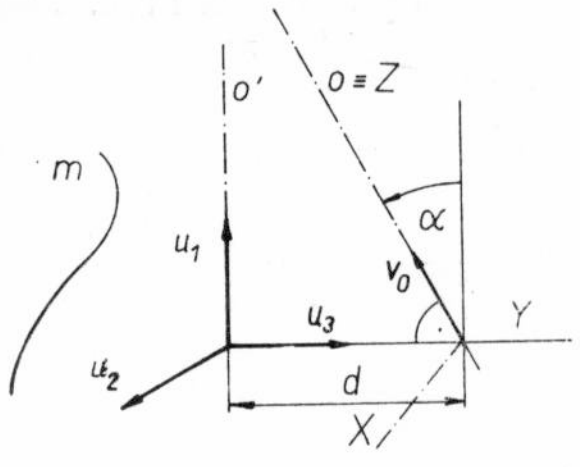

Fig. 45.

(See Fig. 45.) This meridian forms a surface of revolution with the parametric equations

$$\begin{aligned} \mathfrak{F}(a, \varphi) \equiv x^1 &= m_1(a), \\ x^2 &= m^2(a) \cos \varphi, \\ x^3 &= m_2(a) \sin \varphi, \end{aligned}$$

φ is the angle of rotation about axis o'. The velocity vector at point B of surface $\mathfrak{F}(a, \varphi)$ is

$$\boldsymbol{v}_B = \boldsymbol{x} \times B + \boldsymbol{t} = \begin{pmatrix} \cos \alpha \\ \sin \alpha \\ 0 \end{pmatrix} \times \begin{pmatrix} m_1 \\ m_2 \cos \varphi \\ m_2 \sin \varphi \end{pmatrix} + \begin{pmatrix} v_0 \cos \alpha - d \sin \alpha \\ v_0 \sin \alpha + d \cos \alpha \\ 0 \end{pmatrix}$$

$$= \begin{pmatrix} m_2 \sin \alpha \sin \varphi + v_0 \cos \alpha - d \sin \alpha \\ -m_2 \cos \alpha \sin \varphi + v_0 \sin \alpha + d \cos \alpha \\ m_2 \cos \alpha \cos \varphi - m_1 \sin \alpha \end{pmatrix}.$$

Further,

$$\frac{\partial \mathfrak{F}(a,\varphi)}{\partial \varphi} = \begin{pmatrix} 0 \\ -m_2 \sin \varphi \\ m_2 \cos \varphi \end{pmatrix}; \quad \frac{\partial \mathfrak{F}(a,\varphi)}{\partial a} = \begin{pmatrix} m_1' \\ m_2' \cos \varphi \\ m_2' \sin \varphi \end{pmatrix}.$$

The prime indicates a derivative with respect to a, the similar holds below.

The equation of the characteristic is

$$\left| v_B, \frac{\partial \mathfrak{F}}{\partial \varphi}, \frac{\partial \mathfrak{F}}{\partial a} \right| = 0,$$

which yields

$$0 = m_2 \cdot \begin{vmatrix} m_2 \sin \alpha \sin \varphi + v_0 \cos \alpha - d \sin \alpha, & 0, & m_1' \\ -m_2 \cos \alpha \sin \varphi + v_0 \sin \alpha + d \cos \alpha, & -\sin \varphi, & m_2' \cos \varphi \\ m_2 \cos \alpha \cos \varphi - m_1 \sin \alpha, & \cos \varphi, & m_2' \sin \varphi \end{vmatrix}.$$

Upon computation we obtain the equation

$$m_2\{m_1'[(v_0 \cos \varphi - m_1 \sin \varphi) \sin \alpha + d \cos \varphi \cos \alpha] - m_2'[v_0 \cos \alpha + (-d + m_2 \sin \varphi) \sin \alpha]\} = 0.$$

If $m_2 = 0$, the point is located on the axis of the surface of revolution, the tangent plane at this point is not determined, and the condition is thus satisfied trivially ($\partial \mathfrak{F}/\partial \varphi = 0$). Therefore, we may exclude these points. Thus we have the following equation for determining the characteristic:

$$m_1'[(v_0 \cos \varphi - m_1 \sin \varphi) \sin \alpha + d \cos \varphi \cos \alpha] - m_2'[v_0 \cos \alpha + (-d + m_2 \sin \varphi) \sin \alpha] = 0;$$

we can also write

$$(m_1 m_1' + m_2 m_2') \sin \alpha \sin \varphi - m_1'(v_0 \sin \alpha + d \cos \alpha) \cos \varphi + m_2'(v_0 \cos \alpha - d \sin \alpha) = 0.$$

For $v_0 = 0$ we get

$$(m_1 m_1' + m_2 m_2') \sin \alpha \sin \varphi - m_1' d \cos \varphi \cos \alpha - m_2' d \sin \alpha = 0 \tag{12.1a}$$

and under translation of surface $\mathfrak{F}$ along line o, i.e. for $v_0 = \infty$:

$$-m_1' \cos\varphi \sin\alpha + m_2' \cos\alpha = 0\,. \tag{12.1b}$$

Let us again investigate the characteristic on the surface $\mathfrak{F}$ under helical motion. First, let the axes o and o' not be parallel, i.e. $\sin\alpha \neq 0$.

Put $M = v_0 + d\cot\alpha$, $L = v_0 \cot\alpha - d$. Equation (12.1) can then be rewritten into the form

$$(m_1 m_1' + m_2 m_2')\sin\varphi - m_1' M \cos\varphi + m_2' L = 0\,, \tag{12.2}$$

where M and L are constants, m_1 and m_2 are functions of the parameter a which are given if the surface of revolution is given. We can compute φ from equation (12.2) as a function of the parameter a (i.e. $\cos\varphi$ and $\sin\varphi$ as functions of a, which is sufficient for our purpose). Then the characteristic has the form

$$\mathbf{c}(a) = \begin{pmatrix} m_1(a) \\ m_2(a)\cos\varphi(a) \\ m_2(a)\sin\varphi(a) \end{pmatrix}$$

and its helical motion about the axis o generates the desired average envelope.

If we require the explicit representation of the characteristic, we must solve equation (12.2). For this purpose put $m_1' M = F$, $m_2' L = G$, $m_1 m_1' + m_2 m_2' = \varrho$, $\varepsilon \pm 1$. Equation (12.2) then becomes

$$\varrho \sin\varphi - F\cos\varphi + G = 0\,. \tag{12.3}$$

Its solution is

$$(\cos\varphi)_{1,2} = \frac{FG + \varepsilon\varrho(F^2 + \varrho^2 - G^2)^{1/2}}{F^2 + \varrho^2}\,,$$

$$(\sin\varphi)_{1,2} = \frac{-\varrho G + \varepsilon F(F^2 + \varrho^2 - G^2)^{1/2}}{F^2 + \varrho^2}\,.$$

Now, the characteristic has the form

$$x^1 = m_1\,,$$

$$x^2 = \frac{m_2}{F^2 + \varrho^2}\left[FG + \varepsilon\varrho(F^2 + \varrho^2 - G^2)^{1/2}\right],$$

$$x^3 = \frac{m_2}{F^2 + \varrho^2}\left[-\varrho G + \varepsilon F(F^2 + \varrho^2 - G^2)^{1/2}\right].$$

Below we consider several special types of surfaces as given in the table.

Type of surface $\mathfrak{F}$

	I.	II.	III.	IV.
m_1	0	$A + au_1$	$A + \delta \cos a$	a
m_2	a	$B + au_2$	$B + \delta \sin a$	δ/a
m'_1	0	u_1	$-\delta \sin a$	1
m'_2	1	u_2	$\delta \cos a$	$-\delta/a^2$
ϱ	a	a	$-A\delta \sin a + B\delta \cos a$	$a - \delta^2/a^3$
F	0	$u_1 M$	$-\delta M \sin a$	M
K	L	$u_2 L$	$\delta L \cos a$	$-L\delta/a^2$

Type of surface $\mathfrak{F}$

	V.	VI.	VII.
m_1	$a + A$	$a^2 + A$	$A + \delta \operatorname{ch} a$
m_2	$a^2 + B$	$a + B$	$B + \mu \operatorname{sh} a$
m'_1	1	$2a$	$\delta \operatorname{sh} a$
m'_2	$2a$	1	$\mu \operatorname{ch} a$
ϱ	$a + A + 2a(a^2 + B)$	$2a(a^2 + A) + a + B$	ϱ_7
F	M	$2aM$	$M\delta \operatorname{sh} a$
G	$2La$	L	$L\mu \operatorname{ch} a$

$$\varrho_7 = A\delta \operatorname{sh} a + B\mu \operatorname{ch} a + (\delta^2 + \mu^2) \operatorname{sh} a \operatorname{ch} a$$

Upon computation we find that:

I. The surface of revolution is a plane perpendicular to the axis of rotation. We obtain $x^1 = 0$, $x^3 = L$, the characteristic is a line-tangent at the point $-L\boldsymbol{u}_3$.

II. The surface of revolution is a conical surface. The parametric equations of the characteristic are

$$x^1 = A + au_1\,,$$

$$x^2 = \frac{B + au_2}{u_1^2M^2 + a^2}\left[u_1u_2LM + \varepsilon a\left(u_1^2M^2 - u_2^2L^2 + a^2\right)^{1/2}\right],$$

$$x^3 = \frac{B + au_2}{u_1^2M^2 + a^2}\left[-au_2L + \varepsilon u_1M\left(u_1^2M^2 - u_2^2L^2 + a^2\right)^{1/2}\right].$$

III. The surface of revolution is a torus. The parametric equations of the characteristic are

$$x^1 = A + \delta\cos a\,,$$

$$x^2 = \frac{B + \delta\sin a}{M^2\sin^2 a + (B\cos a - A\sin a)^2}\,.$$
$$\left\{-LM\cos a\sin a + \varepsilon(B\cos a - A\sin a)\,.\right.$$
$$\left.\left[M^2\sin^2 a - L^2\cos^2 a + (B\cos a - A\sin a)^2\right]^{1/2}\right\},$$

$$x^3 = \frac{B + \delta\sin a}{M^2\sin^2 a + (B\cos a - A\sin a)^2}\,.$$
$$\left\{L\cos a(B\cos a - A\sin a)\right.$$
$$\left.- \varepsilon M\sin a\left[M^2\sin^2 a - L^2\cos^2 a + (B\cos a - A\sin a)^2\right]^{1/2}\right\}.$$

For $A = B = 0$ we obtain a sphere. The axis o may be chosen such that $L = 0$, and the parametric equations of the characteristic are then

$$x^1 = \delta\cos a\,,\quad x^2 = 0\,,\quad x^3 = \varepsilon\delta\sin a\,.$$

It follows that the characteristic is a circle.

For $A = 0$ we obtain a special position of the torus. The centre of its meridian lies on the shortest transversal of the axes o and o'. If $M = 0$, the characteristic is a plane curve.

We proceed analogously for the remaining types of surfaces of revolution:

IV. The surface of revolution is generated by the rotation of a hyperbola about its asymptote.

V. and VI. The surface of revolution is generated by the rotation of a parabola.

VII. Hyperboloid of revolution.

Finally, we ask whether one of the meridians of the surface of revolution can be the characteristic. In this case, (12.2) would have to be satisfied for $\varphi = const.$, i.e.

$$(m_1 m_1' + m_2 m_2') \sin \varphi - m_1' M \cos \varphi + m_2' L = 0 .$$

By integration we obtain

$$(m_1^2 + m_2^2) \sin \varphi - 2 m_1 M \cos \varphi + 2 m_2 L = K , \quad K = const. ,$$

and after some manipulation we have

$$\sin \varphi \left[(m_1 - M \cot \varphi)^2 + \left(m_2 + \frac{L}{\sin \varphi} \right)^2 \right] = K' , \quad K' = const. ,$$

so that the characteristic may be identical with the meridian only if a torus is involved, with the centre of its prime meridian m having the coordinates $[M \cot \varphi, -L/\sin \varphi]$.

If the meridian is a line, it is more convenient to solve equation (12.2) directly. Thus, let

$$m_1 = A + a u_1 , \quad m_2 = B + a u_2 ,$$

where $A u_1 + B u_2 = 0$, $u_1^2 + u_2^2 = 1$, a is the parameter of the line, and the point with the coordinates $[A, B]$ is the foot of the perpendicular dropped from the origin on to the meridian. Equation (12.2) then becomes

$$a \sin \varphi - u_1 M \cos \varphi + u_2 L = 0 .$$

Under the assumption that $u_1 M = u_2 L = 0$ does not hold, equation (12.2) can be solved and we have

$$a = \frac{u_1 M \cos \varphi + u_2 L}{\sin \varphi} ,$$

so that the characteristic is

$$x^1 = A + \frac{1}{\sin \varphi}(u_1 M \cos \varphi - u_2 L),$$

$$x^2 = B \cos \varphi + \frac{u_2 \cos \varphi}{\sin \varphi}(u_1 M \cos \varphi - u_2 L),$$

$$x^3 = B \sin \varphi + u_2(u_1 M \cos \varphi - u_2 L).$$

Let us investigate whether the characteristic may be a conic section (i.e. a plane curve) in this case. Then

$$\alpha x^1 + \beta x^2 + \gamma x^3 + \delta = 0$$

must hold for some $\alpha, \beta, \gamma, \delta \in \mathbf{R}$ and for all φ.

Upon substitution we obtain the equation

$$\begin{aligned} &\alpha(A \sin \varphi + u_1 M \cos \varphi - u_2 L) \\ &\quad + \beta[B \cos \varphi \sin \varphi + u_2 \cos \varphi(u_1 M \cos \varphi - u_2 L)] \\ &\quad + (\beta u_1 u_2 M - \gamma B) \cos^2 \varphi + (\gamma B - \alpha u_2 L) = 0, \end{aligned}$$

which yields the equations

$$\begin{aligned} \alpha u_1 M - \beta u_2^2 L &= 0, \\ \gamma B - \alpha u_2 L &= 0, \\ \beta B + \gamma u_1 u_2 M &= 0, \\ -\gamma B + \beta u_1 u_2 M &= 0. \end{aligned}$$

We may assume that $u_1 \neq 0$ because the surface of revolution is a plane for $u_1 = 0$ and we know the characteristic. The last two equations yield

$$(\beta^2 + \gamma^2) B = 0, \quad (\beta^2 + \gamma^2) u_2 M = 0.$$

Further, if $B = 0$, we may always put $A = 0$ (for $u_2 \neq 0$ it follows, for $u_2 = 0$ it can be chosen).

Finally, if $u_2 = 0$ and $B = 0$, we obtain the axis of rotation which is a trivial case and can be excluded.

Let us solve our system of equations. We distinguish two cases:

1. $\beta^2 + \gamma^2 \neq 0$. Then $B = 0$, $A = 0$, $u_2 \neq 0$, so that $M = 0$ and the equations can always be solved.

2. $\beta = \gamma = 0$. Then $\alpha \neq 0$ and we have $u_2 L = u_1 M = 0$, which is the excluded case.

In the first case we obtain

$$x^1 = \frac{-u_2 L}{\sin\varphi}, \quad x^2 = \frac{-u_2^2 L\cos\varphi}{\sin\varphi}, \quad x^3 = -u_2^2 L,$$

and these are the equations of a hyperbola; moreover, $M = v_0 + d\cot\alpha = 0$ must hold.

In the second case equation (12.2) is

$$a\sin\varphi = 0,$$

which has the following solutions:

$$a = 0 - \text{circle};\ \sin\varphi = 0 - \text{two lines}.$$

Let us now discuss the case of parallel axes, i.e. of $\sin\alpha = 0$. The equation for the characteristic (12.1) then reads

$$m_1' d\cos\varphi = v_0 m_2'.$$

For $dm_1' \neq 0$ we get

$$\cos\varphi = \frac{v_0 m_2'}{dm_1'}, \quad \sin\varphi = \varepsilon\left[1 - \left(\frac{v_0 m_2'}{dm_1'}\right)^2\right]^{1/2},$$

so that the parametric equations of the characteristic are

$$x^1 = m_1,$$

$$x^2 = m_2\frac{v_0 m_2'}{dm_1'},$$

$$x^3 = \varepsilon m_2\left[1 - \left(\frac{v_0 m_2'}{dm_1'}\right)^2\right]^{1/2}.$$

Let us study this characteristic more closely. Choose any point $W = [w^1, 0, 0]$ on the axis o' and compute the coordinates of its velocity vector $\boldsymbol{v}_W$ under a given helical motion

$$\boldsymbol{v}_W = \boldsymbol{x} \times W + \boldsymbol{t} = \begin{pmatrix} 1 \\ 0 \\ 0 \end{pmatrix} \times \begin{pmatrix} w^1 \\ 0 \\ 0 \end{pmatrix} + \begin{pmatrix} v_0 \\ d \\ 0 \end{pmatrix} = \begin{pmatrix} v_0 \\ d \\ 0 \end{pmatrix}.$$

Now, if the surface $\mathfrak{F}$ is subject to translation in the direction of $\boldsymbol{v}_W$, we obtain by (12.1b) the equation for the characteristic in the form

$$-m_1' d \cos\varphi + m_2' v_0 = 0$$

since the coordinates of the unit vector which determines the direction of translation are

$$\cos\alpha = \frac{v_0}{(v_0^2 + d^2)^{1/2}}, \quad \sin\alpha = \frac{d}{(v_0^2 + d^2)^{1/2}}.$$

Since the equations for determining the characteristic on the surface of revolution $\mathfrak{F}$ under helical motion and under translation are the same, the investigated characteristic is the real contour of the surface $\mathfrak{F}$ in parallel projection in the direction of $\boldsymbol{v}_W$.

The equation for determining the characteristic under rotation about the axis o parallel with o'

$$m_1' d \cos\varphi = 0$$

implies that the characteristic is formed, for $d \neq 0$, by the meridian of the surface $\mathfrak{F}$ in the plane $x_2 = 0$ and by the crater circles of surface $\mathfrak{F}$.

Under translation in the direction of axis o the characteristic is formed by the equatorial and jugular circles of surface $\mathfrak{F}$, which follows from the equation

$$m_2' = 0.$$

If we subject the obtained characteristic c to helical motion, we obtain the required average envelope Φ. It remains to construct its axial section d_0.

Let us choose another Cartesian coordinate system with the axes X, Y, Z so that $Y \equiv x^3$, $Z \equiv o$, and that the orientation of the axes Y and Z is identical with the orientation of lines x^3 and o. We choose the plane of the axial section in the plane XZ and investigate the section of one thread of surface Φ in the halfplane $X \geq 0$, $Y = 0$ (Fig. 45). In the new system of coordinates point $C = [x^1, x^2, x^3]$ of characteristic c has the coordinates

$$X = -x^1 \sin\alpha + y^1 \cos\alpha\,,$$

$$Y = x^3 - d\,,$$

$$Z = x^1 \cos\alpha + x^2 \sin\alpha\,.$$

The parametric equations of the helix of point C in the new coordinate system are

$$X_s = X\cos\psi - Y\sin\psi\,,$$

$$Y_s = X\sin\psi + Y\cos\psi\,,$$

$$Z_s = Z + \psi v_0\,,$$

and its point of intersection D with the plane $Y = 0$ is the point of the required halfmeridian d_0. Therefore, $Y_s = X\sin\psi + Y\cos\psi$ and, consequently, $\psi = -\arctan(Y/X)$. By substituting ψ into (12.4) we obtain the coordinates of point D in the plane XZ. We could compute the points of the front section of surface Φ, i.e. the section through the plane perpendicular to axis Z, in a similar way. This is left for the reader to tackle.

We see that the actual computation of the points of the characteristic and of the section is very tedious. That is why a computer was used to compute the coordinates of these points and the results were represented graphically with the aid of an $X-Y$ plotter. To enable the general application of the program for various surfaces of revolution and various rotations and helical motions, the OBAL procedure was elaborated for computing and plotting the meridian of the surface of revolution $\mathfrak{F}$, the characteristic c and the halfmeridian d_0 of the average envelope Φ. The procedure is written in FORTRAN and makes use of the plotting subroutines of the FELGRAF system developed at the Computer Department of the Faculty of Electrical Engineering of the Czech Technical University in Prague.

Let us briefly describe the input to the OBAL procedure, and its graphical outputs. The surface $\mathfrak{F}$ is an arbitrary surface of revolution which is defined by its meridian m and axis o'. The meridian is given by the parametric equations

$$m^1 = m^1(\mathrm{t})\,,\quad m^2 = m^2(\mathrm{t})\,,\quad \mathrm{t} \in \mathbf{I}\,.$$

Further, the first derivatives of the functions m^1 and m^2 have to be given, i.e. $\mathrm{d}m^1/\mathrm{dt}$, $\mathrm{d}m^2/\mathrm{dt}$, for which we assume that they are continuous in the interval $\mathbf{I}$. These functions must appear in the EXTERNAL description.

The motion of the surface $\mathfrak{F}$ is determined by the axis o and by the reduced thread height, i.e. by the values α, d, v_0 as in Fig. 45.

The surface $\mathfrak{F}$ is mapped in the Monge projection so that its axis o' is identical with the ground line. In this way both its apparent contours are identified with the prime meridians of the surface. The axis of motion o is considered in a position parallel with the second plane of projection. Its second projection o_2 is plotted, generated by rotation of the positive semi-axis x^1 through the angle α, together with the first projection P_1 of its first trace which determines the distance d (to enable verification).

The OBAL procedure then maps the characteristic c by its first projection c_1 (its points are circled) and second projection c_2 (full line). In the next figure the required halfmeridian d_0 of the average envelope Φ is drawn. Under helical motion the arrow on the axis o indicates the reduced thread height v_0, which is equal, in absolute value, to the distance of the top of the arrow from the ring on axis o. The interpolation procedure of the FELGRAF system is used to plot the curves.

Now, let us describe the parameters of the OBAL procedure:

obal (pj, pd, dj, dd, n, ti, ta, al, vo, d, xi, yi, xa, ya)

pj $= m^1$,

pd $= m^2$,

dj $= dm^1/dt$,

dd $= dm^2/dt$,

n = number of division points of interval **I** (including the endpoints),

ti = minimal value of t $\in$ **I**,

ta = maximal value of t $\in$ **I**,

al $= \alpha$,

vo $= v_0$,

d = distance between axes o and o',

xi, yi and xa, ya are the minimal and maximal x and y coordinates in the user's coordinate system on the plotting surface (the x axis is identical with the ground line in the Monge projection); ($0 < \text{ya} - \text{yi} \leqq 300$, $0 < \text{xa} - \text{xi} \leqq 1000$).

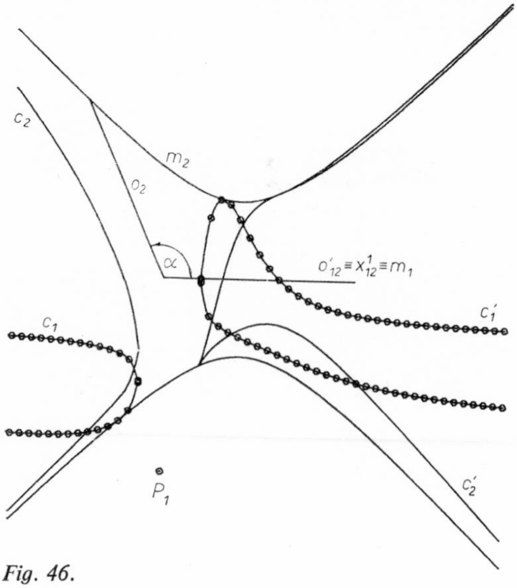

Fig. 46.

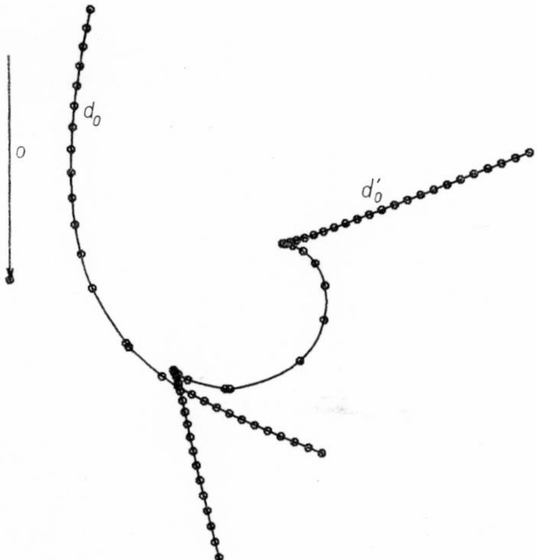

Fig. 47.

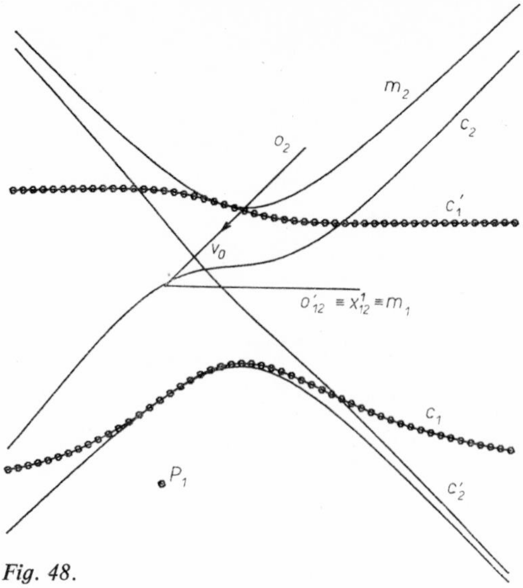

Fig. 48.

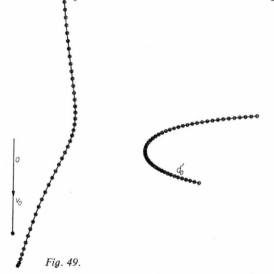

Fig. 49.

Below we present the entire program including the **OBAL** procedure for constructing the average envelope, formed by a one-sheet hyperboloid of revolution generated by the rotation of the equiangular hyperbola

$$m^1 = \mathrm{t}\,, \quad m^2 = [400 + (\mathrm{t} - 20)^2]^{1/2}\,, \quad \mathrm{t} \in \langle -40, 90\rangle\,,$$

about the axis o'.

In the first example the hyperboloid revolves about the axis $o : \alpha = 2$, $v_0 = 0$, $d = 50$, in the second it is subject to helical motion: $\alpha = 0.785\,39$, $v_0 = 20$, $d = 50$. The results of the first and second examples are shown in Figs. 46–47, and 48–49, respectively. In both cases the average envelope is formed by two lateral areas created by the halfmeridians d_0 and d_0' under the given rotation and helical motions, respectively.

Program

```
external m1, m2, dm1, dm2
  call obal (m1, m2, dm1, dm2, 51, -40., 90., 2., 0., 50., -100., -120., 400.,
    /180.)
  call bend (30.)
  call obal (m1, m2, dm1, dm2, 51, -40., 90., 0.78539,20., 50., -100., 120.,
    /400., 180.)
  call bend (30.)
stop
end
real function m1(t)
m1 = t
return
end
real function m2(t)
m2 = sqrt(400. + (t - 20.) ??2)
return
end
function dm1(t)
dm1 = 1.
return
end
```

```
function dm2(t)
dm2 = (t - 20.)/sqrt(400. + (t - 20.) ??2)
return
end
```

```
subroutine obal(pj, pd, dj, dd, n, ti, ta, al, vo, d, xi, yi, xa, ya)
dimension x(n), y(n), oyz(n), yz(n), oy(n), z(n), xv(n)
logical vr
   call benson(3, 1, 1, 0, 0, xi, yi, xa, ya, 0., 0., xa - xi, ay - yi)
   call bplot(0., 0., 0)
   call bsign(10., 1.5)
   call bplot(50., 0., 1)
   call bplot(0., 0., 0)
   call bplot(50. ?cos(al), 50. ?sin(al), 1)
if(vo.eq.0)goto 97
   call bsign(1, 2.)
   call bplot(vo?cos(al), vo?sin(al), 3)
   call bsign(10, 1.5)
97call bplot(0., -d, 2)
p = (ta - ti)/float(n - 1)
mk = 2
ja = 1
vr =. false.
do 1 jj = 1, n
t = ti + float(ji - 1) ?p
x(jj) = pj(t)
z(jj) = pd(t)
1 yz(jj) = -z(jj)
   call bopint
   call bplotn(x, z, 1, n, 1)
   call bclint
   call bopint
   call bplotn(x, yz, 1, n, 1)
   call bclint
bm = vo + d?cos(al)/sin(al)
bl = vo ?cos(al)/sin(al) - d
```

```
7 ma = 1
do 3 i = ja, n
ku = 1
m = 1
t = ti + float(i - 1) ?p
8 f = bm ?dj(t)
g = bl ?dd(t)
r = pj(t) ?dj(t) + pd(t) ?dd(t)
u = f??2 + r??2 - g??2
if(u.lt.0.)goto(9, 5), ma
xx = pj(t)
w = sqrt(u)
yy = pd(t) ?(f?g + r?w)/(f??2 + r??2)
zz = pd(t) ?(f?w - r?g)/(f??2 + r??2)
dy = pd(t) ?(f?g - r?w)/(f??2 + r??2)
dz = -pd(t) ?(f?w + r?g)/(f??2 + r??2)
if(mk.ne.1.or.m.eq.1)goto 32
x(i - 1) = xx
y(i - 1) = yy
yz(i - 1) = -zz
oy(i - 1) = dy
oyz(i - 1) = -dz
xv(i - 1) = xx
vr =. true.
12 ma = 2
goto(13, 3, 2), mk
13 if(p.gt.0,)p = -p
2 if(m.ge.10)goto 11
t = t + p/2.??m
ku = ku + 1
19 m = m + 1
goto 8
32 x(i) = xx
y(i) = yy
yz(i) = -zz
oy(i) = dy
```

```
oyz(i) = -dz
xv(i) = xx
goto 12
9 ku = 2
11 if(p.gt.0.)goto 10
p = -p
mk = 2
3 continue
10 if(ku.le.1)i = 1 - 1
if(vr)ja = ja - 1
vr = .false.
kl = i - ja
if(kl.le.0)goto 4
if(kl.lt.3)goto 16
   call bopint
   call bplotn(x, y, ja, 1)
   call bclint
   call bopint
   call bplotn(x, yz, ja, i, 3)
   call bclint
   call bopint
   call bplotn(y, oy, ja, i, 1)
   call bclint
   call bopint
   call bplont(x, oyz, ja, i, 3)
   call bclint
goto 14
16 call bplotn(x, y, ja, i, 1)
   call bplotn(x, oy, ja, i, 1)
   call bplotn(c, yz, ja, i, 3)
   call bplotn(x, oyz, ja, i, 3)
goto 14
4 if(i.ge.n - 2)goto 23
ja = i + 1
mk = 1
goto 7
```

```
5 if(p.gt.0.)mk = 3
if(m.ge.10)goto 11
t = t - p/2.??m
goto 19
md = 1
50 do 17 jk = ja, i

yn = -yz(jk) - d
zn = x(jk)?cos(al) + y(jk)?sin(al)
xn = -x(jk)?sin(al) + y(jk)?cos(al)
if(xn.eq.0.)goto 28
fi = -atan2(yn, xn)
21 x(jk) = xn?cos(fi) - yn?sin(fi) + 200.
z(jk) = zn + fi?vo
17 continue
if(kl.lt.3)goto 35
   call bclint
35 call bplotn(x, z, ja, i, 3)
if(kl.lt.3)goto 44
   call bolint
44 goto (45, 4), md
45 do 48 it = ja, i
yz(it) = oyz(it)
y(it) = oy(it)
48 x(it) = xv(it)
md = 2
goto 50
28 fi = 1.5708
if(yn.gt.0)fi = -fi
goto 21
23 call bplot(200., 0., 2)
   call bplot(200., 50., 1)
   call bsign(1, 2.)
   call bplot(200., vo, 2)
return
end
```

Let us now discuss the solution of the second fundamental problem. Assume that the conditions for the surfaces $\mathfrak{F}$ and Φ, and for the motion g, are the same as in the first problem solved above. As shown in the introduction to this section, under inverse motion $g^{-1} = (g_1 g_2)^{-1} = g_2^{-1} g^{-1}$ the surface Φ envelops the active part $\bar{\mathfrak{F}}$ of the surface of the tool, identified now with the desired surface $\mathfrak{F}$. Since the surface Φ transforms under motion g_1^{-1} into itself, the average envelopes of the systems of surfaces $g^{-1}\Phi$ and $g_2^{-1}\Phi$ are identical. It is therefore sufficient to investigate the average envelope of surface Φ under rotation g_2^{-1}. However, we have thus come back to the first fundamental problem, with the roles of the surfaces $\mathfrak{F}$ and Φ interchanged.

The second fundamental problem is solved in two steps again:

1. the construction of the characteristic c on surface Φ,
2. the construction of the meridian d_0 of the surface of revolution $\mathfrak{F}$ of the tool.

If Φ is a surface of revolution, it is possible to use directly the previous computer solution of the first fundamental problem. In order to solve the second fundamental problem also for the helical surface Φ, we have to present the solution of the first fundamental problem for the helical surface $\mathfrak{F}$, i.e. we have to derive the equation for determining the characteristic on the helical surface $\mathfrak{F}$. The procedure is similar to that for a surface of revolution, and this fact will be exploited in the following exercise.

12.1 Exercise. Derive the equation for determining the characteristic on the helical surface $\mathfrak{F}$ which is subject to helical motion. The parametric equations of surface $\mathfrak{F}$ are

$$x^1 = m_1(a) + \varphi\tilde{v}_0\,, \quad x^2 = m_2(a)\cos\varphi\,, \quad x^3 = m_2(a)\sin\varphi\,,$$

where $\tilde{v}_0$ is the reduced thread height of the helical motion which generates surface $\mathfrak{F}$. Start with the same assumptions as in deriving this equation for the surface of revolution $\mathfrak{F}$ at the beginning of this subsection.

As we have found out in the exercise, the equation for the determination of the characteristic c on the helical suriace $\mathfrak{F}$ is

$$\begin{aligned}&\left[m_2(m_2' m_2 + m_1 m_1') + \tilde{v}_0(M m_2' + m_1' m_2 \varphi)\right]\sin\varphi + \\ &\left[\tilde{v} m_2'(m_1 + \tilde{v}_0\varphi) - m_1' m_2 M\right]\cos\varphi + m_2' m_2(L - \tilde{v}_0 \operatorname{cotan}\alpha) = 0\end{aligned}$$

when using the same notation as for the surface of revolution (12.2). This transcendent equation is solved by finding a root φ of this equation for a selected parameter a by iteration. This means that for a chosen point B of the meridian m we find the rotational component φ of its helical motion so that it moves, by helical motion, into point C of the desired characteristic c. The meridian d_0 of the desired average envelope is then constructed in the familiar manner with the aid of the characteristic c.

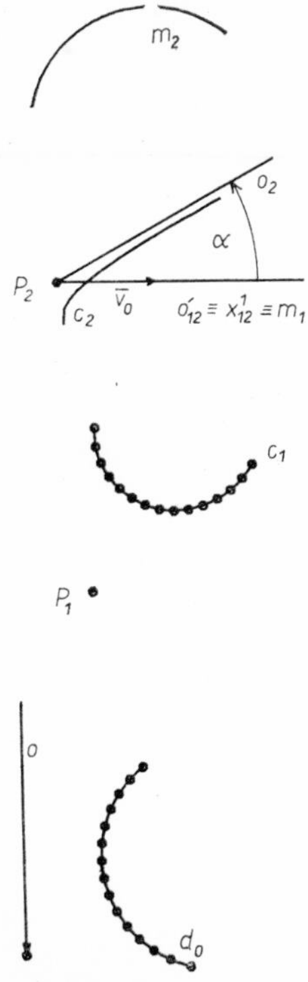

Fig. 50.

Evidently this problem calls for a computer. Fig. 50 shows the graphical output of the computer solution of the second fundamental problem in which a part of the cyclical helical surface Φ is given by the halfmeridian m, the axis o', and the reduced height of thread $\tilde{v}_0$, together with the axis of the desired surface of revolution.

In the Monge projection, in which the same projection planes are chosen as in the first fundamental problem, the given geometrical objects and the characteristic c are plotted on surface Φ. The meridian d_0 of the desired surface $\mathfrak{F}$ of the tool is represented in the bottom part of the figure.

B. *Average Envelopes in Gearing Theory*

One of the fundamental problems of gearing theory is to generate the surfaces of the sides of the cogs of cog wheels, which transform the rotation of one wheel into the rotation of another by mutual contact. If the surfaces of the sides of the cogs are tangent along a curve in the course of the motion (the so-called *curve of instantaneous contact*), under relative motion with respect to one another, the two surfaces envelop each other and form conjugate surfaces.

This relative motion consists of two rotations which, in general, have skew axes. A number of problems of gearing theory are thus reduced to problems of average envelopes generated by double-rotational motion of a surface.

Further, we deal with average envelopes of ruled surfaces under double-rotations with perpendicular concurrent axes. Practical application can be found in a worm and worm gear, whose worm-side surface is a ruled surface. Under motion relative to the worm wheel this surface envelops

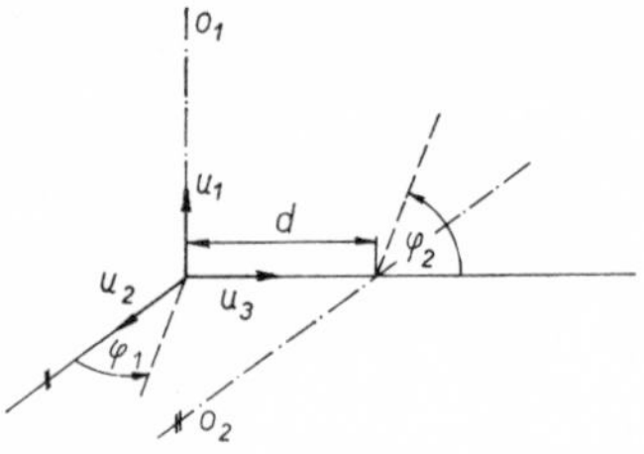

Fig. 51.

the sides of the wheel cog. The relative motion consists of the rotation of the worm about its axis o_2 under simultaneous carrying rotation of the worm about the axis o_1 of the wheel.

Denote by $g_1(\varphi)$ and $g_2(\varphi_2)$ the rotation about the axes $o_1 = \lambda \boldsymbol{u}_1$ and $o_2 = d\boldsymbol{u}_3 + \lambda \boldsymbol{u}_2$, respectively, through the angles φ_1 and φ_2 ($d \in \mathbf{R}$,

$\lambda \in \mathbf{R}$ is the parameter) – Fig. 51. Then we have

$$g_1 = \begin{pmatrix} 1, & 0, & 0, & 0 \\ 0, & 1, & 0, & 0 \\ 0, & 0, & \cos\varphi_1, & -\sin\varphi_1 \\ 0, & 0, & \sin\varphi_1, & \cos\varphi_1 \end{pmatrix},$$

$$g_2 = \begin{pmatrix} 1, & 0, & 0, & 0 \\ -d\sin\varphi_2, & \cos\varphi_2, & 0, & \sin\varphi_2 \\ 0, & 0, & 1, & 0 \\ d(1-\cos\varphi_2), & -\sin\varphi_2, & 0, & \cos\varphi_2 \end{pmatrix}.$$

Further, we have

$$g_2^{-1}(\varphi_2) = \begin{pmatrix} 1, & 0, & 0, & 0 \\ d\sin\varphi_2, & \cos\varphi_2 & 0, & -\sin\varphi_2 \\ 0, & 0, & 1, & 0 \\ d(1-\cos\varphi_2), & \sin\varphi_2, & 0, & \cos\varphi_2 \end{pmatrix},$$

because $g_2^{-1}(\varphi_2) = g_2(-\varphi_2)$.

Consider double-rotation $g_1(\varphi_1)\, g_2(\varphi_2)$ with angular velocities $\omega_1 = \mathrm{d}\varphi_1/\mathrm{dt}$, $\omega_2 = \mathrm{d}\varphi_2/\mathrm{dt}$ (we may assume that $\varphi_2 = \omega_2\omega_1^{-1}\varphi_1$). For the moving directing function $\bar{\boldsymbol{r}}(\mathrm{t})$ of motion $g(\mathrm{t}) = g_1(\varphi_1(\mathrm{t}))\, g_2(\varphi_2(\mathrm{t}))$ we obtain

$$\bar{\boldsymbol{r}}(\mathrm{t}) = (g_1 g_2)^{-1} \frac{\mathrm{d}}{\mathrm{dt}} (g_1 g_2) = g_2^{-1} g_1^{-1} \left[\omega_1 \frac{\mathrm{d}g_1}{\mathrm{d}\varphi_1} g_2 + \omega_2 g_1 \frac{\mathrm{d}g_2}{\mathrm{d}\varphi_2} \right]$$

$$= \omega_1 g_2^{-1} \left(g_1^{-1} \frac{\mathrm{d}g_1}{\mathrm{d}\varphi_1} \right) g_2 + \omega_2 g_2^{-1} \frac{\mathrm{d}g_2}{\mathrm{d}\varphi_2}.$$

Computation yields

$$g_1^{-1} \frac{\mathrm{d}g_1}{\mathrm{d}\varphi_1} = \begin{pmatrix} 0, & 0, & 0, & 0 \\ 0, & 0, & 0, & 0 \\ 0, & 0, & 0, & -1 \\ 0, & 0, & 1, & 0 \end{pmatrix}, \quad g_2^{-1} \frac{\mathrm{d}g_2}{\mathrm{d}\varphi_2} = \begin{pmatrix} 0, & 0, & 0, & 0 \\ -d, & 0, & 0, & 1 \\ 0, & 0, & 0, & 0 \\ 0, & -1, & 0, & 0 \end{pmatrix},$$

$$g_1^{-1} g_1^{-1} \frac{dg_1}{d\varphi_1} g_2 = \begin{pmatrix} 0, & 0, & 0, & 0 \\ 0, & 0, & -\sin\varphi_2, & 0 \\ d(\cos\varphi_2 - 1), & \sin\varphi_2, & 0, & -\cos\varphi_2 \\ 0, & 0, & \cos\varphi_2, & 0 \end{pmatrix},$$

so that

$$\bar{\boldsymbol{r}}(\mathrm{t}) = \begin{pmatrix} 0, & 0, & 0, & 0 \\ -d\omega_2, & 0, & -\omega_1 \sin\varphi_2, & \omega_2 \\ d\omega_1(\cos\varphi_2 - 1), & \omega_1 \sin\varphi_2, & 0, & -\omega_1 \cos\varphi_2 \\ 0, & -\omega_2, & \omega_1 \cos\varphi_2, & 0 \end{pmatrix}.$$

If we write $\bar{\boldsymbol{r}}(\mathrm{t})$ as $\bar{\boldsymbol{r}}(\mathrm{t}) = [\boldsymbol{x}; \boldsymbol{t}]$, we obtain

$$\boldsymbol{x} = \begin{pmatrix} \omega_1 \cos\varphi_2 \\ \omega_2 \\ \omega_1 \sin\varphi_2 \end{pmatrix}, \quad \boldsymbol{t} = \begin{pmatrix} -d\omega_2 \\ d\omega_1(\cos\varphi_2 - 1) \\ 0 \end{pmatrix}.$$

For $\omega_2 \neq 0$, the equation of the moving axoid is

$$(x^1)^2 + (x^3 - d)^2 - \left(\frac{x^2\omega_1}{\omega_2}\right)^2 = \left(\frac{d\omega_1^2}{\omega_1^2 + \omega_2^2}\right)^2,$$

which is a ruled hyperboloid of revolution with the axis o_2. (If $\omega_2 = 0$, the axis is o_1). Now, assume given the ruled surface $\mathscr{P}(\kappa, \lambda)$ as a function of the parameters κ and λ:

$$\mathscr{P} = A(\kappa) + \lambda\, \boldsymbol{u}(\kappa), \quad \text{where} \quad (\boldsymbol{u}, \boldsymbol{u}) = const.$$

Denote by $c(\varphi_2)$ the characteristic of surface $\mathscr{P}$ at time $g_1(\varphi_1)\, g_2(\varphi_2)$. Let $\bar{c}(\varphi_2)$ be its inverse, i.e. $c(\varphi_2) = g_1(\varphi)\, g_2(\varphi_2)\, \bar{c}(\varphi_2)$. For $\bar{c}(\varphi_2)$ we then have the equation

$$\left| \bar{\boldsymbol{r}}(\varphi_2)\, \bar{c}(\varphi_2), \frac{\partial \mathscr{P}}{\partial \kappa}, \frac{\partial \mathscr{P}}{\partial \lambda} \right| = 0.$$

Moreover, $\partial\mathscr{P}/\partial\lambda = \boldsymbol{u}(\kappa)$, $\partial\mathscr{P}/\partial\kappa = A'(\kappa) + \lambda\, \boldsymbol{u}'(\kappa)$, where the prime indicates differentiation wfth respect to κ.

If C is a point of the inverse $\bar{c}(\varphi_2)$ to characteristic $c(\varphi_2)$, then the following must hold:

$$|\bar{\boldsymbol{r}}(\varphi_2)\, C,\ \boldsymbol{u}(\kappa),\ A'(\kappa) + \lambda\, \boldsymbol{u}'(\kappa)| = 0,$$

i.e.

$$\left| \boldsymbol{x} \times C + \boldsymbol{t},\ \boldsymbol{u}(\kappa),\ A'(\kappa) + \lambda\, \boldsymbol{u}'(\kappa) \right| = 0\,.$$

Naturally, point C lies on surface $\mathscr{P}$; thus, let us write it in the form

$$C = A(\kappa) + \lambda(\kappa)\, \boldsymbol{u}(\kappa)\,.$$

We obtain one straight line of the surface for each κ, and the points of the characteristic lying on it are determined by the value of the parameter λ. Therefore, λ is a function of κ (although not single-valued, as we shall soon see). The condition for the function $\lambda(\kappa)$ then reads

$$\left| \boldsymbol{x} \times (A + \lambda \boldsymbol{u}) + \boldsymbol{t},\ A' + \lambda \boldsymbol{u}' \right| = 0\,,$$

i.e.

$$(\boldsymbol{x} \times A + \lambda \boldsymbol{x} \times \boldsymbol{u} + \boldsymbol{t},\ \boldsymbol{u} \times A' + \lambda \boldsymbol{u} \times \boldsymbol{u}') = 0\,.$$

If we develop this equation, we have

$$\begin{aligned}
&\lambda^2(\boldsymbol{x} \times \boldsymbol{u},\ \boldsymbol{u} \times \boldsymbol{u}') \\
&+ \lambda\left[(\boldsymbol{x} \times \boldsymbol{u},\ \boldsymbol{u} \times A') + (\boldsymbol{x} \times A,\ \boldsymbol{u} \times \boldsymbol{u}') + \left|\boldsymbol{t}, \boldsymbol{u}, \boldsymbol{u}'\right|\right] \\
&+ (\boldsymbol{x} \times A,\ \boldsymbol{u} \times A') + \left|\boldsymbol{t}, \boldsymbol{u}, A'\right| = 0\,, \\
&- \lambda^2(\boldsymbol{x}, \boldsymbol{u}')\, \boldsymbol{u}^2 \\
&+ \lambda\left[(\boldsymbol{x}, \boldsymbol{u})(\boldsymbol{u}, A') - (\boldsymbol{x}, A')\, \boldsymbol{u}^2 + (\boldsymbol{x}, \boldsymbol{u})(A, \boldsymbol{u}') - (\boldsymbol{x}, \boldsymbol{u}')(A, \boldsymbol{u}) \right.\\
&\left. + \left|\boldsymbol{t}, \boldsymbol{u}, \boldsymbol{u}'\right|\right] + (\boldsymbol{x}, \boldsymbol{u})(A, A') - (\boldsymbol{x}, A')(A, \boldsymbol{u}) + \left|\boldsymbol{t}, \boldsymbol{u}, A'\right| = 0\,,
\end{aligned}$$

i.e.

$$\begin{aligned}
&\lambda^2(\boldsymbol{x}, \boldsymbol{u}')^2\, \boldsymbol{u}^2 + \lambda\left[(\boldsymbol{x}, A')\, \boldsymbol{u}^2 + (\boldsymbol{x}, \boldsymbol{u}')(A, \boldsymbol{u}) - (\boldsymbol{x}, \boldsymbol{u})(A, \boldsymbol{u}) - \right. \\
&\left|\boldsymbol{t}, \boldsymbol{u}, \boldsymbol{u}'\right| + (\boldsymbol{x}, A')(A, \boldsymbol{u}) - (\boldsymbol{x}, \boldsymbol{u})(A, A') - \left|\boldsymbol{t}, \boldsymbol{u}, A'\right| = 0\,. \qquad (12.5)
\end{aligned}$$

We compute λ as a function of κ from the quadratic equation (12.5) and, for every φ_2, upon substitution into the relation

$$C(\kappa) = A(\kappa) + \lambda(\kappa)\, \boldsymbol{u}(\kappa)$$

we obtain the inverse of the characteristic $\bar{c}(\varphi)$ as a function of the parameter κ.

Now, assume that a ruled helical surface $\mathfrak{F}$ is given with the axis o_2, the jugular helix of radius D, and reduced thread height v_0. The angle of the ruling of the surface with the axis o_2 is assumed to be ψ. Put $\delta = \cot \psi$.

The parametric equations of the surface are

$$\mathfrak{F}(\kappa, \lambda) = A + \lambda \boldsymbol{u}\,, \quad \text{where} \quad A = \begin{pmatrix} D \sin \kappa \\ v_0\kappa + y_0 \\ d + D \cos \kappa \end{pmatrix}, \quad \boldsymbol{u} = \begin{pmatrix} \cos \kappa \\ \delta \\ -\sin \kappa \end{pmatrix}.$$

y_0 represents the second coordinate of the points of the jugular helix at which the line of the surface is parallel with the plane defined by $\boldsymbol{u}_1$ and $\boldsymbol{u}_2$;

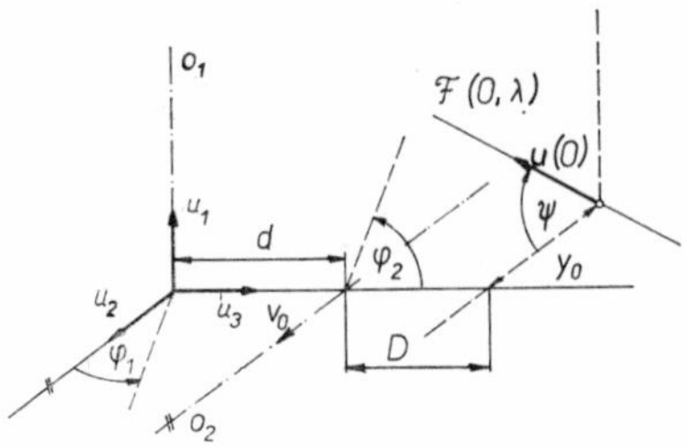

Fig. 52.

y_0 is thus the translation of the given surface in its "fundamental position" in the direction of the axis o_2 (Fig. 52). Let us put $v_0\kappa + y_0 = y$ for short; then $y' = v_0$. Upon differentiation we obtain

$$A = \begin{pmatrix} D \sin \kappa \\ y \\ d + D \cos \kappa \end{pmatrix}, \quad \boldsymbol{u} = \begin{pmatrix} \cos \kappa \\ \delta \\ -\sin \kappa \end{pmatrix},$$

$$A' = \begin{pmatrix} D \cos \kappa \\ v_0 \\ -D \sin \kappa \end{pmatrix}, \quad \boldsymbol{u}' = \begin{pmatrix} -\sin \kappa \\ 0 \\ -\cos \kappa \end{pmatrix},$$

$$\boldsymbol{x} = \begin{pmatrix} \omega_1 \cos \varphi_2 \\ \omega_2 \\ \omega_1 \sin \varphi_2 \end{pmatrix}, \quad \boldsymbol{t} = \begin{pmatrix} -d\omega_2 \\ d\omega_1(\cos \varphi_2 - 1) \\ 0 \end{pmatrix},$$

$$\boldsymbol{u}^2 = 1 + \delta^2\,, \quad (A, \boldsymbol{u}) = y\delta - d \sin \kappa\,, \quad (A, \boldsymbol{u})' = v_0\delta - d \cos \kappa\,,$$

$$(A, A') = yv_0 - dD \sin \kappa\,,$$

$$(\boldsymbol{x}, \boldsymbol{u}') = -\omega_1 \sin(\varphi_2 + \kappa)\,, \quad (\boldsymbol{x}, \boldsymbol{u}) = \omega_1 \cos(\varphi_2 + \kappa) + \delta\omega_2\,,$$

$$(\boldsymbol{x}, A') = \omega_1 D \cos(\varphi_2 + \kappa) + v_0\omega_2\,,$$

$$|\boldsymbol{t}, \boldsymbol{u}, A'| = \begin{vmatrix} -d\omega_2, & \cos\kappa, & D\cos\kappa \\ d\omega_1(\cos\varphi_2 - 1), & \delta, & v_0 \\ 0, & -\sin\kappa, & -D\sin\kappa \end{vmatrix}$$

$$= d\omega_2(\delta D - v_0)\sin\kappa\,,$$

$$|\boldsymbol{t}, \boldsymbol{u}, \boldsymbol{u}'| = \begin{vmatrix} -d\omega_2, & \cos\kappa, & -\sin\kappa \\ d\omega_1(\cos\varphi_2 - 1), & \delta, & 0 \\ 0, & -\sin\kappa, & -\cos\kappa \end{vmatrix}$$

$$= \delta d\omega_2 \cos\kappa + d\omega_1(\cos\varphi_2 - 1)\,.$$

Upon computing we obtain the equation for λ in the following form:

$$\begin{aligned} &\lambda^2\omega_1(1 + \delta^2)\sin(\varphi_2 + \kappa) \\ &+ \lambda\{\omega_1[\delta(v_0 - D\delta) - (D + d\cos\kappa)]\cos(\varphi_2 + \kappa) \\ &+ \omega_1(y\delta - d\sin\kappa)\sin(\varphi_2 + \kappa) - v_0\omega_2 + d\omega_1(\cos\varphi_2 - 1)\} \\ &+ \omega_1 y(v_0 - \delta D)\cos(\varphi_2 + \kappa) = 0\,. \end{aligned}$$

Upon further manipulations this yields

$$\begin{aligned} &\lambda^2\omega_1(1 + \delta^2)\sin(\varphi_2 + \kappa) \\ &+ \lambda\{\omega_1[\delta v_0 - D(1 + \delta^2)]\cos(\varphi_2 + \kappa) \\ &+ \omega_1 y\delta\sin(\varphi_2 + \kappa) - v_0\omega_2 - d\omega_1\} \\ &+ \omega_1 y(v_0 - \delta D)\cos(\varphi_2 + \kappa) = 0\,. \end{aligned}$$

If $\omega_1 = 0$, we have the equation $\lambda v_0\omega_2 = 0$, which has the only solution $\lambda = 0$ for $v_0 \neq 0$, and all λ for $v_0 = 0$ – the trivial solutions. Therefore, we may assume that $\omega_1 \neq 0$. Let us introduce the following notation:

$$v_0 - \delta D = s\,, \quad v_0\omega_2\omega_1^{-1} + d = m\,, \quad \varphi_2 + \kappa = \psi\,.$$

This yields

$$\begin{aligned} &\lambda^2(1 + \delta^2)\sin\psi + \lambda[(\delta s - D)\cos\psi + y\delta\sin\psi - m] \\ &+ ys\cos\psi = 0\,. \end{aligned} \tag{12.6}$$

By special choice of constants particular cases of the ruled surface are obtained:

$D = 0$ – closed helical surface,
$\delta = 0$ – rectangular helical surface,

$s \;\; = 0$ – developable helical surface,
$v_0 = 0$ – ruled hyperboloid of revolution ($y = const.$),
$v_0 = 0,\; y = 0$ – ruled hyperboloid of revolution in a special position,
$v_0 = 0,\; D = 0$ (then also $s = 0,\; y = const$) – conic surface of revolution,
$v_0 = 0,\; D = 0,\; y = 0$ – conic surface of revolution in a special position.

If we put $m = d$, $\varphi_2 = 0$, $\psi = \kappa$, we have the case of a single rotation $g_1(\varphi_1)$ about the axis o_1.

Note that the case of the cylindrical surface of revolution is not included among the special cases; however, here everything is easy: for $\lambda = 0$ and for $\varphi_2 + \kappa = \frac{1}{2}\pi$ the characteristic is a circle and a ruling.

From the quadratic equation (12.6) we now compute λ as a function of φ_2 and κ; the equation of the inverse $\bar{c}(\varphi_2)$ of the characteristic for position φ_2 reads

$$\bar{c}(\varphi_2) = A(\kappa) + \lambda(\kappa)\, \boldsymbol{u}(\kappa)\,, \quad \varphi_2 = const\,;$$

we obtain the characteristic in the form:

$$c(t) = g_1(\varphi_1)\, g_2(\varphi_2)\, [A(\kappa) + \lambda(\varphi_2, \kappa)\, \boldsymbol{u}(\kappa)]\,,$$

$\varphi_1, \varphi_2 = const.$, $\varphi_2\varphi_1^{-1} = \omega_2\omega_1^{-1}$, and the average envelope is

$$\mathscr{A}(\varphi_1, \kappa) = g_1(\varphi_1)\, g_2(\varphi_2)\, [A(\kappa) + \lambda(\varphi_2, \kappa)\, \boldsymbol{u}(\kappa)]\,,$$

$\varphi_2 = \varphi_1\omega_2\omega_1^{-1}$, φ_1, κ are parameters. The curve of instantaneous contact is

$$z(\kappa) = g_2(\varphi_2)\, [A(\kappa) + \lambda(\varphi_2, \kappa)\, \boldsymbol{u}(\kappa)]\,, \quad \varphi_2 = const\,,$$

and the *mesh surface*, as the set of all curves of instantaneous contact, is

$$\mathscr{Z}(\varphi_2, \kappa) = g_2(\varphi_2)\, [A(\kappa) + \lambda(\varphi_2, \kappa)\, \boldsymbol{u}(\kappa)]\,,$$

φ_2, κ are parameters. Computation yields the following expression for the curve of instantaneous contact z in position φ_2:

$$z(\kappa) = \begin{pmatrix} D\sin\psi + \lambda(\varphi_2, \kappa)\cos\psi \\ y_0 + v_0\kappa + \lambda(\varphi_2, \kappa)\,\delta \\ d + D\cos\psi - \lambda\sin\psi \end{pmatrix}, \quad \varphi_2 = const.$$

For the mesh surface we have

$$\mathscr{Z}(\psi, \kappa) = \begin{pmatrix} D\sin\psi + \lambda(\psi, \kappa)\cos\psi \\ y_0 + v_0\kappa + \lambda(\psi, \kappa)\,\delta \\ d + D\cos\psi - \lambda(\psi, \kappa)\sin\psi \end{pmatrix},$$

where we have transformed the parameters by putting $\varphi_2 + \kappa = \psi$. Naturally, λ satisfies (12.6) everywhere.

The characteristic now reads

$$c = \begin{pmatrix} D \sin \psi + \lambda \cos \psi \\ (y + \lambda\delta) \cos \varphi_1 - (d + D \cos \psi - \lambda \sin \psi) \sin \varphi_1 \\ (y + \lambda\delta) \sin \varphi_1 + (d + D \cos \psi - \lambda \sin \psi) \cos \varphi_1 \end{pmatrix},$$

where, of course,

$$y = y_0 + v_0\kappa\,, \quad \psi = \varphi_2 + \kappa = \omega_2\omega_1^{-1}\varphi_1 + \kappa\,,$$

$$\lambda = \lambda(\varphi_2, \kappa) = \lambda(\varphi_1, \kappa)\,;$$

with $\varphi_1 = const.$

Letting φ_1 and κ run through the respective chosen intervals, we obtain the representation of the desired envelope surface.

The curves of instantaneous contact play an important role in assessing the possibility of creating a carrier oil layer between the meshing sides of gear cogs according to the hydrodynamic theory of lubrication. The optimal position of the curve of instantaneous contact is such that at its points we have $\vartheta = \frac{1}{2}\pi$, where ϑ is the angle between the tangent of the curve of instantaneous contact and the vector of the relative velocity of the tangent point (see Fig. 43).

Let us compute the angle ϑ. We determine the instantaneous velocity vector with the aid of the axis of skew motion, the model of which we transfer to the appropriate position, i.e. we rotate it through the angle φ_2 about the axis o_2. Therefore, put $\boldsymbol{s} = g_2 \bar{\boldsymbol{r}} g_2^{-1}$. Then

$$\boldsymbol{s} = g_2 \left[\omega_1 g_2^{-1} \left(g_1^{-1} \frac{\mathrm{d}g_1}{\mathrm{d}\varphi_1} \right) g_2 + \omega_2 g_2^{-1} \frac{\mathrm{d}g_2}{\mathrm{d}\varphi_2} \right] g_2^{-1}$$

$$= \omega_1 g_1^{-1} \frac{\mathrm{d}g_1}{\mathrm{d}\varphi_1} + \omega_2 \frac{\mathrm{d}g_2}{\mathrm{d}\varphi_2} g_2^{-1} = \omega_1 g_1^{-1} \frac{\mathrm{d}g_1}{\mathrm{d}\varphi_1} + \omega_2 g_2^{-1} \frac{\mathrm{d}g_2}{\mathrm{d}\varphi_2},$$

so that

$$\boldsymbol{s} = \begin{pmatrix} 0, & 0, & 0, & 0 \\ -d\omega_2, & 0, & 0, & \omega_2 \\ 0, & 0, & 0, & -\omega_1 \\ 0, & -\omega_2, & \omega_1, & 0 \end{pmatrix}.$$

If point C with coordinates $[x^1, x^2, x^3]$ is given, instantaneous velocity $\boldsymbol{v}_C$ at point C is $\boldsymbol{v}_C = \boldsymbol{s}C$, i.e.

$$\boldsymbol{v}_C = \begin{pmatrix} \omega_2(x^3 - d) \\ -\omega_1 x^3 \\ -\omega_2 x^1 + \omega_1 x^2 \end{pmatrix}.$$

The velocity at the points of the curve of instantaneous contact $z(\kappa)$ will then be

$$\boldsymbol{v}_{z(\kappa)} = \begin{pmatrix} \omega_2(D\cos\psi - \lambda\sin\psi) \\ -\omega_1(d + D\cos\psi - \lambda\sin\psi) \\ \omega_1(y + \lambda\delta) - \omega_2(D\sin\psi + \lambda\cos\psi) \end{pmatrix}, \quad \varphi_2 = const.$$

It is also necessary to determine the tangent vector of the curve of instantaneous contact. If we differentiate equation (12.6) with respect to κ, we obtain $[\lambda = \lambda(\kappa)]$

$$\lambda' = \frac{\cos\psi[\lambda y\delta + sv_0 - \lambda^2(1+\delta^2)] + \sin\psi[\lambda D(1+\delta^2) - sy]}{\cos\psi(\delta s - D) + \sin\psi[y\delta + 2\lambda(1+\delta^2)] - m}.$$

The tangent vector of the curve of instantaneous contact is $z'(\kappa)$, where

$$z'(\kappa) = \begin{pmatrix} D\cos\psi + \lambda'\cos\psi - \lambda\sin\psi \\ v_0 + \lambda'\delta \\ -D\sin\psi - \lambda'\sin\psi - \lambda\cos\psi \end{pmatrix}.$$

The angle ϑ is then computed from the equation

$$\cos\vartheta = \frac{(\boldsymbol{z}', \boldsymbol{v}_z)}{\|\boldsymbol{z}'\| \cdot \|\boldsymbol{v}_z\|}.$$

12.2 Example. For the sake of simplicity let us consider the average envelope of a closed rectangular helical surface, i.e. $\delta = 0$, $D = 0$, in the special case $y_0 = 0$, $m = v_0\omega_2\omega_1^{-1} + d = 0$. Equation (12.6) for λ then reads

$$\lambda^2 \sin\psi + v_0^2\kappa\cos\psi = 0,$$

so that $\lambda_{1,2} = \pm v_0(-\kappa\cot\psi)^{1/2}$. There are two characteristics and only for certain values of the angle κ, as it is easy to ascertain. We consider the plus sign, in which case

$$\lambda = v_0(-\kappa \cot \psi)^{1/2}, \quad \lambda' = \frac{v_0(\kappa - \cos \psi \sin \psi)}{2(-\kappa \cot \psi)^{1/2} \sin^2 \psi}.$$

Upon substitution ($m = 0$ yields $\omega_2/\omega_1 = -d/v_0$) we obtain

$$\omega_1^{-1} \boldsymbol{v}_{z(\kappa)} = \begin{pmatrix} dv_0^{-1}\lambda \sin \psi \\ \lambda \sin \psi - d \\ v_0\kappa + dv_0^{-1}\lambda \cos \psi \end{pmatrix}, \quad \boldsymbol{z}' = \begin{pmatrix} \lambda' \cos \psi - \lambda \sin \psi \\ v_0 \\ -\lambda' \sin \psi - \lambda \cos \psi \end{pmatrix},$$

$$\lambda^2 \sin \psi + v_0^2 \kappa \cos \psi = 0.$$

$$\left(\frac{1}{\omega_1} \boldsymbol{v}_{z(\kappa)}, \boldsymbol{z}^1\right) =$$

$$v_0\left[d\kappa \cot \psi - d + v_0\kappa \frac{2\kappa \cos^2 \psi - \kappa - \cos \psi \sin \psi - \kappa}{2(-\kappa \cos \psi \sin \psi)^{1/2}} + v_0(-\kappa \cos \psi \sin \psi)^{1/2}\right],$$

so that

$$\begin{aligned} \cos \vartheta = \sin \psi[&2d(-\kappa \cos \psi \sin \psi)^{1/2} (\cot \psi - 1) \\ &+ v_0\kappa(2\kappa \cos^2 \psi - \kappa - \cos \psi \sin \psi] \left[(\kappa - \cos \psi \sin \psi)^2\right. \\ &\left.+ 4\kappa^2 \cos \psi \sin \psi - 4\kappa \cos \psi \sin^3 \psi\right]^{-1/2} \left[d^2(1 - \cot \psi)\right. \\ &+ v_0^2\kappa(\kappa - \cos \psi \sin \psi) \\ &\left.+ 2dv_0(\kappa \cos \psi - \sin \psi)\left(-\kappa \cot \psi\right)^{1/2}\right]^{-1/2}. \end{aligned}$$

If the ruled surface is developable, then $s = 0$, i.e. $v_0 = \delta D$. For λ we then obtain the linear equation

$$\lambda(1 + \delta^2) \sin \psi + (y\delta \sin \psi - D \cos \psi - m) = 0$$

($\lambda = 0$ is the trivial solution), which yields

$$\lambda = \frac{D \cos \psi + m - y\delta \sin \psi}{(1 + \delta^2) \sin \psi} = \frac{1}{1 + \delta^2}\left(D \cot \psi + \frac{m}{\sin \psi} - y\delta\right),$$

so that

$$\lambda' = \frac{1}{1 + \delta^2}\left(-\frac{D}{\sin^2 \psi} - \frac{m \cos \psi}{\sin^2 \psi} - v_0\delta\right).$$

Then the curve of instantaneous contact is

$$x^1 = D\sin\psi + \frac{1}{1+\delta^2}\left(D\cot\psi + \frac{m}{\sin\psi} - y\delta\right)\cos\psi\,,$$

$$x^2 = y + \frac{\delta}{1+\delta^2}\left(D\cot\psi + \frac{m}{\sin\psi} - y\delta\right),$$

$$x^3 = d + D\cos\psi - \frac{1}{1+\delta^2}(D\cos\psi + m - y\delta\sin\psi)\,.$$

Upon some manipulation

$$x^1 = \frac{1}{1+\delta^2}\left(D\delta^2\sin\psi + \frac{D}{\sin\psi} + m\cot\psi - y\delta\cos\psi\right),$$

$$x^2 = \frac{1}{1+\delta^2}\left[y + \delta\left(D\cot\psi + \frac{m}{\sin\psi}\right)\right],$$

$$x^3 = d + \frac{1}{1+\delta^2}(D\delta^2\cos\psi + y\delta\sin\psi - m)\,,$$

where $\psi = \varphi_2 + d$, $y = v_0\kappa + y_0$, $\varphi_2 = const.$ We also obtain

$$z' = \frac{1}{\sin^2\psi}\begin{pmatrix} -D\cos\psi - m + \delta y\sin^3\psi \\ -\delta D\cos^2\psi - \delta m\cos\psi \\ y\delta\cos\psi\sin^2\psi \end{pmatrix},$$

$$\frac{1}{\omega_1(1+\delta^2)}\,\mathbf{v}_z =$$

$$\begin{bmatrix} \frac{\omega_2}{\omega_1}(D\delta^2\cos\psi - m + y\delta\sin\psi \\ -d(1+\delta^2) - \delta^2 D\cos\psi + m - y\delta\sin\psi \\ -\frac{\omega_2}{\omega_1}\left[D\delta^2\sin\psi + \frac{D}{\sin\psi} + m\cot\psi - y\delta\cos\psi\right] + y + \delta\left(D\cot\psi + \frac{m}{\sin\psi}\right) \end{bmatrix}$$

Now, these formulas can be exploited to express $\cos\vartheta$; the resulting formula is rather complicated and will, therefore, be omitted.

Let us consider the special case of a conic surface of revolution under double-rotation. This case is obtained by putting $D = v_0 = 0$ in the equations for the average envelope of a developable surface. We then have

$$\lambda = \frac{1}{1 + \delta^2}\left(\frac{d}{\sin\psi} - y_0\delta\right)$$

and the curve of instantaneous contact is represented by the equations

$$x^1 = \frac{1}{1 + \delta^2}\left(d \cot\psi - y_0\delta\cos\psi\right),$$

$$x^2 = \frac{1}{1 + \delta^2}\left(y_0 + \frac{\delta d}{\sin\psi}\right),$$

$$x^3 = \frac{1}{1 + \delta^2}\left(y_0\delta\sin\psi - d\right) + d\,.$$

Understandably, there is only one curve of instantaneous contact because the conic surface of revolution does not change as it rotates about axis o_2, only its parametric representation changes. If we put

$$\bar{x}^2 = x^2 - \frac{y_0}{1 + \delta^2}, \quad \bar{x}^3 = x^3 - d + \frac{d}{1 + \delta^2} = x^3 - \frac{d\delta^2}{1 + \delta^2},$$

we obtain the following formulae for the projection of the characteristic into the plane $x^1 = 0$:

$$\bar{x}^2 = \frac{d\delta}{1 + \delta^2}\frac{1}{\sin\psi}, \quad \bar{x}^3 = \frac{y_0\delta\sin\psi}{1 + \delta^2},$$

i.e.

$$\bar{x}^2\bar{x}^3 = \frac{\delta^2 d y_0}{(1 + \delta^2)^2}\,.$$

This is the equation of a hyperbola with the asymptotes

$$x^2 = \frac{y_0}{1 + \delta^2}, \quad x^3 = \frac{d\delta^2}{1 + \delta^2},$$

which passes through the origin. This case was solved with the aid of the OBAL procedure – Figs. 53 and 54 (here the axes or rotation o_1, o_2 are denoted by o, o').

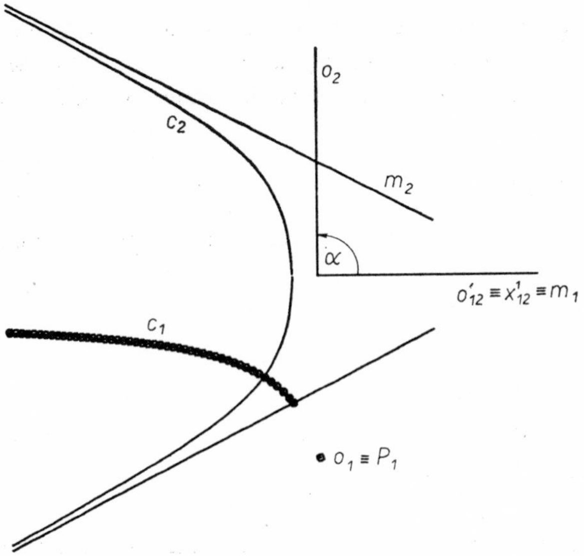

Fig. 53.

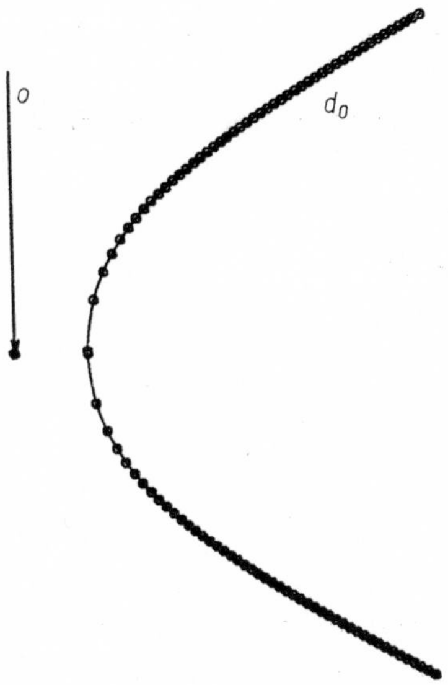

Fig. 54.

If $y_0 = 0$, we have the equation $x^3 = d\delta^2/(1 + \delta^2)$, i.e. the projection of the characteristic is a line parallel with the axis x^2, i.e. the characteristic itself is a hyperbola in a plane parallel with the plane x^1x^2.

Let us add a note on the constructive solution of the characteristic. The conjugate chord of contact of axis o_2 (relative to the zero complex) of the moving directing cone has the parametric equations

$$\begin{aligned} x^1 &= \lambda \cos \varphi_2 - d \sin \varphi_2(\cos \varphi_2 - 1)\,, \\ x^2 &= 0\,, \\ x^3 &= \lambda \sin \varphi_2 + d \cos \varphi_2(\cos \varphi_2 - 1)\,. \end{aligned}$$

Therefore, the conjugate chord of contact is a tangent of the circle $(x^1)^2 + (x^3 - d)^2 = d^2$, with o_2, the inverse of the instantaneous axis of skew motion, and the conjugate chord of contact having a common shortest transversal. For the construction of the curve of instantaneous contact, the axis o_1 is the conjugate chord of contact of axis o_2.

Indeed, the inverse of the axis of skew motion has the Plücker coordinates

$$\bar{\boldsymbol{r}} = \begin{bmatrix} \omega_1 \cos \varphi_2; & -d\omega_2 \\ \omega_2; & d\omega_1(\cos \varphi_2 - 1) \\ \omega_1 \sin \varphi_2; & 0 \end{bmatrix} (\omega_1^2 + \omega_2^2)^{-1/2}\,.$$

Axis o_2 has the Plücker coordinates

$$\begin{bmatrix} 0; & -d \\ 1; & 0 \\ 0; & 0 \end{bmatrix},$$

so that its conjugate chord of contact has the Plücker coordinates

$$\begin{bmatrix} \cos \varphi_2; & 0 \\ 0; & d(\cos \varphi_2 - 1) \\ \sin \varphi_2; & 0 \end{bmatrix},$$

which yields the desired representation.

For the curve of instantaneous contact we know that $\boldsymbol{s} = \bar{\boldsymbol{r}}$ for $\varphi_2 = 0$.

12.3 Theorem. *Let a ruled helical surface with axis o_2 and reduced thread height v_0 be given. Its normals at the points of the curve of instantaneous contact intersect the line with the equation*

$$h = -v_0\omega_2\omega_1^{-1}\boldsymbol{u}_3 + \lambda\boldsymbol{u}_1 ,$$

where λ is the parameter.

Proof. The curve of instantaneous contact has the equation

$$z(t) = \begin{pmatrix} D\sin\psi + \lambda\cos\psi \\ y + \lambda\delta \\ d + D\cos\psi - \lambda\sin\psi \end{pmatrix},$$

where λ satisfies the quadratic equation (12.6). The helical surface has the same equation, only λ is now an arbitrary parameter. The normal at any point $A + \lambda\boldsymbol{u}$ of the helical surface has the equation

$$n = A + \lambda\boldsymbol{u} + \mu(A' + \lambda\boldsymbol{u}') \times \boldsymbol{u} ,$$

μ is the parameter. We obtain

$$\begin{pmatrix} D\cos\psi - \lambda\sin\psi \\ v_0 \\ -D\sin\psi - \lambda\cos\psi \end{pmatrix} \times \begin{pmatrix} \cos\psi \\ \delta \\ -\sin\psi \end{pmatrix} = \begin{pmatrix} -v_0\sin\psi + \delta D\sin\psi + \lambda\delta\cos\psi \\ -\lambda \\ \delta D\cos\psi - \lambda\delta\sin\psi - v_0\cos\psi \end{pmatrix}.$$

Therefore, the normal to the surface is

$$\begin{pmatrix} D\sin\psi + \lambda\cos\psi \\ y + \lambda\delta \\ d + D\cos\psi - \lambda\sin\psi \end{pmatrix} + \mu\begin{pmatrix} -v_0\sin\psi + \delta D\sin\psi + \lambda\delta\cos\psi \\ -\lambda \\ \delta D\cos\psi - \lambda\delta\sin\psi - v_0\cos\psi \end{pmatrix}.$$

Now, compute the point of intersection of the normal constructed at the point of the curve of instantaneous contact with the plane $x^2 = 0$. This yields

$$y + \lambda\delta - \mu\gamma = 0, \quad \text{i.e.} \quad \mu = \frac{y + \lambda\delta}{\lambda}.$$

Upon substitution for μ, we obtain the following for the third coordinate:

$$\begin{aligned}
&\lambda^{-1}\{\lambda(d + D\cos\psi - \lambda\sin\psi) + \\
&\quad (y + \lambda\delta)(\delta D\cos\psi - \lambda\delta\sin\psi - v_0\cos\psi)\} \\
&= \lambda^{-1}\{-\lambda^2(1 + \delta^2)\sin\psi - \\
&\quad \lambda[(\delta v_0 - \delta^2 D - D)\cos\psi + y\delta\sin\psi - d] - \\
&\quad y(v_0 - \delta D)\cos\psi\} \\
&= \lambda^{-1}\{-\lambda^2(1 + \delta^2)\sin\psi - \\
&\quad \lambda[(\delta s - D)\cos\psi + y\delta\sin\psi - m + v_0\omega_2\omega_1^{-1}] - ys\cos\psi\} \\
&= \lambda^{-1}\{-\lambda v_0\omega_2\omega_1^{-1}\} = -v_0\omega_2\omega_1^{-1},
\end{aligned}$$

where we have used (12.6) for λ.

References

Chapter I.

[I] Birkhoff G., Mac Lane S.: A survey of modern algebra. III. ed., Macmillan, New York, 1965.

[II] Braun M.: Differential equations and their applications. Springer Verlag, New York, Heidelberg, Berlin, 1975.

[III] Bishop R. L., Crittenden R. J.: Geometry of manifolds. Acad. Press, New York and London, 1964.

[IV] Bishop R. L., Crittenden R. J.: Tensor analysis on manifolds. Macmillan, New York, 1968.

[V] Chevalley C.: Theory of Lie groups I. Princeton Univ. Press 1946.

[VI] Helgasson S.: Differential geometry and symmetric spaces. Acad. Press, New York and London, 1962.

[VII] Favard J.: Cours de géométrie différentielle local. Gauthier-Villars, Paris, 1957.

[VIII] Kelley J. L.: General Topology. Van Nostrand, New York, Toronto, London, 1957.

[IX] Kreyszig E.: Differential geometrie. II. ed., Akad. Verlag, Leipzig, 1968.

[X] Martin H. G.: Mathematics for engineering, technology and computing science. Pergamon Press, Oxford, 1966.

[XI] Silverman R. A.: Modern calculus and analytic geometry. Macmillan, New York, 1969.

[XII] Sternberg S.: Lectures on differential geometry.

Chapters II and III.

[1] Altmann F. G.: Sonderformen räumlicher Koppelgetriebe und Grenzen ihrer Verwendbarkeit. *Zeitschr. Konstruktion* **4** (1952), 97—106.

[2] Altmann F. G.: Über räumliche sechsgliedrige Koppelgetriebe. *Zeitschr. VDI* **96** (1954), 245—249.

[3] Altmann F. G.: Räumliche fünfgliedrige Koppelgetriebe. *Zeitschr. Konstruktion* **6** (1954), 254—259.

[4] Ballif L.: Sur les surfaces engendrées de deux manieres différentes par le mouvement d'une courbe indéformable. *Comptes rendus Acad. Paris* **158** (1914), 1484—1485.

[5] Beggs J. S.: Ein Beitrag zur Analyse räumlicher Mechanismen. Diss. T.U. Hannover, 1959.

[6] Bennet G. T.: A new mechanism. *Engineering London,* **76** (1903), 777—778.

[7] Bennet G. T.: Deformable octahedra. *Proc. London Math. Soc., second ser.,* **10** (1912), 309—343.

[8] Bennet G. T.: The skew-isogram mechanism. *Proc. London Math. Soc., second ser.,* **13** (1914), 151—173.

[9] Bereis R., Brauner H.: Über koaxiale euklidische Schraubungen. *Monatsh. Math.* **61** (1957), 225—245.

[10] Beyer R.: Neue Wege zur zeichnerischen Behandlung der räumlichen Mechanik. *Zeitschr. angew. Math. Mech.* **13** (1933), 17—31.

[11] Beyer R.: Über die Totlagen eines räumlichen Gelenkviereks. *Zeitschr. Maschienenbau-Betrieb,* (1938), 593—595.

[12] Beyer R.: Zur Analyse räumlicher Kurbeltriebe. *Zeitschr. Maschienenbau-Betrieb, Beilage Getriebetechnik* (1944), 253—259.

[13] Beyer R.: Zur Geometrie und Synthese eigentlicher Raumkurbelgetriebe. *VDI-Berichte* **5** (1955), 5—10.

[14] Beyer R.: Zur Synthese und Analyse vom Raumkurbelgetrieben. *VDI-Berichte,* **12** (1956), 5—20.

[15] Beyer R.: Wissenschaftliche Hilfsmittel und Verfahren zur Untersuchung räumlicher Gelenkgetriebe. *Zeitschr. Konstruktion* **9** (1957), 224—230, 285—290.

[16] Beyer R.: Das Matrizenkalkül als Hilfsmittel zur Untersuchung räumlicher Getriebe. *Zeitschr. Feinwerktechnik* **61** (1957), 318—327.

[17] Beyer R.: A survey of techniques analyzing motion properties of all types of 3-D mechanisms. Trans. of the 5th Conf. on mechanisms. Penton Publ. Co., Cleveland - Ohio, 1958, 141--163.

[18] Beyer R.: Technische Raumkinematik. Springer Verl., Berlin, 1963.

[19] Beyer R., Schrörner E.: Raumkinematische Grundlagen. München, 1953.

[20] Blaschke W.: Nicht-Euklidische Geometrie und Mechanik I, II, III. Hamb. Math. Einzelschr. 34. Heft, 1942.

[21] Blaschke W.: Zur Bewegungsgeometrie auf der Kugel. Sitz. Ber. Heidelberger Akad. Wiss. (1948).

[22] Blaschke W.: Kinematische Begründung von S. Lie's Geraden-Kugel Abbildung. Sitz. Ber. Bayer. Akad. Wiss. (1948), 291—297.

[23] Blaschke W.: Anwendung dualer Quaternionen auf Kinematik. *Ann. Acad. Sci. Fennicae Helsinki,* AI250/3 (1958).

[24] Blaschke W.: Zur Kinematik. *Abh. math. Sem. Univ. Hamburg* 22 (1958), 171—175.

[25] Blaschke W.: Euler und die Kinematik. Sammelband Leonhard Euler, Deutsche Akad. Wiss. Berlin, 1959, 35—41.

[26] Blaschke W.: Kinematik und Quaternione. Deutscher Verl. der Wiss., Berlin, 1960.

[27] Borel E.: Mémoire sur les déplacements à trajectoires sphériques. Mém. savants étrangers, Paris (2) 33 (1908), 1—128.

[28] Bottema O.: Acceleration axes in spherical kinematics. *Trans. ASME* (1964), 1—4.

[29] Brauner H.: Quadriken als Bewegflächen. *Monatsh. Math.* **59** (1955), 45—63.

[30] Brauner H.: Erzeugung eines gleichseitigen hyperbolischen Paraboloids durch Bewegung einer gleichseitigen Hyperbel. *Arch. Math.* 6 (1955), 330—334.
Brauner H.: See also [9].

[31] Bricard R.: Géométrie cinématique I. Paris, 1926.

[32] Buka F.: Über das sphärische Kurbelgetriebe und einen Spezialfall, das Hooke'sche Gelenk. Diss. Göttingen, 1876.

[33] Burmester L.: Kinematische Flächenerzeugung vermittelst zylindrischer Rollung. *Zeitschr. Math. Phys.* **33** (1888), 337—348.

[34] Burmester L.: Lehrbuch der Kinematik. Leipzig, 1888.

[35] Clifford W. K.: Preliminary sketch on biquaternions. *Proc. London Math. Soc.* **4** (1873), 381—395.

[36] Coriolis G. G.: Mémoire sur les équations du mouvement relatif. *Journ. l'école polyt.* **24** (1835), p. 142.

[37] Darboux G.: Sur le déplacements d'une figure invariable. *Comptes rendus Acad. Paris* **92** (1881), 181—121.

[38] Denavit J.: Displacement analysis of mechanisms based on 2 × 2 matrices of dual numbers. *VDI-Berichte* **29** (1958), 81—89.

[39] Denavit J.: Description and displacement analysis of mechanisms based on dual matrices. Diss. Northwestern Univ., Evanston Ill., 1956.
Denavit J.: See also [60].

[40] Denavit J., Hartenberg R. S.: A kinematic notation for lower pair mechanisms based on matrices. *J. Appl. Mech.* **22** (1955), 215—221.

[41] Dijksman E. A.: Motion geometry of mechanisms. Cambridge Univ. Press, 1976.

[42] Dimentberg F. M.: Obščij metod issledovanija konečnych peremeščenij prostranstvennych mechanismov i někotoryje slučai passivnych svjazej. *Trudy sem. po teorii mašin i mech.* **17** (1948), 5—39. *Abstract in Applied Mech. Reviews* **4** (1951), (9).

[43] Dimentberg F. M.: Opredelenije položenij prostranstvennych mechanismov. AN SSSR (1950), 142 p. *Abstract in Applied Mech. Reviews* **5** (1952), (1).

[44] Disteli M.: Über das Analogen der Savaryschen Formel und Konstruktion in der kinematischen Geometrie des Raumes. *Zeitschr. Math. Phys.* **62** (1914), 261—309.

[45] Disteli M.: Über die Verzahnung der Hyperboloidräder mit geradlinigem Eingriff. *Zeitschr. Math. Phys.* **56** (1911).

[46] Dittrich G.: Über die momentane Bewegungsgeometrie eines sphärisch bewegten starren Systems. Diss. TU Aachen, 1964.

[47] Dizioglu B.: Einfache Herleitung der Euler-Savaryschen Konstruktion der räumlichen Bewegung. *Mech. Mach. Theory* **9** (1974), 247—254.
Dragan K.: See [91].

[48] Egger H.: Das Raumgetriebe von Bennet. *Zeitschr. Masch.-Bau/Betrieb* **15** (1936), 42—43.

[49] Elliot E. B.: Some theorems of kinematics on a sphere. *Proc. London Math. Soc.* **12** (1881), 47—57.

[50] Euler L.: Formulae generales pro translatione qualunque corporum rigidorum. *Novi Comm.* **20** (1776).

[51] Federhofer K.: Über die Beschleunigungen bei der Bewegung des starren Körpers. *Zeitschr. angew. Math. Mech.* **7** (1927), 290—298.

Freudenstein F.: See [163].

[52] Gambier B.: Sur les surfaces susceptibles d'être engendrées de plusieurs façons différentes par de déplacements d'une courbe invariable. *Comptes rendus Acad. Aci. Paris* **158** (1914), 1155—1157.

[53] Gambier B.: Etude des surfaces de translation de Sophus Lie. *Nouv. Ann. Math.* **79** (1920), 401—423, 454—479.

[54] Garnier M. R.: Extension de la formule de Savary. *Ann. Sc.* **57** (1940), 113—200.

[55] Garnier M. R.: Cours de Cinématique I, II. 3rd ed. Paris, 1954, 1956.

[56] Goldberg M.: New five-bar and six-bar linkages in three dimensions. *Trans. of the ASME* **65** (1943), 649—656.

[57] Grünwald A.: Darstellung der Manheim-Darbouxs'chen Umschwungsbewegung eines starren Körpers. *Zeitschr. Math. Phys.* **54** (1907), 154—220.

[58] Grünwald J.: Über duale Zahlen und ihre Anwendung in der Geometrie. *Monatsh. Math. Phys.* **17** (1906), 81—136.

[59] Hacisalihoglu H. H.: Study map of a circle. *Journ. Fac. Sc. KTU* **1** (1977), 69—80.

[60] Hartenberg S., Denavit J.: Kinematic synthesis of linkages, Mc Graw-Hill, New York, 1964.

Hartenberg S.: See also [40].

[61] Hohenberg F.: Über die Zusammensetzung zweier gleichförmiger Schraubungen. *Monatsh. Math.* **54** (1950), 222—234.

[62] Horninger H.: Über Trochoidenschraublinien und die durch Trochoidenschraubungen erzeugbaren Kreisschraubflächen. *Monatsh. Math.* **58** (1954), 193—212.

[63] Horninger H.: Über Konchoidenschraublinien und die durch Konchoidenschraubungen erzeugbaren Regelschraubflächen. *Monatsh. Math.* **58** (1954), 226—240.

[64] Horninger H.: Über Trochoidenschraublinien und zyklische Schraubflächen. *Monatsh. Math.* **63** (1959), 39—58.

[65] Horninger H.: Schraubungen und Schraublinien n-ter Stufe. *Monatsh. Math.* **73** (1969), 46—62.

[66] Hoschek J.: Zur Ermittlung von Hüllflächen in der räumlichen Kinematik. *Monatsh. Math.* **63** (1959), 231—240.

[67] Hoschek J.: Eine Verallgemeinerung des Satzes von Holditch. *Monatsh. Math.* **80** (1975), 93—99.

[68] Hoschek J., Lübbert Ch.: Gleitkurven von sphärischen Radlinien 2. Stufe. *Math. Nachr.* **40** (1969), 191—200.

[69] Kalitzin G. S.: Die Begründung der Getriebelehre durch die Mengenlehre. *Acta technica Acad. Sci. Hung.* **11** (1955), 442—448.

[70] Kalitzin G. S.: Gruppentheoretische Eigenschaften der Getribe und Anwendung der Matrizenrechnung zur Berechnung von Getrieben. *Acta technica Acad. Sci. Hung.* **11** (1955), 449—460.

[71] Karger A.: Grundlagen der räumlichen kinematischen Geometrie. I, *Appl. Math.* **14** (1969), 87—93. II, *Appl. Math.* **25** (1980), 161—180.

[72] Karger A.: Kinematic geometry in n-dim. Euclidean and spherical space. *Czech. Math. Journ.* **22** (97) (1972), 83—107.

[73] Karger A.: Kinematic geometry of regular motions in homogeneous space. *Czech. Math. Journ.* **28** (103) (1978), 327—338.

[74] Koenigs G.: Leçons de cinématique. Paris, 1897.

[75] Koenigs G.: Sur les mouvements doublement décomposables et sur les surfaces quiont le lieu de deux familles de courbes égales. *Comptes rendus Acad. Paris* **157** (1913), 988—991.

[76] Koenigs G.: Mémoire sur les courbures conjugées dans le mouvement relatif le plus général de deux corps solides. Mémoires présentés par divers savants à l'Académie des Sciences 35 (1914), second ser., 1—215.

[77] Kotelnikoff A.: Screw calculus and some applications of the same to geometry and mechanics. Annals of the Imperial University of Kazan (1895/96).

[78] Krames J.: Über Fusspunktkurven von Regelflächen und eine besondere Klasse von Raumbewegungen. (Über symmetrische Schrotungen I.) *Monatsh. Math. Phys.* **45** (1937), 394—406.

[79] Krames J.: Zur Bricardschen Bewegung, deren sämtliche Bahnkurven auf Kugeln liegen. (Über symm. Schrotungen II.) *Monatsh. Math. Phys.* **45** (1937), 407—411.

[80] Krames J.: Zur aufrechten Ellipsenbewegung des Raumes. (Über symm. Schrotungen III.) *Monatsh. Math. Phys.* **46** (1937), 38—50.

[81] Krames J.: Zur kubischen Kreisbewegung des Raumes. (Über symm. Schr. III.) *Sb. Akad. Wiss. Wien, math.-nat. Kl. IIa,* **146** (1937), 145—158.

[82] Krames J.: Zur Geometrie des Bennetschen Mechanismus. (Über symm. Schr. V.) *Sb. Akad. Wiss. Wien, math.-nat. Kl. IIa,* **146** (1937), 159—173.

[83] Krames J.: Die Borel-Bricard Bewegung mit punktweise gekoppelten orthogonalen Hyperboloiden. (Über symm. Schr. VI.) *Monatsh. Math. Phys.* **46** (1937), 172—195.

[84] Krames J.: Sur une classe remarquable de mouvements de l'espace. Viration symétrique. *Comptes rendus Akad. Paris* **204** (1937), 1102—1104.

[85] Krames J.: Über die durch aufrechte Ellipsenbewegung erzeugten Regelflächen. *Jber. DMV* **50** (1940), 58—65.

[86] Krames J.: Zur Konstruktion der Krümmungskreise von ebenen und sphärischen Radlinien. *Getriebetechnik* **10** (482), (1942).

[87] Krames J.: Darstellende und kinematische Geometrie für Maschinenbauer. Wien, 1952.

[88] Krames J.: Über kubische Schraublinien und Cayleysche Strahlflächen dritten Grades. Sitz.-Ber. *Öster. Akad. Wiss.* **168** (1959), 239—248.

[89] Krames J.: Koppelgetriebe für windschiefe Drehachsen und konstantes Übersetzungsverhältnis 1 : 1. *Anz. Akad. Wiss. Wien, math.-nat. Kl.* (1965), 145—153.

[90] Lebedev P. A.: Kinematika prostranstvennych mechanismov. Moskva, 1966. (With extensive bibliography of 115 titles in Russian.)

Lübbert Ch.: See [68].

[91] Mangeron D., Dragan K.: Kinematic study with new matrix-tensor methods for four-link spatial mechanisms. *Rev. Mecanique Appl.* **7** (1962), 539—551.

[92] Mangeron D., Topentcharov V. V.: Sur le problème de la permanence des champs vectoriels intéressant la théorie des mécanismes et des machines. *Bul. Inst. Politehn. Iaşi* **11** (15) (1965).

[93] Marbach O.: Die Pohlbahnen des Hooke'schen Gelenks. Diss. Jena, Berlin, 1880.

[94] Meyer zur Capellen W.: Über räumliche Kurbelschleifen und ihre gleichwertigen ebenen Getriebe. *VDI-Berichte* **29** (1958), 91—101.

[95] Meyer zur Capellen W.: Die Extreme der Übersetzungen in ebenen und sphärischen Kurbeltrieben. *Ingenieur-Archiv* **27** (1960), 352—364.

[96] Meyer zur Capellen W.: The instantaneous distribution of acceleration of a spherical moving system. *Journ. Mech.* **1** (1966), 23—42.

[97] Meyer zur Capellen W., Rath W.: Kinematik der sphärischen Schubkurbel. Westdeutscher Verlag, Köln und Opladen, 1960, pp. 1—37.

[98] Müller E.: Über Schiebflächen, deren eine Erzeugendschaar aus gewöhnlichen Schraublinien besteht. *Sitz.-Ber. Akad. Wiss. Wien* (*nat.-wiss. Kl. IIa*), **118** (1909), 3—13.

[99] Müller E.: Die achsiale Inversion. *Jber. DMV* **25** (1916), 209—251.

[100] Müller H. R.: Über zwangläufige Bewegungsvorgänge. *Collect. Mathem.* **3** (1950), 3—10.

[101] Müller H. R.: Über eine infinitesimale kinematische Abbildung. *Monatsh. Math.* **54** (1950), 108—129.

[102] Müller H. R.: Die Bewegungsgeometrie auf der Kugel. *Monatsh. Math.* **55** (1951), 28—42.

[103] Müller H. R.: Über geschlossene Bewegungsvorgänge. *Monatsh. Math.* **55** (1951), 206—214.

[104] Müller H. R.: Zur Ermittlung der Hüllflächen in der räumlichen Kinematik. *Monatsh. Math.* **63** (1959), 231—240.

[105] Müller H. R.: Die kinematische Abbildung in 3-dim. Raum. *Monatsh. Math.* **65** (1961), 252—258.

[106] Müller H. R.: Sphärische Kinematik. VEB Deutscher Verl. Wiss., Berlin, 1962.

[107] Müller H. R.: Kinematische Geometrie. *Jber. DMV* **72** (1970), 143—164.

[108] Prager W.: Beitrag zur Kinematik des Raumfachwerks. *Zeitschr. angew. Math. Mech.* **6** (1926), 341—355.

[109] Raven F. H.: Velocity and acceleration analysis of plane and space mechanisms by means of independent-position equation. *Journ. Appl. Mech.* **25** (1958), 1—6.

[110] Reuleaux F.: Kinematics of machinery. London, 1876.

[111] Reuleaux F.: Theoretische Kinematik. Braunschweig, 1900.

[112] Rodrigues O.: Des lois géométriques qui régissent le déplacement d'un système solide dans l'espace. *Liouville Journ. de math.* **5** (1840).

[113] Shigley J. E.: Kinematic analysis of mechanism. Mc Graw-Hill, New York, 1959.

[114] Schoenflies A.: Geometrie der Bewegung. Leipzig, 1886.

[115] Schoenflies A., Grübler M.: Kinematik in Enzykl. math. Wiss. IV, 190—278, Teubner, Leipzig, 1902.

Schörner E.: See [19].

[116] Schur F.: Über die Bewegung eines starren Körpers durch Abschroten. *Zeitschr. Math. Phys.* **55** (1907), 408—415.

[117] Smeenk D. J.: Rational motions of special spatial fourbars. Diss. Delft, 1973.

[118] Stephanos C.: Mémoire sur la représentation des homographies binaires par des points de l'espace avec application à l'étude des rotations sphériques. *Math. Ann.* **22** (1883), 299—367.

[119] Stephanos C.: Sur la théorie des quaternions. *Math. Ann.* **22** (1883), 589—592.

[120] Strubecker K.: Komplexe Geometrie und aufrechte Ellipsenbewegung. *Jber. DMV* **50** (1940), 43—58.

[121] Strubecker K.: Über Komplexflächen bei euklidischen Schraubungen. *Monatsh. Math.* **62** (1958), 297—323.

[122] Study E.: Geometrie der Dynamen. Leipzig, 1901.

[123] Study E.: Grundlagen und Ziele der analytischen Kinematik. *Sitz.-Ber. Berl. Math. Gesselschaft* **12** (1913), 36—60.

[124] Stübler E.: Das Beschleunigungssystem bei der Bewegung eines starren Körpers. *DMV-Berichte* **19** (1910), 177—185.

[125] Topentcharov V. V.: Sur les éléments cinématiques et certaines propriétés du mouvement euclidien à trois dimensions. *Mathematica* **7** (30) (1965), 163—190.

Topentcharov V. V.: See also [92].

[126] Tölke J.: Spezielle Bewegungsvorgänge III. *Arch. Math.* **21** (1970), 650—659.

[127] Tölke J.: Der sphärische Bresse-Kegelschnitt. *Math. Nachr.* **47** (1970), 345—353.

[128] Tölke J.: Eine Kennzeichnung der sphärischen Trochoidenbewegung. *Element. Math.* **27** (1972), 57—79.

[129] Tölke J.: Oskulierende sphärische Hüllkurvenpaare. *Sb. Akad. Wiss. Wien (math.-nat. Kl.)* **181** (1973), 71—75.

[130] Tölke J.: Kennzeichnung der sphärischen Bahnscheitel. *Sb. Akad. Wiss. Wien (math.-nat. Kl.)* **181** (1973), 253—259.

[131] Tölke J.: Zur Strahlkinematik I. *Sb. Akad. Wiss. Wien (math.-nat. Kl.)* **182** (1973), 177—202.

[132] Tölke J.: Kennzeichnung der sphärischen symmetrischen Rollung. *Sb. Akad. Wiss. Wien (math.-nat. Kl.)* **181** (1973), 261—267.

[133] Tölke J.: Der Schmiegkegelschnitt in der sphärischen Kinematik. *Strojnický Čas.* **25** (1974), 284—287.

[134] Tölke J.: Ein sphärisches Analogon der Affinnormalen. *Sb. Akad. Wiss. Wien (math.-nat. Kl.)* **183** (1974), 39—49.

[135] Tölke J.: Zur Konstruktion des Krümmungsmittelpunktes einer sphärischen Bahnkurve. *Monatsh. Math.* **80** (1975), 61—65.

[136] Tölke J.: Erzeugungsmöglichkeiten der sphärischen Kreispunktkurve. *Mech. Mach. Theory* **10** (1975), 207—215.

[137] Tölke J.: Elementare Kennzeichnungen der symmetrischen Schrotung. *Manuscr. Math.* **15** (1975), 309—321.

[138] Tölke J.: Ein Problem der sphärischen Kinematik. *Abh. math. Sem. Hamburg* **43** (1975), 229—233.

[139] Tölke J.: Zum sphärischen Analogon der Ellipsen-Bewegung. *Sb. Akad. Wiss. Wien (math.-nat. Kl.)* **184** (1975), 403—425.

[140] Tölke J.: Contribution to the theory of axes of curvature. *Mech. Mach. Theory* **11** (1976), 123—130.

[141] Tölke J.: Zum Integrationsproblem sphärischer Hüllkurvenpaare. *Manuscr. Math.* **19** (1976), 189—194.

[142] Tölke J.: The roll-sliding number of associated curves. *Mech. Mach. Theory* **11** (1976), 419—423.

[143] Tölke J.: Eine Charakterisierung sphärischer Trochoidenbewegungen vermöge eigentlicher rollgleitender Hüllkurvenpaare. *ZAMM* **58** (1978), 114—115.

[144] Tölke J.: Ebene euklidische und sphärische symmetrische Rollungen. *Mech. Mach. Theory* **13** (1978), 187—198.

[145] Tölke J.: Eine kinematische Kennzeichnung der gleichseitigen Paraboloide. *Monatsh. Math.* **87** (1979), 69—80.

[146] Tölke J.: Eine Bemerkung zum gleichseitigen hyperbolischen Paraboloid als Bewegfläche bei Zylinderschrotung. *Elemente Math.* **35** (1980), 118—120.

[147] Trinkl F.: Analytisches und zeichnerisches Verfahren zur Untersuchung eigentlicher Raumkurbelgetriebe, dargestellt an einem räumlichen viergliedrigen Kurbelgetriebe mit einem Drehgelenk und drei Drehschubgelenken. Diss. TU München, 1958. Abstract in *Zeitschr. Konstruktion* **11** (1959), 349—359.

[148] Uicker J. J.: On the dynamic analysis of spatial linkages using 4×4 matrices. PhD Thesis, Northwestern Univ. Evanston, Ill., 1965.

[149] Veldkamp G. R.: Canonical systems and instantaneous invariants in spatial kinematics. *Journ. Mech.* **3** (1967), 329—388.

[150] Veldkamp G. R.: An approach to spherical kinematics using tools suggested by plane kinematics. *Journ. Mech.* **2** (1947), 437—450.

[151] Weckert M.: Analytische und graphische Verfahren für die Untersuchung "eigentlicher" Ramkurbelgetriebe, dargestellt an einem räumlichen viergliedrigen Kurbelgetriebe mit einem Drehgelenk und drei Drehschubgelenken. Diss. TU München, 1953.

[152] Weinnoldt E.: Kinematische Erzeugung von Regelflächen 4. Ordnung. *Zeitschr.*

Math. Phys. **52** (1905), 299—330.

[153] Weiss E. A.: Einführung in die Liniengeometrie und Kinematik. Leipzig, Berlin, 1935.

[154] Winter H.: Geschwindigkeitspläne räumlicher Getriebe. *Zeitschr. angew. Math. Mech.* **10** (1930), 274—284.

[155] Wörle H.: Getriebeanalytische und getriebesynthetische Verfahren für den Entwurf zwangläufiger, insbesondere viergelenkiger Raumkurbelgetriebe allgemeiner und spezieller Art. Diss. TU München, 1955.

[156] Wörle H.: Sonderformen zwangläufiger viergelenkiger Raumkurbelgetriebe. *VDI-Berichte* **12** (1956), 21—28.

[157] Wörle H.: Getriebeanalytische und getriebesynthetische Unterlagen für den Entwurf zwangläufiger und übergeschlossener viergelenkiger Raumkurbelgetriebe. *Buletinul Institutului politehnic din Iaşi* **6** (10) (1960), 303—322, **7** (11) (1961), 269—298.

[158] Wörle H.: Untersuchungen über Koppelkurven viergelenkiger räumlicher Kurbelgetriebe. *Zeitschr. Konstruktion* **14** (1962), 390—392.

[159] Wunderlich W.: Kubische Strahlflächen, die sich durch Bewegung einer starren Parabel erzeugen lassen. *Monatsh. Math.* **71** (1967), 344—353.

[160] Yang A. T., Freudenstein F.: Application of dual-number quaternion algebra to the analysis of spatial mechanisms. *Journ. Appl. Math. Mech.* **31** (1964), 300—308.

INDEX